Benchmark Papers
in Geology

Series Editor: Rhodes W. Fairbridge
Columbia University

A selection from the published volumes in this series

Benchmark Papers
in Geology / 64

A BENCHMARK® Books Series

GEOSYNCLINES
Concept and Place
Within Plate Tectonics

Edited by

F. L. SCHWAB

Washington and Lee University

Hutchinson Ross Publishing Company

Stroudsburg, Pennsylvania

Copyright © 1982 by **Hutchinson Ross Publishing Company**
Benchmark Papers in Geology, Volume 64
Library of Congress Catalog Card Number: 81–6807
ISBN: 0-87933-410-X

84 83 82 1 2 3 4 5
Manufactured in the United States of America.

LIBRARY OF CONGRESS CATALOGING IN PUBLICATION DATA
Main entry under title:
Geosynclines, concept and place within plate tectonics.
  (Benchmark papers in geology; 64)
  Includes bibliographical references and indexes.
  1. Geosynclines—Addresses, essays, lectures.
  2. Plate tectonics—Addresses, essays, lectures.
I. Schwab, F. L. (Frederic L.) II. Series.
QE607.G46   551.8'6   81-6807
ISBN 0-87933-410-X   AACR2

Distributed world wide by Academic Press,
a subsidiary of Harcourt Brace Jovanovich,
Publishers.

CONTENTS

Contents

Contents

PART IX: ANALYTICAL KEYS FOR DECIPHERING ANCIENT GEOSYNCLINES PRODUCED BY PLATE TECTONIC MECHANISMS

PART X: GEOSYNCLINES AND PLATE TECTONICS: THE SCIENTIFIC REVOLUTION

SERIES EDITOR'S FOREWORD

The philosophy behind the Benchmark Papers in Geology is one of collection, sifting, and rediffusion. Scientific literature today is so vast, so dispersed, and, in the case of old papers, so inaccessible for readers not in the immediate neighborhood of major libraries that much valuable information has been ignored by default. It has become just so difficult, or so time consuming, to search out the key papers in any basic area of research that one can hardly blame a busy person for skimping on some of his or her "homework."

This series of volumes has been devised, therefore, as a practical solution to this critical problem. The geologist, perhaps even more than any other scientist, often suffers from twin difficulties—isolation from central library resources and immensely diffused sources of material. New colleges and industrial libraries simply cannot afford to purchase complete runs of all the world's earth science literature. Specialists simply cannot locate reprints or copies of all their principal reference materials. So it is that we are now making a concerted effort to gather into single volumes the critical materials needed to reconstruct the background of any and every major topic of our discipline.

We are interpreting "geology" in its broadest sense: the fundamental science of the planet Earth, its materials, its history, and its dynamics. Because of training in "earthy" materials, we also take in astrogeology, the corresponding aspect of the planetary sciences. Besides the classical core disciplines such as mineralogy, petrology, structure, geomorphology, paleontology, and stratigraphy, we embrace the newer fields of geophysics and geochemistry, applied also to oceanography, geochronology, and paleoecology. We recognize the work of the mining geologists, the petroleum geologists, the hydrologists, and the engineering and environmental geologists. Each specialist needs a working library. We are endeavoring to make the task of compiling such a library a little easier.

Each volume in the series contains an introduction prepared by a specialist (the volume editor)—a "state of the art" opening or a summary of the object and content of the volume. The articles, usually some twenty to fifty reproduced either in their entirety or in significant extracts, are selected in an attempt to cover the field, from the key papers of the last century to fairly recent work. Where the original works are in foreign languages, we have endeavored to locate or commission translations. Geologists, because of their global subject, are often acutely aware of the oneness of our world. The selections cannot therefore be restricted to any one country, and whenever possible an attempt is made to scan the world literature.

To each article, or group of kindred articles, some sort of "highlight commentary" is usually supplied by the volume editor. This commentary should serve to bring that article into historical perspective and to emphasize its particular role in the growth of the field. References, or citations, wherever possible, will be reproduced in their entirety— for by this means the observant reader can assess the background material available to that particular author, or, if desired, he or she too can double check the earlier sources.

A "benchmark," in surveyor's terminology, is an established point on the ground that is recorded on our maps. It is usually anything that is a vantage point, from a modest hill to a mountain peak. From the historical viewpoint, these benchmarks are the bricks of our scientific edifice.

RHODES W. FAIRBRIDGE

PREFACE

Geosynclinal theory is a comprehensive scheme that genetically relates regionally extensive, subsiding segments of the earth's crust that fill with sediment *and* the intensely deformed, metamorphosed, and intruded fold belts found around the globe. The concept was first postulated by an American geologist, James Hall, in the 1850s. He pointed out the apparent causal relationship between folded mountain belts and the thickness of the sedimentary rock succession of which they were composed. Shortly thereafter, James Dwight Dana of Yale University attributed the deformation of the thick sediment fill of these *geosynclinals* to the effects of lateral horizontal compression occurring on a global scale. Dana argued that such global compression was essentially localized at the site of the linear geosynclinal. For a century following its initial postulation, geosynclinal theory has probably remained the single most important unifying conceptual scheme in the geological sciences, despite serious differences of opinion as to what the concept actually entails.

In the 1960s and 1970s, geosynclinal theory has been overtaken and largely incorporated into plate tectonics theory, a framework of conceptual ideas or geological paradigm that although rooted in the suggestions of several earlier workers, has now emerged to explain the present-day tectonic activity of the earth: the global network of earthquake belts and volcanic activity, the physiographic features of the ocean basins, and the origin and distribution of modern mobile belts. Plate tectonics theory relates all of these present-day tectonic phenomena to the movement of internally rigid plates of the lithosphere relative to one another. Much of the tectonic activity visible today occurs coincident with the margins of individual plates, although mid-plate disturbances are also observed.

The forty-six papers selected for this volume trace the historical development of the geosynclinal concept from its birth in the middle of the nineteenth century until the present, emphasizing how the concept has been constantly reshaped and broadened in order to conform to changing ideas and new observational data. This historical reassessment of classical geosynclinal theory is particularly appropriate at this point in time because of the growing evidence and consensus within the geological community that plate tectonics mechanisms also operated in the geological past to produce ancient mountain systems, at least as far back as the Proterozoic and perhaps even as early as during Archean time.

The papers included in this volume cover a period from 1859 to 1977. These papers make it clear that during most of this time interval, what was inferred and included within the context of classical geosynclinal theory depended largely on personal biases conditioned by geological and geographical perspective. Nevertheless, geosynclinal theory proved flexible enough to tolerate widespread differences of interpretation, and papers explaining ancient mobile belts in terms of geosynclinal evolution dominate the geological literature throughout the late nineteenth century and well into the twentieth. It is particularly appropriate to examine the extent to which this classical theory can be retained within a geological community now dominated by the paradigm of "the new global tectonics."

I am happy to acknowledge the constructive suggestions I have received from Rhodes Fairbridge, series editor of the Benchmark Papers in Geology series. I am also grateful to Susie White and Peggy Riethmiller for secretarial assistance, and to my wife Claudia who ably provided general help and encouragement.

F. L. SCHWAB

CONTENTS BY AUTHOR

GEOSYNCLINES

INTRODUCTION

The first geologist to formally propose that a direct *causal* relationship exists between thick linear belts of sediment and the formation of mountain systems was James Hall (1859, 1883). Ever since this view was initiated in the middle of the nineteenth century, most geologists have used "geosynclinal theory" to explain the origin of mountain systems, the accretion of continents, and the distribution in time and space of thick lithological assemblages of sediment.

Like most of his contemporaries, Hall subscribed to the view that mountains were essentially produced by vertical uplift and subsidence, and he accepted the then popular notion that the earth's interior was pliable and consequently readily responsive to gravitational loading. Such an assumption was apparent in the actual mechanism invoked by Hall to create mountains. Downwarping of the earth's crust was assumed to have occurred in response to the accumulation of a thick mass of sediment. Subsequent crumpling of the layers of sediment produced local increases in elevation. However, Hall implied that *most* of the elevation of a mountain chain was due to the buoyancy of the sediment-thickened crustal segment; he directly stated that thicker accumulations of sediment produced correspondingly higher mountain ranges. This statement represents one of the earliest assertions of the existence of isostasy, although the term was not to be introduced until some three decades later (by Dutton, 1889).

Although Hall's arguments genetically relating linear fold belts to precursory sediment-filled troughs were convincing, many geologists did not agree with the mechanisms invoked by Hall to downwarp, deform, and uplift the geosynclinal belt. Consequently, they soon began to modify the original geosynclinal concept, asserting that geosynclines were produced by the effects of lateral global compression localized along elongate zones of crustal weakness. This modification of geosynclinal theory, the first of many to occur over the course of the next hundred years,

1

was crafted largely by James Dana (1866; 1873) in the 1860s and 1870s. Dana also focused attention on the position of geosynclines relative to the continental blocks and ocean basins, on the possible role of geosynclines in the growth of continental crust, and on the relative importance of various kinds of geosynclinal source areas.

Early differences between the opinions of Hall and Dana touched off decades of disagreement regarding the relative importance of simple lateral compression versus vertical uplift due to isostasy in producing mountainous topography. This debate, essentially an argument over the internal strength of the crust and its susceptibility to both vertical uplift and subsidence as well as horizontal compression, continued well into the twentieth century. At times, the original concept of the geosyncline was almost completely overshadowed or modified beyond recognition. Nevertheless, again and again geologists returned to the basic premises on which geosynclinal theory was based, reshaping and redefining the concept to best suit their own prejudices or type examples. The generous flexibility allowed within the constraints of geosynclinal theory would nearly be its undoing.

This volume is a collection of classic papers tracing the development of geosynclinal theory from its birth (Part I) to the present time. These papers serve to illustrate that the concept, while useful in a variety of ways as a means of explaining mountain belts, has proved to be an elusive, almost illusionary, difficult-to-define conceptual scheme. Declarations of the definitive characteristics attributed to geosynclines have almost always been with reference to specific examples. Only in the past two decades, with the advent of the theories of the new global tectonics, has it become clear that a wide variety of ancient geosynclines existed, each analogous to the various modern types of ocean basins and continental margins.

Until 1960, the geosynclinal concept produced almost as much confusion as clarity. The geosynclinal characteristic most consistently (but not always) incorporated into the scheme over the past one hundred years is the relationship originally postulated by Hall: the coincidence between mountain belts and linear tracts of thick sedimentary rocks. Otherwise, all aspects of what constitutes a geosyncline have been continually reshaped and redefined, particularly the causal mechanisms invoked to produce and deform geosynclinal basins, the classification schemes used to differentiate among geosynclinal varieties, the depositional environments thought to typify geosynclinal sedimentation, and the position of geosynclinal belts relative to continental blocks and ocean basins.

For example, European geologists (notably in France, Germany, and Switzerland) had enthusiastically embraced geosynclinal theory before the turn of the century, but eventually, led by Haug (1900), they came to fundamentally disagree with the notions of Hall and Dana regarding the shallow-water nature of geosynclinal sediments and the position of geosynclines relative to the continental blocks (Part II). The Alpine belt, unlike the Appalachians, did not straddle a continental margin, nor did it seem to have a floor of continental (sialic) crust. Furthermore, the Europeans perceived geosynclinal development as an orderly process, attested to by systematic lateral and vertical distributions in the sedimentary and volcanic rock infilling (Part III). This view was enthusiastically adopted by American geologists who began to consider the geosynclinal mechanism as deterministic: the geosyncline is the father of the mountain. They believed that geosynclines *necessarily* produced deformed mountain systems and in the process followed a nearly identical, predetermined course of development. Later acceptance of plate tectonics theory has strongly modified this notion and has also explained the many departures from any uniform evolutionary scheme.

Much of the historical development of geosynclinal theory as a unifying theme in geology has centered upon attempts to more clearly define and differentiate the various types of geosynclinal basins. These efforts have resulted in a vast array of classification schemes and terminology, intentionally broad enough to encompass all possible variations from the general case, yet unfortunately too complex to be easily applicable and analytically useful (Part IV). A lively debate over the various kinds of clastic source areas for geosynclinal fill also stemmed from these efforts.

Other workers were less concerned with classification, terminology, and the characterization of individual varieties of geosynclines, considering instead the mechanisms responsible for producing geosynclines in general and for causing their eventual conversion into deformed mountain belts. Some of these early attempts (Argand, 1916, 1920, 1922; Wegener, 1912, 1922, 1966; Holmes, 1931, 1945) were remarkably "modern" in the sense that they interpreted geosynclinal evolution in terms of global mechanisms not unlike those incorporated into modern plate tectonics theory: continental drift, sea-floor spreading, subcrustal convection, and the existence of ocean basins that periodically open and close thus deforming continental margin, trench, and abyssal plain sediments into mobile belts (Part V).

Despite such perceptive foresight, geosynclinal theory has also generated considerable confusion and controversy, and by 1950, the concept had raised as many questions as it had resolved (Part VI). However, just as this general atmosphere of pessimism and disenchantment began to descend over conventional geosynclinal theory, renewed interest in continental drift was generated by the discovery of sea-floor spreading and strengthened by new revelations about paleomagnetism and observations on the physiography and geology of the modern ocean basins. These revitalized mobilist theories immediately provided an entirely different structural framework within which the classical schemes for ancient geosynclinal evolution and mountain-making might also have operated (Part VII). The varied setting of modern mobile belts demonstrates that their ancient equivalents must have been equally as diverse. Consequently, many of the problems concerning ancient mobile belts that classical geosynclinal theory was unable to resolve became understandable when classical theory was modified in light of plate theory (Part VIII). The encouraging success of these efforts led directly to numerous further attempts to better understand the structural and lithological signatures imprinted on the various modern mobile belts. Analytical keys useful for deciphering ancient geosynclinal mobile belts could now be effectively linked to modern, actualistic analogues (Part IX). Again, by using the present as a key to the past, traditional Lyellian and Huttonian philosophy was triumphant.

No problem in geology has been more fascinating or challenging than that of explaining the origin and evolution of the globe's major mountain systems. The continued viability of the geosynclinal explanation for mountain building over the course of the last century has now been extended by the incorporation of conventional geosynclinal theory within the context of modern plate tectonics theory, and probably represents the single most exciting revolution in geological thinking (Part X).

Plate tectonics theory requires that geologists now view geosynclinal origin and evolution from a radically different analytical perspective than that initially used by Hall and Dana. Not only do the depositional, tectonic, and paleogeographical settings for geosynclines vary greatly from one another, they also contrast decidedly with the standardized models originally suggested by

4

Haug, Hall, Stille, Kay, and others. Plate theory indicates that geo-synclines and their component parts can be juxtaposed spatially and temporally to produce an almost endless variety of mobile belts. The resultant complex organization visible in ancient mobile belts stands in stark contrast to the simplistic, systematic character-ization once applied universally to geosynclines within the con-straints of a generalized geotectonic cycle.

Because our models of what ancient geosynclines represent and where their modern analogues now exist have changed, it is also necessary to modify the terminology and classification schemes applied to geosynclines. This volume demonstrates that geosyn-clinal theory repeatedly has been flexible enough to be adopted to new analytical models of the earth. The changes made impera-tive by the universal adoption of plate tectonics mechanisms are neither cosmetic or apocalyptic, although they are probably more fundamental in nature than any of the earlier modifications. The basic aspects of geosynclinal theory—the single most unifying con-cept in geology—can certainly be accomodated within the premises of sea-floor spreading, continental drift, and plate tectonics. For the future, the concept will probably continue to be reshaped to conform to changing modern perceptions.

REFERENCES

Argand, E., 1916, Sur l'arc des Alpes occidentales, *Ecolgae Geol. Helvetiae* **14**:145–191.

Argand, E., 1920, Plissements précurseurs et plissements tardifs des châines de montagnes, *Soc. Helvetique Sci. Nat. Actes* **31**:13–39.

Argand, E., 1922, La Tectonique de l'Asie, *13th Internat. Geol. Congr. Compt. Rend.* **5**:171–372

Dana, J. D., 1866, Observations on the Origin of Some of the Earth's Fea-tures, *Am. Jour. Sci.* ser. 2, **42**:205–211, 252–253.

Dana, J. D., 1873, On Some Results of the Earth's Contraction from Cool-ing, Including a Discussion of the Origin of Mountains and the Nature of the Earth's Interior, *Am. Jour. Sci.* ser. 3, **5**:423–443; **6**:6–14, 104–115, 161–171.

Dutton, C. E., 1889, On Some of the Greater Problems of Physical Geol-ogy, *Philos. Soc. Washington Bull.* **11**:51–64.

Hall, J., 1859, Description and Figures of the Organic Remains of the Lower Helderberg Group and the Oriskany Sandstone, *Natural History of New York*, Part 6, vol. 3, Van Benthusen, New York, 532 p.

Hall, J., 1883, Contributions to the Geological History of the American Continent, *Am. Assoc. Adv. Sci. Proc.* **31**:29–69.

Haug, E., 1900, Les Geosynclinaux et les aires continentales, *Soc. Géol. France Bull.* **28**:617–711.

Holmes, A., 1931, Radioactivity and Earth Movements, *Geol. Soc. Glasgow Trans.* **18**:559–606.

Holmes, A., 1945, *Principles of Physical Geology*, Ronald Press, New York, 532 p.

Wegener, A., 1912, Die Entstehung der Kontinente, *Geol. Rundschau* **3**:276–292.

Wegener, A., 1922, *Die Entstehung der Kontinente und Ozeane*, 3rd ed., Vieweg, Braunschwieg, 144 p.

Wegener, A., 1966, *The Origin of Continents and Oceans*, I. Biram, trans., Dover Publications, New York, 246 p.

Part I

THE GEOSYNCLINAL CONCEPT

Editor's Comments
on Papers 1 and 2

1   **HALL**
Excerpts from *Description and Figures of the Organic Remains of the Lower Helderberg Group and the Oriskany Sandstone*

2   **DANA**
Excerpt from *On Some Results of the Earth's Contraction from Cooling, Including a Discussion of the Origin of Mountains and the Nature of the Earth's Interior*

The geosynclinal concept is at least arguably an American innovation, because the geosynclinal explanation for mountain systems was largely the result of the efforts of two American geologists working in the middle of the nineteenth century: James Hall and J. D. Dana. It was James Hall (1859; 1883) who initially outlined the essential elements of geosynclinal theory. J. D. Dana (1866; 1873) provided the name *geosynclinal*; he also refined and clarified the concept. Together Hall and Dana introduced a novel mechanism explaining the origin of mountain systems that was to dominate geological thought in the second half of the nineteenth century and well into the twentieth.

Hall's concept was based on his own survey work in the northern Appalachian Mountains of New York state and was later expanded by some general observations made subsequently in the continental interior of the American midwest. Hall was struck by the stratigraphic and structural differences that existed between Paleozoic strata in the Appalachian Mountains and equivalent strata in the states around the Great Lakes. In particular, Hall noted that the folded sedimentary succession in the Appalachians was up to ten times thicker than the flat-lying, equivalent-aged sequence in the continental interior. Hall incorporated this observation into his greatest inference: a causal relationship existed between great thicknesses of sediment and mountain building.

Hall first formally summarized his view relating mountain belts to thick, synclinal-shaped, sediment-filled basins in a presidential address to the American Association for the Advancement of Science meeting at Montreal in 1857. This address was not well received and was poorly organized and unclear. (It was finally published in the original version only in 1883). However, shortly after the 1857 speech, the State Geological Survey of New York published a third volume in the series by Hall summarizing the stratigraphy and structure of the northern Appalachians, particularly in southern New York (Hall, 1859). Within this work (Paper 1), Hall clearly and exhaustively described the characteristic features of the Appalachian system and asserted that such characteristics probably applied generally to other major mountain systems around the world.

Hall emphasized the major features of the Appalachian mountain system: an elongate chain of folded, shallow-water, Paleozoic clastic sedimentary rocks, immensely thicker within the mountain chain than in areas west of the belt in the Mississippi Valley. On this basis, Hall concluded that the Appalachian Mountains, *and mountain systems in general*, were preceded by subsiding *synclinal* belts that slowly filled with sediment. Because the sediment fill was dominantly clastic and of shallow-water origin, Hall asserted that the rate of subsidence must have closely matched the rate of sediment supply. This observation lead him in turn to conclude that it was the weight of the sediment load that somehow produced the subsidence of the elongate basin, a concept that owed something to theoretical deductions by Charles Babbage and British physicist J. F. W. Herrschel (Zittel, 1901, p. 305).

Hall was not an advocate of lateral compression as a cause of mountain building, favoring instead the traditional plutonist view that mountains (and lowlands) were due to simple vertical uplift and subsidence. Consequently, Hall interpreted the folds in the sediment fill of his basins as the direct result of the slow subsidence of the syncline—as a result of continued subsidence, the diminished width of the upper surface would produce the smaller scale anticlines and synclines. The subsequent crumpling of the upper surface would produce local increases in elevation, hence some mountainous topography. Continued subsidence would necessarily result in faulting, fracturing, and igneous intrusion. The effects of metamorphism and igneous intrusion, noted by Hall to characterize the innermost zones of the Appalachian system, were rationally related by him to the elevated temperatures and pressures presumably existing within the interior, lower portions of the synclinal. Finally, much of the significant vertical uplift of the resulting folded, synclinal mountain system was inferred to have occurred in conjunction with the general emergence of the continental block of which it formed a part. This emergence produced the bulk of a mountain system's topography and was related by Hall to the lateral migration of material displaced from beneath the subsiding syncline and its subsequent injection beneath the immediately adjacent continental areas.

Hall (Paper 1) also pointed out that the Appalachians occupied a critical position relative to the North American continent. Sediment dispersal apparently occurred mainly by currents flowing parallel to the coastline, approximately coincident with the synclinal axis. (Bruce Heezen, in recent decades, demonstrated the existence of such a mechanism; he called them *contour currents* because they run parallel to the contours of the basin; Heezen, Hollister, and Ruddiman, 1966). Hall emphasized that synclines tended to be situated along continental margins, straddling the continental-oceanic boundary.

Hall's views were given a mixed reception and were hotly debated during and immediately following the American Civil War, particularly in a series of articles published in the *American Journal of Science* (Dana, 1866, 1873, Paper 2; Hunt, 1861, 1873; LeConte, 1872, 1873). T. S. Hunt (1861) largely agreed with Hall that mountain chains coincided with preexisting synclines within which great thicknesses of sediment accumulated, and that sediment loading caused the synclinal subsidence. However, he attributed the folding not to subsidence but to yielding due to global contraction; "hence we conceive that the subsidence invoked by Mr. Hall, although not the sole nor even the principal cause of the corrugations of the strata, is the one which determines their position and direction, by making the effects produced by the contraction, not only of sediments, but of the earth's nucleus itself, to be exerted along the lines of greatest accumulation" (Hunt, 1861, p. 412–413).

J. LeConte was also dissatisfied with some aspects of Hall's theory, particularly Hall's mechanism for folding and vertical uplift. LeConte felt that Hall's theory essentially left "the sediment just after the whole preparation had been made, but before the actual mountain formation has taken place. . .a theory of mountains with the mountains left out" (LeConte, 1873, p. 450). LeConte suggested that the uplift of folded mountain chains was due to the folding itself; "mountain chains are the lines along which the yielding of the surface to the horizontal thrust has taken place. . .this yielding is not by upbending into an arch leaving a hollow space beneath, nor such an arch filled and supported by an interior liquid; but *a mashing or crushing together horizontally like dough or plastic clay, with foldings of the strata, and an upwelling and thick-ening of the whole squeezed mass"* (LeConte, 1872, p. 354).

Among the many opponents who questioned the validity of Hall's scheme, the most prominent was James Dwight Dana. In 1866, Dana wrote a brief paper in which he gave some support to Hall, agreeing that mountain chains in general were character-ized by great thicknesses of shallow-water, folded sedimentary rock sequences. However Dana felt that Hall had, if anything, under-emphasized the close correspondence between rates of sedi-ment accumulation and synclinal subsidence, arguing that such balance extended to a "foot for foot" basis (Dana, 1866, p. 208). Furthermore, Dana, a strong advocate of the view that mountains represented zones of compression produced by a contracting earth, was unconvinced by Hall's mechanism for producing the actual elevation in mountain systems: "it is a theory [Hall's] for the origin of mountains, with the origin of mountains left out" (Dana, 1866, p. 210).

Dana again summarized Hall's theory (and its inadequacies) early in 1873, reiterating his view that the theory failed to satis-factorily explain the vertical uplift of mountain systems. Dana's opus magnum appeared later in 1873 as a series of papers spread through several issues of the *American Journal of Science*. Por-tions of this masterpiece appear as Paper 2. It was in this series of articles that Dana proposed the name *geosynclinal* for Hall's *syn-clinal axis*, a name which continued Anglo-Saxon usage would simplify to geosyncline.

Unlike Hall, Dana regarded the crust as relatively rigid. Warp-ing of the crust was the result of the internal contraction of a cool-ing earth. Geosynclinal subsidence was not due to sediment load-ing, but instead was produced by the active downwarping of crustal segments in response to the tangential stress due to global con-traction. Thus, to Dana, the thick geosynclinal sedimentary se-quence was a *consequence* of subsidence rather than the cause thereof. Lateral compression, together with the lateral displace-ment of material as a result of geosynclinal subsidence, also pro-duced linear *geanticlinal* uplifts adjacent to geosynclinals. These geanticlinals served as the principal source areas for geosynclinal sediments.

Dana agreed with Hall that, based on the example of North America, geosynclines were apparently more likely to develop along the margins of continents at the critical junction between continental and oceanic crust. Dana attributed geosynclinal de-formation to the lateral thrusting of oceanic crust against the continental blocks, which is an interesting precursor of the twentieth century plate model. The actual deformation and uplift of mountain systems was in Dana's understanding clearly separated in time from the earlier stage of geosynclinal subsidence.

In the same paper, Dana anticipated continental accretion (termed *annexation* by a quarter of a century; "each epoch of mountain-making ended in annexing the region upturned, thickened and solidified, to the stiffer part of the continental crust, and that consequently the geosynclinal that was afterward in progress occupied a parallel region more or less outside the former" (1873, p. 171). Dana also vaguely suggested the existence of a systematic tectonic cycle, which was later developed by Bertrand and Haug. He believed that some geosynclinal belts underwent multiple series of deformations (polygenetic) as was the case for the Appalachians, whereas others, such as the Alps, exhibited only a single episode of deformation (monogenetic).

For the remainder of the nineteenth century, controversy and debate over the Hall-Dana geosynclinal theory dominated the geological literature. Variations, modifications, and refinements would develop, but the essential aspects of the theory as originally formulated by Hall and Dana would survive and predominate well into the twentieth century.

REFERENCES

Dana, J. D., 1866, Observations on the Origin of Some of the Earth's Features, *Am. Jour. Sci.* ser. 2, **42**:252–253.

Dana, J. D., 1873, On the Origin of Mountains, *Am. Jour. Sci.* ser. 3, **5**:347–350.

Hall, J., 1859, Description and Figures of the Organic Remains of the Lower Helderberg Group and the Oriskany Sandstone, *Natural History of New York*, vol. 3, Van Benthusen, New York, 532 p.

Hall, J., 1883, Contributions to the Geological History of the American Continent, *Am. Assoc. Adv. Sci. Proc.* **31**:29–69.

Heezen, B. C., D. C. Hollister, and W. F. Ruddiman, 1966. Shaping of the Continental Rise and Geostrophic Contour Currents, *Science* **151**:502–508.

Hunt, T. S., 1861, On Some Points in American Geology, *Am. Jour. Sci.* ser. 2, **31**:392–414.

Hunt, T. S., 1873, On Some Points in Dynamical Geology, *Am. Jour. Sci.* ser. 3, **5**:264–270.

LeConte, J., 1872, On the Formation of Features of the Earth's Surface, *Am. Jour. Sci.* ser. 3, **4**:345–355, 460.

LeConte, J., 1873, On the Formation of Features of the Earth's Surface. Reply to Criticisms of T. Sterry Hunt, *Am. Jour. Sci.* ser. 3, **5**:448–453.

Zittel, K. von, 1901, *History of Geology and Paleontology*, M. M. Ogilvie-Gordon, trans., Walter Scott, London, 562 p.

1

Reprinted from pages 20–21, 23, 49, 68, 69–73, 82–83 of *Natural History of New York, Paleontology*, Part 1, vol. 3, Van Benthusen, New York, 1858, 532 p.

DESCRIPTION AND FIGURES OF THE ORGANIC REMAINS OF THE LOWER HELDERBERG GROUP AND THE ORISKANY SANDSTONE

J. Hall

[*Editor's Note:* In the original, material precedes this excerpt.]

The Hudson-river group, though spreading far to the westward, nevertheless maintains its greatest thickness in the direction of the Appalachian chain. In this direction have accumulated the immense amount of its coarser materials; and we may conceive of that range as indicating the pre-existence of a long coast line from which these materials were abraded, forming a submarine belt of sediments in some degree parallel with the outline of an ancient continent on the east. The force of the current, which was sufficient to bring in this vast quantity of sedimentary matter, extended westward with diminishing force, precipitating the finer mud so slowly as to permit the incipient growth of coral reefs along an equal extent of the ocean bed. Thus from the St. Lawrence on the north, through the Appalachian chain, the coarse sandstones and conglomerates indicate the close of this period; while the same geognostic line, from the northern side of Lake Huron, by the course of the Cincinnati axis, quite to the centre of Tennessee and still farther to the south, is marked by bands of coral limestone.

[*Editor's Note:* Material has been omitted at this point.]

From whatever source, however, we are to look for the sediments of this period, it is clear that the existence of a large part of the western slope of this mountain barrier, the Appalachian chain, in Canada, Vermont, Western Massachusetts and Eastern New-York, is due to the original accumulation of materials during this period, rather than to any subsequent influence which has broken up and dislocated the successive beds of the formations composing it. In proof of this we have only to look at the enormous thickness of the sediments in their normal condition; and we shall be forced to admit that however much broken and plicated and degraded by subsequent denudation, the great mass or quantity of these materials must still remain a strong feature in the line of their accumulation.

[*Editor's Note:* Material has been omitted at this point.]

In considering the distribution of the masses of the formations which we have here described, we find that the greatest accumulations have been along the direction of the Appalachian chain. The original current, transporting the material, has been in the same direction, and consequently a greater deposition of the coarser sediment has marked the lines of the transporting force, which, necessarily diminishing on either side of the centre of this great current, the fine calcareous mud would be gradually conveyed to greater distances and slowly deposited. The material thus transported would be distributed, precisely as in an ocean traversed by a current, like our present Gulf stream ; and in the gradual motion of the waters during that period, to the west and southwest, the finer materials would be spread out in gradually diminishing quantities, till, finally, the deposit from that source must cease altogether.

[*Editor's Note:* Material has been omitted at this point.]

We are accustomed to believe that mountains are produced by upheaval, folding and plication of the strata ; and that from some unexplained cause, these lines of elevation extend along certain directions, gradually dying out on either side, and subsiding at one or each extremity. In these pages, I believe I have shown conclusively that the line of accumulation of sediments has been along the direction of the Appalachian chain ; and, with slight variations at different epochs, the course of the current has been essentially the same throughout. The line of our mountain chain, and of the ancient oceanic current which deposited these sediments, is therefore coincident and parallel ; or, the line of the greatest accumulation is the line of the mountain chain. In other words, the great Appalachian barrier is due to original deposition of materials, and not to any subsequent action or influence breaking up and dislocating the strata of which it is composed.

[*Editor's Note:* Material has been omitted at this point.]

At this point of our inquiry, several questions of importance present themselves : First, what has been the cause of this folding and plication of the strata ; secondly, having been thus folded and plicated, what influence has this action exerted upon the elevation of the parts, or of the whole ; and thirdly, what effects are due to the metamorphism which accompanies this mountain chain ?

It has been long since shown that the removal of large quantities of sediment from one part of the earth's crust, and its transportation and deposition in another, may not only produce oscillations, but that chemical and dynamical action are the necessary consequences of large accumulations of sedimentary matter over certain areas. When these are spread along a belt of sea bottom, as originally in the line of the Appalachian chain, the first effect of this great augmentation of matter would be to produce a yielding of the earth's crust beneath, and a gradual subsidence will be the consequence. We have evidence of this subsidence in the great amount of material accumulated ; for we cannot suppose that the sea has been originally as deep as the thickness of these accumulations. On the contrary, the evidences from ripplemarks, marine plants, and other conditions, prove that the sea in which these deposits have been successively made was at all times shallow, or of moderate depth. The accumulation, therefore, could only have been made by a gradual or periodical subsidence of the ocean bed; and we may then inquire, what would be the result of such subsidence upon the accumulated stratified sediments spread over the sea bottom?

The line of greatest depression would be along the line of greatest accumulation; and in the direction of the thinning margins of the deposit, the depression would be less. By this process of subsidence, as the lower side becomes gradually curved, there must follow, as a consequence, rents and fractures upon that side; or the diminished width of surface above, caused by this curving below, will produce wrinkles and foldings of the strata. That there may be rents or fractures of the strata beneath is very probable, and into these may rush the fluid or semifluid matter from below, producing trap-dykes; but the folding of strata seems to me a very natural and inevitable consequence of the process of subsidence.

The sinking down of the mass produces a great synclinal axis; and within this axis, whether on a large or small scale, will be produced numerous smaller synclinal and anticlinal axes. And the same is true of every synclinal axis, where the condition of the beds is such as to admit of a careful examination*. I hold, therefore, that it is impossible to have any subsidence along a certain line of the earth's crust, from the accumulation of sediments, without producing the phenomena which are observed in the Appalachian and other mountain ranges†.

That this subsidence was periodical, we have the best possible evidence in the unconformability of the Lower Helderberg group upon the Hudson-river group; showing that previous to the deposition of these

---

* I am indebted to Sir WILLIAM LOGAN for this latter suggestion, as the result of his very accurate and extensive observations on the relations of anticlinal and synclinal axes.

† To have an idea of this folding, it is only necessary to take a package of flat sheets of paper, and hold the edges firmly in the same position and relation they had when in a horizontal position, depressing the centre, and as the lower sheets assume the curved direction the upper ones will curve

limestones, there were already foldings and plications, the consequence of a subsidence along the line of accumulation. Subsequently to the deposition of the latter formations, or at intervals during their accumulation, there have been other periods of subsidence, and consequently of folding and plication; so that these are not synchronous, nor are they conformable with each other.

This successive accumulation, and the consequent depression of the crust along this line, serves only to make more conspicuous the feature which appears to be the great characteristic, that the range of mountains is the great synclinal axis, and the anticlinals within it are due to the same cause which produced the synclinal; and as a consequence, these smaller anticlinals, and their correspondent synclinals, gradually decline towards the margin of the great synclinal axis, or towards the margin of the zone of depression which corresponds to the zone of greatest accumulation*.

This affords a partial explanation of the fact already observed; that the mountain elevations in the disturbed regions bear in their altitude a much smaller proportion to the actual thickness of the formations, than do the hills in undisturbed regions. Furthermore it so happens that so soon as disturbance takes place and anticlinals are formed, the beds are weakened at the arching, and become more liable to

---

upwards or wrinkle. This is an illustration after a different manner of the old elementary process of producing foldings in sheets of paper, as illustrative of folded strata by lateral presure. Now, as a set of strata one or two hundred miles in width cannot slide over each other, as sheets of paper do if left to themselves during the process of depression, the beds on the lower side must either become extremely broken, or the higher portions become folded and plicated. That some fractures will take place below there can be no doubt, and these are probably such as we see filled with trappean matter. But the greater movement would undoubtedly take place in the higher beds, which necessarily assume positions and relations such as have been pointed out. This condition and manner of movement offers, moreover, an explanation of the form of trap-dykes, which are often narrower above in the synclinals and on synclinal slopes, the matter filling a fracture opened from below; while in the case of such matter penetrating an anticlinal, it would necessarily widen above from the reversed conditions attending the fracture.

* This mode of depression, which is the result of accumulation, and the production of numerous synclinal and anticlinal axes offers a satisfactory explanation, as it appears to me, of the difference of slope on the two sides of the anticlinals which have been so often pointed out as occurring in the Appallachian range, where the dips on one side are uniformly steeper than on the other.

denuding action. Thus the anticlinals are often worn down to such an extent as to form low grounds or deep valleys; while the synclinal, protected in the downward curving of the beds, remains to form the prominent mountain crest. This is very generally true in many parts of the Appalachian range; and it is only where some heavier or stronger bedded rock occurs, protecting the anticlinals, that they form the higher mountain elevations. Similar features will be observed in other mountain ranges*.

It nowhere appears that this folding or plication has contributed to the altitude of the mountains : on the other hand, as I think can be shown, the more extreme this plication, the more it will conduce to the general degradation of the mass, whenever subjected to denuding agencies. The number and abruptness of the foldings will depend upon the width of the zone which is depressed, and the depth of the depression, which is itself dependent on the amount of accumulation.

We have, therefore, this other element of depression to consider, when we compare mountain elevations with the thickness of the original deposition.

It is possible that the suggestion may be made, that if the folding and plication be the result of a sinking or depression of the mass, then these wrinkles would be removed on the subsequent elevation; and the beds might assume, in a degree at least, their original position. But this is not the mode of elevation. The elevation has been one of continental, and not of local origin; and there is no more evidence of local elevation along the Appalachian chain, than there is along the plateau in the west. As it is, a large mass of the matter constituting the sediments of this mountain range still remain below the sea level, as a necessary consequence of the great accumulation; while in the

---

* The sections of the Geological Survey of Great Britain exhibit numerous examples of this kind. On the geological map of Great Britain, a section across the country presents us with Snowden summit as a synclinal, the height of which is much less than the thickness of the strata from the Longmynd to the Caradoc; while, had the bedded trap of Moel Wyn and Aran Mowddwy, and its superincumbent strata, been sufficiently strong to have resisted denudation, the anticlinal axis would have presented a mountain far higher than Snowden.

plateau of the west, we have a much greater proportion above the level of the sea.

So far, therefore, as our observation extends, we are able to deduce some general principles in regard to the production of this mountain range. To explain its existence, we are to look to the original accumulation of matter along a certain line or zone, the direction of which will be the direction of the elevation. The line of the existing mountain chain will be the course of the original transporting current. The minor axes or foldings must be essentially parallel to the great synclinal axis and the line of accumulation. The present mountain barriers are but the visible evidences of the deposits upon an ancient ocean bed ; while the determining causes of their elevation existed long anterior to the production of the mountains themselves. At no point, nor along any line between the Appalachian and Rocky mountains, could the same forces have produced a mountain chain, because the materials of accumulation were insufficient ; and though we may trace what appears to be the gradually subsiding influence of these forces, it is simply in these instances due to the paucity of the material upon which to exhibit its effects. The parallel lines of elevation, on the west of the Appalachians, are evidenced in gentle undulations, with the exception of the Cincinnati axis, which is more important, extending from Lake Ontario to Alabama, and is the last or most western of those parallel to the Appalachian chain.

[*Editor's Note:* Material has been omitted at this point.]

The facts here adduced relative to the strata composing the Appalachian range and their extension to the west and southwest, are all capable of verification; and the deductions hence drawn seem to me perfectly legitimate. I believe, moreover, that this mountain chain, in its component parts, and in its mode of accumulation, and the process by which it has assumed its present position, does not differ materially from other mountain ranges.

The direction of any mountain chain, I would infer, corresponds with the original line of greatest accumulation, or that line along which the coarser and more abundant sediments were deposited. The changes consequent upon the accumulation of such a mass of sediments would, often at least, prevent the immediate deposition of another series of beds of consecutive age in the same direction. Neither is it probable that distinct ranges of mountains, though composed of sediments of the same age, would have a corresponding direction. The Rocky mountains, though perhaps fundamentally composed of deposits of the same age as the Appalachians, have had their materials derived from a different source, and distributed by a current having a different direction. Moreover the greater height of the Rocky mountains appears to be due to later deposits than those constituting the Appalachian range; and if we may credit all the facts stated and their verification by collections of fossils, the strata of newer age than the Coal measures, with the limestones of that age, constitute a large part of the mass producing the altitude of that range of mountains*.

If it be true that original deposition or accumulation has given origin to mountain ranges, then, the greater that accumulation, the higher will be the mountain chain ; and if, after the formation of the older strata along certain lines, there shall supervene conditions allowing the deposition of later formations above the older ones, we may, on the final elevation of the continent, have mountains of greater altitude composed of strata of successive ages. I can conceive, moreover, that under analogous circumstances, the direction of the later currents in the transportation of material may not always have coincided with that of the former ones ; and we may have diverging or cross ranges of mountains with higher summits, where the greater accumulation or the combined accumulations from several sources have taken place.

[*Editor's Note:* In the original, material follows this excerpt.]

---

* See the Reports of Nicollet, Fremont; of Emory, Abert, Cooke and Johnston*; of Captain Stansbury, Captain Marcy, Dr. D. D. Owen : Reports of Explorations and Surveys for a railroad route to the Pacific, Marcou, Blake, Newberry and others : Emory's Report on the United States and Mexican Boundary Survey, Geology, etc., by the writer. Also, results derived from Explorations on the Upper Missouri in 1853, made under my direction, by F. B. Meek and F. V. Hayden†. In addition to all these, may be cited the facts acquired by, or the results derived from, the observations of all the explorers of this mountain range.

* Ex. Doc. No. 41, Thirtieth Congress : Notes upon the Minerals and Fossils, etc., by Prof. J. W. Bailey.

† A communication made by the writer to the American Association for the Advancement of Science, at the meeting of 1855. (Not printed.)

2

Reprinted from pages 429–431 of *Am. Jour. Sci.* ser. 3, 5:423–443 (1873)

ON SOME RESULTS OF THE EARTH'S CONTRACTION FROM COOLING, INCLUDING A DISCUSSION OF THE ORIGIN OF MOUNTAINS AND THE NATURE OF THE EARTH'S INTERIOR

J. D. Dana

[*Editor's Note:* In the original, material precedes and follows this excerpt.]

3. *Kinds and Structure of Mountains.*

While mountains and mountain chains all over the world, and low lands, also, have undergone uplifts, in the course of their long history, that are not explained on the idea that all mountain elevating is simply what may come from plication or crushing, the *component parts* of mountain chains, or those simple mountains or mountain ranges that are *the product of one process of making*—may have received, *at the time of their original making*, no elevation beyond that resulting from plication.

This leads us to a grand distinction in orography, hitherto neglected, which is fundamental and of the highest interest in dynamical geology ; a distinction between—

1. A simple or *individual* mountain mass or range, which is the result of *one process of making*, like an individual in any process of evolution, and which may be distinguished as a *monogenetic* range, being *one in genesis ;* and

2. A composite or *polygenetic* range or chain, made up of two or more monogenetic ranges combined.

The Appalachian chain—the mountain region along the Atlantic border of North America—is a *polygenetic* chain ; it consists, like the Rocky and other mountain chains, of several *monogenetic* ranges, the more important of which are : 1. The Highland range (including the Blue Ridge or parts of it, and the Adirondacks also, if these belong to the same process of making) pre-Silurian in formation ; 2. The Green Mountain range, in western New England and eastern New York, completed essentially after the Lower Silurian era or during its closing period ; 3. The Alleghany range, extending from southern New York southwestward to Alabama, and completed immediately after the Carboniferous age.

The making of the Alleghany range was carried forward at first through a long-continued subsidence—a *geosynclinal*[*] (not a *true* synclinal, since the rocks of the bending crust may have had in them many true or simple synclinals as well as anticlinals), and a consequent accumulation of sediments, which occupied the whole of Paleozoic time; and it was completed, finally, in great breakings, faultings and foldings or plications of the strata, along with other results of disturbance. The folds are in several parallel lines, and rise in succession along the chain, one and another dying out after a course each of 10 to 150 miles; and some of them, if the position of the parts which remain after long denudation be taken as evidence, must have had, it has been stated, an altitude of many thousand feet; and there were also faultings of 8,000 to 10,000 feet, or, according to Lesley, of 20,000 feet.[†] This is one example of a *monogenetic* range.

The Green Mountains are another example in which the history was of the same kind: first, a slow subsidence or geosynclinal, carried forward in this case during the Lower Silurian era or the larger part of it; and, accompanying it, the deposition of sediments to a thickness equal to the depth of the subsidence; finally, as a result of the subsidence and as the climax in the effects of the pressure producing it, an epoch of plication, crushing, etc. between the sides of the trough.

In the Alleghany range the effects of heat were mostly confined to solidification; the reddening of such sandstones and shaly sandstones as contained a little iron in some form ;[‡] the coking of the mineral coal; and probably, on the western outskirts where the movements were small, the distillation of mineral oil, through the heating of shales or limestones containing carbohydrogen material, and its condensation in cavities among overlying strata; with also some metamorphism to the eastward; while in the making of the Green Mountains, there was metamorphism over the eastern, middle, and southern portions, and imperfect metamorphism over most of the western side to almost none in some western parts.

Another example is offered by the Triassico-Jurassic region of the Connecticut valley. The process included the same stages in kind as in the preceding cases. It began in a geosyn-

---

[*] From the Greek γῆ, *earth*, and *synclinal*, it being a bend in the earth's crust.

[†] See an admirable paper on these mountains by Professors W. B. and H. D. Rogers, in the Trans. Assoc. Amer. Geol. and Nat., 1840–42. J. P. Lesley gives other facts in his "Manual of Coal and its Topography," and in many memoirs in the Proceedings of the American Philosophical Society. A brief account is contained in the author's Manual of Geology.

[‡] Oxide of iron produced by a wet process at a temperature even as low as 212° F. is the red oxide $Fe_2 O_3$, or at least has a red powder. (Am. Jour. Sci., II, xliv, 292.)

clinal of probably 4,000 feet, this much being registered by the thickness of the deposits; but it *stopped short of metamorphism*, the sandstones being only reddened and partially solidified; and *short of plication or crushing*, the strata being only tilted in a monoclinal manner 15° to 25°; it ended in numerous great longitudinal fractures, as a final catastrophe from the subsidence, out of which issued the trap (dolerite) that now makes Mt. Holyoke, Mt. Tom, and many other ridges along a range of 100 miles.*

These examples exhibit the characteristics of a large class of mountain masses or ranges. A geosynclinal accompanied by sedimentary depositions, and ending in a catastrophe of plications and solidification, are the essential steps, while metamorphism and igneous ejections are incidental results. The process is one that produces final stability in the mass and its annexation generally to the more stable part of the continent, though not stable against future oscillations of level *of wider range*, nor against denudation.

It is apparent that in such a process of formation elevation by direct uplift of the underlying crust has no necessary place. The attending plications may make elevations on a vast scale and so also may the shoves upward along the lines of fracture, and crushing may sometimes add to the effect; but elevation from an upward movement of the downward bent crust is only an incidental concomitant, if it occur at all.

We perceive thus where the truth lies in Professor LeConte's important principle. It should have in view alone *monogenetic* mountains and these only *at the time of their making*. It will then read, plication and shovings along fractures being made more prominent than crushing:

Plication, shoving along fractures and crushing are the true sources of the elevation that takes place *during the making* of geosynclinal monogenetic mountains.

And the statement of Professor Hall may be made right if we recognize the same distinction, and, also, reverse the order and causal relation of the two events, accumulation and subsidence; and so make it read:

Regions of monogenetic mountains were, previous, and preparatory, to the making of the mountains, areas each of a slowly progressing geosynclinal, and, *consequently*, of thick accumulations of sediments.

---

* This history is precisely that which I have given in my Manual of Geology, though without recognizing the parallelism in stages with the history of the Alleghanies.

Part II

THE EUROPEAN PERSPECTIVE:
A QUESTION OF GEOGRAPHICAL BIAS

Editor's Comments
on Papers 3, 4, and 5

The Hall-Dana geosynclinal concept was initially applied with enthusiasm by European geologists to describe and genetically explain the geological framework of their continent. However, while they concurred that European mountain belts developed, like those of North America, from elongate basins filled with abnormally thick sedimentary successions, it soon became apparent that substantial differences existed between the Caledonian, Hercynian, and Alpine belts and those of North America. The differences in opinion were at first subtle and minor (Bertrand, 1887), but the intensity of the debate became progressively more bitter and fundamental (Haug, 1900). Translated excerpts of these two papers appear here as Papers 3 and 4, respectively.

Brief excerpts from Marcel Bertrand's 1887 paper constitute Paper 3. In this early work, Bertrand was so eager to substantiate Dana's conclusion that geosynclinal belts straddle continental margins that he almost abstractly rearranged the complex tectonic framework of Europe to emphasize its similarity with the symmetrical tectonic framework of North America. Bertrand formally proposed the concept of continental accretion, tracing the growth of Europe by the systematic folding of three laterally-adjacent marginal geosynclinal belts against an unyielding northern foreland.

Bertrand also supported Dana's contention that the development of mobile belts occurred in two distinct stages: an early stage of geosynclinal filling followed by a later stage of orogeny. Bertrand suggested that these stages proceeded in an orderly, evolutionary

26

fashion—a theme to which he would enthusiastically return in less than a decade (Paper 6). Bertrand did not specify whether deformation occurred in a continuous, essentially regular fashion, or whether it was sporadic and pulsatory. He emphasized rather the continuity and systematic distribution in time and space of distinctive lithological-structural *zones* (such as the flysch and molasse belts) within individual geosynclinal belts. Finally, almost parenthetically, he noted that the most striking Alpine terrain, the flysch belt, consisted of thick, slowly deposited, fine-grained, *deep-water* sediments, quite in contrast to the usual shallow-water facies described by Hall and Dana.

In an interesting aside made even more prophetic since the acceptance of modern plate tectonics theory, Bertrand pointed out the curious similarities in the distribution and orientation of major tectonic boundaries visible on both sides of the Atlantic. He speculated that such similarities supported the earlier inferences of Suess (1885–1901) and de Lapparent (1883) that the two continents were once linked together prior to the opening of the modern Atlantic Ocean.

Emile Haug (Paper 4) categorically stated that the geosynclinal model proposed by Hall and Dana was inaccurate, especially with respect to two characteristics: the depositional environments of geosynclinal sediment and the position of geosynclinal basins with respect to continental blocks.

Haug pointed out the contradiction between the insistence by Hall and Dana that geosynclinal sediments were by definition shallow-water littoral deposits and the claims by Neumayr (1875) and Suess (1875) that geosynclinal sediments were deeper-water, abyssal pelagic deposits. (Neumayr and Suess [1920] later collaborated on a study of the bathymetry of Alpine sediments.) Haug concluded that both interpretations were incorrect, contending instead that geosynclinal sediments were deposited at great depth, but at a shallower depth than the abyssal domain. He defined this characteristic geosynclinal depth of deposition as the *bathyal zone*, a zone extending from 80 meters to 900 meters, essentially comprising the outer continental shelf and upper part of the continental slope. Haug also believed that bathyal deposits represented almost continuous sedimentation over long periods of time. Therefore to Haug, unlike to Hall and Dana, the original geosyncline was actually a deep, elongated *trough*. Haug recognized that rates of subsidence and sedimentation might differ appreciably, leading in some cases to starved geosynclines and in other cases to filled geosynclines.

Haug firmly opposed the conviction of Hall and Dana that geosynclines were necessarily confined, like the Appalachians, to continental margins. Other European geologists, led by Steinmann (1905), concluded that the ophiolite terrains of greenstone, basalt, and chert beneath the normal geosynclinal assemblage actually represented oceanic rather than continental crust. These characteristic lithologies of a true geosyncline (abyssal sediments, cherts, and ophiolites) became known as the *Steinmann trinity*. The European school consequently argued with increasing confidence that geosynclines were *intercontinental* features developed at least in part within oceanic crustal areas. This idea constituted a dramatic turnabout from the American view.

The seeds of many other ideas appear in Haug's classic treatise. He recognized that geosynclinal development was evolutionary— older deposits of deep-water, muddy oozes were overlain by progressively coarser-grained sediment. A transition from purely marine to mixed marine and continental deposits usually occurred. The geosynclinal stage of development was succeeded by an orogenic stage. During the early stages of development, small embryonic geanticlines developed within the subsiding geosyncline as folding began. This concept of sediment source areas *within* the geosyncline not only anticipated Argand's concept of *cordilleras*, but stood in sharp contrast to the rather vague American concept of external *borderlands* as the major source of geosynclinal sediment.

The European characterization of geosynclinal sedimentation as largely deep-water persisted well into the twentieth century. In a widely read classical paper, O. T. Jones of Cambridge University reinforced this contention as well as the growing European consensus that geosynclinal belts developed between rather than adjacent to continental blocks (Jones, 1938). He carried out extensive studies of Lower Paleozoic rocks of Britain, characterizing the sedimentary cover of areas *bordering* the Caledonian geosyncline as relatively undeformed "deposits of limited thickness . . . they constitute Haug's neritic facies . . . the seas in which they were laid down have been termed 'platform' or 'shelf seas'" (Jones, 1938, p. 63). Geosynclinal sediments, to Jones, were in sharp contrast to the shallow shelf sediments: "deposition appears to have been continuous, apart from occasional indications of bottom disturbance . . . bedding is uniform . . . sometimes with an appearance of rhythmic deposition. . . . the graded, bedded sandstones which Bailey (Geol. Mag., 1930, p. 84) has suggested should be named 'graywacke' are essentially deposits of the geosyncline and are characteristically found near one or the other of its margins" (Jones, 1938, p. 63).

Hans Stille, a professor in Berlin, is best known for his schemes of geosynclinal classification and nomenclature (1936a, 1936b, 1940). Paper 5 is included because it summarizes Stille's earlier convictions about the nature of geosynclines. Like Dana and unlike Hall, Stille believed that geosynclines were zones of permanent (or long-term) subsidence (that is, generally troughs). However, Stille contended that not all geosynclines would be deformed eventually into mountain ranges, an argument directly opposed to the views of Hall, Dana, and Haug. For Stille, geosynclines and long-lived sedimentary basins were identical. The folding that occurred in those geosynclinal troughs destined to become mountain ranges (*muttergeosynklinalen*) did so in a series of brief, episodic revolutions separated by long anorogenic intervals.

REFERENCES

Bailey, E. B., 1930, New Light on Sedimentation and Tectonics, *Geol. Mag.* **67**:71–92.

Bertrand, M., 1887, La châine des Alpes et la formation du continent européan, *Soc. Geol. France Bull.* 15:423–447.

de Lapparent, A., 1887, Conference sur le sens des mouvements de l'écorce terrestre, *Soc. Géol. France Bull.* ser. 3, **15**:215–241.

Haug, E., 1900, Les Geosynclinaux et les aires continentales, *Soc. Géol. France Bull.* **28**:617–711.

Jones, O. T., 1938, On the Evolution of a Geosyncline, *Geol. Soc. London Quart. Jour.* **94**:60–110.

Neumayr, M., 1875, *Erdgeschichte*, vol. 1, Bibliographische Inst., Leipzig, p. 364.

Neumayr, M., and E. Suess, 1920, *Erdeschichte*, vol. 1, 3rd ed., Bibliographische Inst., Leipzig, 543 p.

Steinmann, G., 1905, Die geologische bedeutung der Tiefseeabsätze und der ophiolitischen Massengestein, *Naturf. Gesell. Freiberg Ber.* **16**:44–65.

Stille, H., 1936a. The Present Tectonic State of the Earth, *Am. Assoc. Petroleum Geologists Bull.* **20**:848–880.

Stille, H., 1936b, Wege und Ergebnisse der geologisch-tektonischen Forschung, *Kaiser Wilhelm Gesell. Jahr. 25,* **2**:77–97.

Stille, H., 1940, *Einfuhrung in den Bau Amerikas*, Gebruder Borntraeger Berlin, 169 p.

Suess, E., 1875, *Die Entstehung der Alpen*, Braumuller, Vienna, 168 p.

Suess, E., 1885–1900, *Das Antlitz der Erde*, 3 vols., Tempsky, Wien. (English translation: *The Face of the Earth*, 1904–1924, 5 vols., Clarendon Press, Oxford.)

3

THE ALPINE CHAIN AND THE FORMATION OF THE EUROPEAN CONTINENT

M. Bertrand

These excerpts were translated expressly for this Benchmark volume by C. A. Schwab and F. L. Schwab, Washington and Lee University, from pages 443–444 and 445–447 of "La châine des Alpes et la formation du continent européen" in Soc. Géol. France Bull.
15:423–447 (1887)

Comparison with North America The Allegheny chain, well-known for the regularity of its folds and whose features recall those of the Jura, stretches out in a direction parallel with the Atlantic coast. All the formations up through the Carboniferous (and as the study of flora has shown, even to the Permian[1]) are involved in the folding. The intensity of folding decreases toward the west and folding disappears. except for broad undulations, as one comes to the great plain of the Mississippi where the same formations are encountered in a nearly horizontal attitude. As for the age of the entire system, the Alleghenies are comparable to the Hercynian chain and, like it, they are bordered by a zone of Carboniferous units (which in the Alleghenies also includes Permian units).

To the north of the Alleghenies is the Green Mountain chain in which Lower and Middle Silurian units are folded, faulted, and overturned, whereas to the west, the Upper Silurian is horizontal and rests with unconformity on the older Silurian units.[2] As the Alleghenies are to the Hercynian chain, the Green Mountains would correspond to the Caledonian chain.

However, if one examines the respective positions of these different chains on the map (Fig. 5) [not reproduced here], one sees that they face one another from across the two sides of the Atlantic Ocean; the Carboniferous border of the Hercynian chain, if continued, runs into that of the Alleghenies, and the Caledonian chain, if extended, runs into the Green Mountains. The possibility of a former connection is suggested on this basis.

Undoubtedly, this relationship cannot be considered as demonstrated. In order to guard against such premature conclusions, it is sufficient to consider the Antilles and the two chains of the western Mediterranean, the Betic chain and the Atlas chain; these two face one another in like manner from the two sides of the Atlantic, like two limbs of a broad anticlinal fold whose arch has slumped. Yet along these coasts it is nearly certain that underwater extensions of the structures do not exist; it would be unlikely, in light of their recent origin, that such continuity existed or that any trace of it would survive violent deformation. However, for the more ancient chains this kind of argument is even less valid; but some argue strongly for such a former junction, particularly those who make the comparison on the basis of similar flora and fauna.

In fact, the continental flora of Devonian and Carboniferous ages are much more similar in the United States and Europe than the marine fauna of the same age. Moreover, until Miocene time, coastal species seem to have been easily propagated from Europe to America; the Cretaceous fauna from the Antilles with *Polypiers, Acteonelles, Nerinees,* and *Rudistids* show striking similarities to the fauna of Gosau in the Alps. The Eocene *Polypiers* of Jamaica and Cuba are identical to those of the Castel-Gomberto reefs in the Vincentian, and Miocene sea urchins of Antigua are found on Malta.[3] These comparisons have suggested Suess's hypothesis, adapted by de Lapparent,[4] that a former continent or chain of islands, Plato's Atlantis, connected Europe and America until Miocene time. This continent or chain of islands would have been something other than fragments of the Hercynian chain.

[*Editor's Note:* Material has been omitted at this point.]

Summary In spite of its apparent complexity, the formation of the present European continent seems to have resulted from a series of remarkably regular and relatively simple movements: three large fold belts formed successively, each encircling the former and inverted toward the northern margin. The rule posed by Dana, that successive zones of folding are formed on the border of oceans and developed against an ancient continental nucleus, is perhaps more applicable to Europe than to America for which it was formulated.

The fact that these simple folding episodes have resulted in such a complicated figure for Europe (which seems to contrast with the symmetry of North America) may have been what has caused it to be said that the two continents were constructed on two different plans; this complexity is largely the result of the piling up and fragmentation of the European chains after their initial formation. But after rational analysis, the reconstruction of the primitive continuity of our little continent shows that it is designed with as much simplicity as North America, with which it has otherwise been blended from the start.

Naturally, the history of sedimentary deposits is closely linked with that of orogenic phenomena, and consideration of the three successive chains permits the particularities of the sedimentary phenomena of the different periods to be grouped into a single, collective scheme.

In the Silurian, the land is to the north, the sea covers the largest portion of Europe and North America. A fold belt is formed from Norway to the Saint Lawrence with the southern limit more or less coincident with the present-day Alps (that is to say, with the Mediterranean peninsula) and also corresponding to the unconformities seen in Shropshire and the Ardennes.

This first chain is modified: atmospheric actions wear it down, some masses of sandstone and puddingstone (Old Red Sandstone) fill the empty depressions located at its feet, while pelagic deposits extend to the south.

A new fold belt is uplifted behind the first, forming a sinuous belt across the continent from the Alleghenies to Westphalia, from Silesia to the Dniester and the Oural. The isolated basin in which coal is deposited is located between this fold belt and the ancient continent. Away from this zone, Carboniferous formations are no more productive than other more recent terrains; coal formation, at least in marine deposits, seems to have been closely linked to the uplift of the Hercynian chain.

The second chain is modified like the first; the New Red Sandstone fills in portions of the depressions formed around it. Triassic lagoons, Jurassic straits and gulfs, and Jurassic coral banks are also established in these depressions. They receive a series of continental and coastal deposits while the open ocean is restricted to the south, in the Alpine region. Then the Alps in their turn are uplifted, producing a third great fold belt that embraces the entire Mediterranean zone from the Pyrenees to the Himalayas. Then, the only evidence that marks the former continuity of the Appalachian and Hercynian chains and explains the similarity of coastal fauna in Europe and America through Miocene time disappears into the Atlantic.

Without doubt only one overall interpretation exists in this case, and some details are still missing. For the Alps, I am limited to considering the northern margin, because it is simplest and most comparable to older chains; but if one studied the entire belt of Tertiary folds one would see it closely linked to the Alps as a series of southern extensions across the Mediterranean. Likewise, the older chains with their relatively simple northern border might have had more or less complex southern extensions whose shape and position, difficult to decipher, undoubtedly had some influence on the more recent folding producing their irregularities. If the conclusions that I have proposed here are accepted, we find that the succession of movements corresponds quite well to what a hypothetical observer at the summit of the primitive arctic continent would see. He would first have observed a large solid wave forming in the sea that extended at his feet, rising up slowly to conceal the horizon from him, then congealing while unfurling on its edges. Later, some openings would be made in the large, continuous high barrier, and he might see a second wave, then a third, each forming successively farther to the south, and each, like the first, unfurling (breaking) in turn. It is probable that he would have to wait today to see a fourth wave formed behind the Alps (that is, in the Mediterranean region). But *rules of three* are inapplicable to these subjects, and we will never know if the waiting of our hypothetical observer will come true or if he will be disappointed.

NOTES

[1]White and Fontaine, Permian flora of West Virginia, 1880.
[2]Dana, Manual of Mineralogy, p. 211.
[3]Suess, Das Antlitz der Erde, p. 365.
[4]Traité de Géologie, p. 1188 and 1280.

4

GEOSYNCLINES AND CONTINENTAL AREAS

E. Haug

*These excerpts were translated expressly for this Benchmark volume by C. A. Schwab and F. L. Schwab, Washington and Lee University, from pages 618–623, 626–627, and 630–632 of "Les Géosynclinaux et les aires continentales" in Soc. Géol. France Bull. **28**:617–711 (1900)*

FIRST PART--GEOSYNCLINES

I. General Concepts

Definition. The concept of the geosyncline is undoubtedly due to James Hall. In fact it is this famous paleontologist who, after explaining the enormous accumulation of sediment coincident with certain zones of the earth's surface as a consequence of gradual subsidence of the sea floor, showed (1859)[1] that "the axis of the greatest subsidence coincides with the axis of greatest accumulation," thus establishing the direct relationship in each case of sediment thickness and regional subsidence. He also contended that the subsidence of the sediment mass produced a "large synclinal axis." It is to this concave downfolding of the earth's crust that Dana (1875)[2] gave the name "geosynclinal." Moreover, he correctly attributed its formation to lateral compression, not as Hall did, to the weight of sediment.

At the same time that he outlined the mechanics of sediment accumulation, Hall also postulated the conclusions that were to form the fundamental basis for subsequent theories of orogeny; i.e., fold belts occur coincident with zones of great sediment thickness. We have often given to this law the following form: mountain chains form on the sites of geosynclines.

The classic example is the Appalachian chain, in which folding has affected a thickness of sediment that American geologists estimate at 40,000 feet. In the central Himalayas, Diener,[3] basing his calculations on the observations of Griesbach, estimates that the sedimentary succession deposited across a single region increases from 900 to 14,000 feet without any important unconformities that would indicate significant interruptions in sedimentation.

Bathymetric Character of Geosynclines. Suess reserved judgement on the relationship between cause and effect existing between folding and formation of geosynclines, but he did propose another conclusion (1875)[4] that amplifies the principle originally stated by James Hall and can be stated as follows: In the folded zones, the sedimentary series is generally complete and possesses a "pelagic" character; conversely, in relatively undeformed zones, the sedimentary succession is incomplete and is interbedded with brackish (shallow) water deposits. The Triassic and the Portlandian stage of the Jurassic are well-known examples supporting this conclusion.

A certain contradiction apparently exists between James Hall's original principle and this important statement by Suess. In effect, American authors, and James Hall himself, have often insisted very forcefully on the littoral (coastal) or at least "shallow water" character of geosynclinal sediment; on the other hand, Suess understands by "pelagic" deposits, relatively deep water deposits, and Neumayr went so far as to consider certain Jurassic deposits of the Alpine region as "abyssal."

Evidently, exaggeration exists on both sides. In the Appalachians, Cambrian and Silurian deposits at the base of the folded series are coarse detritus indicating very shallow water deposition; younger deposits (Devonian and Carboniferous) were certainly deposited in a sea of relatively greater depth. On the other hand, with regard to the supposedly "abyssal" deposits of the eastern Alps, the silicious radiolarian limestones of Jurassic and Lower Cretaceous (Neocomian) age that Neumayr[5] has compared to radiolarian oozes forming at abyssal depths in modern ocean basins, Johannes Walter[6] has shown that these two deposits certainly do not have similar origins. As to the assimilation of white chalk into Globigerina muds, Cayeux and Walter have both adequately explained the phenomenon.[7]

Personally, I have reached the conclusion that geosynclines, in most cases, correspond to relatively deep seas but not abyssal depths. It is necessary to adapt a special term for designating the bathymetric zone existing between the *abyssal zone* (correctly named) and the shallow zone (sometimes incorrectly referred to as "littoral"). I have proposed the name *neritic zone*[8] for the shallow zone for lack of a corresponding term for "shallow water" in French or "Seichtwasser" in German, and while waiting for a better term, I name (with Renevier) the intermediate zone between the neritic and abyssal zones as the bathyal zone, and I assign to it as specific boundaries the depth from 80 or 100 meters to 900 meters.

In modern seas, surface marine currents can be felt throughout this bathyal zone. Temperature there is nearly constant and organisms living there are essentially temperature-sensitive. Light only penetrates slightly into the upper layers and chlorophyl activity is consequently negligible so that algae and herbivorous organisms are absent. The fauna consists mainly of carnivorous organisms and mud eaters (limivores) that nourish themselves on the organic remains accumulating from the bodies of surface-dwellers. Ornate shells are rarer and colors less vivid than in the neritic zone.

However, because the bottom is swept by the passage of strong marine currents, certain benthonic organisms not ordinarily found at such great depths are remarkably abundant and diverse thanks to the abundant nourishment furnished by the supply of surface plankton.

A large number of geological formations are without doubt deposited in the bathyal zone and it is precisely these that largely constitute geosynclinal sediments. These include graptolite-bearing schists, muds, and nodular ammonitic marls (chalky muds) and limestones. The red ammonitic limestones and *Aptychus* schists seem to have been formed under the same bathymetric conditions, although in areas where most fossiliferous detritus was swept away and where the lack of subsidence prevented accumulation of thick masses of sediment.

Geosynclinal formations of questionable origin have received the name "pelagic facies," which is misleading in terms of the kinds of organisms preserved there. One knows in effect that Murray groups sediments in modern seas according to their mode of formation as either "pelagic deposits" or "terrigenous deposits." The first include "those that are formed toward the center of the big oceans and which are composed principally of the remains of pelagic organisms associated with the final products of the decomposition of rocks and minerals."[9] However, sediments of the bathyal zone do not at all conform to this definition, belonging for the most part, instead, to the category of terrigenous deposits. Despite this drawback, the term "pelagic facies" could be applied with good reason to some sediments rich in ammonites, even if earlier ideas regarding the habitat of cephalopods are still valid. Based on present knowledge, it no longer seems likely that these molluscs [ammonites] can be interpreted as good swimming organisms living on the ocean surface in the manner of the *Argonaut*. Like *Spirula* and *Nautilus,* they had to live near the bottom and were members of the benthonic community. Apart from the rather rare cases where ammonite shells remained floating after death to be carried by currents into littoral regions (as proposed by Alcide d'Orbigny[10]), it seems that most often ammonites lived on the ocean bottom in the very place where they are found today. Besides the excellent state of preservation of the "peristome" and other delicate ornaments in certain deposits, the hypothesis of ammonites as "floating organisms" cannot adequately explain the close association of numerous individuals of the same type in one spot, suggesting the existence of a collection of like organisms living together near the bottom. Most of the belemnites and *Nautilus* seem to have lived generally at lesser depths than the ammonites (i.e., in the deeper part of the neritic zone), and of the various types of ammonites, species such as *Trachyceras, Reineckia, Hoplites,* etc. were probably eurythermic (i.e., they tolerated rather large variations in temperature and could therefore live at variable depths as opposed to other species), particularly *Phylloceras* and *Lytoceras,* which were sternothermic, requiring a constant temperature and therefore living only in the bathyal zone. They characterize the deeper portion of geosynclines whereas exclusively eurythermic species generally live in the shallower, littoral part of geosynclines. Thus is explained the contrast between the "province of central Europe" and the "Mediterranean province," which in reality correspond to the neritic zone and bathyal zone of the same zoological "epoch."[11]

A close analogy exists between the modes of distribution of fossils in clays and marls of the Mesozoic and the distribution of animals on the clay bottom of the bathyal zone in modern seas. In both cases, the organisms are distributed together in a completely sporadic manner. Localities where they are abundant are often separated by vast regions where life is almost absent, and in all these latter localities one type predominates, frequently to the exclusion of all other types.

If the paleontological characteristics speak in favor of attributing geosynclinal sediments to a relatively deep bathymetric zone, it is the same with lithological characteristics. Whenever the lithological changes in geosynclinal sediment fill are traced across (and perpendicular to) its axis (that is, across the line of greatest sediment thickness), we see first a gradual thinning, then an abrupt lateral passage produced by multiple alterations of deep water sediment with the shallower water sediments deposited in the neritic zone. I will review an example that I have studied myself.[12]

[*Editor's Note:* Material has been omitted at this point.]

Relations Between Geosynclines and Folds. Relations between geosynclines and folds are of two orders: those that exist between the
geosyncline and subsequent folds; and those that exist between the
geosyncline and folds developed prior to its formation. We have seen
that it is James Hall who first showed that folds originated on the
site of geosynclines. To the same author[13] are due the notions that
folds form at depth and that folding occurs completely independent of
the elevation of mountainous terrain.

The first step in the formation of folds on the site of a geosyncline is the birth of an anticline or median geanticline that divides
the primitive geosyncline into two secondary geosynclines. At any rate,
this is what takes place in two places that have particularly attracted
my attention: in the Prealps of Chablais and Roman Switzerland, and
in the Alps of Dauphinois. In both of these cases the median geosyncline exists beginning with Lias time. It persists during most of the
Mesozoic, and it is this median axis that will control the median
axis of later folds, that is, the line on both sides of which folds
will lean in opposite directions in such a way as to form a compound
"fan."

I do not at all wish to claim that a fold located along the geosynclinal axis will always play an identical tectonic role, but I won-
der if inversely, the disposition of folds in a fanlike pattern found
in many chains or portions of chains is not due precisely to the existence of such a geanticline separating the two geosynclines. Thus the
many exceptions to the law of unilateral thrusts, first formulated by
Suess in 1875 and studied since for verification around the globe,
might be explained. . . . I have no intention of studying the mechanism of folding here and I will restrain myself from stating certain
facts. We note, in numerous cases, that there exists an almost perfect
parallelism between fold axes and the axes of multiple geosynclines
whose formation preceded the episodes of folding.

[*Editor's Note:* Material has been omitted at this point.]

Siting of Geosynclines with Respect to Continental Masses. The
American authors credited with the concept of the geosyncline have
always taken as the point of departure for their orogenic theories
the fundamental idea that mountain chains are formed on the boundary
between continents and *oceans,* and that continents grow by the addition of more and more recent chains. In this hypothesis, because geosynclines originate at the borders separating continents and oceans,
the sediments that would accumulate there would be exclusively coastal
(littoral) deposits, and the zone of submergence where thick sedimentation occurs would be separated from the deep sea (abyssal zone) by
a simple slope.

It is easy to demonstrate that it is not in such conditions that
geosynclines are formed and that far from originating at the margins
of continents, they are always found between two continental masses
and constitute mobile zones found between two relatively stable masses. A few examples prove this concept.

The largest mountain range of the globe, the Himalayas, corresponds to a vast geosyncline where sediments reach immense thicknesses. Neither in the Paleozoic nor in the Mesozoic do the deposits
show a littoral character there, and at no time was the region found
situated on the edge of a large ocean. We can accept the fact that
some more recent chains have been successively added to the extremely

ancient continent of central Asia to the south and that the Himalayas were the last to be formed, but this very mobile belt has always been bounded on the south by the stable region of the Indian peninsula, a fragment of a much vaster continent.

Likewise, the chains of central Europe, folded at the end of the Paleozoic era and toward the end to the Tertiary, when taken as an entity occur between the older chains of northern Europe and the former African continent. Here, as in Asia, the condition implied by the geo-synclinal theory of American geologists is found to be untrue. Like-wise if one considers the European chains in detail. The Pyrenees are crushed between the Massif Central of France and the Iberian Meseta; the Carpathians are compressed against the Russian platform and their hinterland (foreland) is composed entirely of ancient massifs; the Dinarides are inserted between the "terre orientale" eastern land of Mojsisovics and the Adriatica massif of which Monte Gargano is a frag-ment; and the Atlas chain is likewise pinched between the old African continent and the crystalline chain on the coast, today in large part collapsed beneath the Mediterranean. In every one of these well-known examples, the folded chain always corresponds to a geosyncline. The subsided mobile zones form a series of sinuous channels that outline the ancient consolidated cores.

Even the position of the Appalachian geosyncline no longer indi-cates that one is in the presence of a continental-ocean margin zone. American geologists agree that the "Piedmont Plateau," an Archean chain extending from the Hudson to Alabama parallel with the Atlantic coast, constituted a southeastern coast that supplied sediment to the Appalachian geosyncline. Therefore the ocean cannot be found to the southeast on the present-day site of the Atlantic. Following James Hall, many authors have admitted that the deep sea was toward the northwest in a little-folded region where facies contemporaneous with those of the Appalachians are quite different from those of the Appa-lachian geosyncline. I do not believe that this interpretation is justified.

In fact, as far as the Silurian deposits, one can hardly contend that the coral-bearing limestones of Cincinnati were formed in shallow-er water than the graptolitic schists of the Appalachians; regarding the Devonian, it includes well-developed, cephalopod-bearing schists (Marcellus-Naples formations) that are replaced in the Midwestern states by thinner limestones bearing the remains of coelenterates, crinoids, and brachiopods, all of the neritic type.

In addition, one knows that the "Algonkian continent," which at the beginning of the Cambrian occupied the center of North America, had been covered only intermittently by Paleozoic seas. This makes it difficult to reconcile the temporary existence of an "ocean" on this site.

From all these examples, one can attribute the character of a general law based on the following two observations: geosynclines, essentially mobile regions of the earth's surface, are always situated between two continental masses--regions that are relatively stable; and geosynclines constitute, before filling up, marine depressions of rather considerable depth. Continental areas are, on the other hand, either emergent areas or areas only temporarily invaded by shallow seas.

[*Editor's Note:* In the original, material follows this excerpt.]

NOTES

[1]James Hall. Natural History of New York. Vol. III, p. 70, Albany, 1859.

[2]James D. Dana. Manual of Geology. 2nd edit., p. 748.

[3]In E. Suess. Are great ocean depths permanent? Nat. Science, vol. II, p. 18.

[4]Eduard Suess. Die Entstehung der Alpen. Vienne, 1875, p. 98.

[5]Neumayr. Erdeschichte, t. I, p. 364.

[6]J. Walter. Ueber die Lebenweise fossiler Meeretheire, Zeitschr. d. D. Geol. Ges., vol. XLIX, p. 214, 1897.

[7]I will not say anything about the ideas of Renevier concerning the abyssal type muds.

[8]Annual review of Geology. Revue gen. des Sciences, 30 June, 1898, p. 496.

[9]John Murray and A. F. Renard. Report on deep-sea deposits, Report on the Scientific Research of the Voyage of H.M.S. Challenger, p. 185, 1891.

[10]Cours élémentaire de Paléontologie, t. I, p. 85, 1849.

[11]Pompeckj and I both arrived independently at this interpretation, published a few weeks apart from one another. Pompeckj postulated it first in connection with the Lias of Anatolia (Paleontologische und stratigraphische Notizen aus Anatolie, Zeitschr. d. D. Geol. Ges., vol. XLIX, p. 826, 1898). I was lead to the same conclusion from the upper Jurassic and lower Cretaceous, based in part on some observations by Kilian (V. Revue annuelle de Géologie. Revue gen. des Sciences, 30 June, 1898, p. 497).

[12]Les chaînes subalpines entre Gap and Digne, p. 55, Bull. Sérv. Carte géol., no. 21, 1891.

[13]Hall, 1859, p. 72. "It nowhere appears that this folding or plication has contributed to the altitude of mountains. . . It is possible that the suggestion may be made, that if the folding and plication be the result of a sinking or depression of the mass, then the beds might assume, in a degree at least, their original position. But this is not the mode of elevation. The elevation has been one of continental, not local origin."

5

BASIC PROBLEMS OF COMPARATIVE TECTONICS

H. Stille

This excerpt was translated expressly for this Benchmark volume by R. B. Youngblood, Washington and Lee University, from pages 6–9 of Grundfragen der verleichenden Tektonik, Gebruder Borntraeger, Berlin, 1924, 443 p.

SOME CONCEPTS OF TECTONICS

I. The Terms Geosyncline and Geanticline

A. The Term Geosyncline. The term geosyncline is not used in a uniform manner in the geological literature. In its broadest sense, a geosyncline is defined as a subsiding area of sedimentation that persists in time (Stille, v. 5, p. 7). Perhaps one should say, "an area sinking through time," because even though a geosyncline usually contains a thick succession of sediment, the amount of sediment can be minimal under special circumstances--for example, when the geosyncline is located at a fairly great distance from land, or when there is only a small influx of terrigenous material. In general, however, the rate of subsidence is matched by the rate of sedimentation. As a result, a great thickness of strata results, as is shown by analyses of ancient geosynclinal belts. For example, one is reminded of the succession of strata up to 6,000 meters thick in the Appalachians, of the 14,000 foot thick concordant succession in the Himalayas, of the 5,500 meter thick succession of carbonate in the California Valley between the Sierra Nevada and the Coastal Range, and according to Schuchert, of the continuous, 76,000 foot Algonkian-Paleozoic-Mesozoic succession deposited in the Rocky Mountain geosyncline of Idaho, western Montana, and southeastern British Columbia. The post-Variscan sedimentary succession in portions of the Lower Germany Basin, which accumulates up to a thickness of about 7,000 meters, indicates the usual deep subsidence of these types of sedimentary basins. However, there are also examples of deeply subsiding basins that are exclusively or almost exclusively continental in nature. I point, for example, to the 6,000 foot thick Tertiary deposits in the central Cascade Mountains of the western United States, or to the 3,500 meter thick continental Triassic deposits of New Jersey.

The term geosyncline is often applied, however in a restricted sense, only to those particularly deep depressions of our Earth that as a consequence of deep subsidence are generally marine (covered by the sea) and that subsequently become the sites for folding. However, there is no basic difference between these "true" geosynclines and other depressions--for example, in geosynclines where sedimentation

is mainly continental in nature, remaining so despite strong subsidence
of the basin floor either as a consequence of high rates of sediment
supply or the fact that the subsiding trough is a portion of a larger,
rising continental block, the sea either cannot gain access into it or
can do so only temporarily. In addition, there are many other varieties
of geosynclines.

The term geosyncline was first used by Dana, but the full concept
on which it was based goes back to Babbage (1833) and Herschel (1836),–
who assumed that great depressions of the Earth were a prerequisite
for the emergence of thick bodies of strata. The concept is further
rooted in the ideas of James Hall, who in 1857 recognized folds to be
natural features that accompany and are a consequence of subsidence
and sedimentation. Hall spoke of great "synclinal" axes along which
smaller scale anticlines and synclines are formed. What Hall labeled
large "synclines" in this sense, Dana subsequently called "geosynclines"
and assigned the term "synclinorium" (i.e., "mountain range of synclines") to the systems of mountain chains that developed from them.
The typical standard reference model is the geosyncline from which
the Applachian mountain system developed.

According to Schuchert, it is a characteristic of the Appalachians,
as well as North American geosynclines in general, to be situated within a continental mass and separated from the ocean by a "borderland."
Conversely, there are geosynclines situated between continents--"mesogeosynclines" or "mediterraneans," according to Schuchert. The Euro-
pean-Asian Tethys is a typical example. E. Haug, in particular, pointed
to this intercontinental situation as apparently the basic characteristic of geosynclines. He assumed that there probably was a great Pacific continent off the western coast of North America, giving an intercontinental character to the geosyncline from which the Cordilleran
belt of North and South America developed.

Incidently, in expanding my use of the term geosyncline to all
basins that subside over long intervals of time, I am merely following
the example of Dana, who included even the great ocean basins as geosynclines. Also, Schuchert, after placing the North American geosynclines situated within a continental block in their proper perspective
and describing them in detail as the standard type, thought that the
concept of geosynclines should, in the final analysis, apply to "all
the regionally-extensive, time-persistent, downflexured parts of the
lithosphere."

One might use the designation sunken basin or sunken trough as
the German translation for geosyncline, referring mainly to marine
basins or troughs. Often a continental trough becomes a marine trough
as a result of marine transgression, but under the influence of regression, a marine trough can also become a continental trough.

In accordance with the Dana-Hall premise that sedimentation and
folding are interconnected, the idea that mountain ranges are born
out of their laps goes hand in hand with the term geosynclines. However, we also know of areas of long-term, appreciable subsidence accompanied by long-term, thick sedimentation in which actual folding either
did not subsequently follow or has not yet occurred, or in which orogeny was largely restricted to tensional movements. I am thinking, for
example, of the sedimentary basins of the North European Old Red Sandstone, of the great Neopaleozoic (Permo-Carboniferous) continental
geosynclines in Central Europe, of the Newark-type basins in Eastern
America, and in a certain sense also of the Triassic basins of Germany.

Even if the occurrence of large folds may be the rule in subsiding basins, it cannot be the defining characteristic for the term geosyncline, contrary to frequent statements in the geological literature. It would be equally as logical to deny the term "woman" to a woman who had not borne a child.

In summary, it certainly must be admitted that there is a certain extreme type of subsiding basin that corresponds to what in many cases is commonly designated a geosyncline, that is: a regionally extensive, relatively narrow zone of subsidence, wholly or at least predominantly the site of marine sedimentation that subsequently becomes the location from which mountain ranges are formed. But a basic distinction between geosynclines and other areas of perhaps less subsidence (areas of shorter-term or perhaps long-term but continental sedimentation, or from irregularly-shaped sedimentary basins perhaps only partially deformed) is not possible. Thus, we must either extend the term geosyncline to all basins undergoing long-term subsidence--which seems to me the most practical solution and has also found a wide application-- or we must introduce a new term for this totality from which "real" geosynclines would form a subcategory that would not be sharply delineated.

B. The Term Geanticline. Even Dana contrasted geanticlines, which we use to mean areas undergoing long-term uplift, with geosynclines. If geosynclines are sites of sedimentation, in contrast, geanticlines are mainly sites of erosion. And just as sedimentation in geosynclines generally matches the rate of subsidence, the progressive uplift of geanticlines matches the rate of erosion, thus maintaining geanticlines permanent source areas for the adjacent geosynclines.

Because geanticlines border geosynclines, they frame or enclose them, at least initially, in a paleogeographical sense. However, during times of subsequent folding, mountain chains rise from within the geosyncline itself, and the geanticlines become unable to maintain their role as forelands sitting in front of the fold belt, although they continue to control to a large extent the type of folding that occurs in the neighboring geosyncline.

REFERENCES

[*Editor's Note:* Stille did not actually cite Babbage (1833) or Herschel (1836) in his original paper, that is, no specific references. I have been able to trace Herschel to two 1836 letters written to Lyell and Murchison, later published in 1837 in a book by Charles Babbage. The Babbage reference, I believe, is to a talk given in 1833, rather general remarks, made to the Geological Society, and published by them in 1834. I list these references below, plus the additional one from this section on Schuchert.]

Babbage, C., 1833-1834, *Geol. Soc. Proc.* 2:72.
Herschel, J. F. W., 1836, personal communications, two letters, Feb., 1836 to Lyell and Nov., 1836 to Murchison. Subsidence of the Earth's Crust as a Consequence of the Weight of the Strata, in C. Babbage, *The Ninth Bridgewater Treatise: A Fragment,* 1837, John Murray, London, pp. 216-217.
Schuchert, C., 1923, Sites and Natures of the North American Geosynclines, *Geol. Soc. America Bull.* 34:151-260.

Part III

THE GEOTECTONIC CYCLE: THE CASE FOR SYSTEMATIC GEOSYNCLINAL EVOLUTION

Editor's Comments
on Papers 6 Through 10

Hall, Dana, and Haug, among others had all stated their general conviction that the processes by which geosynclines were initially formed, filled, and eventually deformed into uplifted mountain ranges seemed to occur in an orderly, sequential fashion. This intuitive conviction implied that geosynclinal sedimentation on the basis of lithology must also proceed in a systematic, evolutionary pattern. Marcel Bertrand (1897) first formally proposed this idea, a concept that has influenced most subsequent stratigraphic analyses of geosynclinal mobile belts.

Bertrand (Paper 6) recognized that a consistent relationship existed among the various major lithological facies within a geosynclinal sedimentary prism. The secular variation, coarsening upward, was manifested by older, interbedded shale and chert sequences (often resting on ophiolites) that were overlain by younger, coarser flysch and still younger molasse. Based on this apparently cyclical, unidirectional pattern of geosynclinal filling, geologists began to conclude that geosynclines exhibited an overall deterministic behavior—geosynclinal evolution was characterized by the sequential occurrence of distinctive igneous, metamorphic, and sedimentary rock suites. From such a fixed, time-dependent pattern, there could be little variation. Such views were enthusiastically backed by Stille (1913), Tyrell (1933) von Bubnoff, (1925, 1931), Kossmat (1921) and Kraus (1927, 1928) among others. Kraus (1927) refined Bertrand's terminology, proposing four discrete stages of geosynclinal development: *preorogen* (pre-orogenic), *tieforogen* (deep orogenic), *hochorogen* (full orogenic), and *nachorogen* (postorogenic).

North American geologists became aware of Bertrand's ideas largely due to the efforts of Paul Krynine (Papers 7, 8, 9, and 10), who would later remark, "As is usual with great pioneering work done before its time, Bertrand's ideas were promptly forgotten" (Krynine, 1951, p. 745). Krynine's brief but convincing case for geotectonic cycles and their commanding influence over geosynclinal sedimentation won the endorsement of many sedimentologists and stratigraphers. The close link Krynine established between sedimentary and tectonic environments would fundamentally govern the approaches used to analyse geosynclinal terrains during much of the twentieth century. Later classic papers expanding these ideas include those by Pettijohn (1943) Tercier (1946) Krynine (1948) and Krumbein, Sloss, and Dapples (1949).

In retrospect, it is perhaps surprising that such consensus could exist among geologists everywhere as to the reality of *simplistic* geotectonic cycles (Cady, 1950). For following World War II, detailed knowledge of several mobile belts made it apparent that the geosynclinal cycle, at least as formulated, was far too simple an approximation of reality. The type of geosynclinal sedimentation, the kinds of igneous and metamorphic rocks within mobile belts, and the various stages of tectonism experienced—even within portions of the same mobile belt—bore no consistent relationship to one another in time or space despite enthusiastic attempts to forcibly fit pieces of individual mobile belts into identical secular and spatial molds. Such inconsistencies, initially denied and subsequently deplored (Paper 19) would eventually kindle pleas that the entire geosynclinal concept be scrapped (Paper 34).

REFERENCES

Bertrand, M., 1897, Structure des Alpes Francais et Récurrence de Certain Facies Sédimèntaires, *6th Internat. Geol. Congr. Compt. Rend.* pp. 161–177.

Cady, W. M., 1950, Classification of Geotectonic Elements, *Am. Geophys. Union Trans.* **31**:780–785.

Kossmat, F., 1921, Die mediterranen Kattengebirge in ihrer Berichtigung zum Gleichgewichtszustande der Erdrinde, *Akad. Wiss. (Leipzig) Math.-Naturw. kl. Abh.* **38**:46–68.

Kraus, E., 1927, Der orogene Zyklus und seine Studien, *Zentralbl. Mineralogie Paläontologie, Abt. B,* pp. 216–233.

Kraus, E., 1928, Das Wachstum der Kontinente nach der Zyklustheorie, *Geol. Rundschau* **19**:353–386, 481–493.

Krumbein, W. C., L. L. Sloss, and E. C. Dapples, 1949, Sedimentary Tectonics and Sedimentary Environments, *Am. Assoc. Petroleum Geologists Bull.* **33**:1859–1891.

Krynine, P. D., 1948, The Megascopic Study and Field Classification of Sedimentary Rocks, *Jour. Geology* **56**:130–165.

Krynine, P. D., 1951, A Critique of Geotectonic Elements, *Am. Geophys. Union Trans.* **32**:743–748.

Pettijohn, F. J., 1943, Archean Sedimentation, *Geol. Soc. America Bull.* **54**:925–972.

Stille, H., 1913, *Evolutionen und Revolutionen in der Erdgeschichte,* Gebruder Borntraeger, Berlin, 32 p.

Tercier, J., 1946, Problèmes de Sédimentation dans l'Insulinde, *Ver. Schweizer. Petroleum-Geologen u. -Ingenieure Bull.* **44**:7–19.

Tyrell, G. W., 1933, Greenstones and Graywackes. *Réunion internat. pour l'étude de Precambrien et des vielles châines de montagnes en Finlande, 1931, Compt. Rend.* pp. 24–26.

von Bubnoff, S., 1925, Ueber Werden und Zerfall Knotinental-Schollen, *Fortschr. Geologie Palaontologie* **10**:1–84.

von Bubnoff, S., 1931, *Grundprobleme der Geologie,* Gegruder Borntraeger, Berlin, 237 p.

6

THE STRUCTURE OF THE FRENCH ALPS AND THE RECURRENCE OF CERTAIN SEDIMENTARY FACIES

M. Bertrand

This excerpt was translated expressly for this Benchmark volume by C. A. Schwab and F. L. Schwab, Washington and Lee University, from pages 174–177 of "Structure des Alpes Francais et Récurrence de Certain Facies Sédimèntaires" in 6th Internat. Geol. Congr. Compt. Rend., 1897, pp. 161–177.

In conclusion, with the reservations and uncertainties pointed out above, one can consider the following four cycles:

Huronian Chain	Silurian Chain	Carboniferous Chain	Alpine Chain
A_1 Laurentian Gneiss	A_2 Cambrian Gneiss	A_3 Devonian Gneiss	A_4 Permian Gneiss
B_1 ?	B_2 Fine-grain schistose Flysch (Hudson schists, Ordovician?)	B_3 Fine-grain schistose Flysch (Culm)	B_4 Fine-grain schistose Flysch (Schistes lustres)
C_1 ?	C_2 Coarse Flysch (Upper Silurian)	C_3 Coarse Flysch (Carboniferous)	C_4 Coarse Flysch (Cretaceous and Eocene)
D_1 Red sands (Precambrian or Cambrian)	D_2 Red sands (Devonian)	D_3 Red sands (Permian)	D_4 Puddingstones and molasse (Oligocene and Miocene)

Each cycle, I repeat, includes four terms that correspond to one another; the importance of each stage of sedimentation is of relative importance in the different chains, but each is systematically related to the others in terms of their positions in the cycle. This timetable permits us to outline a general scheme for the history of mountain chains. First is the formation of a large geosyncline prior to central thrusting: fine-grained flysch (B) accumulates within the geosyncline. This original basin is deformed and divided by a progressively more sharply defined central thrust. Along its margins, two narrower basins deepen, and in them coarser flysch (C) accumulates. A period of more intense deformation piles a series of folds onto the same spot, giving birth to large, layered folds; following this period of violent rupture of stability, molassic sandstones and puddingstones (D) accumulate at the foot of the chain.

This series of successive phenomena produces an internal, equally gradual uplift; granitic magmas rise up particularly in the central portion and transform the terrain into gneiss before absorbing it into its mass. Gneissic metamorphism can reach into the terrain immediately underlying the finer schistose flysch (B), as far as the units of the preceding chain that are red beds (D); granite penetrates in isolated bodies that intrude the finer-grained flysch (B) (Silurian in Scotland, Culm in Brittany and Saxony, Triassic in the Alps).

If one extended this sequence of events and applied it to *a future chain,* it would be seen that in this future chain, the gneiss facies would reach into the Tertiary, and granites would intrude as far as the Cretaceous. We know of no part of the globe where this occurs, and it probably does not exist. However, I cannot help comparing this inferred consequence with a recent remark by Lawson that struck me vividly. In the Sierra Nevada and in the Pacific coastal chains as far as the Andes, Lawson believes that he has found traces of a very recent intrusion of granite that has cut at the base to the Triassic and Jurassic. This could mark the intrusion of a large laccolith accompanying the slow uplift of the entire coast. Becker has described some Cretaceous gneiss. Can one see there the indications of orogenic movements comparable to the earlier movements? In any case, this same Pacific coast has been the site of uplift of considerable magnitude until Tertiary time. Russel pointed out some stratified formations that do not appear to be terraces at the foot of Mount Saint Elias in Alaska; they are 2000 meters above sealevel and contain as fossils, shells of exclusively marine organisms. In the Barbados (which despite the Isthmus of Panama, is in the same zone) some exposed beds contain radiolaria identical to modern forms that live only at ocean depths between 3000 and 4000 meters. Exceptional mobility of the earth and exceptional uprising of granitic magmas are clearly indicated. These are important inferences to grasp and follow without rushing into an unrealistic conclusion; one can say that based on all we know of the globe, this region is the one that exhibits the most similar geological situation to past cyclical sedimentary series and may be the site of a future chain, now in the state of formation.

We are quite far from the Alps; I shall return to conclude. The facts that I believe are established for the Savoy Alps are the following: the existence of Permian gneiss; the existence on the same site of schist deposits, which are crystalline at their base and include distinctive facies of Triassic and Liassic age; and the association of these schists with large masses of greenstones.

There exists, in fact, a connection between the zone where these particular features occur and the location of the original geosynclinal basin that preceded the era of Alpine uplifts. I propose a hypothesis that views this coincidence of facts as being directly related by cause and effect, rather than an accidental coincidence. I would say that nearly a *necessary* connection exists within the history of the chain. The point of examining sedimentary sequences elsewhere is to make similar relationships evident in other chains.

In the Alpine belt, the similarities include the occurrence of flysch resting on a basement of uncertain age. Also, the position of younger red sands is comparable, and if one accepts the evidence existing in the Green Mountains and the Taunus, gneiss of progressively younger age occurs toward the center of successive chains; this is no less clearly and conclusively demonstrated than in the Alps. One cannot deny that the three examples, Cambrian gneiss, Devonian gneiss, and Permian gneiss, collectively support the postulated scheme, and a comparison using these examples is significant.

As regards the postulated cycles of sedimentation, the framework of the table above seems justified to me; the details remain debatable. This is not simply a case of groping, for this searching reveals a conclusion in which I believe: that is, an ordered, systematic relationship between the various tectonic, sedimentary, and eruptive phenomena characterizes the different stages of evolution of mountain chains; these stages constitute the four great chapters, the four essential units in the history of the globe.

7

Reprinted from *Geol. Soc. America Bull.* **52**:1915 (1941)

DIFFERENTIATION OF SEDIMENTS DURING THE
LIFE HISTORY OF A LANDMASS

BY PAUL D. KRYNINE

The tectonic and geomorphic processes which shape the successive stages of the development of a landmass also control on a broad regional scale the sediments related to each stage. Such differentiation is most easily studied in medium-grained clastics ("sandstones").

These processes operate: (1) directly by controlling both relief and topography within source area and basin of deposition; and also controlling depth and intensity of metamorphism and igneous activity which shape the material to be brought into the region of erosion; and (2) indirectly by influencing climate through topography.

The following tectonic stages and their corresponding "sandstones" seem to be well defined in the central part of the Appalachian Trough, and their counterparts appear to be also present in the Alpine and Himalayan regions:

(1) *Peneplanation (or early geosynclinal) stage:* cyclic deposition on fluctuating flat surface after much weathering, characterized by *first-cycle quartzites* and quartzose sandstones (Gatesburg).

(2) *Geosynclinal stage:* basinal deposition interrupted by local vertical buckling, and regional marginal upwarping which shifts earlier sediments to the center of the basin after low-rank metamorphism. Typical sediment: *graywackes* (normal dark gray or red). Examples: Oswego, Juniata, Chemung, Catskill, Alpine molasse, Siwalik of India.

Reworked second cycle quartzites (Tuscarora, Oriskany) and unusual tectonic arenites (Bellefonte) may be interbedded with graywackes.

(3) *Post-geosynclinal stage:* uplift (frequently by faulting) after folding and magmatic intrusion of geosyncline. Typical sediments: *arkoses* and similar first-cycle, fresh clastics from deep-seated high-rank sources (Triassic of Connecticut).

These types may be complicated locally.

8

Reprinted from *Geol. Soc. America Bull.* **52**:1915–1916 (1941)

PALEOGEOGRAPHIC AND TECTONIC SIGNIFICANCE
OF SEDIMENTARY QUARTZITES

BY PAUL D. KRYNINE

Sedimentary orthoquartzites and highly quartzose sandstones originate either as first-cycle deposits following intense chemical weathering of source area or as second-cycle affairs through reworking of pre-existing quartzose sediments.

First-cycle quartzites generally follow prolonged and intense chemical decay in peneplaned regions (Gatesburg, Uinta). After passage through beach or dune stages deposition proceeds on flat surfaces in subaerial or subaqueous environments. Marine deposition in typically shallow seas, highly charged with dissolved silica and alkalies, results in large-scale authigenesis. In portions of such seas inaccessible to clastic detritus cherts appear in abundance. Beside authigenic minerals first-cycle quartzites are characterized by concentration of rounded stable detrital mineral species like tourmaline and zircon, resulting in great complexity of these species: 13 varieties of tourmaline in Gatesburg, 9 in Uinta. Locally abnormal contamination from monadnocks may bring floods of angular unstable minerals. First-cycle quartzites are related to beginning or end of a geosynclinal cycle. They follow arkoses and precede graywackes.

Second-cycle quartzites are formed through reworking of pre-existing quartzose sediments. Petrographically they contain in their coarser-grade sizes fragments of older quartzites and at places much detrital chert (Tuscarora-Shawangunk, Oriskany, Weber). Authigenesis is less abundant than in first-cycle quartzites. Tectonically second-cycle quartzites are related to mature portions of the geosynclinal stage and are formed on large scale during periods of regional marginal upwarping of geosynclines which bring into the zone of erosion older quartzites with resulting shift of sediments toward the center of the geosyncline. Second-cycle quartzites are hence interbedded with graywackes.

9

Reprinted from *Geol. Soc. America Bull.* **52**:1916 (1941)

PALEOGEOGRAPHIC AND TECTONIC SIGNIFICANCE
OF GRAYWACKES

BY PAUL D. KRYNINE

Graywackes, which in the classical sense are "dirty" sandstonelike rocks containing an abundance of slate, shale quartzite, and chert fragments, are typical of the medium and late depositional stages of a geosyncline. These stages are characterized by local vertical buckling along zones of weakness in the central part of the geosyncline and by large-scale regional tilting, warping, and minor revolutions (Taconic) in the marginal parts. This brings older sediments of the early stages again into the zone of erosion, either without alteration or after subjection only to low-rank metamorphism (slates and phyllites) due to shallow burial and moderate orogenesis. After reworking, these sediments are shifted toward the center of the geosyncline. Since the central parts of the geosyncline are preserved best and the margins generally eroded, graywackes produced by reworking of mildly metamorphosed marginal deposits become the typical clastic sediment of ancient geosynclines. The term as used here is a general one covering both typical dark graywackes and red schist arenites. Within one geosynclinal cycle low-rank metamorphic material generally predominates in the earlier (older) graywackes, sedimentary material in the later.

Typical examples of early graywackes are: Oswego, Chemung graywackes, and red schist arenites of Juniata and Catskill; of late graywackes, Pennsylvanian of Appalachian region, Alpine molasse, and Siwalik of India.

Reworking of older quartzites may produce second-cycle quartzites interbedded with graywackes such as the Tuscarora and Oriskany.

10

Reprinted from *Geol. Soc. America Bull.* **52**:1918–1919 (1941)

PALEOGEOGRAPHIC AND TECTONIC SIGNIFICANCE OF ARKOSES

BY PAUL D. KRYNINE

Arkose deposits are produced when a granitic (or highly feldspathic) terrane is broken up, eroded, and deposited under conditions when mechanical weathering proceeds faster than chemical decay. Relative rather than absolute figures are important. Ultra-rapid (linear or vertical) erosion proceeding through impact on steep slopes in a tropical humid climate masks and neutralizes chemical decay and produces large-scale arkose deposits quicker than relatively slow erosion in semiarid, arid, or glacial regions. So-called "deserts" of the American Southwest are quite abnormal in this respect since to a large extent they depend upon high rainfall on mountain tops and also they rework fluvial deposits inherited from pluvial Pleistocene.

A restudy of arkoses in the geologic column would assign most of them to ultra-humid climates operating on very steep (not necessarily high) relief.

Tectonically arkoses are par excellence related to granitic terranes in regions of steep youthful topography. This places them at the very end of the geosynclinal stage after magmatic intrusion, uplift, and block faulting, and before peneplanation (Triassic of Eastern United States).

The common mistake in the field of confusing arkoses (rich in feldspar) with graywackes (rich in nonfeldspathic rock fragments) is sure to lead both to petrologic and to interpretative disaster.

Part IV

GEOSYNCLINES: CLASSIFICATION, NOMENCLATURE, AND A DEBATE OVER SOURCE AREAS

Editor's Comments
on Papers 11 and 12

11 KAY
Excerpt from *Development of the Northern Allegheny Synclinorium and Adjoining Regions*

12 KAY
Excerpts from *North American Geosynclines*

In the decades between the American Civil War and the Great Depression of the 1930s, papers detailing the definitive characteristics by which geosynclines could be recognized grew in number and diversity almost as rapidly as the number of geologists authoring them. It became obvious that there were geosynclines and *geosynclines.*

By what criteria did one recognize geosynclinal belts? If bathymetry of depositional environments is diagnostic, we must ask whether geosynclinal sediments were shallow-water (neritic) deposits (Hall and Dana) or deep-water (bathyal) deposits (Suess, Neumayr, Jones, and Bailey). Was the geosyncline a physiographic entity, essentially a trough (Dana and Haug) that as a consequence filled with sediment, or did proximity to continental clastic source areas produce rates of sediment supply sufficient to produce linear belts of subsidence thanks to sediment loading (Hall)? Could entire ocean basins be considered geosynclines (Haug)? Were all sedimentary basins geosynclines (Stille and Kay)? Was folding and uplift of mountainous chains an integral and inevitable result of the geosynclinal process (Hall, Dana, and Haug)?

It was apparent by the 1920s that geosynclines could not be identified on the basis of any one single characteristic. In fact, many individual "definitive" criteria were ambiguous at best, downright contradictory at worst. One's choice of representative examples of both modern and ancient geosynclines depended on regional and national perspective, even on hemisphere. For example, the North American school gravitated toward the original Appalachian belt as the classical ancient geosyncline. For

North American geologists, therefore, the best modern analogue was often taken to be the Gulf Coast (Barton, Ritz, and Hicke, 1933; Russel, 1936). Conversely, European geologists generally cited the Alpine belt as the most typical ancient geosynclinal belt; modern analogues included the present Mediterranean as well as the Indonesian area in the western Pacific (Kuenen, 1935; Tercier, 1936).

Unfortunately everyone who followed Hall and Dana committed the same basic error: extending the detailed characteristics of a specific geosyncline to the general, worldwide case. In an attempt to bring a degree of order to the resulting chaos, several workers attempted to systematically classify geosynclines within a scheme that would tolerate the obvious worldwide diversity. Unfortunately, cumbersome and often confusing nomenclature accompanied these classifications.

Charles Schuchert's uneven and overly long treatment essentially conceded that the American-European schism was irreparable (Schuchert, 1923). Consequently he incorporated the basic differences into two major categories of geosynclines. *Mesogeosynclines* were essentially Mediterranean-type geosynclines: intercontinental, situated on oceanic crust, and filled dominantly with deep-water (bathyal and abyssal) sediment. Mesogeosynclines conformed with most western European perceptions. Conversely, geosynclinal belts like those of North America (that is, developed near the outer margins of, but within continental areas) were called *monogeosynclines, polygeosynclines,* and *parageosynclines.* Monogeosynclines were simple troughs, like the "original" Appalachian belt studied by Hall. Polygeosynclines were divided by a geanticlinal axis into two or more parallel troughs. Parageosynclines were situated right at the periphery of continental blocks. Schuchert cited the island arcs of Asia as typical examples. Schuchert succeeded in providing a new terminology, but his categories were not clearly defined and he did not succeed in defining the unique characteristics by which geosynclinal belts could be distinguished from other geotectonic units.

Hans Stille (Paper 5) was far more successful and devoted much of his life to analyzing geosynclines and providing for them a useful nomenclature. Because much of Marshall Kay's monumental work (Paper 11 and 12) retains premises that are fundamentally those of Stille, it is appropriate to recapitulate some of Stille's ideas.

Throughout his career, Stille (1924) applied the term geosyncline to all sedimentary basins that showed notable thickening (Hall's criterion). In 1936 he elaborated on an earlier contention that not all geosynclines would eventually become mountain ranges. Geosynclines that produced mountain ranges were referred to as *muttergeosynklinalen* or *mother geosynclines*. (Stille, 1936a, 1936b). Geosynclines in general could therefore be subdivided into *orthogeosynclines* and *parageosynclines*, a scheme later adhered to by Kay and his followers. Orthogeosynclines were geosynclines in the sense that Hall, Dana, and Haug had intended: mountainous belts of thick sedimentary rock exhibiting plastic folding and lateral thrusting and an intimate association with deep-seated plutonic and metamorphic rocks. Orthogeosynclines were invariably intercratonic. However, in Stille's thinking, they could lie either between two stable continental blocks (the Alpine belt) or between stable continental crust and a segment of stable oceanic crust that he referred to as *tiefkratone* (deep-sea cratons; translated as *thalassocratons*), as is the case of the Appalachian belt. On the other hand, parageosynclines were located only within a single stable continental area and hence could be considered intracontinental basins. The Alsace (Rhine) graben and the Paris basin are typical examples. Stille made it clear that even in his tolerant scheme, parageosynclines were not truly geosynclines in the sense originally intended by Hall and Dana.

Stille later (1941) analysed the igneous activity and metamorphism associated with the evolution of orthogeosynclinal belts. He documented the successive occurrence of preorogenic ophiolite-greenstone suites, synorogenic plutonism and migmitization, late orogenic andesitic volcanism, and a final phase of post orogenic simatic (basaltic) volcanism. Because the bulk of this activity—particularly the ophiolites, greenstones, and synorogenic igneous rocks—were confined largely to the internal zones of geosynclines, Stille differentiated two major subdivisions of orthogeosynclines: an external amagmatic (or miomagmatic) zone and an internal pliomagmatic zone. The names *miogeosyncline* (lesser geosyncline) and *eugeosyncline* (true or wholly geosynclinal) were formally substituted for the miomagmatic and pliomagmatic zones in a series of letters written by Stille in 1940 and 1941 in response to questions posed by Marshall Kay. Stille's categorization reflected his contention that though both eugeosynclines and miogeosynclines underwent deep subsidence, active volcanism in the preorogenic and synorogenic stages occurred only in the eugeosynclinal zones. Consequently field work could allow the two to be easily distinguished.

Portions of an article by Kay dealing with the northern Allegheny synclinorium are included as Paper 11 to illustrate Kay's ideas in the formative stages. He obviously wrestled with the problem of distinguishing between intracratonic geosynclines (Stille's parageosynclines) and intercratonic and extracratonic (continental margin) orthogeosynclines. Kay coined the name *autogeosyncline* for basins such as the Michigan Basin (or Paris Basin) whose development appeared to be unrelated to orogeny. The Allegheny synclinorium was first named a *deltageosyncline*, a variety of parageosyncline that was essentially a belt of thick, clastic terrigenous material deposited in front of rising mountainous areas as a complex of alluvium. This variety was later renamed *exogeosyncline* (Kay, 1947). Kay then proposed an *epieugeosyncline*, which was a geosyncline superimposed on an orogenic belt (like the Gulf of California or Great Basin); a *taphrogeosyncline*, essentially a rift trough (like the Rhine Graben or Triassic basins of eastern North America); a *zeugogeosyncline*, a geosyncline yoked to adjacent intracratonic source areas (the Denver Basin); and finally a *paraliageosyncline*, a geosyncline located along coastal margins like the northern Gulf of Mexico. Kay also acknowledged the *idiogeoysncline* proposed by Umbgrove (1933), a late cycle geosyncline located between stable continental crust and offshore island arcs (such as the marginal basins of the East Indian island arc, although Kay contended that it was simply a late-cycle miogeosyncline).

As to orthogeosynclines in general, Kay incorporated Stille's notion that they were indeed separable into two distinct zones, the eugeosyncline and miogeosyncline, each characterized largely by distinctive sedimentary suites. Kay emphasized that the miogeosyncline contained a sedimentary sequence almost identical to that of the craton, only of much greater thickness.

In 1944 Kay presented a brief outline of the ideas he would detail later in his *Geological Society of America Memoir* (Paper 12). In the earlier paper, he not only summarized some of the distinguishing characteristics of miogeosynclines, eugeosynclines, deltageosynclines, autogeosynclines, and taphrogeosynclines, but also attempted to settle the controversy dealing with geosynclinal source areas. From the first, American geologists had recognized that much of the clastic material in the Appalachian belt thickened and coarsened toward the east. The first seeds of the *borderland* hypothesis were planted even by Hall and Dana. Hall (1843) pointed out that Paleozoic sandstones in the Appalachians thickened eastward, therefore requiring a source area to the east. The

Rogers brothers and others even made vague references to an Atlantic continent off the coast of eastern North America. Dana (1875), advocating the permanency of continental blocks and ocean basins, suggested instead a geanticlinal upwarping on the seaward side of geosynclines as a mechanical complement to the downwarped geosyncline. T. C. Chamberlain (1882) published the first paleogeographic map of North America, on which he showed an "Archean Appalachian Highlands" offshore in the Atlantic Ocean. Schuchert (1910) showed such borderlands offshore all North American geosynclines, gave them distinctive names (Appalachia, Cascadia, etc.), and argued that they provided the bulk of the sediment fill. This view dominated the American view until about World War II.

Quite a contrary view evolved in western Europe. Most Europeans favored a view first expressed by Argand (1916) that the bulk of sediment within geosynclines was derived from the erosion of embryonic ridges (cordilleras) that rose within geosynclinal belts during the early phases of compression and crumpling. Polite acknowledgement was given to the possible influence of the craton and offshore borderland sources, but far more significant for Europeans were these intrageosynclinal, cannibalized source areas. Unhappily, again a polarization developed across the Atlantic.

Kay helped reconcile these differences of opinion in a series of papers published between 1937 and 1951. In 1937 Kay suggested that much of the clastic material in the Appalachian belt was derived from the erosion of *tectonic lands*, uplifted terrains of pre-existing sediment produced by episodic deformation. Tectonic lands were analogous to Argand's cordilleras, although Kay did not specify that they had to be linear fold belts. Meanwhile, Hess (1939) had been emphasizing the importance of volcanic rocks and ophiolites in geosynclinal belts and suggested that volcanic island arcs might also be important soure areas for the geosynclinal fill. Kay was receptive to the idea that geosynclinal source areas might indeed be multiple in kind and origin, as is apparent in a series of articles written by him in the 1940s (see especially, Kay, 1945).

Paper 12 consists of critical excerpts from Kay's *Geological Society of America Memoir 48*. This classic nicely summarizes the controversy that existed over geosynclinal source areas and details the major types of geosynclines discriminated by Kay. The *Memoir* settled some arguments regarding source areas by being conciliatory and receptive to various schemes. Kay's classification scheme won a very mixed reception, however, and would be followed by two decades of debate as to whether Kay had helped or hindered the original geosynclinal concept. The long, Greek-based terms were unpalatable to many practical geologists who regarded them as bizarre, easily forgotten, and imprecise. In response to the global search to meet oil needs, a similar scheme was developed by Weeks (1952) that employed Anglo-Saxon terminology. The final classification scheme generated from this global approach was simpler to understand. Instead of *taphrogeosyncline*, we had *rift basin*, and so on. Also, some important modifications and additions were proposed. For example, the Swiss geologist Trümpy (1955) proposed the useful suggestion that the abyssal plain stage of geosynclinal development should be separately designated with the label *leptogeosynclinal*.

REFERENCES

Argand, E., 1916, Sur l'arc des Alpes occidentales, *Eclogae Geol. Helvetiae* **14**:145–191.

Barton, D. C., C. H. Ritz, and M. Hickey, 1933, Gulf Coast Geosynlines, *Am. Assoc. Petroleum Geologists Bull.* **17**:1146–1158.

Chamberlain, T. C., 1882, *Geology of Wisconsin*, vol. 4, Madison, Wis., 779 p.

Dana, J. D., 1875, *Manual of Geology*, 2nd ed., Ivison, Blakeman, and Taylor, New Tork, 828 p.

Hall, J., 1843, *Geology of New York. IV. Comprising the Survey of Fourth Geological District*, State of New York, Albany, 683 p.

Hess, H. H., 1939, Island Arcs, Gravity Anomalies, and Serpentine Intrusions: A Contribution to the Ophiolite Problem, *17th Internat. Geol. Congr. Report* **2**:263–282.

Kay, G. M., 1937, Stratigraphy of the Trenton Group, *Geol. Soc. America Bull.* **48**:233–302.

Kay, G. M., 1944, Geosynclines in Continental Development, *Science* **99**:461–462.

Kay, G. M., 1945, North American Geosynclines—Their Classification, *Geol. Soc. America Bull.* **56**:1172.

Kay, G. M., 1947, Geosynclinal Nomenclature and the Craton, *Am. Assoc. Petroleum Geologists Bull.* **31**:1289–1293.

Kuenen, P. H., 1935, *Geological Interpretation of the Bathymetric Results: Snellius Expedition in the East, Parts of Netherlands East Indies*, vol. 5, part 1, Brill, Leyden, 124 p.

Russel, R. J., 1936, Physiography of Lower Mississippi River Delta, *Louisianna Geol. Survey, Dept. Conserv. Geol. Bull.* **8**:1–199.

Schuchert, C., 1910, Paleogeography of North America, *Geol. Soc. America Bull.* **20**:427–606.

Schuchert, C., 1923, Sites and Natures of the North American Geosynclines, *Geol. Soc. America Bull.* **34**:151–229.

Stille, H., 1936a, Wege und Ergebnisse der geologisch-tektonishen Forschung, *Kaiser Wilhelm Gesell. Jahr. 25,* **2**:77–97.

Stille, H., 1936b, Present tectonic state of the Earth, *Am. Assoc. Petroleum Geologists Bull.* **20**:849–880.

Stille, H., 1941, *Einfuhrung in den Bau Nordamerikas.* Gebruder Borntraeger, Berlin, 443 p.

Tercier, J. 1936, Dêpots marim actuel et series géologiques, *Eclogae Geol. Helvetiae* **32**:1304–1335.

Trümpy, R., 1955, Wechselbeziehungen zwischen Paleogeographie und Deckenbau, *Naturf. Gesell, (Zürich) Vierteljahrschr.* **100**:217–231.

Umbgrove, J. F., 1933, Veschillende Typen van teriare Geosynclinalen Indischen Archipel, *Leidse Geol. Meded.* **5**:33–43.

Weeks, J. G., 1952, Factors of Sedimentary Basin Development that Control Oil Occurence, *Am. Assoc. Petroleum Geologists Bull.* **36**:2071–2124.

11

Reprinted from pages 1639–1644 of *Geol. Soc. America Bull.* **53**:1601–1658 (1942)

DEVELOPMENT OF THE NORTHERN ALLEGHENY SYNCLINORIUM AND ADJOINING REGIONS

G. M. Kay

[*Editor's Note:* In the original, material precedes this excerpt.]

CLASSIFICATION OF THE ALLEGHENY BELT

APPALACHIAN "GEOSYNCLINES"

The term geosyncline has been applied in the broadest sense to those regions of the earth that have undergone relatively great depression or sinking. Many authors include the Allegheny belt in the Appalachian geosyncline. The stratigraphy of the Paleozoic systems evidences the changing characters of the belts of downward movement and permits their classification.

The deformation of the Allegheny belt is separable into two principal classes of movements—those when terrigenous sediments failed to reach the belt and those when there were highlands to the east whose clastic detritus crossed the Adirondack axis. There was never significant terrigenous materials gained from the west and north; minor quantities are present in the base of overlapping sequences. When terrigenous rocks were absent or quite subordinate within the belt, the margining axes at times formed relatively low arches, as in early and earliest middle Trenton along the Adirondack axis and in Clinton time along the Cincinnati axis. When the troughs bordering the axes were sinking rapidly, the axes became flexures, and the Allegheny belt was relatively depressed along the margins, as in middle Trenton time along the Adirondack, and in Cayugan and Ulsterian times along the Cincinnati.

When there were highlands to the east, the belt was depressed in a semifunneled form, with the apex along its eastern margin, the greatest depression in the area near the terrigenous source and the deformation crossing the Adirondack axis with no appreciable influence. The deformation of the later Ordovician (Cincinnatian) (Fig. 6a) and later Devonian (Erian to Conewangan) (Fig. 10) is of this sort and affects the entire belt, whereas that of the Silurian involved only a part of the breadth. This deformation, attended by thick deposition, in the main has led to the reference of the belt to the Appalachian geosyncline. The belt contrasts with the geosynclines to the east and west, which have undergone great depression irrespective of the character of their sediments; as a whole, the Allegheny belt was never a trough.

The Michigan basin is an elliptical geosyncline (Newcombe, 1933, p. 101). During the Silurian to middle Devonian, it subsided considerably, though terrigenous sediments are practically absent (Pl. 2b). In the later Paleozoic, terrigenous sediments entered the area from the Allegheny belt across the Chatham sag in the Cincinnati arch; but independent subsidence continued during middle Mississippian (Meramecan) when detrital sediments were few.

The Michigan basin seems to have received the centrally thickening sediments because it was sinking, whereas the Allegheny belt was tilted to a considerable degree because of sinking synchronous with the production of terrigenous sediments to the east, beyond its margin. The former is a geosyncline defined by causes within its area; the latter was depressed in complement to uplift near by. The limits of the former are similar for a long time; the gross sinking of the latter is the sum of periodic movements. Both are within the craton or "shield."

The extracratonic Champlain trough was periodically sinking during earlier Ordovician, when terrigenous sediments were relatively unimportant, its western limit being sinuous. During later Ordovician, when thick terrigenous sediments were laid within, it continued to sink more rapidly than loading alone would cause if isostatic adjustment were immediate and complete. The eastern margin lies beneath the Taconic allochthone but can be interpreted from evidence in the easternmost autochthonous and westernmost allochthonous sections. It has been considered a submarine slope, a flexure with greater subsidence southeastward than in the Champlain belt (Bailey, Collet, and Field, 1928), or as a barrier separating two troughs (Ulrich and Schuchert, 1902, p. 639).

The boulder beds in the northwest marginal Magog trough sediments have been interpreted as submarine landslip deposits by those who believe the Quebec axis was a flexure (Bailey, Collet, and Field, 1928, p. 603-604). The writer concurs that in early Ordovician time there were no highlands along the axis, for, if there had been, there should be thick terrigenous sediments in the Champlain belt; but the presence (Raymond, 1913, p. 30) in local abundance (C. Faessler, personal communication, 1941) of crystalline boulders with exotic blocks of Lower Cambrian to Lower Ordovician limestone (Clark, 1924) such as would be found beneath the Taconic allochthone (Bailey, Collet, and Field, 1928, p. 601-602) require uplift along the axis with southeast sedimentary transfer of the rocks. They may have gained their sedimentary position by landslips but must have been raised before being transported and have been deposited originally in the margin of the Champlain trough. The Cow Head breccia of Newfoundland (Schuchert and Dunbar, 1934, p. 84-86) had similar paleogeographic position and has been attributed to "talus and landslides formed along a fault scarp that came into existence during mid-Ordovician orogeny . . . formed at the nose of thrust sheets"; it has been thought to be autochthonous. The post-lower Trenton Rysedorph conglomerate in the Magog belt southeast of Albany (Ruedemann, 1901; 1930) contains exotic boulders of Champlain trough facies (Kay, 1937, p. 276-277) and has been attributed to the "erosion of anticlinal ridges" (Ruedemann, 1930, p. 113).

The Quebec axis probably did not separate faunas, but facies, which in turn influenced faunas. There were uplifts along the axis, some of which may have risen above sea level, any of which must have directed currents and thus controlled distribution of facies. The conglomerates at Quebec were formed by movements antedating the Vermontian disturbance, for they are in Canadian beds and evidence the prolonged activity along the axis. The comparable "exotic" blocks in the northern Ouachita allochthone have been attributed to thrusting of rocks toward the geosynclinal margin (Hendricks, 1940); or landslipping from an archipelago, possibly bordered by normal faults (Kramer, 1933; Moore, 1934). The Haymond conglomerate of west Texas has boulders supposedly "derived from the erosion of thrust sheets advancing from the southeast" (King, 1937, p. 91), that is from the Ouachita geosyncline; their deposition antedates the important Marathon orogeny and thrusting. Those in the Ordovician Woods Hollow formation of the same region (King, 1937, p. 24-25) are of foreland facies. These are analogous to those along the Quebec axis.

A structurally active and relatively higher though perhaps interrupted barrier formed the eastern margin of the Champlain trough. To the west lay the Adirondack belt, alternately an arch and flexure. The trough sank endogenetically during Cambrian and earlier Ordovician time and became deeply depressed as it filled with terrigenous sediments from the Vermontian geanticline in the western Magog belt in middle Trenton.

CLASSIFICATION OF GEOSYNCLINES

Schuchert (1923, p. 195-197) has classed troughs such as the Champlain and Magog as monogeosynclines, formed by the partition of the earlier polygeosyncline, the "St. Lawrence Trough," by an intrageosynclinal geanticline, the Quebec "barrier." Stille (1936a, p. 517) has separated geosynclines into the greater intercontinental belts, orthogeosynclines, and those lying within the shield areas or cratons, "parageosynclines," the latter a term having a different earlier definition (Schuchert, 1923, p. 199). Inasmuch as the character of areas is transitory, any classification is fundamentally temporal. The writer adopts an Ordovician basis of reference for definition of paleogeographic belts; this may reflect his interest in the system, but it seems advantageous because the record is extensive, the belts had become distinct, and had not been altered by orogenies.

The Magog geosyncline has graptolitic slates, formed in at least moderate depth, and associated cherts and volcanic rocks, as in northern Newfoundland (Sampson, 1923; Heyl, 1936) and southeastern Quebec (Cooke, 1935; Clark and Fairbairn, 1936); they are in a belt later intruded (Keith, 1928). Lower Ordovician graptolites are of distinctive genera, and practically limited to the shaly and cherty sequences of the Magog belt, reflecting a structurally controlled facies. Similar rocks in the Ouachita area contrast with carbonates on the foreland, and they are found in the western belt of the "Cordilleran geosyncline" (Ferguson, 1924, p. 20; Ross, 1934, p. 942-945). That facies rather than land barriers limited their distribution is shown by thin graptolitic shales intercalated within the carbonates in Oklahoma (Decker, 1936), West Texas (King, 1937, p. 26-30), Utah (Clark, 1935), and southeastern Idaho (Mansfield, 1927); that local conditions changed, by the intercalation of thick graptolitic shales within limestone formations, as in British Columbia (Walker, 1926, p. 24-31; Evans, 1933).

The Magog belt lies in a pliomagmatic zone (Stille, 1936a), and the trough is a eugeosyncline.[3]   Eugeosynclines are characterized by a prevalence of shales and cherts, the presence of volcanic rocks, ophiolites or greenstones (Stille, 1936a, p. 517; Bailey, 1936, p. 1717-1723); the rocks have been interpreted as formed at great depths (Ruedemann, 1936, p. 1545-1563) to an extreme that does not seem warranted by the coarser texture of some of the interbeds. They have been the locus of principal subsequent plutonic invasions.

In contrast to the Magog eugeosyncline, the Champlain belt contains dominant carbonates of shallow-water origin, unaffected by subsequent volcanism; it is a miogeosyncline. The lithologies are more like those on the craton than in the eugeosyncline, and with the craton, the belt is essentially amagmatic. But the miogeosyncline is extracratonic, inasmuch as the sediments in most regions have been thrust subsequently on the foreland.

Just as the Magog and Champlain troughs comprise the two parts of the St. Lawrence orthogeosyncline,[4] there are two similar stratigraphic belts in the Cordilleran orthogeosyncline. The western, eugeosynclinal belt has been classed as pliomagmatic (Stille, 1936b, p. 139-143) and called "Nevadan." Inasmuch as the name Nevadian has been applied to the intrusive disturbance at the close of the Jurassic period, it is

---

[3] In reply to the writer's letters of November 8, 1939 and April 23, 1941, in which he requested a definition of orthogeosyncline and emphasized the contrasts between the stratigraphy of the pliomagmatic and amagmatic zones of Stille and the similarities of the stratigraphy of the latter with that of the craton, Stille, after discussion in a letter of March 20, 1940, proposed the following classification in a letter of May 24, 1941:

"1. Orthogeosynklinalen (alpinotype Geosynklinale) zerfallend in
    a) eugeosynklinale Zonen (pliomagmatische Mutterzonen der Interniden)
    b) miogeosynklinale Zonen (miomagmatische Mutterzonen der Externiden).

"2. Parageosynklinalen (germanotype Geosynklinalen auf schon konsolidiertem Untergrunde)."
The terms eugeosyncline and miogeosyncline have been defined recently (Stille, 1941, p. 15).

[4] An orthogeosyncline is defined on the character of the rocks and their history and is preferred to polygeosyncline, which is a geometrically descriptive term. Orthogeosynclines are polygeosynclines, but not all polygeosynclines are orthogeosynclines.

confusing to use Nevadian or Nevadan for the stratigraphic belt. The lower Ordovician (Deepkill) graptolitic shales in Idaho (Umpleby, Westgate, and Ross, 1930; Ross, 1934) may be in the belt; equivalent rocks in southeastern Alaska (Buddington and Chapin, 1929, p. 72-79) are typically eugeosynclinal, and volcanic and cherty slates are prevalent in the late Paleozoic, Triassic, and Jurassic of western British Columbia. The eastern, miogeosynclinal belt has the well-described sequence of southeastern Idaho (Mansfield, 1927) and is the Cordilleran geosyncline of most authors. Keith (1928) emphasized the importance of subsequent plutonic intrusions in the western belt; Stille illustrates (1936b) the movements along the eastern border as westward, but the belt was thrust on the eastern belt (Umpleby, Westgate, and Ross, 1930, map; Merriam, 1940, Pl. 1), much as the Magog was thrust on the Champlain. The overthrust sediments of the eugeosynclinal belt form an unnamed allochthone inasmuch as the orogeny that produced the structure has not been dated; it is certainly post-Carboniferous and probably pre-Nevadian. Nolan (1929) has described the geosynclinal nature of the eastern belt in Nevada.

The Allegheny belt seems comparable to the Coloradan, to the east of the Cordilleran belts, each having had at some times depressions filled with detrital materials gained from outside the belt; it is proposed that these intracratonic, terrigene-filled depressions be called deltageosynclines (Fig. 6; Pl. 2), in contrast to the endogenetic geosynclines like that in the Michigan basin that are termed autogeosynclines (Pl. 2b) in reference to their independent sinking and isolation from those near the more active tectonic zones.

In eastern North America, the sequence has been development of the orthogeosynclines, uplift of an intraeugeosynclinal geanticline Vermontia, folding and thrust of the eugeosynclinal Magog belt on the miogeosynclinal Champlain (Taconian revolution), intrusion within the eugeosynclinal (Acadian disturbance) and folding and thrust of the miogeosynclinal Champlain on its foreland, the Allegheny belt of the craton (Appalachian revolution). A comparable sequence seems represented in the Cordilleran belts in the Mesozoic and early Tertiary history. In the Ouachita region, magmatic invasions are not recognized, though the observable part of the belt is quite marginal; the thrusting is not dated with certainty, though it is at least in part intra-Pennsylvanian, and two belts of thrusting are not known. Exotic boulder conglomerates are present on the foreland margin of eugeosynclines—the source in the miogeosynclinal belts—and the deposits preceded the major thrusting. In a discussion of similar stratigraphy in the Ural region of eastern

Russia the writer (Kay, 1941b) misapplied the term orthogeosyncline to the more restricted eugeosynclinal belt.

The Ouachita and Virginian belts illustrate the confusion if one does not recognize the temporal nature of the belts. If one grant that the earlier Paleozoic rocks of the former are eugeosynclinal, though they lack characteristic volcanics, they are overlain by sediments of the Pushmataha series (Harlton, 1938, p. 853) that are like and presumably continuous with the lower Pennsylvanian sediments of the Virginian geosyncline (Van der Gracht, 1931) that is marginally intracratonic with reference to the earlier Paleozoic definition but is miogeosynclinal. On the other hand, the early Paleozoic "Ouachita" eugeosyncline must trend southeastward from Arkansas if it joins the extension of the Magog belt. It may be that the Pushmataha geosyncline is related to a different cause from the earlier trough and that the "Ouachita" geosyncline is composite. The great thrusting in the Virginian-Pushmataha belt may be limited to the area in which it lies on eugeosynclinal facies. That the Virginian miogeosyncline trended diagonally to the Ordovician Champlain trough has been stated (Fig. 10); whatever factors determined the trend and character of early Paleozoic belts had been superseded by others in late Paleozoic time.

Belts that had been sinking independent of terrigenous loading sank more when orogenic events produced quantities of detritus in neighboring belts. The Champlain trough developed this contrast in middle Trenton but retained its linear limits; and the later, transverse Virginian trough likewise maintained an essentially linear northwest boundary whether terrigenous rocks were unimportant or preponderant. These miogeosynclines contrast with the great semilenticular depressions of the simple deltageosyncline, or the more sinuous and varied outlines' when the terrigenous sediments were distributed from a number of centers in other deltageosynclinal belts.

Foreland folds within the eastern border of the Allegheny belt failed to gain the magnitude of those of the comparable Coloradan belt of the late Paleozoic and particularly the Cretaceous; even during the late Gulfian epoch, uplifts within the Coloradan deltageosynclinal belt yielded significant quantities of clastic sediment (Knight, 1938); nor were they as important as the uplifts along the Wichita-Amarillo axis in the Ouachita foreland.

[*Editor's Note:* In the original, material follows this excerpt. Only the references cited in the preceding excerpt are reproduced here.]

REFERENCES

Bailey, E. B. (1936) *Sedimentation in relation to tectonics*, Geol. Soc. Am., Bull., vol. 47, p. 1713–1726.

Bailey, E. B., Collet, L. W., and Field, R. M., (1928) *Paleozoic submarine landslips near Quebec City*, Jour. Geol., vol. 36, p. 577–614.

Buddington, A. F., and Chapin, Thomas (1929) *Geology and mineral deposits of southeastern Alaska*, U. S. Geol. Survey, Bull. 800.

Clark, T. H., (1924) *The paleontology of the Beekmantown series at Levis, Quebec*, Bull. Am. Paleont., vol. 10, no. 41.

Clark, T. H., (1935) *A New Ordovician graptolite locality in Utah*, Jour. Paleont., vol. 9, p. 239–246.

Clark, H. T. and Fairbairn, H. W. (1936) *The Bolton igneous group of southern Quebec*, Royal Soc. Canada, Tr., vol. 30, sec. 4, p. 301–311.

Cooke, H. C. (1937) *Thetford, Disraeli and eastern half of Warwick map-areas, Quebec*, Geol. Survey Canada, Mem. 211.

Decker, C. A., (1936) *Some tentative correlations on the basis of graptolites of Oklahoma and Arkansas*, Am. Assoc. Petrol. Geol., Bull., vol. 20, p. 301–311.

Ferguson, H. G. (1924) *Geology and ore deposits of the Manhatten district, Nevada*, U. S. Geol. Survey, Bull. 723.

Harlton, B. H. (1938) *Stratigraphy of the Bendian of the Oklahoma salient of the Ouachita Mountains*, Am. Assoc. Petrol. Geol., Bull., vol. 22, p. 852–914.

Hendricks, T. A., (1940) *Structure of the western part of the Ouachita Mountains* (abstract), Am. Assoc. Petrol. Geol., Program 25th Ann. Meeting, p. 34–35.

Heyl, G. R., (1936) *Geology and mineral deposits of the Bay of Exploits area*, Newfoundland Dept. Nat. Res., Bull. 3.

Kay, G. M., (1937) *Stratigraphy of the Trenton group*, Geol. Soc. Am., Bull., vol. 48, p. 233–302.

Kay, G. M., (1941b) *Classification of the Artinskian series in Russia*, Am. Assoc. Petrol. Geol., Bull., vol. 25, p. 1396–1404.

Keith, Arthur (1928) *Structural symmetry in North America*, Geol. Soc. Am., Bull., vol. 39, p. 321–385.

King, P. B. (1937) *Geology of the Marathon region, Texas*, U. S. Geol. Survey, Prof. Paper 187.

Knight, S. H., (1938) *Origin of Late Upper Cretaceous sediments of the Laramie and Hanna basins, Wyoming* (abstract), Geol. Soc. Am., Pr. 1937, p. 94.

Kramer, W. B. (1933) *Boulders from Bengalia*, Jour. Geol., vol. 41, p. 590–621.

Mansfield, G. R. (1927) *Geography, geology, and mineral resources of southeastern Idaho*, U. S. Geol. Survey, Prof. Paper 152.

Merriam, C. W. (1940) *Devonian stratigraphy and paleontology of the Roberts Mountain region, Nevada*, Geol. Soc. Am., Spec. Paper 25, 114 pages.

Moore, R. C. (1934) *The origin and age of the boulder-bearing Johns Valley shale in the Ouachita Mountains of Arkansas and Oklahoma*, Am. Jour. Sci., 5th ser., vol. 27, p. 432–453.

Newcombe, R. J. B. (1933) *Oil and gas fields of Michigan*, Mich. Dept. Conserv., Geol. Survey, Publ. 38.

Nolan, T. B. (1929) *A Late Paleozoic positive area in Nevada*, Am. Jour. Sci., 5th ser., vol. 16, p. 153–161.

Raymond, P. E. (1913) *Ordovician of Montreal and Ottawa*, Internat. Geol. Congr., XII Sess., Canada, Guidebook 3, p. 137–160.

Ross, C. P. (1934) *Correlation and interpretation of Paleozoic stratigraphy in south-central Idaho*, Geol. Soc. Am., Bull., vol. 45, p. 937–1000.

Ruedemann, Rudolf (1901) *Trenton conglomerate of Rysedorph Hill, Rensselaer County, New York, and its fauna*, N. Y. State Mus., Bull. 49, p. 3–114.

Ruedemann, Rudolf, (1930) *Geology of the Capitol district*, N. Y. State Mus., Bull. 285.

Ruedemann, Rudolf, and Wilson, T. Y. (1936) *Eastern New York Ordovician cherts*, Geol. Soc. Am., Bull., vol. 47, p. 1535–1586.

Sampson, Edward (1923) *The ferruginous chert formations of Notre Dame Bay, Newfoundland*, Jour. Geol., vol. 31, p. 571–598.

Schuchert, Charles (1934) *Stratigraphy of western Newfoundland*, Geol. Soc. Am., Mem. 1.

Stille, Hans (1936a) *Wege und Ergebnisse der geologisch-tektonischen Forschung*, Festschr. Kaiser-Wilhelm Gesellsch. Förd. Wissen., Band 2, Berlin.

Stille, Hans (1936b) *Die Enwicklung des amerikanischen Kordillerensystems in Zeit und Raum*, Preuss. Akad. Wissen., physo-math. Kl., Sitzungsber., vol. 25, p. 134–153.

Ulrich, E. O. and Schuchert, Charles (1902) *Paleozoic seas and barriers in eastern North America*, N. Y. State Mus., Bull. 52, p. 633–663.

Umpleby, J. B., Westgate, L. G., and Ross, C. P. (1930) *Geology and ore deposits of the Wood River region, Idaho*, U. S. Geol. Survey, Bull. 814.

van der Gracht, W. A. J. M van W. (1931a) *The Permo-Carboniferous orogeny in south-central United States*, K. Akad. Wetensch. Amsterdam, Afd Naturk., 2d Sec., Deel 27, no, 3, p. 1–170.

van der Gracht, W. A. J. M. van W. (1931b) *Permo-Carboniferous orogeny in south-central United States*, Am. Assoc. Petrol. Geol., Bull., vol. 15, p. 991–1057.

Walker, J. F. (1926) *Geology and mineral deposits of Windermere map-area, British Columbia*, Geol. Survey Canada, Mem. 148.

12

Reprinted from pages 1, 3–6, 29–33, 105–107 of *Geol. Soc. America Mem. 12*, 1951, 132 p.

NORTH AMERICAN GEOSYNCLINES

G. M. Kay

ABSTRACT

Geosynclines are surfaces of regional extent that subside considerably during formation of their included surficial rocks; they have limits in time and space. Early Paleozoic North America had a rather stable center (hedreocraton) margined by deeper sinking belts (miogeosynclines) that initially received carbonate rocks, and quartz sands from the interior; neither area had appreciable volcanism. The continental borders have distinctive volcanic flows and fragmentals, which with associated sediments show deep subsidence (eugeosynclines) and development of associated tectonic welts. Lands raised in the eugeosynclinal belts yielded sediments to the adjoining miogeosynclines; with deformation of the latter, terrigenous detritus spread into subsiding areas (exogeosynclines) in the margin of the hedreocraton. The craton periodically gained basin or trough-shaped depressions isolated from highland source areas (autogeosynclines) or receiving debris from associated intracratonal elevations (zeugogeosynclines).

The stratigraphy of the rocks in Paleozoic eugeosynclines shows that they were formed in regions much like modern island arcs. Some of the ancient eugeosynclines are continuous in trend with modern volcanic archipelagoes, which lie on the continental side of the ocean basins. The eugeosynclines became consolidated by orogeny and by intrusion of plutonic rocks; ultrabasic intrusions are restricted to them, and great batholiths are predominant within their areas. Deeply-subsiding, relatively non-volcanic troughs (epieugeosynclines) developed later, and were succeeded by fault-bounded depressions (taphrogeosynclines). The belts became rather stable additions to the craton. They finally developed coastal plains passing marginally into coastal geosynclines (paraliageosynclines).

It is axiomatic that nearly all kinds of surficial rocks are thickest in geosynclines; geosynclinal facies are characterized by thickness rather than kind. The better sorted sediments virtually are restricted to the cratons and adjoining geosynclines, and volcanic detritus is prevalently in eugeosynclines and the proximal parts of miogeosynclines. Because of the mobility of lands in eugeosynclinal belts, poorly sorted detritus, such as conglomerate and graywacke, is most abundant, but carbonate and siliceous rocks prevailed over large areas for long times.

Plutonic rocks are extensive in the pre-Paleozoic and volcanic rocks are widespread. Thus, by analogy, each part of the continent has in some past time been eugeosynclinal. Eugeosynclines and modern island arcs are thought to be in areas of thin sial. Sialic nuclei grew through orogenic consolidation of eugeosynclinal belts, formed at the expense of an originally universal simatic crust, of which the ocean basins, the low cratons, are relics. The nuclei coalesced into a single continental mass by earliest Paleozoic, when thin-sialic eugeosynclinal belts extended to the present continental margin. Stratigraphic evidence contradicts hypotheses that assume an original floating sialic block on surrounding and subjacent sima; the continent has developed through dynamic processes.

[*Editor's Note:* Material has been omitted at this point.]

INTRODUCTION

DEFINITION OF GEOSYNCLINE

The term "geosynclinal" was introduced by Dana (1873, p. 430) for a "long continued subsidence" evidenced in a "consequent accumulation of sediments," though the concept had been anticipated by Hall (1859, p. 66); Knopf (1948) recently published a summary of geosynclinal theory. If one accepts original usage as a guide, many of the earth features to which the name has been applied (Glaessner and Teichert, 1947) are not geosynclines. The term was not intended for subsequently gained structure, for Dana described the "catastrophe of plications and solidifications" ending "a geosynclinal accompanied by sedimentary depositions" as forming "synclinoria." The concept was associated with theories of mountain building, but he did not restrict it to a subsidence containing rocks later folded, for he included the Triassic of Connecticut among three examples. The geosyncline is the surface developed at the base of extensive surficial rocks that subsided deeply during their deposition or accumulation.

The form of the geosynclinal surface is a measure of subsidence through time. Thickness of accumulation only approximates the deformation, records it only if both the lower and upper surfaces were horizontal planes, or of identical relief, at their respective times of formation (Kay, 1945, p. 429–433). Buried sedimentary surfaces warped through a long time to form deeply subsided troughs are geosynclines even though they contain deepening water rather than thick accumulations; if later filled without further depression, the thick sequence will have accumulated in the antedating geosyncline.

The term has been applied, particularly by some European geologists (Haug, 1900; Tercier, 1940) to physiographic surfaces of sedimentation, such as are measured by depth of water or surface elevation. This usage does not accord with the original definition, and has led to such confusion that some have suggested the abandonment of the word (J. W. Evans, 1926, p. lxxv).

Great thickness of rocks, generally considered evidence of a geosyncline, is the result of time as well as of rates of depression and deposition; slow but long sinking and sedimentation can result in as great thickness as rapid depression and deposition of shorter span. At a stage in the subsidence of the geosyncline, the rocks at the surface of sedimentation are dependent on source, transporation, and depth of water, not with how much previously has been laid or is to come. The magnitude of subsidence that limits designation as a geosyncline is ill-defined, as in the classification of the surface below sediments thickening toward a center or axis but evidencing depression of less than thousands of feet. Insofar as process is concerned, the basin, trough, or furrow beneath such a deposit during or at the close of deposition is geosynclinal; it has the quality of a geosyncline, but not the magnitude because the differential movements proceeded too slowly or for too short a time or both.

In each accompanying section and map, the approximate span of time is indicated in the legend; illustrations record the thickness of rocks, but may not truly represent the deformation in the time; deformation in representative geosynclines will be compared and the duration alloted to stratigraphic intervals, presented.

A requisite applied by many is linearity, extension along a trend. Deep furrows (Bucher, 1933, p. 424), filled as they deformed, are generally classed as geosynclines, but there is hesitancy in accepting thick accumulations of non-terrigenous detrital sediments in basins, structural depressions with essential circularity, developed as deposition progressed. The writer recognizes these qualifications and includes as geosynclines deep sediment-filled depressions that have geosynclinal attributes but are doubtful geosynclines. The matter will be discussed more fully in reference to the intracratonic geosynclines.

The term geosyncline should be restricted to a surface of regional extent subsiding through a long time while contained sedimentary and volcanic rocks are accumulating; great thickness of these rocks is almost invariably the evidence of subsidence, but not a necessary requisite. Geosynclines are prevalently linear, but non-linear depressions can have properties that are essentially geosynclinal.

CLASSIFICATION

The classification of geosynclines preferred by the writer is based on the form and origin of the contained rocks, the "accumulations." The contents evidence the nature of the associated or complementing tectonic environments. The rocks are not the geosyncline, but are the basis for determining the conditions during formation of the geosyncline.

The central part of North America is so well known that there is agreement with respect to its principal form and development in the early Paleozoic. The comparatively stable interior was bordered by more mobile geosynclinal belts. Such a shield is a craton ("shield") (Stille, 1936, p. 84–85) and the bordering linear or arcuate geosynclines are orthogeosynclines ("straight" or "real geosynclines") (Stille, 1936). Orthogeosynclines were classified as eugeosynclines ("truly or wholly geosynclines") and miogeosynclines ("lesser geosynclines"), by Stille (1941, p. 15), who used the terms quite informally, but clearly implied a volcanic basis of distinction. An eugeosyncline is a surface that has subsided deeply in a belt having active volcanism, a miogeosyncline in a belt lacking active volcanism.

As volcanic rocks are practically absent in the orthogeosynclines that adjoin the North American early Paleozoic craton, they are thus miogeosynclines. A craton is transitory, expanding as orogenies add rocks of former orthogeosynclines, contracting as new orthogeosynclines reduce its area. The early Paleozoic craton of North America which had persisting influence on continental development and has close correlation with present structures is an hedreocraton ("steadfast craton") (Kay, 1947), a term introduced to retain a constant reference in discussing prolonged history. Although the hedreocraton was comparatively stable, geosynclines or basins subsiding during deposition formed in many areas at several times. The volume of the contained rocks is great, but relatively small compared to that of the rocks in the orthogeosynclines that surround the hedreocraton.

The Paleozoic paleogeography of North America has been interpreted in the past on the theory that the continent was margined from the beginning of Paleozoic time by great lands of ancient crystalline rocks, Cascadia, Appalachia, and Llanoria on west, east, and south that persisted long as dominant sources of detritus and ultimately foundered beneath seas advancing from the bordering oceans. The writer comes to the contrary conclusion that the borders of North America are dominated by geosynclines, deposited beneath paleogeographic troughs adjoining and including linear tectonic and volcanic islands, analogous to the present island arcs (Pl. 1). This is the theory of marginal volcanic troughs and island arcs (Kay, 1944). The volcanic rock-bearing geosynclines lying beyond the miogeosynclines in the orthogeosynclinal belts, designated eugeosynclines, developed through long spans of time, were severely deformed and intruded, and superseded by other sorts of geosynclines, some of which are still forming.

SEDIMENT TERMINOLOGY

The classification of sedimentary rocks has been considered recently in several publications (Krynine, 1948; Lombard, 1949; Pettijohn, 1949). The terms used will depend on the purposes for which they are applied; classifications emphasize texture, chemical or mineralogical composition, or a combination of the two. The writer prefers the terms rudaceous, arenaceous, silty and lutaceous for textures, applying an "-ite" suffix to the roots for consolidated rocks—thus rudite, arenite, siltite and lutite. Compositionally defined rocks similarly compound the chemical or mineral root, as in argillite, carbonatite, quartzite, calcitite and dolomitite for rocks having dominant argillaceous, carbonate, quartz, calcite, and dolomite constituents. This extends Grabau's terminology (1913) for what he called clastic to all detritus, eliminating the subjective determination of origin of particles from the definition.

Some names connote both texture and composition, such as graywacke for poorly sorted silty to arenaceous rocks having quartz and argillaceous constituents with plagioclase and rock fragments abundant in some phases, and arkose, arenaceous rocks having quartz and feldspar as principal constituents. Rocks also can be classed with compositional prefixes to textural terms, as argillicalcilutite, argillisilicisiltite, calcisilicarenite, or the less formidable argillaceous calcite lutite, argillaceous quartz siltite, or calcareous quartz arenite. Primary and secondary structural terms such as shaly, laminated, cross-bedded and stylolitic and concretionary are added as modifying adjectives. In the present paper, concerned with general summaries of lithologies, carbonatite, calcitite, dolomitite, quartzite, graywacke, and arkose are used commonly; limestone is synonomous with carbonatite. The origin and significance of the rocks are summarized in the discussion of geosynclines and sedimentation; the writer deplores use of the terms geosynclinal and foreland for lithic facies, as they confuse tectonic and surficial concepts.

Clastic and nonclastic are used frequently in classifying sediments, with varying definitions of the terms. A clastic rock is one composed of particles broken from larger masses, fragments. Detritus is of discrete particles that can have drifted in currents, thus may have rubbed one another.

Clastic and detrital sediments that are of fragments of pre-existing rocks are termed terrigenous (earth-derived) though these are but the physically broken constituents and the precipitated chemical sediments can also gain their substance from earlier rocks. "Tectonic land" is applied to lands raised by tectonic movements, in contrast to volcanic lands and deltal plains that rise by accumulation.

Detritus can include such varied constituents as fragments broken from consolidated rocks or organic masses—terrigenous or organic clastic detritals, or such nonclastic materials as organic skeletons, precipitates from fluids, or discrete authigenic minerals formed prior to burial. Calcite limestones—calcitites—are almost invariably detrital, with the exception of such rock as travertine, crystallized into a solid mass. Thus virtually all sedimentary calcitites are calcirudites, calcarenites, calcisiltites or calcilutites, or mixtures of considerable range. They may become calcite marbles with recrystallization. On the other hand, if dolomitites are preponderantly metasomatic rocks (Pettijohn, 1949, p. 309), as the writer believes, the rocks are rarely of dolomite detritus, but are of dolomitized calcite detritus. Textures can be detrital or crystalline. Thus dolomitite may be dolomitized calcarenite, or calcilutite, but is rarely dolomarenite of dolomilutite; the texture can be described as aphanic or phaneric (De Ford, 1946) without implication of original detrital texture.

[*Editor's Note:* Material has been omitted at this point.]

PALEOGEOGRAPHY OF THE CONTINENTAL BORDERS

INTRODUCTION

The continent in early Paleozoic had an interior platform, the hedreocraton, separated by flexures from bordering belts of earlier and greater depression, the miogeosynclines. The limits of these are not wholly defined, but their essential characters are evidenced in abundant records. The nature of the more peripheral parts of the continent is more obscure. There are few areas of known early Paleozoic sediments beyond the belts of Cambrian miogeosynclines. The rocks in these belts are generally younger, so that they have synclinorial form; early Paleozoic rocks if present are deeply buried. Extensive intrusions have destroyed much of the early record, and so metamorphosed the rocks as to obscure their original sections.

The paleogeography of the continental borders has been a subject of hypotheses and theories. The first hypothesis, that of marginal reefs, was expressed by Dana nearly a century ago. Dana's succeeding hypothesis of Archaean protaxes, popular later in the nineteenth century, considered present crystalline ranges to have persisted from early times (Pl. 12). The widely accepted theory of borderlands, developed principally by Schuchert, placed long-lived Cascadia and Appalachia on the opposite sides of the continent as great lands of crystalline, principally granitic rocks, extending from well within the present coasts to beyond the oceanic shores. The theory of marginal volcanic geosynclines and island arcs (Kay, 1944) considers regions beyond the miogeosynclines to have had deep-sinking belts of sediments and marine volcanic rocks—eugeosynclines—with smaller areas in volcanoes and linear tectonic welts rising within them.

THEORY OF MARGINAL REEFS

The conception of the eastern and known part of North America of 90 years ago was expressed by Dana (1856).

"The region toward the Atlantic border, afterwards raised into the Appalachians, was already, even before the Lower Silurian [Ordovician] era closed, the higher part of the land" lying "as a great reef or sand-bank, partly hemming in a vast continental lagoon, . . . thus separating more or less perfectly the already existing Atlantic from the interior waters."

This land was not considered due to mountain making, nor considered the source of sediments.

"The Atlantic border, from Labrador to Alabama, long in preparation, was at last folded into mountains" at the close of the Paleozoic; "no such event had happened since the revolution closing the Azoic Period." "On the idea that the rocks of our continent have been supplied by sands and gravel from a continent now sunk in the ocean, . . . the whole system of progress . . . is opposed to it. The existence of an Amazon on any such Atlantic continent in Silurian, Devonian, or Carboniferous times is too wild an hypothesis for a moment's indulgence." "The Continent in those early times received the northern Laborador current,—which would have kept by the shore as now, along the eastern border of the Azoic," and thence "over the Appalachian region, where the sandstones and shales were extensively accumulated; and therefore its aid in making these deposits can scarcely be doubted."

The lands in the north remained from the pre-Cambrian to serve as a source; at the close of the Azoic, "some parts of the continent were left as dry land, which appear to have remained so, as a general thing, in after times; for no subsequent strata cover them." This may be called the theory of marginal reefs.

THEORY OF ARCHAEAN PROTAXES

The increase in knowledge in the succeeding years led to the development of Dana's theory of Archaean protaxes (1890, p. 36–48).

"It appears that Archaean operations first established the boundaries, and that Paleozoic Mesozoic rock-making went on in the troughs between these boundary ranges . . . The boundaries separating the Atlantic and Pacific borders from the Continental interior should be drawn, as far as possible, along the ranges of Archaean ridges . . . These were boundaries at the beginning of Paleozoic time; and they have been ever since the most important division-lines for noting progress. On account of the Archaean origin of these axial lines in the two mountain chains, and the fact that in their elevation the existence of the Appalachian and Rocky Mountain chains had their beginning, I propose to call each the Archaean protaxis of the chain . . . Much the larger part of the later fragmental rocks, limestones excepted, are made out of what the Archaean ridges have lost.

"But after the Lower Silurian [Ordovician] era had passed . . . the Green Mountain area . . . emerged . . . Observe here what a blow the fact of this closed Northeast Bay gives the old theory that . . . the coarse and fine sediment for Appalachian rock-making, during the Upper Silurian era and afterwards, came in, period after period, from the northeast, through Labrador currents." It is "the unavoidable conclusion that all the sedimentary beds of New York and the Alleghenies, through the Upper Silurian, Devonian, and Carboniferous eras . . . were made within the Interior sea out of material derived, as far as non-calcareous, from the wear of rocks about it, and that the tidal and other currents of the interior sea distributed the material."

Walcott was the first who could base interpretation on substantial knowledge of earliest Paleozoic stratigraphy. In the text and illustrations of a report published in 1891 (Walcott, 1891, p. 363–369) he conformed to Dana's theory of Archaean protaxes.

"As the continent was slowly depressed and the waters advanced upon the land, the sediments now forming the Lower and Middle Cambrian series were accumulated in the various interior bodies of water to the eastward and westward of the main land area and between it and the outlying ridges. As the sea was transgressing over the surface of the continent on its way northward across the broad interior in the late Middle or early Upper Cambrian time, it was also working along the base of the border ridges and depositing the sediments derived from them, conformably upon those deposited while the main mass of the continent was above the water. Secular disintegration had prepared an immense amount of material for the advancing sea to assort and rearrange in early Cambrian time. . . . The great accumulation of sediment resulted from the distribution of material worn from the shore by waves and brought into the sea by the rivers of the Interior Continental region and the outlying ridges."

A new conception of the early Paleozoic nature of the Atlantic border is suggested n a footnote in the same article (p. 365):

"It is not improbable that the area of the great coastal plain of the Atlantic slope was then an elevated portion of the continent and that much sediment deposited during Cambrian and later Paleozoic was washed from it into the seas immediately to the west."

The name Appalachia was given to a land in this region that was the source of later Devonian clastic detritus (H. S. Williams, 1897, p. 395).

The conception of borderlands thus advanced differed from that of the protaxes of Dana in that the lands were not so directly related to the belts of rocks then classified as Archaean or pre-Cambrian.

In the time of Dana, larger areas along the Atlantic seaboard were thought to be pre-Cambrian than at present (1890b, p. 378–383); hence some of his protaxes were based on rocks that are not pre-Cambrian. Subsequent work has shown that the intrusives are dominantly intra-Paleozoic. Thus, in New Hampshire a synclinorium of Devonian has taken the place of an anticlinorium of pre-Cambrian with the recognition that the high-rank metamorphism reflects proximity to intrusions rather than antiquity (Billings, 1934; 1931). The belt of metasedimentary rocks in the western Green Mountains is the principal residue in New England, and these may be Cambrian older than the lowest beds in which fossils have been seen (Clark, 1936; Booth, 1950). Such metasediments as in the Glenarm series of the western Piedmont have similar stratigraphic position (Cloos and Hieranen, 1941, p. 196; Stose and Stose, 1944) and the Virgilina series of southern Virginia and the Carolinas (Jonas, 1932; Laney, 1917; P. B. King *et al.*, 1945). The Ocoee series of the Appalachians of eastern Tennessee seems definitely pre-Cambrian (G. W. Stose and A. J. Stose, 1944).

Dana's western Rocky Mountain protaxis (Dana, 1890c) lies quite within the craton in the south and within the miogeosynclinal belt in Canada, far east of the postulated "Cascadia" of later authors. Walcott placed a second "border ridge" in the Sierras (Walcott, 1891, p. 364). One assumes that this land was retained from Dana's maps, inasmuch as it had been established long before that the plutonic rocks are intra-Mesozoic in the Sierra Nevada of California (Whitney, 1865, p. 226) and the Coast Range of British Columbia (Dawson, 1886, p. B10–13). Among the last "pre-Cambrian" remnants, the Shuswap terrain of British Columbia has like the crystallines of New Hampshire become recognized as of metamorphosed sediments of Paleozoic and Mesozoic age adjoining intrusives (C. E. Cairnes, 1940). The oldest rocks that remain are such as the Siskiyou series of northern California, pre-Devonian (Hinds, 1940).

THEORY OF BORDERLANDS

The theory of borderlands became a necessary successor to that of Archaean protaxes when the theory of permanent pre-Cambrian uplifts became untenable, with recognition that sediments in miogeosynclinal belts had peripheral origin and that the clastic sediments in the Paleozoic of the East could not be attributed to the transportation of the Labrador current. There is abundant evidence that sediments in the Middle and Upper Ordovician, Silurian and Middle and Upper Devonian in eastern United States had an eastern source, for there are eastwardly coarsening textures in synchronous beds. Appalachia of H. S. Williams (1897, p. 345–397) was established in recognition of these relations. The critical question is whether these rocks came from great crystalline borderlands formed prior to the beginning of the Cambrian, lying on each side of the continent, or were from lands raised from earlier Paleozoic geosynclinal belts by intra-Paleozoic mountain building.

The theory of borderlands has been developed in the last half century, particularly by Charles Schuchert (Schuchert, 1910, p. 464–475; 1923, p. 158–164).

"Appalachia, the eastern borderland, extended to unknown eastern limits, occupying the area of the present Piedmont and coastal plains and probably the continental shelf as well. Cascadia, the western borderland, occupied the site of the modern coast ranges and apparently extended some distance beyond the present coastal margin and a corresponding borderland, Llanoria, stretched away to the south into the region which is now the Gulf of Mexico. . . . These elements persisted as the dominant features of the North American continent through all the changes of the Paleozoic era" (Schuchert and Dunbar, 1941, p. 125–126).

There has been little discussion of the evidence on borderlands in earliest Paleozoic time, though some have expressed scepticism.

If the early Cambrian sediments came from belts peripheral to the miogeosynclines, the presence of borderlands of some sort is demanded. The Lower Cambrian being absent on the craton, the entire interior of the continent could supply detritus to the miogeosynclines in the epoch. "Secular disintegration had prepared an immense amount of material for the advancing sea to assort and rearrange in early Cambrian time" (Walcott, 1891, p. 364). The diminishing pre-Cambrian area was the source of the basal sediments of the gradually spreading later Cambrian seas, for the equivalent beds are argillaceous and limy as they pass, thickening, into the bordering miogeosynclinal belts (McKee, 1943; Wheeler, 1943). The Lower Cambrian sands along the inside of the miogeosynclines must have had similar origin, for they are continuous with those basal to the younger series on the craton. "The basal Waucobian deposits are astonishingly similar in lithic composition throughout the Great Basin area and clearly represent the erosion products carried by streams from the isolated granitic areas to the east and north, re-sorted by wave action in the transgressing sea (Deiss, 1941, p. 1090)." Yet one reads that

"The name Cascadia has continued in current use and is the term used here to indicate the land mass whose central part furnished much of the Cambrian sediments in the area under consideration" (Deiss, 1941, p. 1093). "The overwhelming predominance of sandy sediments in these lower Cambrian deposits is remarkable. Much of the sand probably came from the uplands of Cascadia" (Schuchert and Dunbar, 1933, p. 127).

The most comprehensive study that seemed to give evidence of a peripheral source in the Appalachians was that of Barrell (1925) in the south. But the stratigraphic relation of the Ocoee series that seemed coarser and nearer the source than the Chilhowie series on the northwest is in doubt; the Ocoee has been considered to be older (G. W. Stose and A. J. Stose, 1944) and partly of augen gneiss resembling sheared conglomerate. Other conglomerates in the lower Cambrian of the region grade into finer sediments away from the craton (Booth, 1938; 1950). As the Middle Cambrian is predominantly non-terrigenous, Cascadia and Appalachia, under the theory, ceased to supply sands at the very time that seas spread over the craton, increasing the distance from the low lands of pre-Cambrian crystallines.

A few have questioned the prevalent view that lands characterized the marginal part of the continent in Paleozoic time.

"In the west, throughout the Cordilleran belt from the Pacific to the border of Laurentia, the sedimentation of the Paleozoic and Triassic built a great cumulus of deposits, buttressed here and there by volcanics, but otherwise without secondary structures. . . ." "The existence in any period of anything but volcanic land west of the Cordilleran geosyncline is hardly proved." "The areas of the Cascadia of Schuchert's maps was open sea during much of the Paleozoic" (Crickmay, 1931, p. 71, 17). "It has been assumed by some that the Columbia plateau is a 'positive area' or a region that has consistently stood high through geologic time. . . . The most vital evidence bearing on this theory is buried beneath the lava floods, but much evidence contradicts this interpretation" (Waters, 1933, p. 260). "The casual way in which some writers deal with the question of a Paleozoic 'Cascadia' to the west of the marine deposits of known Paleozoic age in California and Nevada is likely to rouse either considerable admiration or doubt. If Cascadia existed, where are the clastic deposits that accumulated near its borders?" (R. D. Reed, 1941, p. 99). "Prevailing American opinion has always considered both the Appalachians and the Ouachitas as mountains that have risen out of intracontinental geosynclines. . . . The hinterlands of these mountains would have been the 'borderlands' Appalachia and Llanoria. Does Llanoria exist at all, if we view it as an ancient rigid block in the earth's crust, acting first as a source region of the clastic flysche sediments?" (Van der Gracht, 1931b p. 1047–1048).

THEORY OF MARGINAL VOLCANIC GEOSYNCLINES AND ISLAND ARCS

The theory that the marginal parts of the present continent had Paleozoic volcanic geosynclines and island arcs (Pl. 1) has developed during the past decade.

Stille (1936a) perceived that western North America had the "Pacific" "pliomagmatic" belt and the "Rocky Mountain" "amagmatic" belt; that the contrast pertained not alone to plutonic intrusions, but also to volcanic extrusions; and that volcanic extrusions prevailed during the early Paleozoic on both continental borders (1940) in what he termed eugeosynclines (1941). Though Stille proposed the term, his use was somewhat informal and incidental to the presentation of the thesis that orogenic movements are periodic; the concept of eugeosynclines as subsiding surfaces is only inferred. However, he clearly intended a volcanic distinction:

"In diesem Sinne lässt sich innerhalb des Begriffes Orthogeosynklinale geradezu der "vollgeosynklinale" ("eugeosynklinale") Zustand, charakterisiert durch eine besonders im initialen Magmatismus zum Ausdruck kommende hohe Geosynklinalität, von einem "mindergeosynklinalen" ("miogeosynklinalen") unterscheiden, in dem der initiale Magmatismus fehlt."

Stratigraphic evidence in the East led the writer (1937) to conclude that there were tectonically raised linear lands in generally subsiding marginal belts; he referred to more peripheral volcanic belts as orthogeosynclines (1941) until clarification of the definition of that term (1942b, p. 1642).

Hess (1939; 1940) noted that the distribution of ultrabasic intrusions in the Atlantic and Pacific borders is similar to that in orogenic zones in other continents and in island arcs. Kay applied Fraser and Millard to the belts on the Pacific border having Paleozoic eugeosynclines and miogeosynclines, respectively, and Magog and Champlain to those on the Atlantic (Fig. 1) (1942b; 1947). Eardley (1947) summarized evidence from the Pacific border that convinced him that there are analogies to modern volcanic archipelagoes in the Paleozoic paleogeography.

The phrase "marginal volcanic geosynclines and island arcs" is somewhat anachronistic, in that it includes both stratigraphic and paleogeographic elements. Paleogeographically, it is conceived that there were bathymetric troughs, partly floored by volcanic rocks, "marginal volcanic troughs" (Kay, 1944), separating festoons of volcano-bearing islands and tectonic welts. Stratigraphically, the volcanic rocks of the islands as well as the sediments and volcanic rocks between and among them are in unit geosynclines lying between and among the tectonic swells, or grading cratonward into the miogeosynclinal belts. The relations are such as have been illustrated for the Welsh Ordovician eugeosyncline (Fearnsides, 1910).

Sedimentary and volcanic rocks in the belts are principally indigenous. Though some of the detritus came from older deposited and intruded rocks raised in the welts, a large part may be of ultimate volcanic source, first-cycle extrusive flows and fragmentals of all textures, or second-cycle sediments derived from erosion of rocks of volcanic origin; and some regions at times have great thicknesses of indigenous carbonatites and cherts—nonterrigenous sediments whose chemical constituents can have had distant ultimate origin. Some of the rocks eroded from tectonic welts are plutonic, as well as also ultimately volcanic, sediments of volcanic origin in earlier cycles. The subsidence in each geosyncline is measured by the thickness of the rocks, but with the addition or subtraction of the differences in original elevations of surfaces of reference at the base and top of the sections.

Discussion of the theory will follow a summary of the stratigraphic sections along the continental borders.

[*Editor's Note:* Material has been omitted at this point.]

SUMMARY AND CONCLUSIONS

The term geosyncline is applied to surfaces of regional extent subsiding deeply during the accumulation of succeeding sediments and volcanic rocks. It does not refer to physiographic surfaces, nor is it restricted to contained sequences having specified lithic suites or having had subsequent orogenic and volcanic histories. Geosynclines are classified on their forms, and on the nature and sources of their contents, reflecting tectonic and volcanic environments within and beyond their borders. The several classes are in three categories. The first, the orthogeosynclines, in linear belts lying outside the relatively stable continental interior or hedreocraton, comprise volcanic eugeosynclines and nonvolcanic miogeosynclines. They are commonly the sites of subsequent orogenies, and plutonic invasions are concentrated in the eugeosynclinal strips. The second category comprises the intracratonal geosynclines, grading into surfaces that are hardly geosynclinal in certain attributes. Exogeosynclines complement extracratonal source lands; autogeosynclines develop without complementing highlands; and zeugogeosynclines adjoin intracratonal elevations. The third category, geosynclines formed in orthogeosynclinal belts subsequent to the orogenic and plutonic consolidation of rocks of the antecedent and subjacent orthogeosynclines, includes epieugeosynclines—deeply subsiding troughs in relatively nonvolcanic environments; taphrogeosynclines—fault-bounded subsiding blocks; and paraliageosynclines—sinking troughs lying between the continents and ocean basins. The terms are defined in the succeeding glossary of definitions.

The interior of North America was rather stable, a craton, in the early Paleozoic and has been called the hedreocraton. Deeply subsiding nonvolcanic belts, miogeosynclines, had varying durations along the west, south, and east borders. Orthoquartzites are thickest in the basal part of sequences in the miogeosynclines along the margin of the hedreocraton, where crystalline rocks were exposed. Carbonatites prevailed when the low interior had been blanketed by sediments, until orogenic welts raised along the opposite side supplied terrigenous detritus. As uplifts affected the miogeosynclinal belts, derived detritus, principally graywacke and argillite, spread into the margin of the hedreocraton to be deposited in exogeosynclines.

The restriction of Paleozoic and early Mesozoic volcanic rocks to the continental margins, and the similarities of their stratigraphy and subsequent volcanic history to that of modern island arcs, leads to the conclusion that Paleozoic North America was fringed by festoons of volcanic and tectonic islands, rather than by borderlands of ancient crystalline rocks. Graywackes are prevalent in eugeosynclines, because frequently there were associated tectonic welts, but conglomerates and argillites are common, and carbonatites accumulated extensively and thickly when lands were elevated only slightly; basaltic and andesitic rocks are more frequent than rhyolitic. There is a suggestion of local interrupted progression from quartz-free to quartz-rich lavas, but such sequences are repeated within the same broad eugeosynclinal belts. Though ultrabasic intrusions seem to progress outward from the hedreocraton, the volcanic belts seem to have retained rather constant limits.

Post-Protozoic sediments and associated volcanic rocks would form a layer almost 2 miles thick if distributed evenly over the present continent. Orthogeosynclines contained the largest proportions of this volume, because they subsided relatively rapidly and long. The thickest sequences of sediments of each principal class are in orthogeosynclines. Sediments are dependent not only on the amount of subsidence but also on the presence or absence of near-by sources and the relative stability of regions of source and transportation. Most post-Protozoic terrigenous detritus came ultimately from Paleozoic and younger volcanic and plutonic rock, much of it raised in welts in orthogeosynclinal belts.

The study of geosynclines involves hypotheses of continental genesis. The advent of eugeosynclines is obscure, but belts for the most part have been consolidated by orogenies and attendant plutonism. Thus island arcs and eugeosynclines are now of limited extent, the residue of former more extensive belts. Volcanic belts have decreased in number through time, with attendant decrease in the extent of volcanism, or else they have developed from previously rather stable ocean basin. The outline of the continent came into being rather late in geologic history, perhaps at the close of the Protozoic and beginning of the Paleozoic, when thin sialic volcanic arcs extended to about its present limits. With consolidation of the marginal belts, volcanism has become less extensive. The continent with its thick sial represents the geographic contraction of broader areas of thin sial, with the addition of interior sialic constituents into the more simatic rocks.

CLASSIFICATION AND DEFINITIONS

A *geosyncline* is a surface of regional extent subsiding deeply during accumulation of succeeding surficial rocks. *Orthogeosynclines* are long and narrow, forming belts adjoining and separating *cratons*, relatively immobile regions of the earth, either of higher, continental level or the lower ocean basins. The *hedreocraton* in North America is the craton of the earliest Paleozoic that strongly influenced subsequent continental structure.

Orthogeosynclines are of two kinds. *Eugeosynclines* lie in belts of active volcanism in association with relatively rapid subsidence, whereas *miogeosynclines* are in nonvolcanic belts. Both persisted along the Pacific and Atlantic borders of North America in the early Paleozoic (Figs. 1, 2; Pls. 1, 2, 9).

Intracratonal geosynclines (parageosynclines of Stille) are of three sorts. *Exogeosynclines*, within cratonal borders, gain sediment from erosion of complementing highlands outside the craton in the orthogeosynclinal belts; the late Ordovician and later Devonian of Pennsylvania and near-by states were laid in typical examples (Pl. 5, and Figs. 4 and 5), the later Cretaceous of Wyoming in a more complex exogeosyncline (Pl. 11). *Autogeosynclines* subside in elliptical basin or trough form without associated highlands; the late Silurian of Michigan (Pls. 6 and 7) and Mississippian of Illinois (Pl. 8) are in examples. *Zeugogeosynclines* are similar, but contain sediment from eroded complementing highlands within the craton; the Pennsylvanian of central Colorado and northern New Mexico was laid in the best example (Fig. 6; Pl. 9).

Geosynclines of three types formed in orthogeosynclinal belts after relative immobilization and consolidation to the hedreocraton through orogeny and plutonic intrusion. *Epieugeosynclines*, deeply subsiding troughs with limited volcanism associated with rather narrow uplifts and overlying deformed and intruded eugeosynclines, contain Mississippian and Pennsylvanian in Maritime Canada (Pl. 14 and Fig. 14) and Cretaceous and Tertiary in California (Pl. 14 and Fig. 11). *Taphrogeosynclines* are sediment-filled deeply depressed rift blocks, bounded by one or more high angle faults; late Triassic geosynclines along the Atlantic seaboard are of this kind (Fig. 16). *Paraliageosynclines* subsided deeply in belts passing into coastal plains along the present continental margin, are exemplified by that of the Paleocene and Eocene along the north coast of the Gulf of Mexico (Pls. 15 and 16).

The several geosynclines are represented in sections to uniform scale in Figure 19, and with scale related to rate of subsidence in Figure 20.

[*Editor's Note:* None of the figures are included here. Only the references cited in the preceding excerpts have been reproduced.]

REFERENCES

Barrell, Joseph (1925) *The nature and environment of the Lower Cambrian sediments of the southern Appalachians*, Am. Jour. Sci., vol. 9, p. 1-20.

Booth, V. H. (1938) *Oak Hill series in Vermont* (abstract), Geol. Soc. Am., Bull., vol. 49, p. 1869.

Booth, V. H. (1950) *Stratigraphy and structure of the Oak Hill succession in Vermont*, Geol. Soc. Am., Bull., vol. 61, p. 1131-1168.

Bucher, W. H. (1933) *The deformation of the Earth's crust*, Princeton Univ. Press.

Cairnes, C. E. (1940) *The Shuswap rocks of southern British Columbia*, Sixth Pac. Sc. Congr., 1939, Pr., vol. 1, p. 259-272.

Clark, T. H. (1936) *A lower Cambrian series from southern Quebec*, Royal Canad. Inst., Tr., vol. 21, p. 135-151.

Cloos, Ernst, and Heitanen, Anna (1941) *Geology of the "Martic overthrust" and the Glenarm series in Pennsylvania and Maryland*, Geol. Soc. Am., Spec. Paper 35.

Crickmay, C. H. (1931) *Jurassic history of North America, its bearing on the development of continental structure*, Am. Philos. Soc., Pr., vol. 70, p. 15-102.

Dana, J. D. (1856) *On the plan of development in the geological history of North America*, Am. Jour. Sci., vol. 22, p. 335-349.

Dana, J. D. (1873) *On some results of the Earth's contraction from cooling including a discussion of the origin of mountains and the nature of the Earth's interior*, Am. Jour. Sci., vol. 5, p. 423-443; vol. 6, p. 6-14; 104-115; 161-172.

Dana, J. D. (1890a) *Areas of continental progress in North America*, Am. Jour. Sci., vol. 1, p. 36-48.

Dana, J. D. (1890b) *Archaean axes of eastern North America* Am. Jour. Sci., vol. 39, p. 378-383.

Dana, J. D., (1890c) *Rocky Mountain protaxis and the post-Cretaceous mountain-making along its course*, Am. Jour. Sci., vol. 40, p. 181-196.

Dawson, G. M. (1888) *Report on an exploration in the Yukon district, Northwest Territory, and adjacent portions of British Columbia*, Geol. Survey Canada, Ann. Rept. 3, pt. B, p. 1-183.

De Ford, R. K. (1946) *Grain size in carbonate rock*, Am. Assoc. Petrol. Geol., Bull., vol. 30, p. 1921-1928.

Deiss, C. F. (1938) *Cambrian geography and sedimentation in the central Cordilleran region*, Geol. Soc. Am., Bull., vol. 52, p. 1085-1115.

Eardley, A. J. (1947) *Paleozoic Cordilleran geosyncline and related orogeny*, Jour. Geol., vol. 55, p. 309-342.

Evans, J. W. (1926) *Regions of compression*, Geol. Soc. London, Quart. Jour., vol. 82, p. lx-cii.

Fearnsides, W. G. (1910) *North and central Wales*, Geol. Ass. London, Geology in the field, vol. 2, p. 786-825.

Glaessner, M. F., and Teichert, Curt (1947) *Geosynclines: a fundamental concept in geology*, Am. Jour. Sci., vol. 245, p. 465-482; 571-591.

Grabau, A. W. (1913) *Principles of Stratigraphy*, A. G. Seiler, New York.

Hall, James, (1859) *Description and figures of the organic remains of the Lower Helderberg group and the Oriskany sandstone*, N. Y. Geol. Survey, Paleont., vol. 3.

Haug, Emile (1900) *Les geosynclinaux et les Aires Continentales*, Soc. Géol. France, vol. 28, p. 617–711.

Hinds, N. E. A. (1940) *Paleozoic section in the southern Klamath Mountains, California*, 6th Pacif. Sci. Cong. 1939. Pr., vol. 1, p. 273–287.

Jonas, A. I. (1932) *Geology of the kyanite belt of Virginia*, Va. Geol. Survey, Bull. 38.

Kay, Marshall (1937) *Stratigraphy of the Trenton group*, Geol. Soc. Am., Bull., vol. 48, p. 233–302.

Kay, Marshall (1941) *Taconic allochthone and the Martic thrust*, Science, vol. 94, p. 73.

Kay, Marshall (1942b) *Development of the Allegheny synclinorium and adjoining regions*, Geol. Soc. Am., Bull., vol. 53, p. 1601–1658.

Kay, Marshall (1944) *Geosynclines in continental development*, Science, vol. 99, p. 461–462.

Kay, Marshall, (1945a) *Paleogeographic and palinspastic maps*, Am. Assoc. Petrol. Geol., Bull., vol. 20, p. 426–450.

Kay, Marshall (1947) *Geosynclinal nomenclature and the craton*, Am. Assoc. Petrol. Geol., Bull., vol 31, p. 1280–1293.

King, P. B. et al. (1946) *Tectonic map of the United States*, Am. Assoc. Petrol. Geol.

Knopf, Adolph (1948) *The geosynclinal theory*, Geol. Soc. Am., Bull., vol. 57, p. 649–670.

Krynine, P. D. (1948) *The megascopic study and field classification of sedimentary rocks*, Jour. Geol., vol. 56, p. 130–165.

Laney, F. B. (1917) *The geology and ore deposits of the Virgilina district of Virginia and North Carolina*, Va. Geol. Survey, Bull. 14.

Lombard, A. E. (1949) *Critères descriptifs et critères génétiques dans l'étude des roches sédimentaires*, Soc. Géol. Belge, Bull., vol. 3, p. 214–271.

McKee, E. D. (1943) *Some stratigraphic principles illustrated by Paleozoic deposits of northern Arizona*, Am. Jour. Sci., vol. 241, p. 101–108.

Pettijohn, F. J. (1949) *Sedimentary Rocks*, Harper and Brothers, New York.

Reed, R. D. (1941) *California's record in the geologic history of the world*, Calif. Dept. Nat. Res., Div. Mines, Bull. 118, pt. 2, p. 99–118.

Schuchert, Charles (1910) *Paleogeography of North America*, Geol. Soc. Am., Bull., vol. 20, p. 427–606.

Stille, Hans (1936a) *Die Entwicklung des amerikanischen Kordillerensystems in Zeit und Raum*, Preuss. Akad. Wiss. Phys.-Math. Kl. Sitsungsber. 15, p. 134–155.

Stille, Hans (1936b) *Wege und Ergebnisse der geologisch-tectonischen Forschung*, Festschr. Kaiser-Wilhelm Gesellsch. Förd. Wiss., Bd. 2.

Stille, Hans (1940) *Wandlungen im Magmatismus unserer Erde*, Naturw., Jg. 28, vol. 21, p. 321–326.

Stille, Hans (1941) *Einführung in den Bau Amerikas*, Borntraeger, Berlin.

Stose, A. J., and Stose, G. W. (1944) *Geology of the Hanover-York district, Pennsylvania*, U. S. Geol. Survey, Prof. Paper 204.

Stose, G. W., and Stose, A. J. (1944) *The Chilhowee group and Ocoee series of the southern Appalachians*, Am. Jour. Sci., vol. 242, p. 367–390; 401–416.

Tercier, J. (1940) *Dépôts marine actuels et séries géologiques*, Ecl. Geol. Helv., vol. 32, p. 47–100.

van der Gracht, W. A. J. M. van W. (1931b) *Permo-Carboniferous oro-geny in south-central United States,* Am. Assoc. Petrol. Geol., Bull., vol. 15, p. 991–1057.

Walcott, C. D. (1891) *Correlation papers; Cambrian,* U. S. Geol. Survey, Bull. 81.

Waters, A. C. (1933) *A summary of the sedimentary, tectonic, igneous and metalliferous history of Washington and Oregon,* Ore deposits of the western states (Lindgren Vol.), Am. Inst. Min. Met. Eng., p. 253–265.

Wheeler, H. E. (1943) *Lower and Middle Cambrian stratigraphy in the Great Basin area,* Geol. Soc. Am. Bull., vol. 54, p. 1781–1824.

Whitney, J. D. (1865) *Report of progress and synopsis of field work from 1860 to 1864,* Geol. Survey Calif., Geol., vol. 1.

Williams, H. S. (1897) *On the southern Devonian formations,* Am. Jour. Sci., vol. 3, p. 393–403.

Part V

CAUSAL MECHANISMS: "OLD GLOBAL TECTONICS" THEORIES, EARLY TWENTIETH CENTURY

Editor's Comments
on Papers 13, 14, and 15

Despite important disagreements over detail, most geologists in the early part of the twentieth century agreed that geosynclines were the necessary precursors of most mobile belts. Furthermore, examination of continental fabric suggested that each continental block was an interwoven mosaic of mobile belts of various ages. Therefore it was crucial to elaborate fully on the mechanism(s) whereby geosynclines were initially developed and later deformed and uplifted into mobile belts that could eventually be welded to ancient continental nuclei.

Long before the new global tectonics was proclaimed, many theories had related geosynclinal development of mobile belts to single, worldwide mechanisms. The visible differences in the geological and geographical distribution of geosynclines, their persistence through time, and their varying evolutionary histories were perceived as possibly due only to variations in the manner in which the same causal mechanism manifested itself.

Early global tectonics theories insisted that the positions of continental blocks, ocean basins, and geosynclines were rigidly fixed in time and space. Dana, like most of his contemporaries, had accepted a model in which the earth was regarded as contracting. This was the "establishment dogma" of most of the nineteenth and early twentieth centuries. The resultant lateral compression was invoked to produce the initial downwarping of a geosynclinal trough, its later compressional deformation, and its eventual uplift. A fundamental concept, still valid today, was developed by Suess (1885–1909), who recognized that oceanic and continental crusts were petrologically distinct (*sima* and *sial*). Dana believed that geosynclines marked this critical junction between contrasting crustal types, and believed that such zones would consequently be particularly susceptible to crustal shrinkage.

However, there were many obvious shortcomings to these early fixist theories and it was their inadequacy that lead to mobilist theories such as continental drift. For example, the varied locations of geosynclines were perplexing: sometimes along the continental margins, sometimes within the interiors of continental blocks, and sometimes in far-flung island arcs. Furthermore, the origin and mysterious disappearance of geosynclinal source areas such as borderlands and volcanic island arcs was puzzling. If continents were allowed to drift, the borderland problem was easily resolved:

> Before the new idea of continental migration was offered to science, it was thought by geologists that these lands had foundered, block by block, under the ocean. This older explanation of the Atlantic and Indian ocean-basins is fatally affected by difficulties arising from the physics of the case. We shall not enter on a discussion of those difficulties.
>
> However, gigantic founderings of the kind are not necessary assumptions, if the two young ocean-basins have been opened by the migration of the continents. For according to the migration theory, the missing land east of North America *was*, in large part, Europe and northern Africa. The missing land west of Europe and Africa *was* in large part, North America and so on (Daly, 1926, p. 279).

Continental drift, as proposed by Taylor (1910) and Wegener (1912a, 1912b, 1915), provided a mechanism capable of producing crustal tension as well as compression. For example, the distension and separation of preexisting supercontinents into two or more blocks could produce small ocean basins (incipient geosynclines?) bordered on both margins by continental landmasses (cratonic and borderland source areas). As continents drifted iceberg-like through the simatic crust, their leading edges were folded (the Cordilleran belt?). Closure of an ocean basin between coverging continental blocks could produce mobile belts like the Alps and Urals. Later continental rifting might bifurcate such a belt (the Appalachian-Caledonian belt?). If one accepted the reality of Gondawanaland, the sharp contrasts between the symmetrical North American craton, rimmed on all sides by mobile belts, and the puzzling one-sided frameworks of South America, Africa, Australia, and Antarctica, could be rationally explained. Significantly, if continental drift was a viable mechanism, geosynclines represented a variety of continental margin types, and each of the various kinds of continental margins might have ancient analogues within mobile belts.

The success with which early theories of continental drift explained the confusing array of observations made about geosynclines is impressive (DuToit, 1937a; 1937b). The degree to which a number of geologists anticipated the ideas that today constitute the basic format of the new global tectonics has been recounted at length (see, for example, Davis et al., 1974). Various theoretical aspects of plate tectonics were described with almost eerie foresight by Emile Argand (Paper 13), Reginald Daly (1926), David Griggs (Paper 14), and Arthur Holmes (Paper 15) among others, beginning in the 1920s. While the decades that followed would refine these early speculations and provide for them a firmer factual footing, it is enlightening to read how remarkably accurate a picture they gave.

Argand (1924) published his monumental work on the tectonics of Asia (the paper was actually read in 1922 before the 13th International Geological Congress). He was particularly interested in the evolution of the Alpine-Himalayan belt, but felt that this zone was generally representative of many geosynclinal belts. He related the then popular Taylor-Wegener version of continental drift to the geosynclinal mechanism for generating mountain systems. Argand's work is important because it allowed for a somewhat different mechanism of continental drift and introduced the idea of orogeny by continental collision and gravitational sliding (Paper 13).

Conventional continental drift required the differential movement of iceberg-like continental blocks through simatic ocean crust, a process vigorously opposed by geophysicists on mechanical grounds (Jeffreys, 1929). Argand allowed for the rifting and movement of continents as a consequence of the injection and growth of new sima between continental blocks, with the continents themselves essentially passively conveyed passengers. Continental blocks torn apart and forcibly separated by the injection of new ocean crust from below formed new ocean basins. To Argand, such ocean basins were geosynclines. Subsequent closing of ocean basins by convergence of the bordering continental fragments produced a deformed geosynclinal belt composed largely of marine and continental margin sediments.

Therefore, to Argand, the Alpine-Himalayan belt was the product of the opening and closure of an ancient ocean basin, Tethys, like a vise between the jaws of the Indo-African and Eurasian continents. Similarly, the Appalachian-Caledonian belt was generated by the early Paleozoic rifting of a Laurasian supercontinent, the opening of a proto-Atlantic Ocean, and subsequent ocean closure and continental collision. Argand argued that the present-day Atlantic Ocean represented a re-opening of this proto-Atlantic:

> the mid-Atlantic Ridge, similar to a wreck of sial tossed around and deformed in the sima, seems to be the heritage of the time in which the New World was beginning to separate from the Old. It seems to consist, as previously said, of the remains of the sial stretched at the bottom of the S-shaped furrow along which the disjunction was being prepared; of masses of sial collapsed from both vertical margins, with sliding and dragging of some terrigenous deposits, even of melted sial (Argand in Carozzi, 1977, p. 147).

Reginald Daly, one of the most imaginative and influential geologists in the early part of the twentieth century, also accepted the reality of continental drift, but employed a mechanism not unlike sea-floor spreading. According to Daly (1926), continental drift occurred in response to the doming of a primitive Pangean continent that was progressively stretched and fragmented. Slabs broken from this supercontinent slid off the domal uplift into a series of low-lying, east-west and north-south ocean basins. If domal stretching continued, new ocean crust developed on the site of the dome as a result of the vertical migration of basaltic magma through a series of dikes and fissures. Compression occurred only on the leading edges of continental blocks; trailing edges were subject to tension or passive subsidence. Thus, two fundamentally different types of continental margins could occur: *Atlantic-type* and *Pacific-type* (so named by Suess, 1885–1909).

> The new hypothesis automatically accounts for the general absence of mountain chains bordering the Atlantic, Arctic, and Indian Oceans. For in the more recent geological time these have been regions of tension, not of horizontal compression. On the other hand, the Pacific should be, and is, well framed in mountain chains. In brief, the Atlantic and Pacific types of coast-line, fundamental features of the earth, are adequately explained by the hypothesis of the migration of continents (Daly, 1926, p. 290).

Daly's remarks were certainly prophetic (see Papers 29 and 30) and he proved to be equally ahead of his time in his views on the possible relationship between mobile belts on the adjacent sides of the North Atlantic:

> If North America slid westward and Europe eastward, having been formerly joined together, then individual rock-structures of the original great mass of land were torn apart. The sundered

parts of the structures should be visible in the existing continents. . . . One can hardly fail to be impressed with the remarkable matching of rock structures on the two sides of the Atlantic.
Thus, the Appalachians of Newfoundland and Nova Scotia correspond in composition, structure, and date of formation with
the Hercynian mountain-mass of Britain and Brittany" (Daly,
1926, p. 284–285).

Daly considered the mid-Atlantic ridge as a continental crustal
remnant situated roughly at the site of the original junction between the Old and New Worlds: "a long strip of the original continent, a strip left behind when that continent was torn into fragments, which slid away, respectively to westward and eastward"
(Daly, 1926, p. 280).

In 1931, Arthur Holmes published a paper (originally read in
1929) that improved Daly's model of continental drift considerably.
Holmes was quite prepared to accept the evidence for continental
drift, but he sought to provide for it a more plausible causal mechanism than Daly's presumed crustal doming. Holmes was keenly
interested in the possible ways in which the earth dissipated the
internal heat generated by radioactive decay. He postulated the
existence of a relatively brittle, crystalline crustal layer (essentially the lithosphere as presently defined) that rested on a denser
comparatively plastic substratum (earlier labelled *asthenosphere*
by Barrel, 1914). Holmes postulated that convective movements,
developed within the underlying substratum in order to dissipate
radioactive heat, would progessively stretch and fragment the
rigid surface crust. This would eventually lead to the drift of continental blocks as new areas of ocean crust (derived from the substratum) grew between the separating continental blocks. Downward moving currents compressed geosynclinal belts located
along continental margins into fold belts. The Atlantic Ocean was
an "opening" ocean situated over ascending convection limbs,
while the Pacific Ocean was bordered largely by descending limbs.

Griggs (Paper 14) accepted Holmes's hypothesis that convection currents might occur within a plastic zone beneath a rigid
outer shell of the earth (Holmes, 1929, 1932). To test the implications of this hypothesis, Griggs conducted a series of experiments
to investigate the effects of such convection on the crust. Extracts
of his paper are included here to document how thoroughly such
earlier workers understood a possible mechanism for geosynclinal
development of mobile belts, well before modern plate tectonics
theory.

Holmes popularized and refined his model explaining continental drift and the origin of mobile belts by thermal convection

in the 1945 edition of his classic textbook, *Principles of Physical Geology*. This later version did not differ appreciably from the version originally published in 1931. In the brief excerpts reprinted here (Paper 15), one can recognize the seeds of such future concepts as the low velocity zone, subduction zones, and spreading axes, all of which would later form the conceptual core of plate tectonics. Holmes characterized his model as very speculative. Interestingly, Holmes's ideas concerning the existence of a global series of rising and descending convection cells postdated the work of Kober (1923), who published a map showing a global submarine mountain chain that largely coincided with the present-day distribution of the ocean ridge and rise system. What is most significant about Holmes's mantle convection concept is that it provided a "geophysically acceptable " mechanism for continental drift. Convective lateral flow of the mantle provided "an endless travelling belt" on which the continental blocks could passively drift; no longer were they forced to push, iceberg-like, through a solid sima.

REFERENCES

Argand, E., 1924, La Tectonique de l'Asie, *13th Internat. Geol. Congr. Compt. Rend.* **5**:171–372.

Barrell, J., 1914, The Strength of the Earth's Crust, *Jour. Geol.* **22**:289–314, 441–468, 655–683

Carozzi, A. V., ed. and trans, 1977, *Tectonics of Asia*, E. Argand, Hafner Press, New York, 218 p.

Daly, R. A., 1926, *Our Mobile Earth*, Charles Scribner, New York, 342 p.

Davis, G. A., B. C. Burchfield, J. E. Case, and G. W. Viele, 1974, A Defense of "An Old Global Tectonics," in C. F. Kahle, ed., *Plate Tectonics—Assessments and Reassessments*, American Association of Petroleum Geologists Memoir 23, pp. 16–23.

DuToit, A. L., 1937a, A Geological Comparison of South America with South Africa, *Carnegie Inst. Washington Publ. 381*, pp. 1–157.

DuToit, A. L., 1937b, *Our Wandering Continents*, Oliver & Boyd, Edinburgh, 366 p.

Holmes, A., 1929, Radioactivity and Earth Movements, *Geol. Soc. Glasgow Trans.* **18**:559–606.

Holmes, A., 1932, The Thermal History of the Earth, *Washington Acad. Sci. Jour.* pp. 169–195.

Jeffreys, H., 1929, *The Earth: Its Origin, History, and Physical Constitution*, Cambridge University Press, Cambridge, 364 p.

Kober, L., 1923, *Lehrbuch der Geologie*, Holder-Ichler-Tempsky, Vienna, 425 p.

Suess, E., 1885–1909, *Das Antlitz der Erde*, 3 vols., Tempsky, Vienna. (English translation: *The Face of the Earth*, 5 vols., Clarendon Press, Oxford.)

Taylor, F. B., 1910, Bearing of the Tertiary Mountain Belt on the Origin of the Earth's Plan, *Geol. Soc. America Bull.* **21**:179–226.

Wegener, A., 1912a, Die Enstehung der Kontinente, *Petermann's Geog. Mitt.* **58**:185–195, 253–257.

Wegener, A., 1912b, Die Enstehung der Kontinente, *Geol. Rundschau* **3**:276–292.

Wegener, A., 1915, *Die Enstehung der Kontinente und Ozeane,* Vieweg, Brauschwieg, Germany, 135 p. (English translation: *The Origin of Continents and Oceans,* Dover, Toronto, 246 p.)

13

Reprinted with permission from pages 32–33, 134–135, 138–140, and 162 of
Tectonics of Asia, A. V. Carozzi, ed. and trans., Hafner Press, New York, 1977, 218 p.

TECTONICS OF ASIA

E. Argand

[*Editor's Note:* In the original, material precedes this excerpt.]

In general, a large portion of the original energy will be consumed by deformations within the continental block itself; it cannot be otherwise because this derivation of energy is the *first one* that can be discussed at present with some degree of reliability; this consumption occurs *in situ*—it involves volumes in comparison with which all the geosynclinal chains, and other types that emerge from the seas, weigh very little, at least in the Alpine cycle. The collision of two continents, or the more gradual confrontation between the two portions of the same continent separated by a zone of weakness, is expressed by a certain repercussion of the compression on the jaws themselves. Therefore, within the width of the colliding objects, a kind of backlash occurs whose deforming effects—all other features being equal—gradually decrease from the zone of collision toward the rear areas: in the case of very powerful compression, this *second* derivation of energy may become extremely efficient. Thus, the first derivation, complicated or not by intense secondary compression, always produces within the mass of continents very powerful effects. As we shall see soon, these effects consist essentially of a folding, with a large radius of curvature, of old indurated frames: these *basement folds*, through the internal tensions resulting from their formation, generate important fractures. Within continental areas, fractures will most probably not occur unless such conditions exist or unless there is distensional traction.

The other still important derivative effect is, in the case of a vise, the folding of geosynclinal chains: arcs are formed whose distribution and deformation depend, originally, on the first fragmentation, to which is added later a division in much shorter and more specialized segments, under the influence of the rupturing through fractures that reaches the jaws and divides them into high and low sectors of unequal resistance, not to mention other factors that will become more evident later. (203) The amount of energy diverted by this *third* derivation to the benefit of geosynclinal chains and of other new chains can increase, particularly when the jaws have a small amount of heterogeneity and are therefore not much deformed.

Geosynclinal chains usually are overturned *backward*, that is, toward the jaw that is more directly responsible for their formation. When they happen to be in close contact with the jaw or to be thrust over it, these chains restitute to the upper parts of the continent a small portion of the energy they acquired from it.

This restituted energy, a *fourth* component of the energies in a distributed state, is naturally very much lower than the others. Its effects usually are limited to that part of the continental margin that is most exposed to the backward pressure of geosynclinal chains. The restituted energy propagates from this margin toward the rear zones, and its effects decrease rapidly in that direction. The propagation within the continental block takes place through the unconformable covers or the basement consisting of old indurated folds. In the first case, *cover foldings* are generated; in the second, the energy restituted to the old frames is added to those energies that are liberated directly within them and that participate together in the generation of basement folds and related fractures.

In the distribution of the useful energies that occur within a continent and in its related blocks, it is therefore possible to visualize at least four derivations more and more remote from the original energy. The first one, more important than all the others except in particular cases, is never absent as long as there is any trace of movement of the continent. The second is a consequence of compression; it is really only a portion of the first one and is active similarly inside the continent with a few peculiar features concerning its location. The first, including this secondary derivation whenever it exists, is the energy that can properly be called *intracontinental*. The third, which is borrowed from the first or, which amounts to the same, from the second, remains very low compared with the intracontinental energy; it is consumed in new chains of geosynclinal or Circumpacific type. The fourth, still much lower, is the energy *restituted* to the continent, and in a detailed viewpoint there are other kinds of distribution. When discussing East Asia, we shall indicate to what extent these factors have a chance to occur in the deformation of the Circumpacific chains and of the five continents against which these chains are molded. Meanwhile, we shall look at the Tethys and its two rear zones.

In tectonics, the prevailing practice is to relate the direction of movements to continents assumed to be motionless, to old massifs considered in their passivity as obstacles and not in their active displacements. The rear zones correspond to the forelands of this older nomenclature, (204) and in the remaining part of this work, I shall try as much as possible to use this manner of speaking. The other terminology, as may be seen, has the advantage of better defining the origins, the deformations being expressed in terms of power and not of resistance. It is clear that the fact of the nearing of jaws—with consequences drawn or to be drawn—remains independent of any convention. With respect to initial energies, the media and the conditions under which they are liberated escape our visualization so much so that we are not going to say anything more about them.

The mobilistic theory has somewhat neglected the concept of geosyncline. It is therefore appropriate to sketch a connecting link. A geosyncline will generally result from a horizontal *traction* that stretches the raft of sial. The stretching is at first easier in the deeper part of the sial rather than in the upper part, where extension fractures may develop. While thinning, the sial sinks and develops a depression: the subsidence inherent in the geosynclinal process does not, therefore, stem from an original radial stress; it is only the vertical effect of a horizontal distension. The overburden of the deposits helps of course to accentuate the alveole, but the latter is not necessarily the original feature. Until compensation, the sima rises under the thinned sial; this behavior accounts for the frequent association of green rocks with bathyal and abyssal sediments. The mixture of abyssal with shallower sediments takes place through submarine sliding on the slope. The vertical margins of the thinned zone, which have preserved the normal thickness of the sial, represent the jaws of the geosyncline. Whenever compression replaces traction, the jaws are brought closer and the classical geosynclinal deformation begins with its embryonic folding by means of cordilleras, furrows, and true foredeeps: the conclusion is almost always the formation of two geosynclinal chains with opposite overfolding.

If traction continues, instead of giving way to compression the sial continues to stretch and the sima appears at the bottom of the alveole. Along the transverse alignments where such a situation occurs, the geosynclinal condition is replaced by the oceanic condition; if such a situation becomes generalized, only an ocean is left. If a compression occurs at this stage or just before it, when the sial is really very thin, the lack of synergy will lead to the generation of one or two trends of marginal chains, of Circumpacific type, and not of the double chain of geosynclinal type. If the compression continues, the latter type will establish itself gradually and may perhaps eventually predominate, but the traces of the simple or double marginal condition will persist, although veiled.

The ordinary geosynclinal behavior is therefore, in essence, an incomplete lenticular cutting up of the continent; whenever the cutting is complete, the oceanic condition appears.

In addition, events can occur in a more brutal fashion and proceed by means of high and steep distensional fractures that eliminate the stretchings and result in more massive jaws—hence the differences in the subsequent flexible deformations when they exist.

All these geosynclinal deformations may become complicated in the same manner as the corresponding marginal deformations, namely, by the effect of bow and stern stresses, (300) of oblique and longitudinal behaviors, in reciprocal alternations or in simple succession. The resulting complexity has an appreciable chance to be greater in a geosyncline than in a double marginal arrangement because of an effect of synergy.

Small intracontinental geosynclines can be generated without stretching, through the simple continued effect of an overburden of sediments. As long as the original factors operate without outside interference, these alveoles remain almost full while subsidence operates along the substratum of their axial zone.

There are between all these types of geosynclines transitions and interactions that I shall not discuss here.

The *North American continent* shows the process of drift over lenses on a grandiose scale. This interpretation is immediately suggested by the distribution of the Bouguer anomalies in the United States. The front on the Pacific coast shows positive regions about which one cannot say as yet to what extent the default of gravity due to drift over lenses, the positive perturbation of Helmert related to the upper crest of the continental slope, and the more local distribution of the densities should participate in the explanation of the observed anomalies. But in the rear, (303) where the drift over lenses is very obviously displayed, there is evidence that it would be dangerous to eliminate this factor from the explanations pertaining to the frontal part. The zone of strong negative anomalies extends hundreds of kilometers to the east of the visible Laramide front, beneath the High Plains. In the same direction, the major portion of the anomalies gradually decreases and the lens similarly thins while its lower surface rises and the sima with it. The major part of this conformation ends along a winding line—the general trend being meridian—enclosed between the 103rd meridian to the west and the 97th meridian to the east, with some reservations due to the degree of approximation compatible with the large spacing of the stations. Possibly, a solidary compensation, involving both the region of the chains and that of the High Plains, may lead to a sinking of the latter, thus contributing to its lack of mass. But since nothing similar occurs on the west side of the chains, the unilateral character of the situation is obvious, and I can visualize only drift over lenses accounting for the essentials of it.

Farther east, beneath the center of the United States, weak negative anomalies seem to prevail with noticeable positive exceptions, which I shall not discuss. Farther away, the sub-Appalachian lens appears with respect to the visible intumescence slightly off the axial line, toward the east or southeast. The effect is more moderate than it is for the lens of the great chains of the west, but it is of the same direction, and drift over lenses explains the essentials of both conformations. While drifting to the west, the continent tends to abandon a portion of its deep-seated lenses, which remain behind.

My *Proto-Atlantic,* a Caledonian geosyncline that in some parts is much older and whose traces run from the Arctic domain to the Antarctic regions, is perhaps not irreconcilable, for all we know, with the fixist imagery. The large-scale collapse processes assumed to have generated the present Atlantic would have renewed the general arrangement of the ancient plan and would correspond to the return of a certain number of similar conditions. The Mid-Atlantic Ridge would show a beginning of filling by folds trending across the Hercynian and Alpine plan and would initiate, by following the essentials of the Caledonian plan, a post-Alpine orogenic cycle. The long persistence of geosynclines, whose lifetime can encompass several cycles, with the preservation of the same general plan, is certainly an established fact. But the alternate return of plans that intersect each other on such a scale and for such long durations of time presents quite a different problem whose solutions is difficult in terms of fixist hypotheses unless randomness is assumed.

According to the mobilistic concept, the formation of the Proto-Atlantic becomes simply an old outline of the Atlantic that was generated by the traction and the thinning of a very old continent, both associated with drifts. The geosynclinal condition has been reached, as is well known; whether this has also been the case for the oceanic condition is not known, but fundamentally it does not matter; (304) nor is it important to know whether the geosyncline was simple, branching, or multiple. The Caledonian folding, a product of reverse drifts, of the nearing of systems that had been separated, reestablished the welding. It will probably never be known if it did reestablish it over the entire length: the present Atlantic conceals too many things. The folding of this great Caledonian branch is only the scar of an ancient wound, subsequently reopened with great consequences as shown by the gaping space of the present-day Atlantic. One of the advantages of mobilism in this matter is to reduce to the same concept the explanation of the former and the present Atlantic by using the intersection of the plans instead of being hindered by them.

It has been pointed out—not without reason—that the Caledonian geosyncline of Scotland and of Scandinavia behaved as a double chain with opposite overfoldings. On both sides, one sees along the margin clean-cut thrusts that in fact result from Caledonian basement folds. These folds have reworked the Precambrian of the very margin of the Hebridian and Baltic massifs, including the Cambrian or Cambro-Silurian coverings, and have satisfied their internal tensions by means of low-dipping great fractures and by going beyond any ordering. But the much more flexible style that prevails throughout most of the internal part of the Highlands, as well as in that of the Scandinavian chain, is properly the geosynclinal style of plastic behavior.

The *Mediterranean* and its chains present to the mobilistic theory a difficult field of application and a test that this theory must undergo if it pretends to have more than a passing acceptance. The smallness of the scale allows neither the large statistical approaches by which this theory is successful nor the easy unfolding of the greater deformations that it uses. The type of small-scale complication displayed by these structures, all this *three-dimensional puzzle* of which many pieces are deformed to the extent of being almost unrecognizable, and furthermore the very demanding stratigraphy, ready to lock up in its refined chronology each phase of some importance of the movements, require from the theory a calculated and cautious progress. Besides, it could not, without the help of a great number of operational artifices borrowed from the observed tectonics, penetrate very far into this problem, which is too close to the lower limit of the scales within its interest.

But, if one sees only the difficulties, no positive work will ever be accomplished. Therefore, let us admit, until further notice, that the history of the Tethys and of its Mediterranean segment has exhausted all the complications of the multiple behaviors that I have related to the geosynclinal condition, including the transitions to the oceanic and marginal conditions. Besides, I shall limit myself to a smaller number of sketches (Figures 13 to 27).

[*Editor's Note:* Material, including figures 13 to 27, has been omitted at this point.]

If the original mobilism can easily be considered, following a simple reasoning, as a particular case of the concept of framed folding, this should also be the case, and the more rightly so, of that type of mobilism that I have presented loaded with all the aspects of concrete tectonics. While undertaking this delicate operation, I have rejected, after examination, all the suggestions of the vague eclecticism that would have attempted to reconcile irreconcilable terms in the twilight of powerless combinations or in the seesaw games of a delightful skepticism. These little artifices have never led to anything serious, regardless of the situation considered. I have simply attempted to draw in full light the map of the compatible, and it turns out that such a field was still very extensive.

[*Editor's Note:* In the original, material follows this excerpt.]

14

A THEORY OF MOUNTAIN-BUILDING

D. Griggs

[*Editor's Note:* In the original, material precedes and follows this excerpt.]

THE DYNAMIC CONVECTION MODEL.

The principles of the dynamic model are best illustrated in Plate II, Fig. 1, which is a photograph of the small model. Here the small glass cell holds a clear liquid (glycerine) on which floats a black plastic crust (cylinder oil and fine sawdust). The drums are rotated by means of small pulleys belted to the larger pulleys which are turned by hand. When the rotation of the drums is slow, the plastic crust is thickened and pulled down into a slight trough (in this model the oil sticks to the glass walls so that it is impossible to see the surface configuration in the photograph). As the currents increase in velocity, the crust is thickened more and more until finally a narrow downfold is formed as shown in the photograph. At this time of rapid currents, the surface of the crust is irregularly folded, and its average level is just a little lower than before the currents began. If the currents are slowed down and stopped, the downward component of force disappears and the thickened mass rises to buoyant equilibrium, so that its surface is considerably above its original level. Thus the reaction of the crust to the three phases of the convection current cycle is demonstrated.

Because the inertial forces were not negligible in this small model, and because it was desired to study in more detail the structures produced, a larger model was built with the physical properties given in Table I. A photograph of this model in

105

operation is shown in Plate II, Fig. 1. Here the substratum was very viscous waterglass and the continental crust was a mixture of heavy oil and sand, which possessed a yield point that could be varied by changing the proportion of oil used, and so made to have the proper strength for the model scale. The rotating drums show in the foreground. At the time the photograph was taken, the drums were in motion, and had developed a downfold and the two thrust faults in the crust shown by the dark lines. The actual development of these structures cannot be satisfactorily depicted short of moving picture portrayal.

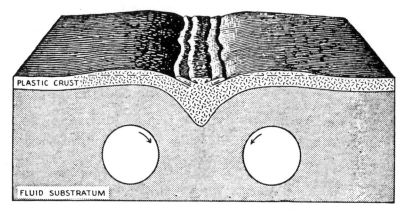

Fig. 14. Stereogram of Large Model with Both Drums Rotating, Showing Tectogene and Surface Thrust Masses with Relations Similar to Kober's Orogen.

The writer has a series of movies of the apparatus and is endeavoring to make arrangements so that anyone interested can obtain them at a nominal cost.

The type of thrusts formed at the junction of two downcurrents is shown in Fig. 14. This sketch illustrates the relations during the height of current velocity. The similarity in type of structure to that of Kober's Orogen is striking. An interesting feature of these thrusts is that they are formed by passive resistance of the overlying mass and active underthrusting of the foundation.

When only one drum is rotated (corresponding to the development of a single convection cell), the effect on the crust is different—the crustal downfold formed is not so narrow, and is asymmetrical in that it is steeper on the side facing the current. The structures are correspondingly asymmetrical. As the current velocity is increased, the crust is markedly thinned above it, and transported into the thickened part of the crust. Finally, the current sweeps all the crustal cover off and piles it up in a peripheral downfold. This stage in the development is illustrated in Fig. 15, drawn from the model.

This indication that a singly active cell may sweep off the superjacent continental crust opens wide avenues for speculation as to the formation of the circum-Pacific mountains and indeed as to the primary segregation of the continental masses

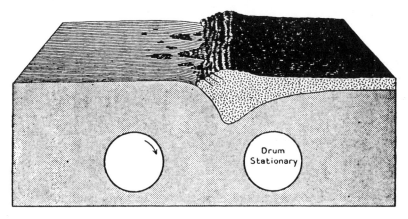

Fig. 15. Stereogram of Large Model with only One Drum Rotating, Showing Development of Peripheral Tectogene.

themselves. Here is a possible deformative force which could effectively counteract the tendency of erosion to distribute the continental material uniformly over the surface of the globe.

Correlation of the Convection and Orogenic Cycles.

We began with a generalized review of the mountain-building cycle; progressed through the hypothesis of thermal convection-current cycles in the earth; saw from model experimentation how subcrustal currents may deform the continental crust; and now we proceed to a synthesis of the facts and inferences gleaned from these varying modes of approach.

Fig. 16 shows the suggested correlation between the convection-current cycle and the mountain-building cycle. During the first phase of the convection-current cycle, when the cur-

THE MOUNTAIN BUILDING CYCLE

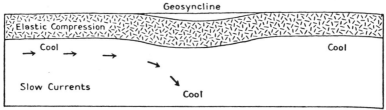

1. First stage in convection cycle — Period of slowly accelerating currents.

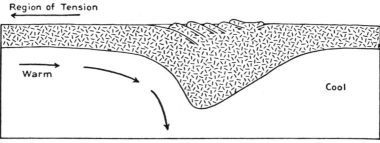

2. Period of fastest currents — Folding of geosynclinal region and formation of the mountain root.

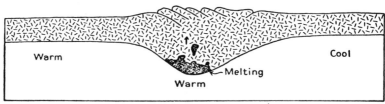

3. End of convection current cycle — Period of emergence. Buoyant rise of thickened crust aided by melting of mountain root.

Fig. 16. Hypothetical Correlation between Phases of the Convection-Current Cycle and Phases of the Mountain-Building Cycle. Structural Relations Drawn from the Model.

rents are slowly accelerating, they will exert an ever-increasing tendency to compress the crust. The compressive force in the crust reaches a maximum where the convection current dives down at the boundary of a cell. At this point there is a vertical component of stress as well as the tangential drag, and the combination of maximum compressive stress and the vertical stress localizes the crustal deformation which in the first stage of the mountain-building cycle takes the form of a geosyncline (Fig. 16-1). If the currents were constant in velocity, the geosyncline would reach an equilibrium position in which no further sinking would occur. With the slowly accelerating currents, however, the compressive stress is constantly increasing and causes continued depression of the geosynclinal trough, in agreement with the geological evidence for continued subsidence of geosynclines. According to the estimate given above, this period of acceleration would be of the order of 25 million years.

When the currents attain high enough velocity, the compressive stress on the crust is sufficient to deform it violently. This increase in compressive stress is accompanied by an increase in the vertical component at the cell boundary, and between the two the crust is downfolded as shown in Fig. 16-2. The thrusts shown in the diagram were added from a study of the model.

In these diagrams and in the discussion of the model, no attempt has been made to reproduce the sedimentary filling of the geosynclinal trough. The thrusts shown in Fig. 16-2 are foundation thrusts, corresponding to similar features in the Alps described by Swiss geologists, and it is to be expected that, as in the Alps, each foundation thrust would be connected with a nappe or thrust in the sedimentary rocks above. The folding and faulting in the sedimentary filling, however, must necessarily be very much more complicated than in the relatively homogeneous and massive granitic crust. This increase in complexity can certainly be better demonstrated in the region of the "roots of the nappes" in the Alps than in any model.

During this period of orogenic folding by the most rapid currents, which probably lasts from five to ten million years, the thickened mass of the crust is kept submerged by the downtow of the sinking current. The vertical component of stress developed by the current acts to prevent the swollen crust from rising to isostatic equilibrium. This persistent deviation from isostatic adjustment which has been shown in the model is corroborated by the common geological observation that mediterranean sedimentation occurred simultaneously with the peak of diastrophism in the mountain systems of the world.

As the currents decelerate in the third phase of the convection cycle, the compressive force in the crust decreases, the downtow decreases, and the thickened mass rises buoyantly, gradually attaining isostatic equilibrium as the currents slow and stop (Fig. 16-3). This produces the elevation which characterizes the third phase of the mountain-building cycle. Observation of the model shows that during this rise, the thickened part of the crust expands laterally and exhibits a tendency to flow away from the center of the folded portion of the crust under the influence of gravity. The absence of compression which characterizes this phase of the cycle, and the tendency toward lateral expansion are in agreement with geological observations that the period of elevation is not a period of dominant compression, but is one of dominant uplift and normal faulting. This change in character of deformation is difficult to explain by the hypothesis of thermal contraction, but can be readily demonstrated in the cyclic-convection theory.

The persistence of major isostatic disequilibria in the East Indies has puzzled Vening Meinesz, Umbgrove, and Kuenen (Kuenen, 1936, pp. 196-7). If the gravity deficiencies here observed are due to crustal downfolds, it is reasonable to suppose that they were formed at the time of the greatest folding of the sediments on the adjacent islands. This conclusion seems unequivocally established by the exact parallelism of regions of most intense folding and the negative anomaly bands. The age of this folding is Miocene, so that one must explain the persistence of the anomalies for something like 20 million years. This presents great difficulty to the thermal contraction theory of orogeny, but is to be expected from the cyclic-convection theory, since we have seen that the period of deceleration is estimated to last 25 million years.

Reference

Kuenen, P. H.: "The Negative Isostatic Anomalies in the East Indies (with Experiments)," Leidsche Geo. Mededeelingen, Vol. 8, 169-214, 1936.

15

Reprinted from pages 505–509 of *Principles of Physical Geology*, Ronald Press, New York, 1945, 532 p.

PRINCIPLES OF PHYSICAL GEOLOGY

A. Holmes

[*Editor's Note:* In the original, material precedes this excerpt.]

THE SEARCH FOR A MECHANISM

It has been shown that in looking for a possible means of " engineering " continental drift we must confine ourselves to processes operating within the earth. To be appropriate, the process must be capable (*a*) of disrupting the ancestral Gondwanaland into gigantic fragments, and of carrying the latter radially outwards as indicated in Fig. 210 : Africa and India towards the Tethys ; Australasia, Antarctica, and South America out into the Pacific ; (*b*) of disrupting Laurasia, though much less drastically, and again with radially outward movements towards the Tethys and the Pacific, as indicated in Fig. 209. We have already seen that the peripheral orogenic belts probably mark the regions where opposing systems of sub-crustal currents came together and turned downwards. The movements required to account for the mountain structures are in the same directions as those required for continental drift, and it thus appears that the sub-crustal convection currents discussed on pages 408 to 413 may provide the sort of mechanism for which we are looking (Fig. 262).

To explain the peripheral orogenic belts three systems of convection currents are called for (or three co-ordinated groups of systems), with their ascending centres situated beneath Gondwanaland, Laurasia, and the Pacific respectively. Incidentally, it should be noticed that the coalescence of the usual chaotic or small convective systems into three gigantic ones involves a coincidence that can rarely have happened in the earth's history, and one that is just as likely to have come about during the Mesozoic era as at any other time. The often-asked question : How is it that Pangæa did not begin to break up and unfold until Mesozoic time ? thus ceases to have any significance. If continental drift could have been caused by the gravitational forces invoked by Wegener, then it should have occurred once and for all very early in the earth's history, since those forces have always been in operation. If convection currents are necessary, continental drift may have accompanied all the greater paroxysms of mountain building in former ages but, if so, it would usually have been on no more than a limited scale. That there was a quite exceptional in-tegration of effort in Mesozoic and Tertiary times is forcibly suggested by eruptions of plateau basalts and building of mountains on a scale for which it would be hard to find a parallel in any earlier age.

There are, therefore, good reasons for supposing that at this critical period of the earth's history the convective circulations became unusually powerful and well organised.   Currents

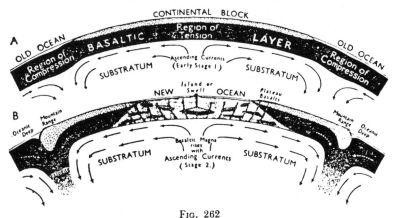

FIG. 262

Diagrams to illustrate a purely hypothetical mechanism for " engineering " continental drift. In A sub-crustal currents are in the early part of the con-vection cycle (Stage 1 of Fig. 215). In B the currents have become sufficiently vigorous (Stage 2 of Fig. 215) to drag the two halves of the original continent apart, with consequent mountain building in front where the currents are descending, and ocean floor development on the site of the gap, where the currents are ascending

flowing horizontally beneath the crust would inevitably carry the continents along with them, provided that the enormous frontal resistance could be overcome. The obstruction that stands in the way of continental advance is the basaltic layer, and obviously for advance to be possible the basaltic rocks must be continuously moved out of the way. In other words, they must founder into the depths, since there can be nowhere else for them to go (Fig. 262).

Now this is precisely what would be most likely to happen when two opposing currents come together and turn downwards beneath a cover of basaltic composition. The latter then suffers intense compression, and like the sial in similar circumstances it is eventually drawn in to form roots (*cf.* Figs. 215 and 216). On the ocean floor the expression of such a down-turning of the basaltic layer would be an oceanic deep. The great deeps bordering the island festoons of Asia and the Australasian arc (Tonga and Kermadec) probably represent the case where the sialic edge of a continent has turned down to form the inner flanks of a root, while the oceanic floor contributes the outer flanks.

It is not difficult to see that a purely basaltic root must have a very different history from one composed of sial. The density of sial is not significantly increased by compression. Consequently, when a sialic root is no longer being forcibly held down, it begins to rise in response to isostasy, heaving up a mountain range as it does so. But when rocks like basalt or gabbro (density 2·9 or 3·0) are subjected to intense dynamic metamorphism they are transformed into schists and granulites and finally into a highly compressed type of rock called *eclogite*, the density of which is about 3·4. Since this change is known to have happened to certain masses of basaltic rocks that have been involved in the stresses of mountain building, it may safely be inferred that basaltic roots would undergo a similar metamorphism into eclogite. Such roots could not, of course, exert any buoyancy, and for this reason it is impossible that tectonic mountains could ever arise from the ocean floor. On the contrary, a heavy root formed of eclogite would continue to develop downwards until it merged into and became part of the descending current, so gradually sinking out of the way, and providing room for the crust on either side to be drawn inwards by the horizontal currents beneath them (Fig. 262).

The eclogite that founders into the depths will gradually be heated up as it shares in the convective circulation. By the time it reaches the bottom of the substratum it will have begun to fuse, so forming pockets of magma which, being of low density, must sooner or later rise to the top. Thus an adequate source is provided for the unprecedented flows of plateau basalt that broke through the continents during Jurassic and Tertiary times. Most of the basaltic magma, however, would naturally rise with the ascending currents of the main convectional systems until it reached the torn and outstretched crust of the disruptive basins left behind the advancing continents or in the heart of the Pacific. There it would escape through innumerable fissures, spreading out as sheet-like intrusions within the crust, and as submarine lava flows over its surface. Thus, in a general way, it is possible to understand how the gaps rent in the crust come to be healed again ; and healed, moreover, with exactly the right sort of material to restore the basaltic layer. To sum up : during large-scale convective circulation the basaltic layer becomes a kind of endless travelling belt on the top of which a continent can be carried along, until it comes to rest (relative to the belt) when its advancing front reaches the place where the belt turns downwards and disappears into the earth.

To go beyond the above indication that a mechanism for continental drift is by no means inconceivable would at present be unwise. Many serious difficulties still remain unsolved. In particular, it must not be overlooked that a successful process must also provide for a general drift of the crust over the interior : a drift with a northerly component on the African side sufficient to carry Africa over the Equator, and Britain from the late Carboniferous tropics to its present position. The northward push of Africa and India, of which the Alpine system and the high plateau of Tibet are spectacular witnesses, could not have been sufficient by itself to shove Europe and Asia so far to the north. To achieve this the aid of exceptionally powerful sub-Laurasian currents directed towards the Pacific is required. The total northward components might then overbalance the southward components, and a general drift of the crust would be superimposed on the normal radial directions of drift.

It must be clearly realised, however, that purely speculative ideas of this kind, specially invented to match the requirements, can have no scientific value until they acquire support from independent evidence. The detailed complexity of convection systems, and the endless variety of their interactions and kaleidoscopic transformations, are so incalculable that many generations of work, geological, experimental, and mathematical, may well be necessary before the hypothesis can be adequately tested.   Meanwhile it would be futile to indulge in the early expectation of an all-embracing theory which would satisfactorily correlate all the varied phenomena for which the earth's internal behaviour is responsible.   The words of John Woodward, written in 1695 about ore deposits, are equally applicable to-day in relation to continental drift and convection currents : " Here," he declared, " is such a vast variety of phenomena and these many of them so delusive, that 'tis very hard to escape imposition and mistake."

SUGGESTIONS FOR FURTHER READING

A. Wegener
The Origin of Continents and Oceans.   Methuen, London, 1924.

A. L. du Toit
Our Wandering Continents.   Oliver and Boyd, Edinburgh, 1937.

T. H. Holland
The Evolution of Continents : A Possible Reconciliation of Conflicting Evidence.   Proceedings of the Royal Society of Edinburgh, Vol. LXI., Part II., No. 13, 1941.

Part VI

GEOSYNCLINAL THEORY IN THE MID-TWENTIETH CENTURY: SOME UNANSWERED QUESTIONS

Editor's Comments
on Papers 16 Through 19

Space limitations prohibit reprinting two papers published in the late 1940s which critically evaluated the state of geosynclinal theory and its impact on explaining the earth's first-order features. A paper by Glaessner and Teichert (1947), "Geosynclines: A Fundamental Concept in Geology," reexamined the concept of the geosyncline as originally conceived in order to judge whether the concept, as it existed in modified fashion in the middle of the twentieth century, still provided a viable explanation for global mobile belts. The paper provided special insight because the authors had read and summarized the original studies of Hall, Dana, Haug, Bertrand, Argand, Stille, and others, and were thus able to historically document the contrasting perspectives of European and American geologists. Glaessner and Teichert concluded that the geosynclinal concept occupied a definite and permanent position "in the theoretical and terminological apparatus of modern geology . . . the concept as employed by modern writers does not differ essentially from that established by its originators. . . . there is a certain vagueness to this concept which makes it impossible at present to define a geosyncline in absolute terms and quantitative relations so as to decide once and for all what is a geosyncline and what is not" (Glaessner and Teichert, 1947, p. 585). They insisted that additional work would lead to such definition.

A second paper, by Knopf (1948), "The Geosynclinal Theory," briefly reviewed the original Hall-Dana concept. However, much of the paper emphasized later modifications of the original concept:

> these additions involve (1) volcanism and intrusion during the growth of the parent geosyncline; (2) isostatic control during the folding consequent upon appression (compression) of the geosynclinal sediments; (3) metamorphism resulting from geosynclinal conditions and the events attending the folding: batholithic intrusion, syntectonic and epitectonic, and the relation between batholithic intrusions and the successive epochs of foldings that comprise a large-scale orogenic revolution; and (4) metalliferous deposition as aftermaths of the successive cycles of igneous activity during the orogenic revolution. This greatly developed and expanded form of the Hall-Dana theory is referred to herein as "the geosynclinal theory". It constitutes a great—probably one of the greatest—unifying principles in geologic science (Knopf, 1948, p. 651).

Despite such optimistic promotions, other studies make it clear that unhappily, nearly a century after Hall first published his geosynclinal hypothesis, no clear consensus existed as to what the concept entailed in a definitive, absolute sense. Similarly, no single classification scheme had been generally accepted by the global geological community as a whole. Despite the longevity of this "single, great, unifying principle," the same questions that had intrigued Hall and Dana continued to nag the geological community in the mid-twentieth century. Were geosynclines genetically related to the development of the two major first-order features of the earth's crust, the sialic continental blocks and simatic ocean basins? Were the continental mobile belts unrelated to the mid-ocean mountain systems? What forces produced the downwarping of linear segments of crust extending across broad regions of the globe? How were lateral compressional stresses produced? What were the major sources for geosynclinal sediment? Where were modern analogues of each of the various types of ancient geosynclinal belts? Was the location and development of geosynclines systematic or random? Papers 16 to 19 were selected to demonstrate the air of ambiguity permeating geological thinking in the years immediately preceding the formulation of global plate tectonic theory.

Eardley (Paper 16) reconstructed a detailed model of the Cordilleran geosyncline in Paleozoic time that remarkably (and purposefully) resembled the abyssal plain-trench-island arc inland sea complexes of the western Pacific. This model differed appreciably from Kay's cross-section across the New England Appalachians and bore little resemblance to the Appalachian geosyncline as originally described by Hall and Dana. Eardley was considerably influenced by Hess (1938), who had emphasized the petrological similarities between modern island arc systems and *some* ancient mobile belts. Eardley's model effectively eliminated the problem of what to do with offshore continental borderlands like Cascadia, replacing them instead with a complex series of volcanic and tectonic archipelagos. Eardley's model was truly actualistic, but why did it seem unacceptable as a model for the Alpine and Appalachian systems? Were there geosynclines and *geosynclines*? Was the concept too all-embracing to be useful?

Wells (Paper 17) clearly enunciated the dichotomy between Atlantic (ensialic) and Pacific (ensimatic) geosynclines but could offer no viable genetic reason for the distinctions. Dickinson (1962) would soon argue that truly oceanic (ensimatic) geosynclines did not really exist.

Gilluly (Paper 18) wrestled with the apparent permanence of large continental blocks in spite of the continued secular loss of sialic material to the oceans via erosion. Persistence of continents confirmed that sial must be constantly regenerated, Gilluly presumed, largely through volcanic activity. Gilluly, like many others, was puzzled by the apparent absence of geosynclinal belts developed within oceanic crustal areas (that is, essentially as marginal additions to continental areas). He weakly suggested that perhaps intense metasomatism erased all traces of the original oceanic condition. A reverse process and equally as necessary, the transformation of continental segments into oceanic segments, seemed to Gilully an even more difficult process to conceptualize. To accomplish it, he speculatively invoked some sort of subcrustal erosion and transport by convection currents to thin the sial. Stille had already proposed the term *quasi-craton* for these subsided areas, a term adopted particularly for such areas in the western Pacific (Fairbridge, 1961). Beloussov and Ruditch (1961) conceived the process of *basification* or *simatization* of thinned (and stretched) continental crust.

Gilluly also prophetically intimated that the same subcrustal flow beneath oceanic crust toward the continents might carry shelf sediments into the continental mass causing marginal coastal orogenic zones, underthrusting, trenches, and (andesitic) igneous activity (not unlike the Cordilleran-type orogeny later proposed by Dewey and Bird, 1970). Still, Gilluly could not bring himself to naively make a direct connection between "degradational transfer, isostatic return at depth, and mountain making," despite his suspicion that some sort of intimate connection indeed existed.

Dott's 1964 treatise (Paper 19) on mobile belts, sedimentation, and orogenesis typifies the disillusioned and skeptical attitude of many geologists regarding some of the fundamental classical principles of geology—including the geosynclinal concept—after the initial acceptance of at least some of the basic premises of continental drift and sea-floor spreading as put forth by Gilluly and Hess, and later by Heezen (1960), and Dietz (1963) among others.

Dott uneasily pointed out that processes such as lateral continental accretion were more enthusiatically supported by geologists than by firm scientific data. Similarly, Dott conceded that despite interpretative efforts to the contrary, geosynclinal belts in particular and mountain building processes in general did not depend on crustal types. Orogenic belts appeared to him to develop almost haphazardly in time and space, with apparent disregard for the distribution of continental blocks and ocean basins on the one hand, or the location, age, and structural grain of other mobile belts on the other. Dott emphasized that the so-called orderly appearance in time of successively distinct igneous and sedimentary rock types was also a gross oversimplification supported more by convention than facts. He could see no definitive, typically geosynclinal lithologies.

To Dott, it was obvious that geosynclines (defined as orogenic belts in which thick sedimentary successions accumulate) were the *result* of crustal mobility, rather than vice versa. Whether already mobile areas of the crust actually evolved as geosynclines into mountain ranges was largely determined by the presence of proximal sediment source areas. Dott recounted the revision in thinking that had allowed the older borderlands concept to be replaced with the more modern concept of island arcs and tectonic lands.

However, Dott doubted that Dietz's hypothesis (Papers 22, 24, and 25) proposing that orthogeosynclines (equated with continental shelf and continental slope-rise sediment prisms) became mountain ranges via sea-floor spreading was generally applicable because few ancient mobile belts resembled Dietz's Atlantic-type continental margin in geological setting, lithology, provenance, or tectonic history, at least from the fixed perspective of the present. Dott, like many others, was puzzled by the abundance of "empty" modern island arc-trench systems that encircle the globe. These arc-trench systems all seemed to be related to the continents, even though some were separated from continental blocks by back arc basins (Heiskanen and Vening Meinesz, 1958; Vening Meinesz, 1955). Whereas simple volcanic arcs such as Hawaii, Tahiti, and the Azores occurred in the middle of ocean basins, no arc-trench systems were mid-ocean features. Finally, many geologists noted the marked contrast between the continent-linked arc-trench systems and the global rift system that obviously traversed both ocean basins and continental blocks, a phenomenon emphasized and explained genetically by Carey (1958) and Heezen (1960).

REFERENCES

Beloussov, V. V., and E. M. Ruditch, 1961, Island Arcs in the Development of the Earth's Structure (Especially in the Region of Japan and the Sea of Okhotsk), *Jour. Geology* **69**:647–658.

Carey, S. W., 1958, A Tectonic Approach to Continental Drift, in S. W. Carey, ed., *Continental Drift: A Symposium*, Hobart University, Tasmania, Australia, pp. 177–355.

Dickinson, W. R., 1962, Petrogenetic Significance of Geosynclinal Andesitic Volcanism Along the Pacific Margin of North America, *Geol. Soc. America Bull.* **73**:1241–1256.

Dewey, J. F., and J. M. Bird, 1970, Mountain Belts and the New Global Tectonics, *Jour. Geophys. Research* **57**:2625–2647.

Dietz, R. S., 1963, Collapsing Continental Rises: An Actualistic Concept of Geosynclines and Mountain Building, *Jour. Geology* **71**:314–333.

Fairbridge, R. W., 1961, The Melasian Border Plateau, a Zone of Crustal Shearing in the S. W. Pacific, *Bur. Cen. Siesmo. Int.* ser. A, **22**:137–149.

Glaessner, M. F., and C. Teichert, 1947, Geosynclines: A Fundamental Concept in Geology, *Am. Jour. Sci.* **245**:465–482, 571–591.

Heezen, B. C., 1960, The Rift in the Ocean Floor, *Sci. American* **203**:98–110.

Heiskanen, W. A., and F. A. Vening Meinesz, 1958, *The Earth and Its Gravity Field*, McGraw-Hill, New York, 470 p.

Hess, H. H., 1938, Gravity Anomalies and Island Arc Structure with Particular Reference to the West Indies, *Am. Philos. Soc. Proc.* **79**:71–96.

Knopf, A., 1948, The Geosynclinal Theory, *Geol. Soc. America Bull.* **57**:649–670.

Vening Meinesz, F. A., 1955, Plastic Buckling of the Earth's Crust: The Origin of Geosynclines, *Geol. Soc. America Spec. Paper* **62**:319–330.

16

Reprinted with permission from pages 334–342 of *Jour. Geology* 55:309–342 (1947)

PALEOZOIC CORDILLERAN GEOSYNCLINE AND RELATED OROGENY

A. J. Eardley

University of Michigan

[*Editor's Note:* In the original, material precedes this excerpt. Figure 1 has not been reproduced here.]

VOLCANIC ARCHIPELAGO

CONCEPT OF A WESTERN LANDMASS

In the opinion of several geologists, the source of the sediments, and especially of the volcanics, of the Cordilleran geosyncline in Paleozoic time was to the west. Schuchert named the landmass from which they came "Cascadis" (*Historical Geology* [1924]) and characterized it as the greatest of all "borderlands." In the Mesozoic era he believed Cascadis was divided into two smaller landmasses, each with shorelines farther west than the Paleozoic one. The Schuchert concept of Cascadis seems to be the most widely accepted theory today among paleontologists and stratigraphers. Deiss[76] recently depicted the shoreline of Cascadis during the Cambrian in about the position that Schuchert illustrates it, namely, at the western margin of the southern Rocky Mountain trough, the Montana dome, and the northern Rocky Mountain trough (see map, Fig. 1).

[76] See ftn. 4.

[*Editor's Note:* The following footnotes appeared in the preceding pages of this article and are referred to in the pages reproduced here. They are reprinted on p. 132 for the reader's convenience.]

Schofield[77] has reviewed the concept of "Cascadia" and agrees that a western landmass existed; but, on the basis of Proterozoic, Silurian, and Triassic conglomerates of British Columbia and southeastern Alaska, he regards it as having lain west of the present west coast of North America, except during the Proterozoic (Beltian), when, he thinks, the shoreline was just west of Kootenay. Cairnes[78] has shown that the Shuswap terrane is not all pre-Cambrian rock and therefore not a former land area, but that, on the contrary, it is the metamorphosed facies of rocks of several periods and that it was the site of deposition during at least part of the Paleozoic. The Shuswap terrane has been depicted as part of Cascadis by Schuchert and others, but this seems now disproved.

The previous pages on the great Cordilleran geosyncline—its divisions, sediments, and deformation—have suggested several stimulating lines of evidence by which the position and constitution of the lands from which the sediments came can be deciphered. The nature of the sediments alone is very informative; and this, together with their distribution, provides enough data to reconstruct a general picture of the western lands.

EVIDENCE OF A WEST-LYING ARCHIPELAGO

Volcanic rocks.—The abundance of volcanic rocks, mostly tuffs and lavas, in the Pacific trough, with none in the eastern, has been noted. The volcanoes were not situated along a landmass between the two troughs because such an elevation probably did not exist except in the south in Carboniferous time, when the Manhattan geanticline rose. If a medial land area had existed and had been the site of volcanoes, the eastern trough would have received as much volcanic

material as the western. It is therefore concluded that the volcanoes were aligned along the western flank of the Pacific trough. They are known to have been especially numerous opposite the Klamath and Alexander troughs and were active at intervals in each period from pre-Ordovician to the Present.

In the modern geologic picture, rows of active volcanoes are on the concave side of island arcs or archipelagos, but on the larger islands, such as Japan, the volcanic rocks are widely distributed on both sides. It therefore seems of some significance that the Klamath and Alexander troughs are on the concave sides of what would be the west-lying island arcs.

The great bulk of the lavas in both the Klamath and the Alexander troughs are andesites. Some basalts occur in the Klamaths, but only andesitic lavas have been found in the Alexander trough. This is especially instructive because the volcanoes of the island arcs, both modern and ancient, have emitted chiefly andesites. W. H. Hobbs[79] has recently reviewed this distribution and has drawn an andesite-basalt line which separates the basalts of the volcanoes of the ocean floor from the andesites of the island arcs in the western and southern Pacific. Rarely do appreciable quantities of andesite occur on the mainland opposite an archipelago. Inspection of the geologic map of eastern Asia (Fig. 7) reveals abundant andesite on the Japanese archipelago but an absence of it on the Asiatic mainland across the Sea of Japan and the Sea of Okhotsk.

It may be concluded, therefore, that the andesitic nature of the preponderance of volcanic material in the western trough of the Cordilleran geosyncline

[79] "Mountain Growth: A Study of the Southwestern Pacific Region," *Proc. Amer. Phil. Soc.*, Vol. LXXXVIII (1944), pp. 221–68.

[77] See ftn. 73.   [78] See ftn. 51.

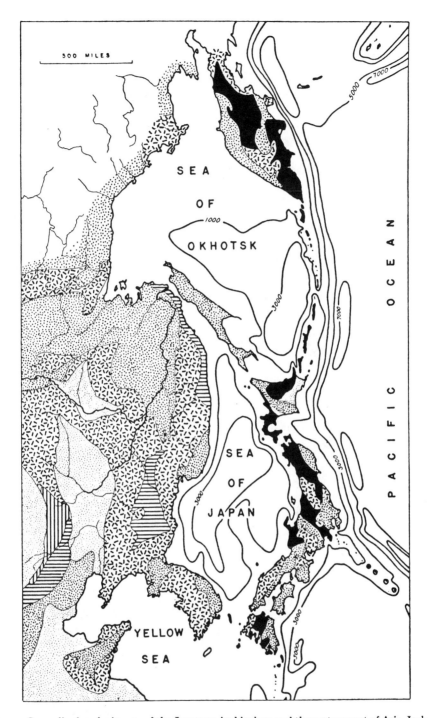

Fig. 7.—Generalized geologic map of the Japanese Archipelago and the eastern part of Asia. Isobaths are in meters. Coarsely stippled areas are those chiefly of sedimentary rocks, but with large areas of Archean gneisses and schists and some smaller areas of intrusive and extrusive rock. Finely stippled areas denote alluvium. Hachured areas as those of plutonic rocks, chiefly granite to granodiorite, but with considerable areas of Archean gneiss and schist and some sedimentary rocks. Solid black areas are andesite. Horizontally ruled is basalt and vertically ruled is trachyte.

125

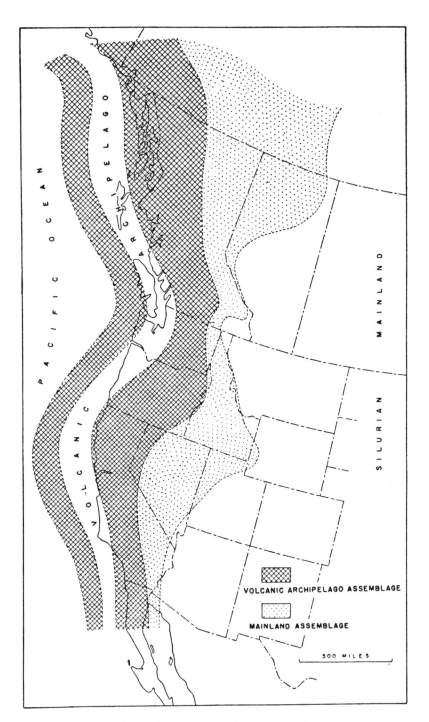

FIG. 8.—The volcanic archipelago and its relation to the sediments of the Paleozoic Cordilleran geosyncline. The map is generalized for the Silurian period on the assumption that the Copley volcanics of the Klamath Mountains are Silurian in age and correlate with the Silurian volcanics and graywackes of the Alexander Archipelago. The volcanic archipelago assemblage consists of andesitic volcanics in many forms, graywacke, conglomerate of andesite and granitoid rocks, massive limestones and coarse intra-formational limestone conglomerates, chert, arkose, dark shale, and siliceous shale (all the above varieties are generally found now in their metamorphosed form). The mainland assemblage consists of orthoquartzite, arkosic sandstone, sandstone, conglomerate (generally pea and pebble size, composed of quartzite and quartz pebbles), limestone and dolomite, cherty limestone and chert, and various shales and siltstones.

126

also indicates that the land to the west was a volcanic archipelago.

Graywackes, arkoses, and conglomerates.—A graywacke is a dirty sandstone in the sense that a clean sandstone is composed largely of quartz grains. Graywackes have an abundance of slate, phyllite, argillite, and chert fragments. The grains come from previously deposited clastics that have suffered low-grade metamorphism.[80] From a study of graywackes in the Appalachian geosyncline, from the Alpine molasse, and from the Siwalik strata of India, Krynine concludes that graywackes are due to erosion of previously deposited sediments in a geosyncline that had been folded and somewhat metamorphosed by widespread orogeny along the margins. He also thinks it is possible that geanticlines with geosynclines could bring the previously deposited sediments to the surface and result in the formation of graywackes.

Buddington[81] states that the Silurian graywackes, in general, of southeastern Alaska are composed of particles of rock similar to the kinds that form the pebbles and cobbles in the conglomerates with which they are interbedded and, in addition, of a considerable percentage of plagioclase, potassic feldspar, and quartz grains. The conglomerates are largely made of andesite pebbles and boulders; but slate, diorite, rhyolite, and limestone pebbles are abundant, if not dominant, in some conglomerates. One specimen of graywacke of Devonian or Silurian age, for example, consisted of particles of andesite, felsite, plagioclase, granophyre, quartz, spherulitic rhyolite, and orthoclase, with a chloritic and slightly calcareous groundmass.

[80] P. D. Krynine, "Paleogeographic and Tectonic Significance of Graywackes" (abstr.), *Bull. Geol. Soc. Amer.*, Vol. LII (1941), p. 1916.

[81] See ftn. 54.

The association of the graywackes and conglomerates that Buddington describes is very revealing of their origin and substantiates Krynine's conclusions. The conglomerates in themselves are indicative of a volcanic archipelago to the west and deserve further mention. The following is a résumé of the Silurian conglomerates according to Buddington. Varieties of conglomerate are as follows:

1. A conglomerate composed almost wholly of well-rounded andesite or andesite porphyry cobbles and boulders; the matrix may be calcareous, and lenses of limestone are intercalated, but limestone cobbles are sparse.

2. A conglomerate composed almost wholly of limestone cobbles or boulders in a limestone or andesitic tufflike matrix; this type is rare, but beds 100 feet thick have been noted.

3. Peculiar conglomerates intermediate between 1 and 2, consisting of pebbles and cobbles of andesite and limestone in a greenish tufflike matrix.

4. A homogeneous-appearing rock composed of fragments of andesite in a matrix of the same material; the structure is that of a conglomerate or waterworn breccia.

The limestone fragments are usually of a dense-textured limestone typical of the Silurian, and many carry fossils of Silurian age. The fossils are the same as from the overlying limestone. It is, therefore, believed that the limestone conglomerates are intra-formational and that the limestone fragments are of practically the same age as the volcanic fragments. Vertical movements of the sea bottom, perhaps local, must have accompanied the volcanism and must have resulted in contemporaneous erosion and submarine slumping of slightly compacted fine lime mud. A part of the volcanic material, at least, must have been erupted from central volcanoes, which were built up above the surface of the ocean and were thus subjected to erosion.

Although recognizing unsolved elements in the problem of origin of the graywackes, conglomerates, and limy argillaceous beds, Buddington visualizes a sedimentary environment as follows; the great lens-shaped beds of conglomerate may be local deposits made by torrential streams, and the graywacke may be in part the more finely comminuted peripheral marine equivalent. The calcareous shale and argillaceous limy beds, which are locally intercalated with the clean, thick-bedded limestone, may be in part the more distant offshore equivalent of the conglomerate and graywacke.

The limestone is in part dense white on fresh surfaces and massive, with only rare, if any, evidence of stratification. Beds as thick as 2,000 feet have been observed. In part it is interbedded with thin-layered limestone, nodular and shaly limestone, calcareous shaly argillite, dense platy siliceous layers, green-gray shale, and sparse buff-weathering sandstone. The massive limestone seems to be due to rapid deposition, and, where clean, the site of accumulation was sufficiently distant from land so as not to have received any clastic material. Volcanic activity has been thought of as contributing to the deposition of the limestone through the activity of magmatic waters or meteoric waters draining from a volcanic terrane, but the chemistry and oceanography of the problem have not been worked out.

Schofield[82] has already been cited as having discussed the problem of granitoid pebbles and cobbles in the conglomerates of several periods, especially the Triassic. Buddington refers to them also. In one locality the Britannia map area of

[82] See ftn. 73.

British Columbia, an arkose is described as composed of irregular grains of quartz, plagioclase, orthoclase, and sericite schist. The lack of rounding of the grains, the freshness of the plagioclase, and the considerable thickness of the unstratified beds prove that the material accumulated rapidly and was transported only a short distance from a source of granitoid plutonic rocks. Buddington failed to trace the granitoid clastics to their source, despite the fact that their size and abundance indicated to him a nearby local origin. It seems necessary, he believes, to assume that granitoid intrusions existed in a land that formerly stood to the west, where only the Pacific Ocean now lies.

Krynine[83] has studied the tectonic significance of arkoses and concludes that they are deposited just after a granitoid terrane has been uplifted and while it is being vigorously dissected. They are derived from the deformed geosyncline into which granitoid rocks have been intruded. The plutons become exposed by erosion of the mountains created by orogeny and then are uplifted in a further stage of deformation and vigorously eroded.

Granite plutons are seldom exposed in arcs of small volcanic islands. We must look to the larger islands of an archipelago for the source of granitoid conglomerates and arkoses. The geologic map of the Japanese Archipelago (Fig. 7) shows extensive areas of granitic intrusions and pre-Cambrian gneisses, which could furnish the necessary material. A major archipelago like the Japanese has had a long orogenic history and is composed of rocks that will make not only graywackes but also arkoses. Such a one seems to have been the source land of

[83] See ftn. 80.

the sediments of the western part of the Cordilleran geosyncline.

Orthoquartzite and chert.—The eastern trough lacks the volcanics, graywackes, and arkoses, but it has a dominant lithology that is exceedingly scarce in the western trough—sandstones cemented by silica. These are commonly called "quartzites," such as the Tintic, Brigham, Bingham, Weber, Wells, Quadrant, Flathead, Deadwood, etc.; but perhaps the names "orthoquartzite" and "quartzose sandstone" would safeguard against the misconception that they are metamorphic rocks. Some of the Proterozoic quartzites in the Rocky Mountain trough are regionally metamorphosed, but it is very doubtful if any of the Paleozoic and Mesozoic quartzites are affected in the least.

According to Krynine,[84] sedimentary orthoquartzites and quartzose sandstones originate either as deposits following prolonged chemical weathering of a terrane or though re-working of pre-existing quartzose sediments. The first group generally develops after chemical decay in peneplained regions. The shallow seas into which the sands were deposited were, at the time, precipitating silica, and, as a result, orthoquartzites were formed. If part of the sea bottom is removed from sand deposition, considerable beds of chert will form. If Krynine's postulate is true, it would appear that the sand grains of the orthoquartzites were derived from the mainland of the continent. It does not seem possible that they could have come from the west, where volcanic activity dominated. Associated with the Paleozoic quartzites are several limestone formations, 500–2,000 feet thick, and several somewhat thinner shale formations. The assemblage

of quartzite, shale, and limestone supports the conclusion of a mainland origin.

Great beds of chert are present in the sediments of the volcanic archipelago. Extensive beds of chert and cherty limestone are present in the Rocky Mountain trough, as well as in the inland basins and shelfs of the mainland, and therefore the factors governing the precipitation of the silica are probably several. Its transportation in solution in marine currents may result in precipitation at a great distance from its source. I find it easy to believe that a large part of the silica originated in the volcanic activity of the archipelago; that some of it was carried by currents across the seas between the archipelago and the mainland free from the area of deposition of volcanic material; and that it was precipitated copiously in the shallow seas of the eastern trough and mainland shelf, where, from place to place and from time to time, clay, lime mud, or sand was accumulating. The problem needs special attention beyond the limits possible here, but the distribution of the cherts and orthoquartzites fits in nicely, if only theoretically, with the postulate of a western volcanic archipelago.

Distribution of troughs and sediments.— The pre-Devonian troughs of the Cordilleran geosyncline are isopached in Figure 1. The spread of data and the amount of interpretation have been discussed on previous pages. In brief review it is clear that the contouring of the western trough between the Alexander Archipelago and the Klamath Mountains is principally a guess, but, at least, a logical one and not at variance with known information. The Klamath trough is contoured on the assumption of a Paleozoic age of the Copley, Chanchelulla, and Salmon formations. The western flank of the northern Rocky Mountain trough is

[84] *Ibid.*

masked by metamorphism, so that its exact configuration is not known. Also the continuation of the trough northward from Jasper is fairly speculative. Elsewhere in the great geosyncline, substantial data support the contouring.

The pattern of the troughs, together with a postulated western volcanic archipelago, reminds one of the seas and archipelagos of the western Pacific, and it is natural to compare the two. In doing so it must be borne in mind that in our own western geosyncline we have the sedimentary record through the geologic periods, whereas in the western Pacific Ocean we have the shorelines of the moment, the distribution of the rocks on the archipelagos and the mainland adjacent to the seas, but practically no information about the sediments that are forming in the seas. A few dredgings have been made in a few places and have been reported on, but these represent at best only the topmost layer. Under the great seas between the archipelagos and the mainland we have little idea where troughs of greatest accumulation are or what the nature is of the sediments in the troughs below the surface. Instead of making a comparison, we find ourselves supplying the lost parts of one with the known parts of the other and making a complete geologic province, for we have the sedimentary record in one and the source of the sediments and the site of their deposition in the other. Knowing what we do of the origin of sediments, it is not difficult to recognize the fit or misfit of the parts.

In the first place, the shape and scale of the geosynclinal elements of the two regions is comparable (cf. Figs. 7 and 8). The Japanese Archipelago including Kamchatka was selected for comparison because it is the closest modern parallel that could be found. The curvature of the arcs, their convexity toward the Pacific, their width and length, are conspicuously alike.

In the second place, the geology of the Japanese Archipelago is somewhat the same as that postulated for the source land of the sediments of the Pacific trough of the Paleozoic Cordilleran geosyncline. If the Tertiary and Cretaceous strata are imagined removed from the Japanese islands, then the proper rocks and terrane are present to supply such sediments as are found in the Pacific trough. The most abundant rocks mapped in the Japanese Archipelago are as follows: andesite, granite, syenite, schistose granite, gneiss, schist, slate, chert, sandstone, limestone, diorite, pyroxenite, amphibolite, gabbro, and trachyte, in approximately decending order of abundance. From the nature and distribution of these rocks it is realized not only that the Japanese Archipelago and related seas are a comparable example but that their history probably ran through the Paleozoic era in a way appreciably like that of the Cordilleran geosyncline. The highly generalized geologic map of the Japanese Archipelago and the adjacent mainland (Fig. 7) is presented to show the sources of the sediments and the depths of the seas into which the sediments drain.

Figure 8 depicts the two great assemblages of sediments in the Cordilleran geosyncline as they were deposited during Silurian time. The Silurian was selected for illustration because it seemed to give a pattern that more closely simulates that of the Japanese region. In other periods, either volcanism was not prominent opposite both the Klamath and the Alexander troughs at the same time, or the seas did not spread over the eastern trough. Deep-water sediments, such as must be accumulating in the Sea of Japan and the

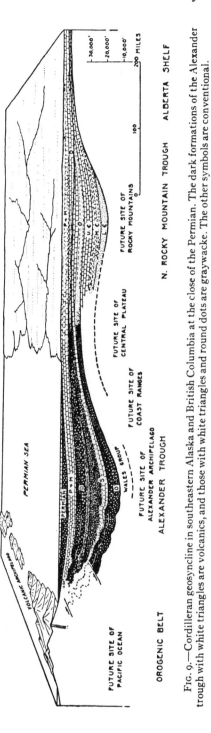

Fig. 9.—Cordilleran geosyncline in southeastern Alaska and British Columbia at the close of the Permian. The dark formations of the Alexander trough with white triangles are volcanics, and those with white triangles and round dots are graywacke. The other symbols are conventional.

Sea of Okhotsk, may be present in the Cordilleran geosyncline, but they have not been recognized.

A consideraton of the sedimentary conditions about volcanic archipelagos and the comparison of these with the mainland assemblage stimulate thoughts that ramify into many aspects of sedimentation but which are outside the objectives of this article. It must suffice to conclude that the distribution of sediments in the Cordilleran geosyncline indicates that a volcanic archipelago existed west of the western trough during the Paleozoic era.

ARCHIPELAGO AND OROGENIC BELT

The sedimentary evidence that establishes a great volcanic archipelago on the west side of the Cordilleran geosyncline at the same time indicates the existence there of an orogenic belt, because, from all we know, similar archipelagos in other parts of the world are the result of crustal movements in zones of active deformation.

Added to the sedimentary evidence of a Paleozoic orogenic belt is the dynamic evidence in the form of numerous unconformities between stratigraphic systems in the Pacific trough. Further dynamic evidence is the metamorphism of the trough's formations and, in addition to this, the convincing observation that the grains of the graywackes and the pebbles of the conglomerates were derived from already dynamically metamorphosed rocks of Paleozoic age. These, in short, are the lines of evidence that support the theory of a great orogenic belt to the west of the Cordilleran geosyncline during Paleozoic time. The theory is illustrated ideally in Figure 9.

FOOTNOTES

[4]C. F. Deiss, "Cambrian Geography and Sedimentation in the Central Cordilleran Region," *Bull. Geol. Soc. Amer.*, Vol. LII (1941), pp. 1085–1116.

[51]C. E. Cairnes, "The Shuswap Rocks of Southern British Columbia," *Proc. 6th Pacific Sci. Cong.*, Vol. I (1939), pp. 259–72.

[54]A. F. Buddington, and T. Chapin, "Geology and Mineral Deposits of Southeastern Alaska," *U. S. Geol. Surv. Bull. 800* (1929).

[73]S. J. Schofield, "Cascadia," *Amer. Jour. Sci.*, Vol. CCXXXIX (1941), pp. 701–14.

17

Reprinted from *Geol. Soc. America Bull.* **60**:1927 (1949)

ENSIMATIC AND ENSIALIC GEOSYNCLINES*
Francis G. Wells

Deposits in geosynclinal prisms are of two types: (1) sediments derived from the weathering of the crust and deposited on a downsinking floor of sial, the classical geosyncline; and (2) effusive rocks and accumulates on a sima floor.

Deposits of (1) consist of argillaceous rocks, quartz sandstones, widespread limestones and dolomites, arkoses, conglomerates, and terrestrial deposits. Open and alpine types of folding and overthrusting predominate, and the igneous rocks are of the Atlantic type.

Deposits of (2) consist of spilitized basaltic flows and pyroclastics overlain by andesitic and dacitic pyroclastics and flows. Shales and mudstones composed of volcanic dust and ash, with little quartz and primary kaolin, are intercalated with the more silicic effusives. These grade into sandy graywacke also composed of volcanic debris and containing little if any quartz. Limestones, except for small reefs, are absent. Chert lentils are associated with graywackes. No terrestrial deposits are present. From top to bottom the deposits are almost entirely volcanic and show no traces of chemical weathering. Isoclinal folding and high-angle reverse faulting rule. Peridotites are characteristic, and the other intrusive rocks are of the Pacific type. Palingenesis is common.

The terms ensialic and ensimatic geosynclines are proposed to designate these distinctly different types. The sialic crust has been built up of increments of ensimatic geosynclines. Complete sections through ensimatic geosynclines are exposed only in Precambrian shields, the Keewatin of Canada being a good example, but what are believed to be younger ensimatic geosynclines are found around the peripheries of the sima basin of the Pacific, and the rocks younger than the Chico formation of the northern Klamath Mountains probably are in a younger ensimatic geosyncline.

* Published with the permission of the Director, U. S. Geological Survey.

18

Reprinted from *Geol. Soc. America Spec. Paper* **62**:7–18 (1955)

Geologic Contrasts between Continents and Ocean Basins

James Gilluly

U. S. Geological Survey, Denver Federal Center, Denver, Colo.

ABSTRACT

Isostasy implies that the differences in surface elevation of continents and ocean basins must reflect differences in density that in turn imply gross lithologic contrasts between these crustal segments. Some petrologists infer that tholeiites and magmas more siliceous are wholly continental. Oceanic rocks, collected from islands, may not fairly represent the oceanic crust, but at any rate they do not differ sharply from continental rocks. All major magma types occur; the Hawaiian magmas are saturated, for instance. Quartzose rocks seem confined to islands that rise from high ridges, but such rocks are sufficiently abundant to suggest that most high ridges are partly sialic. Probably the deeper oceanic basins are essentially free from sial.

The secular loss of sialic material to the pelagic areas requires that the continents should be either lower or smaller than in the past, unless there has been concurrent addition to the sial from the mantle. A rough computation, based on Arrhenius' studies of deep-sea cores, suggests a loss of 1.5 km of material from the continents during 2 billion years. The fact that the continents seem as large or larger than in early geologic history suggests addition of sial during geologic time.

If the continents have grown areally by accretion of shelf geosynclines to a central nucleus, it seems that pelagic sediments and simatic geosynclinal floors should be common. Perhaps, though, their absence is not a conclusive argument against such growth: metasomatism may have changed the original composition of a mafic rock during orogeny, and furthermore many of the Pacific border mountains show no geosynclinal floors older than Paleozoic. A former simatic basement may have been folded downward during mountain making and be now overlain by sial crowded over it. Thus we may have former oceanic segments changed to continental. There are also many areas where stratigraphic and other evidence compels us to assume that former continental segments have been depressed at least as much as 3000 meters.

These changes in level raise isostatic problems as great as those of plateau uplift, but of the same kind. It is suggested that thinning or thickening of sial by subcrustal erosion or deposition is responsible for both.

The contrasts between continents and ocean basins invite attention to the visible processes now operating to modify them. These processes, though powerful, do not seem to account for the diversities.

The shore line—junction of the realms of denudation and deposition—is critical in dynamical geology. Sediment is now being carried across this boundary at a rate great enough, if continued, to erase all the topography above sea level in less than 10 million years, if compensating uplift did not occur. An analysis indicates that subcrustal flow induced by isostatic response to unloading may influence both coastal structures and differentiation of sial, but such flow does not in any way explain the contrast between Pacific and Atlantic structures nor can it be the governing factor in orogeny. These must result from other movements deep within the mantle, perhaps piloted by the shallow movements.

INTRODUCTION

Continental crustal segments differ from suboceanic in elevation and surface form; that they also differ in mean composition is suggested by observation and so strongly implied by indirect evidence as to be generally accepted.

The geophysical inferences from the hypsographic, gravitational, and seismological data as to differences in crustal structure between continents and ocean basins are being discussed by other participants in this symposium; I will therefore launch directly into the geological clues to the problem. These seem to be of three kinds: petrologic, stratigraphic, and structural, as inferred chiefly from morphology.

PETROLOGIC CONTRASTS

The inference of a broad petrographic contrast between continents and ocean basins is several generations old (Fisher, 1889, p. 232–248; Suess, 1909, p. 626). Indeed, since Marshall (1912, p. 28) drew his "andesitic line" such a supposed difference has become for many geologists the main criterion for distinguishing continental from oceanic crustal segments.

The consensus seems to be that, although mafic and even ultramafic intrusive masses are found in both continent and ocean basin, true granites are confined to the continents. Similarly, although basaltic lavas are ubiquitous, andesite and rhyolite are continental rocks. Kennedy (1933, p. 249; Kennedy and Anderson, 1938, p. 38) has contended that basaltic lavas include two distinct suites: olivine basaltic, of worldwide distribution, and tholeiitic, of continental habitat.

So sharp a distinction among the lavas is not supported by the facts—unless ocean and continental are to be redefined in accordance with it. Powers (1935; 1954) and Tilley (1950, p. 40–43) have shown that the Hawaiian magma is tholeiitic and very similar to the continental flood basalt of the Deccan, Columbia River, and Thulean fields. Thus, whether or not the tholeiitic and olivine-basaltic magmas are fundamentally different, as Kennedy, Turner and Verhoogen (1951, p. 181), and others insist, tholeiite is certainly not restricted to continents[1].

Nor can a strong case be made for the restriction of any other common igneous rock species. Few continental basalts are as rich in olivine as the widely distributed oceanites, or in pyroxene as the abundant oceanic ankaramites. Yet picrites are known on the continents, and ankaramite was named from the continental island of Madagascar. Even granite and quartz-bearing gneiss are found on the Seychelles (Lemoine, 1911, p. 1; Reed, 1949, p. 554–555)—truly oceanic islands by any criteria other than petrographic.

A partial list of oceanic occurrences of rocks more siliceous than basalt—and hence often considered continental varieties—is given in Table 1. From it have been excluded the large islands like Cuba, Madagascar, New Zealand, and New Caledonia, all of which are commonly considered continental. Also excluded are the oligoclase andesites of Hawaii (Macdonald, 1949, p. 1571) and other islands.

These records seem to me to show that a generalization that is broadly valid has come, by repetition, to be considered absolute. If the "granitic layer" of seismologists is made of real rocks and is not, like "the zone of flow", a condition, it must be represented by the shield areas. Yet Sederholm's (1925) average sample of the Fennoscandian Shield includes 8.2 per cent mafic rocks—sima in the usual termi-

---

[1] So many transitional and even overlapping characters are found between tholeiites and olivine basalts (Turner and Verhoogen, 1951, analyses p. 180; Williams, *in* Williams, Turner, and Gilbert, 1953, p. 37–48) that their separation seems to me of quite subordinate rather than fundamental petrogenetic value.

TABLE 1.—*Oceanic occurrences of silicic igneous rocks*

Topographic setting	Island	Approx. Lat.	Approx. Long.	Rock varieties	Source
		PACIFIC OCEAN			
Albatross Plateau	Marquesas	9°S	140°W	andesite	Chubb, 1930
Albatross Plateau	Galapagos	0°	91°W	andesite	Chubb, 1934
Albatross Plateau	Easter	26°S	109°W	rhyolite	Bandy, 1937
Fiji-New Zealand ridge	Fiji	17°S	179°E	granite, schist	Pegau, 1934
Fiji-New Zealand ridge	Tonga	19°S	173°E	rhyolite and granitic inclusions	Reed, 1949
Fiji-New Zealand ridge	Kermadec	30°S	178°E	rhyolite and granitic inclusions	Marshall, 1912
Fiji-New Zealand ridge	Chatham Is.	44°S	176°W	andesite	Marshall, 1911
Fiji-New Zealand ridge	Bounty Is.	48°S	179°E	andesite	Marshall, 1911
Fiji-New Zealand ridge	Auckland Is.	51°S	167°E	andesite	Marshall, 1911
Caroline ridge	Truk, Yap, Map	7°N	152°E	andesite, schist	Marshall, 1912
		INDIAN OCEAN			
Ridge NE. of Madagascar	Seychelles	5°S	55°E	granite, gneiss	Lemoine, 1911, p. 1; Reed, 1949, p. 554, 555
Mid-Indian Ridge	Kerguelen	49°S	70°E	andesite	Nordenskiold, 1913, p. 24
Mid-Indian Ridge	St. Paul	39°S	78°E	andesite	Nordenskiold, 1913, p. 25
Mid-Indian Ridge	Mauritius	20°S	57°E	clay slate	Reed, 1949, p. 543
		ATLANTIC OCEAN			
Mid-Atlantic Ridge	Iceland	65°N	18°W	rhyolite, "granite"	Pjeturss, 1910, p. 5
Mid-Atlantic Ridge	Ascension	8°S	14°W	rhyolite	Reed, 1949, p. 227
Mid-Atlantic Ridge	Tristan da Cunha	37°S	12°W	gneiss inclusions	Reed, 1949, p. 230
	Falkland Is.	52°S	60°W	granite, quartzite	Reed, 1949, p. 236
	Cape Verde Is.	17°N	25°W	granitic rocks	Gagel, 1910, p. 3

nology of geophysics. Why should it be surprising that the oceanic crust, though dominantly sima, should contain a roughly comparable amount of sial? After all, there seems to be a range of several per cent in wave speeds recorded from oceanic profiles immediately below the sediment.

Washington's (1922, Fig. 1) plotted values of density against elevation seem to show no discontinuity at sea level, even though, as might be expected, island rocks are rather consistently less dense than they should be to fit the curve appropriate to the mean depth of the surrounding sea floor.

In summary, it seems that an excellent case can be made for a strong statistical difference in the petrologic composition of oceanic and continental crustal segments. But this difference is after all statistical and not exclusive. The best sialic layer we can find on the continents contains some magnesia-rich rocks, and it is probable that the dominant sima of the oceanic floors is here and there contaminated with siliceous ones. There are oceanic granites and continental oceanites.

This suggests that such rises as the Mid-Atlantic, Carlsberg, and Walfisch, or plateaus like the Albatross, intermediate in elevation between deep ocean floors and continental plates, may also be intermediate in the mean density of their component rocks.

Such rises could be adequately explained by an extra thickness of sima or by somewhat lower mean density of sima. But the clear geologic evidence from the southwest Pacific that the sial plates of Fiji, New Caledonia, and a host of other islands in the area between Fiji, New Zealand, and Australia were formerly larger—though much of the area now lies at oceanic depths—seems to demand that at least some sial is also present in the deep ocean floor. True, much of this area is submerged by less than 4 km of water, but the problem of its depression is still comparable to that involved in the uplift of the Tibetan Plateau. Granitic rocks in Macquarie Island glacial drift may have come from a small mass, but it was doubtless near by and hence now at great depth beyond the fault scarp bounding the island (Mawson and Blake, 1943). Madagascar, the Antilles, Falklands, and Southern Antilles all rise from deep water and show geologic characters implying a formerly greater extent.

CHANGES IN DISTRIBUTION OF LAND AND OCEAN

The geologic record is of course clear in showing remarkable changes of former sea areas to land and the converse. But it is not so clear in indicating whether the changes involve merely epicontinental shelves and troughs or whether they also include transformation of truly oceanic crustal segments into part of the continent and the converse.

If the total quantities of sial and sea water had remained constant throughout geologic time and if isostasy generally prevailed, the average continent should now be either smaller or standing lower with respect to sea level than at earlier times. The continents must continually lose sial to the sea floor. Although the bulk of the material is doubtless deposited on the continental embankment whence it may be returned to the land during a later orogeny—itself a wasteful process—nearly all the material reaching pelagic areas should be permanently lost. Even if, as seems unlikely, all the rocks from Timor, the Alps, Barbados, and Cuba that have been thought to be pelagic are really so, their total bulk is trivial compared with that of the secular accumulation of pelagic sediments.

Arrhenius (personal communication, 1954) has estimated on the basis of his studies of eupelagic sediments of the eastern Pacific (1952) that the present rate of inorganic sedimentation in pelagic areas is about 100 mg per sq. cm per 1000 years. In 2 billion years at this rate 900 m of clay would accumulate on the deep-sea floor, as well as variable amounts of biogenic silica and carbonates. Without considering the far greater bulk of shelf sediment, this clay alone would represent an average continental denudation of 1.5 km. Water displaced by the sediment should encroach on the continents, continually decreasing their "freeboard."

Yet nothing in the geologic record supports such a secular change; from the earliest geologic dates we seem to have had huge epicontinental seas. If today's conditions are atypical of the past, it seems more likely that continental relief is slightly greater, rather than less, than formerly. It seems a necessary inference that new sial has been evolving throughout geologic time (Rubey, 1951, p. 1147; Gilluly, 1937, p. 440–441), and both Rubey and Wilson (1952, p. 444) have shown that the present rate of volcanism implies an adequate supply.

POSSIBLE CHANGE OF OCEANIC TO CONTINENTAL SEGMENTS

Dana long ago suggested that the average continent, having a central shield lowland bordered by Paleozoic and younger mountains, has grown laterally by consolidating its own continental embankments or geosynclines, now represented by the mountain chains, with an ancient shield nucleus. Wilson (1951) has interpreted geologic ages inferred from radioactive minerals in different ancient structural trends as supporting such an accretionary outward growth of North America.

Though such a growth may have some statistical support from these radioactive age determinations they are still too few to prove the point, and the record by no means demonstrates that transformation of truly oceanic segments to continental has taken place. Indeed many mountains are intracontinental, not maritime. The older Sierra Nevada lies nearer the continental border than the younger Rockies. Most of the geosynclinal sediments of the Alpine, Appalachian, and Cordilleran troughs were derived from lands oceanward from these geosynclines. And few or no truly pelagic sediments have been recognized, even among the oldest strata of the youngest maritime ranges—and it seems that they should be if truly oceanic segments have been added to the lands. Furthermore, the basement of these geosynclines, where exposed or penetrated by the drill, seems essentially identical in composition to the old shield areas—not to the sima or ultrasima that geophysical data compel us to infer as predominating beneath oceanic sediments. The Indonesian belt of negative gravity anomalies strikes landward into the thick sedimentary piles of the Burma-Himalaya tectonic zones (Evans and Crompton, 1946). But the sediments rest on sial, not sima. Even where the rocks of the negative strip emerge in the Indonesian islands as culminations along structural axes that sink to oceanic depths in both directions, the exposed oldest rocks are sialic. Similar relations seem to prevail in the Aleutian arc and its continuation in the Alaska Peninsula.

These arguments are so persuasive that many geologists consider the possibility of converting oceanic segments to continental so remote as to be negligible. And, of course, most geophysicists would agree with Bullard (1954a, p. 404) that it is practically impossible to turn a segment of continental crust into oceanic.

It does not, however, seem to me quite so definite. Recent work on submarine sediments has thrown doubt on the adequacy of such features as ripple mark and scour-and-fill to prove shallow deposition, and the composition of the sediments of the Mindanao Deep (Neeb, 1943) is more like that of offshore muds than of pelagic oozes, despite the great depth from which the samples were dredged. Some Los

Angeles Basin rocks appear to have been deposited at depths of several hundred fathoms (Woodring, 1938, p. 17), and the supposedly deep-water sediments of Cuba, Barbados, and Timor are in areas where bordering trenches even now descend to great depths so that moderate migration of structural axes might transform deeps into land. Perhaps a re-examination of many eugeosynclinal deposits in the light of what is now known of turbidity currents may lead to revising the interpretation of the depth represented by the association of pillow lavas, graywackes, and pyroclastic slates so commonly found in them.

Furthermore, the granitic basements found beneath many eugeosyclines may not be ancient sial but old sima metasomatically changed to granitic gneisses. In many an ancient terrane, the oldest rock, much replaced and altered, is an amphibolite—perhaps representing an oceanic floor that owes its present composition and density to metasomatism.

Wells (1949) has stressed the contrast between geosynclines developed on a sialic basement and the kind he calls "ensimatic" in which no such basement can be found and in which the deepest rocks exposed are simatic. Despite tremendous surficial shortening in at least five orogenic pulses between Devonian and late Jurassic time, for example, no Precambrian rocks are exposed between Utah and the Pacific Coast along the Fortieth Parallel[2]. The absence of Precambrian rocks cannot be due to lack of deep erosion since the last orogeny because none are exposed near this latitude east of the Sierra, either, where the main deformation was Paleozoic. It seems possible that the absence of exposed Precambrian may be, as Wells implied, because it was all simatic and its excess density favored its depression into the mountain roots during orogeny.

Most of the circum-Pacific mountains show no signs of a pre-Paleozoic floor and no indications of an ancient sial foundation. They may be areas that were formerly truly oceanic, with simatic floors, now added to the continent by orogenic processes. The absence of recognizably pelagic sediments in them may be due in part to depression of abyssal sediment into the roots, partly to metasomatic alteration, and partly to the fact that sediment deposited by density currents even in such extreme depths as that of the Mindanao Deep (Neeb, 1943) are terrigenous muds with few characters that would distinguish them from sediments of much shallower depths, especially after mild metamorphism.

POSSIBLE CHANGES OF FORMER LAND SEGMENTS TO OCEANIC

As metasomatism seems to trend everywhere toward increasing sial, and injections of ultrabasic rock seem everywhere subordinate, it is not so easy to visualize transformation of continental segments into oceanic. Yet it seems quite indisputable that such transformations have taken place, though it must be granted that the oceanic depths concerned are probably nowhere in excess of 3 km. Actually, the transformation does not appear to me to raise any greater geophysical difficulties than the familiarly demonstrable transformation of a low-lying plain into a high plateau.

---

[2] Rocks in the San Gabriel and Klamath mountains of California shown on the geologic map of the State as Precambrian are the only ones so designated within 100 miles of the Pacific Coast of North America. There is real doubt that the age of the Klamath rocks is so great, as no fossils older than Silurian have ever been collected from the region. In any event, the Pacific border ranges contain trivial quantities of Precambrian rocks compared with older ranges.

Along many coasts of Atlantic type a former land surface extends seaward beneath coastal sediments for an unknown distance. The sialic surface rocks have been traced by geophysical techniques along the English Channel beyond the Scilly Islands (Hill and King, 1953), and Ewing and his colleagues have traced the pre-Mesozoic unconformity of the Atlantic coastal plains to the edge of the continental shelf (Ewing et al., 1939; 1950), thereby going far toward the restoration of an *Appalachia* that had been increasingly discredited.

In many parts of the Atlantic coast lines we have comparable evidence of a geologically young change of former land to submarine. On the east coast of Greenland a flexure in the pre-basaltic surface carries a former extension of the island below the sea for a strike length of several hundred miles (Wager and Deer, 1938). The Deccan traps, with their intercalations of plant beds, are similarly drowned in the Arabian Sea (Lees, 1953). Sedimentary features of Karroo sediments in South Africa seem to demand a source in what is now the south Atlantic (Jessen, 1948, p. 51). The list of former continental areas now drowned off structurally quiescent coasts could be very greatly extended. These offer problems of a kind different from those presented by Pacific shores.

The density of Mesozoic and younger sediments of the Coastal Plain and shelf must be far lower than that of any possible subcrust. If isostasy is generally valid over regions so large as the continental shelf, a sial plate so extensive and thick enough to stand above the sea—as it must have done in Triassic time—cannot have had its surface drowned by adding a sedimentary load. Adding ice on top of a floe may submerge the original ice, but the new surface will stand higher rather than lower than before. If the shelf area was formerly and is now in isostatic balance, either the sial plate must have been thinned or a denser material must have been added to it.

Now the upper surface of the sial—the former erosion surface on which the Mesozoic sediments were unconformably deposited—has been protected from erosion by the sedimentary cover everywhere southeast of the Fall Line. Some of the Mesozoic rocks are terrestrial. If drowning has been due to injection of denser material into this crustal segment, one might expect more evidence of volcanism or seismic activity than is found. It seems more likely that the sial has been thinned. If so the thinning was not by surficial but by subcrustal "erosion".

The drowning of such areas as the Atlantic shelf and the raising of such areas as the Colorado Plateau seem to demand changes in thickness of the underlying sial. In plateau uplift one might consider regional metasomatism as a major factor in thickening the underlying sial, except that migmatites seem more common in areas of contemporaneous compression. Even in the plateau problem the lateral transfer of sial at the base of the crust seems more likely. Where a former erosional surface has been depressed so as to receive sediment, however, there is no chance of surficial erosion being adequate. If, as isostasy requires, the sial is thinned or the sima thickened above the Mohorovičić discontinuity, in order to produce the change in level, the process responsible must be subcrustal flowage. A possible modifying factor in such flowage is the shifting of surficial loads by erosion and sedimentation, but, as is discussed later, this cannot alone determine the course and location of such currents.

The loss of land areas to the sea is not, however, confined to coasts of Atlantic type.

The California borderland shows strong evidence of block faulting and subsidence (Shepard and Emery, 1941), and another Pacific border province—the area between Fiji and New Zealand—has a geologic history requiring subsidence on a gigantic scale both in area and depth (Marshall, 1912). The Pleistocene tills of Macquarie Island with their granitic boulders on an island that exposes no granites testify to the very young subsidence along the boundary fault of the island.

In these places, it seems that no direct effect of subcrustal current need be postulated. The crowding together of sialic blocks by any tectonic process must lead to differential yielding and hence differential thickening and thinning of smaller areas—areas so small, indeed, that isostasy need no longer rule. Hence the yielding during periods when underthrusting was inactive may lead to faulting and local drowning as in the California borderland. Such drowned borderlands seem absent from coasts where deeps are actively forming.

Lawson (1932; 1936) has called attention to the indirect but large effects of erosion and sedimentation in orogeny. That these may play a large part in producing subcrustal flow seems to me probable from the quantitative relations involved, though it seems equally clear that the heat supply called for and the world pattern of orogeny demand a more deep-seated motion in the mantle. To bring out some of the factors involved I would like to refer to the quantities we are dealing with.

RATE OF EROSION

The present rate of denudation is difficult to evaluate accurately. Kuenen (1950a, p. 233) has placed the undissolved load annually delivered by streams to the sea at 12 km³. Clarke's figure (1924, p. 121) for the dissolved load is equivalent to 1.5 km³, and Kuenen (p. 234) has estimated the sediment produced by marine abrasion at 0.12 km³. The total is 13.6 km³. If the rate estimated by Dole and Stabler (1909, p. 84) for the whole United States were valid for the entire earth, the volume transferred would be about 5 km³, or if the rate for the Mississippi Basin were applicable, 7.5 km³. Dole and Stabler made no allowance for the traction load. We are probably not far off to accept the total of 13.6 km³ as the volume of solid rock annually transferred from land to sea.

The volume of the land block above the level of the sea is 130×10^6 km³ (Kossinna, 1933, p. 882). If the relief were not renewed, the removal of 13.6 cubic km annually would reduce the continents to sea level in less than 10 million years. Of course isostatic uplift must take place, and even if it did not the rate of erosion would certainly decrease greatly as base leveling was approached.

A more realistic estimate would consider only the mountain areas, defined as those more than 0.2 km in elevation. The volume of the mountains so defined is 69.25 million cubic km (Kossinna, 1933). Allowing a difference in density of 0.6 between root and subcrust the volume of their roots is 380 million cu. km. If erosion were confined to these areas and isostatic uplift were concurrent all the mountains would be removed in about 33 million years—a time perhaps equivalent to that since Early Miocene.

The enormous influence that erosion and sedimentation must have on crustal processes is suggested by some other rough computations. If sial is assumed to be 12 km thick, over all the land masses to the -3000 m isobath, its volume is roughly 3×10^9 cubic km. If the present rate of erosion could be continued for about 225 million years, an equal volume of material would be transferred. This is the time since the Late Paleozoic. Erosion at its present rate would rework a volume eight or nine times that of all the earth's sial, if it proceeded for 2 billion years.

It is, of course, widely assumed that the present rate of erosion is abnormal, and some students consider it 10 times the average rate during geologic time. I have elsewhere given reasons for thinking that perhaps it is a little higher than average, but probably not by more than a few per cent (Gilluly, 1949). The geology of the Gulf Coastal shelf seems to me to support this estimate.

A rough but conservative estimate of the volume of Cenozoic sediment of the Mississippi Embayment, Gulf Coast, and continental shelf between Pensacola and Corpus Christi, using the Tectonic Map of the United States and papers by Weaver (1950) and Carsey (1950) as guides, turns out to be 5 million cubic km. At the present discharge rate of the Mississippi (Fisk *et al.*, 1954) this would require 20 million years to accumulate instead of the 60 million commonly attributed to Cenozoic time. This is a factor of 3. Not much weight can be given so crude an estimate as present data permit, but I submit that many factors make it seem conservative. These include reworking of Tertiary rocks, loss beyond the 6000 foot depth curve I have used as an arbitrary cutoff for the shelf, loss of dissolved material to the open sea, and bypassing of present sediment. Furthermore the Mississippi probably drains a larger area now than during Tertiary time. Contributions of other streams in the area are negligible. I therefore think we are justified in assuming that present erosion rates are surely not twice the average rate during the Cenozoic and, indeed, perhaps no greater than that average.

The main point of this discussion, however, is concerned with the effect of this transfer on the structures of the crust. Continental borders are zones of contrasting crustal density. Along Pacific-type coasts the zone is extremely narrow; along Atlantic, much wider. But the density contrast on the two sides must bring about shear stresses so that this zone is tectonically critical. Secondly this is the zone separating areas of general upward tendency from those of general downward tendency. If the transfer of material during the past was about as it is now, a volume equal to all the sial of the earth has crossed the critical zone six or eight times during the 2 billion years of earth history.

As the total length of earth shore lines is about 370,000 km (Kossinna, 1933, p. 873), sediment is now crossing the shore line at an average rate of 36 cubic km per million years, per km of shore.

Though this is the average per kilometer, it is of course clear that erosion and sedimentation are both unevenly distributed. Climates, courses of streams, accidents of configuration of the marine basin and its current regimens, and many other factors lead to almost capricious irregularities in loading and unloading of crustal segments. One shelf receives highly radioactive sediments, another none; one loses by submarine landslips (Heezen and Ewing, 1952), another gains. Inevitably there are wide differences among localities and regions in thickness of accumulation, rate of depression, mobility, and many other factors.

If isostasy is to be maintained, powerful currents must return an equal mass of subcrustal materials to positions beneath the continents. Were the currents uniformly distributed through a zone 50 km thick, they would be moving at about 1 mm per year. From considerations just mentioned it is obvious that the counterflow cannot be expected to be uniform. The differential motion recorded between crustal blocks on either side of the San Andreas rift of 5 cm per year may thus reflect different rates of flow of subcrustal currents generated in the way just outlined.

Perhaps this subcrustal flow from beneath the oceans carries the shelf sediments back into the continental mass. This may explain both the raised borders of the continents stressed by Dana and by Jessen (1948) (crowding and thickening of the sial plate, even in nonorogenic coastal zones). It may also partly explain the apparent lack of pre-Mesozoic continental shelves referred to by Kuenen (1950b), and it would operate in the right sense to produce the underthrusting of continental borders along tectonically active coasts that is recorded in the deeps, the fault pattern, and the igneous activity there localized. Perhaps, to carry the speculation a bit further, the surprisingly high heat flows measured in the deep oceans (Revelle and Maxwell, 1952; Bullard, 1954b) may be partly explained by the fact that this counterflow controls the shallow subcrustal motion in the mantle. Thus convection cells in the mantle rise beneath the ocean, as the pattern of the basins as a whole suggests, rather than beneath the continents whose blanket of radioactive sial has been considered to control a convection overturn in the opposite sense.

It is clear that the subcrustal counter flow of material in response to shifting supracrustal loads cannot be the cause of mountain making. Sedimentary prisms of the Pacific border ranges may be ensimatic and built into deep water where isostatic movement would permit geosynclinal accumulation. But the sedimentary prisms of most of the older mountain ranges rest on sialic floors. It seems impossible for simple sedimentary loading to depress a sialic basement far enough to allow the accumulation of geosynclinal thicknesses of sediment, and, on the other hand, it seems impossible for the sial floor to have been submerged enough in the beginning to get the process started. In short, the loading and unloading needed to begin the regional subcrustal streaming could hardly involve enough sediment to start a geosyncline. Furthermore, the mouth of the Mississippi where, for several million years, vast sedimentary loads have been added to the crust is precisely an area of essential isostatic equilibrium and shows no signs of either strong seismicity or a bordering uplift. While the accumulated sedimentary thickness there must be approaching the amount commonly found in mountain ranges, there is no sign of unusual subcrustal unrest.

Despite these facts—certainly fatal to a naive direct connection between degradational transfer, isostatic return at depth, and mountain making—I cannot but feel that a less direct but still intimate connection between these processes does indeed exist. The masses involved are comparable, and the time constant of a cycle reasonable. The processes can hardly be independent, even though the current rate of denudation may be twice the average for past geologic time. I see no strong evidence that it is.

REFERENCES CITED

Arrhenius, G. (1952), Sediment cores from the East Pacific: Swedish Deep-Sea Expedition, Repts., v. 5, pts. I–V incl., 227 p., Göteborg.

Bandy, M. C. (1937), Geology and petrology of Easter Island: Geol. Soc. America Bull., v. 48, p. 1589–1609.

Bullard, E. C. (1954a), A comparison of oceans and continents: Roy. Soc. London, Proc., Ser. A, v. 222, p. 403–407.

———— (1954b), The flow of heat through the floor of the Atlantic Ocean: Roy. Soc. London, Proc., Ser. A, v. 222, p. 408–429.

Carsey, J. B. (1950), Geology of Gulf Coastal Area and continental shelf: Am. Assoc. Pet. Geol., Bull., v. 34, p. 361–385.

Chubb, L. J. (1930), Geology of the Marquesas Islands: Bernice P. Bishop Mus. Bull., no. 68, p. 1–71.

———— (1933), Geology of Galapagos, Cocos and Easter Islands: Bernice P. Bishop Mus. Bull., no. 110, p. 1–67.

———— (1934), The structure of the Pacific Basin: Geol. Mag., v. 71, p. 289–302.

Clarke, F. W. (1924), The data of geochemistry: U. S. Geol. Surv., Bull. 770, 841 p.

Dole, R. B., and Stabler, H. (1909), Denudation: U. S. Geol. Surv., Water-Supply Paper 234, p. 78–93.

Evans, P., and Crompton, W. (1946), Geological factors in gravity interpretation: Geol. Soc. London, Quart. Jour., v. 102, p. 211–244.

Ewing, M., Woollard, G. P., and Vine, A. C. (1939), Geophysical investigations in the emerged and submerged Atlantic Coastal Plain, Part III, Barnegat Bay, N. J. section: Geol. Soc. America Bull., v. 50, p. 257–296.

————, Worzel, J. L., Steenland, N. C., and Press, F. (1950), Geophysical investigations in the emerged and submerged Atlantic Coastal Plain, Part V: Woods Hole, New York and Cape May sections: Geol. Soc. America Bull., v. 61, p. 877–892.

Fisher, O. (1889), Physics of the Earth's Crust: 2nd. ed., Macmillan, London, 391 p.

Fisk, H. N., McFarlan, E., Jr., Kolb, C. R., and Wilbert, L. J., Jr. (1954), Sedimentary framework of the modern Mississippi delta: Jour. Sed. Petr., v. 24, p. 76–99.

Gagel, C. (1910), Die Mittelatlantischen Vulkaninseln, in Handbuch der Regionalen Geologie; Bd. 7, T. 10, H. 4, 32 p.

Gilluly, J. (1937), The water content of magmas: Am. Jour. Sci., 5th. ser., v. 33, p. 430–441.

———— (1949), The distribution of mountain building in geologic time: Geol. Soc. America Bull., v. 60, p. 561–590.

Heezen, B. C., and Ewing, M. (1952), Turbidity current and submarine slumps and the 1929 Grand Banks earthquake: Am. Jour. Sci., v. 250, p. 849–874.

Hill, M. N., and King, W. B. R. (1953), Seismic prospecting in the English channel and its geological interpretations: Geol. Soc. London, Quart. Jour., v. 109, pt. 1, p. 1–21.

Jessen, O. (1948), Die Randschwellen der Kontinente: Petermanns Geog. Mitt., Ergänzungsheft 241, p. 1–205.

Kennedy, W. Q. (1933), Trends of differentiation in basaltic magmas: Am. Jour. Sci., 5th. ser., v. 25, p. 239–256.

————, and Anderson, E. M. (1938), Crustal layers and the origin of magmas: Bull. Volcan., Sér. 11, Tome 111, p. 23–82.

Kossinna, E. (1933), Die Erdoberfläche, in Handbuch der Geophysik; Bd. 11, Abt. 6, p. 869–952.

Kuenen, P. H. (1950a), Marine geology: John Wiley & Sons, New York, 568 p.

———— (1950b), The formation of the continental terrace: Brit. Assoc. Adv. Sci., v. 7, p. 76–80.

Lawson, A. C. (1932), Insular arcs, foredeeps and geosynclinal seas of the Asiatic coast: Geol. Soc. America Bull., v. 43, p. 353–381.

———— (1936), The isostasy of large deltas: Geol. Soc. America Bull., v. 49, p. 401–416.

Lees, G. M. (1953), The evolution of a shrinking earth: Geol. Soc. London, Quart. Jour., v. 109, p. 217–257.

Lemoine, P. (1911), Madagascar, *in* Handbuch der Regionalen Geologie; Bd. 7, T. 4, H. 6, 44 p.

Macdonald, G. A. (1949), Hawaiian petrographic province: Geol. Soc. America Bull., v. 60, p. 1571.

Marshall, P. (1911), New Zealand, *in* Handbuch der Regionalen Geologie; Bd. 7, H. 5, Abt. 1, 78 p.

———— (1912), Oceania, *in* Handbuch der Regionalen Geologie: Bd. 7, pt. 2, 36 p.

Mawson, D., and Blake, L. R. (1943), Macquarie Island, its geography and geology: Australasian Antarctic Expedition, Sci. Repts., A, v. 5, 194 p.

Neeb, G. A., *in* Kuenen, P. H., and Neeb, G. A. (1943), Bottom samples: The Snellius Expedition, Brill, Leiden, v. 5, pt. 3, 265 p.

Nordenskiold, O. (1913), Antarktis, *in* Handbuch der Regionalen Geologie; Bd. 8, T. 6, H. 15, 29 p.

Pegau, A. A., *in* Ladd, H. S. (1934), Geology of Vitilevu, Fiji: Bernice P. Bishop Mus. Bull., no. 119, p. 36–42.

Pjeturss, H. (1910), Island, *in* Handbuch der Regionalen Geologie; Bd. 4, H. 1, 22 p.

Powers, H. A. (1935), Differentiation of Hawaiian lavas: Am. Jour. Sci., 5th. ser., v. 30, p. 57–71.

———— (1955), Composition and origin of basaltic magma of the Hawaiian Islands: Geochim. Cosmochim. Acta, v. 7, p. 77–107.

Reed, F. R. C. (1949), The geology of the British Empire: 2nd. ed., E. Arnold and Company, London, 764 p.

Revelle, R., and Maxwell, A. E. (1952), Nature, London, v. 170, p. 199–200.

Rubey, W. W. (1951), Geologic history of sea water: Geol. Soc. America Bull., v. 62, p. 1111–1147.

Sederholm, J. J. (1925), The average composition of the earth's crust in Finland: Comm. géol. Finlande Bull., no. 70, p. 1–20.

Shepard, F. P., and Emery, K. O. (1941), Submarine topography off the California coast: Geol. Soc. America, Spec. Paper 31, 171 p.

Suess, E. (1909), Das Antlitz der Erde: Bd. 3, H. 2, p. 626.

Tilley, C. E. (1950), Some aspects of magmatic evolution: Geol. Soc. London, Quart. Jour., v. 106, p. 37–62.

Turner, F. J., and Verhoogen, J. (1951), Igneous and metamorphic petrology: McGraw-Hill Book Co., New York, 602 p.

Wagner, L. R., and Deer, W. A. (1938), A dike swarm and crustal flexure in east Greenland: Geol. Mag., v. 75, p. 39–46.

Washington, H. S. (1922), Isostasy and rock density: Geol. Soc. America Bull., v. 33, p. 375–410.

Weaver, P. (1950), Variations in history of continental shelves: Am. Assoc. Petr. Geol., Bull., v. 34, p. 351–360.

Wells, F. G. (1949), Ensimatic and ensialic geosynclines: Geol. Soc. America Bull., v. 60, p. 1927 (abstract).

Williams, H., Turner, F. J., and Gilbert, C. M. (1953), Petrography: W. H. Freeman and Co., San Francisco, 406 p.

Wilson, J. T. (1951), On the growths of continents: Roy. Soc. Tasmania, Pap. and Proc. 1950, p. 85–111.

———— (1952), Orogenesis as the fundamental geological process: Am. Geophys. Union, Trans., v. 33, p. 444–449.

Woodring, W. P. (1938), Lower Pliocene mollusks and echinoids from the Los Angeles Basin, Calif.: U. S. Geol. Surv., Prof. Paper 190, 67 p.

19

Reprinted from *New York Acad. Sci. Trans.* ser. 2, **27**:135–143 (1964)

MOBILE BELTS, SEDIMENTATION AND OROGENESIS[*][†]

R. H. Dott, Jr.

Department of Geology, University of Wisconsin, Madison, Wis.

Introduction

Certainly the most gnawing geological question on earth is "Why continents and ocean basins?" The many different geological differences between these initial earth features are so impressive as to compel one to believe that they existed for a very great time. And yet, recent re-evaluation of rates of erosional denudation of North America (Judson and Ritter, 1964) indicates that the everlasting mountains, and the plains as well, could be reduced to sea level in a mere 10 million years. But, even granting that 200 meters per million years probably greatly exceeds the average rate of denudation of continents throughout geologic time, it is an inescapable conclusion that continents have been rejuvenated somehow.

In the course of time, addition of sialic material to continents must have occurred at the edges or bottoms or both. Evidence of a long-continued chemical segregation or differentiation of atmosphere, oceans, crust, mantle and core of the earth is very compelling (Rubey, 1955; Engel, 1963; MacDonald, 1964). As conceived here, differentiation of the earth's interior would add new sial chiefly from below; however it is reasonable to suppose that some lateral spreading would occur, especially in early stages of continent formation when the entire crust presumably was relatively thin. Hurley *et al.* (1962) present impressive evidence from Rb/Sr ratios in rocks of North America that there has been "continuous generation of primary sial from subsialic source regions." They treat this accretion in terms of area, finding that apparent growth has been roughly proportional to areal extent of the age provinces. For North America the apparent average areal growth rate has been approximately 7,000 km.^{2} per million years. Moreover, they conclude that "growth seems to have been operative over most of geologic time," and imply that the rate has been more or less constant. It would seem more conservative to regard growth in terms of volume only, leaving the question of areal increase open. Furthermore, though we have no adequate proof at present, it seems highly probable, as in the case of past average denudation rates, that the process of sialic accumulation has been far from linear through time. It seems dangerous to project apparent average rates very far back in time. Though controversial, there is evidence that continental crust may have accumulated largely in the interval from 3.0 to 1.0 billion years ago (Tilton & Hart, 1963; Gastil, 1960; Tatsumoto & Patterson, 1964).

[*] This paper was presented at a meeting of the Section on November 2, 1964.
[†] Geophysical and Polar Research Center Contribution No. 150. This paper is based on studies which were made possible by grants from the Wisconsin Alumni Research Foundation and the National Science Foundation.

In any event, vertical volumetric accretion of continents has received far less consideration than lateral, and the conventional hypothesis favored by many geologists has been that of lateral growth by orogenesis.

The orogenic or structurally mobile belts of the earth's crust constitute the next most compelling and puzzling geologic features. And, although it has long been thought that mobile belts played a major role in crustal evolution, nonetheless a long history of hypothesis formulation, testing, and discard, checkers the career of geology's most sophisticated, unifying theoretical integration. Nowhere else in the vast labyrinthine corridors of earth studies do we find any concept or group of concepts which have the breadth and depth of impact than the theories of geosynclines and orogenesis. Yet today we still lack a satisfactory explanation. A myriad of relevant hypotheses and speculations, such as pole wandering, continental drift, heat convection in the mantle, and spreading sea floors cannot now be satisfactorily tested in order to allow confident rejection of any or all of them. It is appropriate, then, to sift, winnow and re-examine the historical evidence bearing upon the great genetic questions about mobile belts. Ultimately, geologic history provides one of our most powerful tests of hypotheses, and the historian must have his house in order if he is to be prepared to administer that test when needed.

This paper attempts to resift some of the present knowledge of the sedimentary record in mobile belts to discern if any new or revised generalizations about sediment patterns can be discovered. Ultimately it is hoped, of course, that such reevaluation can strengthen or weaken one or more of the many alternative genetic explanations now extant.

The Issues

Lateral continental accretion. The theory of areal accretion of continental crust throughout time by its continuous generation through orogenesis has several weaknesses as applied to North America where the case has always been most compelling. First, it assumes that virtually all orogenic belts form at margins between continents and ocean basins, yet this is clearly denied by worldwide distribution of post-Precambrian belts. Moreover, many belts crisscross and intersect each other in such complex ways as to raise serious doubts about any simple, orderly sequential development. Many are not marginal, but lie either wholly within (i.e. crosscut) continents and others lie well out in oceanic areas if island arc-trench systems are accepted as akin to true orogenic belts. In this way, similar gross structural characteristics are taken to be the overriding common denominator rather than location, stratigraphy or petrology. Structural patterns also suggest that cratonic "welding" is somewhat reversible, for some mountains seem to represent remobilization of old parts of cratons. And the great rift zones with their commonly associated flood basalts within cratons, though not directly related to the mobile belts, also prove that continental crust is not immune to profound deformation even long after it is formed.

Second, while absolute age data of granitic plutons and high grade metamorphic terrains early suggested, at least statistically, a crudely concentric arrangement of centrifugally younging ages in several continents, the proper interpretation is by no means clear. As there have appeared more examples of discordant ages from a single rock and important younger plutons located centripetal to older ones, some geochronologists have questioned the original explanation (for example: Tilton & Hart, 1963; Zartman, *et al.*, 1964). For at least the last one billion years, it now appears more probable that net lateral volume increase of sial may not have occurred at all (Gastil, 1960), for important granitic bodies of Precambrian and early Paleozoic age occur essentially at the present continental margins. The radiogenic age data might instead be explained by crude lateral spread of orogenic plutonic and metamorphic events simply reworking older sial rather than increasing its area (Tilton & Hart, 1963; Gilluly, 1963). Indeed extensive orogenic reworking seems virtually inescapable for most other continents where the age patterns are far more complex than in North America.

Third, in no known mobile belt has an originally oceanic or quasi-oceanic initial floor or basement been discovered. Wherever the basement is visible in the hearts of the belts it is sialic. The argument that younger orogenesis has simply obliterated and converted initial oceanic material is not wholly satisfactory because in many cases much older, highly discordant structural trends are still preserved through the mask of younger orogenies.

Fourth, the common view that early eruptions in orogenic belts are mafic ophiolites followed by an orderly evolution to more felsic varieties simply does not hold in most cases; therefore the implication that evolution occurred from oceanic to continental crust during the history of volcanism in the "orogenic cycle" is oversimplified. Dickinson (1962) has shown from the composition of volcanic and plutonic rocks of the Pacific margin of North America that the fundamentally sialic nature of the crust beneath this mobile belt has not changed at least since mid-Paleozoic times. As with the sediments of eugeosynclines, so the volcanic products vary enormously through time and show no clear trend.

Geosynclines. The main thesis of this paper is that mobile belts have formed in very complex patterns with respect to continents. A corollary may follow that they bear little on the most fundamental question of the origin of initial earth features, the continents and oceans. Their role may only have been to modify continental and oceanic crust once formed. Where, and only where, mobile belts or portions thereof have lain in proximity to great potential sediment sources, have they tended to become filled with thick sedimentary accumulations. It is such a "filled mobile belt" that James Hall first recognized and J. D. Dana called geosynclinal. To be more specific, we would call it today an *orthogeosyncline* (Kay, 1951); but throughout this paper, orthogeosyncline and geosyncline are used synonymously. The viewpoint of this paper, therefore, places structural mobility first as the most fundamental phenomenon—a cause, and geosynclinal accumulation as secondary thereto—a result. Rather than the geosyncline necessarily preceding orogenesis, as generally assumed, I believe the causal relation to be the reverse. The mobile belt is the "chicken;" the orthogeosyncline a later-to-be-hatched "egg." But, as a hen's new egg is only a *potential* chick, so a mobile belt is only a potential geosyncline. Modern island arc-trench systems are but "hopeful chicks" in this ornithological analogy, still in incubation, but without assurance of hatching into true geosynclines, much less of achieving full maturity as mountain ranges.

Role of continental shelf sedimentary prisms. An apparent consequence of this view is rejection of Dietz's (1963) hypothesis of the origin of orthogeosynclines and mountain ranges through: 1) long "geosynclinal" accumulation of continental shelf ("miogeosynclinal") and continental rise or slope ("eugeosynclinal") sediments, and 2) their final orogenic "collapse" and up-heaval into mountains at the margins of stable cratons by spreading of the sea floors. Presence today of thick late Mesozoic and Cenozoic sedimentary prisms at several margins of the present continents provides a very appealing, apparently analogous genetic link with ancient supposed marginal geosynclinal-orogenic accretions to continental nucleii and seems to answer the embarrass-ing question of "Why are there no pre-Mesozoic shelves?" However, there are several difficulties: 1) as Dietz himself admits, his hypothesis does not account for the many nonmarginal belts; 2) evidence is lacking in ancient geosynclines of initial accumulation upon oceanic crust; 3) the bulk of geo-synclinal sediments even in miogeosynclines are not paralic as claimed for shelves by Dietz, nor are they all craton-derived; and most important, 4) orthogeosynclines reveal much greater structural mobility very early and throughout their histories than seems evidenced in modern continental shelves and rises. Open margined shelves may have quite a different expla-nation than mobile belts. Cram (1962) makes a strong case for phase change at the M discontinuity beneath great sedimentary load on the Gulf Coast. Alternatively, large shelf prisms on "open" coasts may not have formed earlier because previously continents were not so large and high.

Variability. The sediments of mobile belts, particularly of eugeosynclines, are so variable that some simplifying "models" are mandatory. TABLE 1 shows examples of the principal associations in terms of source and process-environment energy factors, the most objectively assessable parameters. Tectonics is considered a constant, for in all cases there is evidence of a high order of structural mobility. The importance of eugeosynclinal volcanism, unconformities and conglomerates, tangible evidences of continuing mobility, has been summarized for North America by Kay (1951). Most foreign belts reveal similar evidences, though ironically the classic Alpine example is exceptional in that volcanic materials are very scarce therein. Space allows discussion and reference to only a few situations with special implications.

Very early inception of volcanism in most eugeosynclines is of prime importance, particularly in considering alledged analogies with continental terrace sediment prisms. Daly as early as 1912 noted (p. 575) that "vulcanism is always or almost always contemporary with geosynclinal sedimentation." He even speculated that transfer of magma upward could account volumetrically for geosynclinal subsidence and sedimentation, an idea currently enjoying revival.

Sedimentary sources. For terrigenous clastic sediments, three distinct source types or mixtures can explain the bewildering array of petrographic hybrids encountered in mobile belts (TABLE 1). These are: 1) stable cratonic source external to the mobile belt and supplying compositionally mature quartz sandstones and shale; 2) internal volcanic sources within the mobile belt, either submarine or subaerial, producing lavas, pyroclastics and tuffaceous sediments; and 3) internal tectonic lands raised in the mobile belt by diastrophism and producing chiefly lithic and feldspathic debris mixed with quartz.

Carbonate rocks. Special circumstances must exist to produce carbonate rocks in eugeosynclines. In all cases, important carbonate rocks represent special subtropical to tropical climatic conditions, shallow agitated water, and relative paucity of terrigenous clastic debris. Reefs and banks are known

TABLE 1

RELATION OF SOURCE AND ENVIRONMENTAL FACTORS
IN EUGEOSYNCLINAL SEDIMENTATION

Dominant Source Factors	Process – Environment Factors	
	Low or Intermittent Energy	High or Constant Energy
Submarine Volcanic Vents Only	1. Acidic pyroclastics (Tertiary, Japan). 2. Mafic flows, bedded chert, fine clastics; minor but widespread quartz arenite (lower Paleozoic, western Nevada-Idaho).	(Example unknown)
Small and/or Ephemeral Volcanic Lands	1. Lavas, volcanic con-glomerate and flysch-like graywacke-mudstone suites (Jurassic, southwest Oregon; Franciscan Formation, Calif.).	1. Lavas, pyroclastics, graywacke-mudstone; rare limestone (Ordovician, Cale-donides). 2. Vitric tuffs and lithic (volcanic) graywackes with oolite and fossil debris (Carboniferous, Queensland, Australia). 3. Lavas, tuff, volcanic conglomerate, reefal limestone, siltstone, quartz arenite (Silu-rian, Gaspe, Canada; Permo-Triassic, Pacific coast).
Large Volcanic Lands	(Example unknown)	1. Lavas, pyroclastics, tuffaceous sediments and conglomerates with nonmarine fossils (Cretaceous-Tertiary, Central Chile; Terti-ary, Columbia Plateau-Cascade region).
Small (or Distant) Tectonic Lands	1. Local conglomerate, lithic graywacke, mudstone (Cretaceous-Tertiary, Alpine flysch; Cretaceous, southern Chile; early Cretaceous, southwest Oregon; Carboniferous, Ouachita Belt).	1. Chert and quartzite conglomerate and sandstone intertongued with calcarenite (Pennsylvanian, cen-tral Nevada). 2. Quartz arenites, con-glomerate, local lime-stone (Jurassic, central Oregon).
Large Tectonic Lands	(Example unknown)	"Molasse": 1. Red bed clastic wedges (conglomerate, sand-stone, shale) (Old Red Sandstone, Britain; Catskill, Appalachians). 2. Non-red coaly wedges (sandstone, coal, shale) (Carboniferous, Appa-lachians; Cretaceous, Rocky Mountains).
Cratonic Land Source	1. Quartz arenites inter-bedded with lava-chert-mudstone-graywacke successions (Ordovi-cian, central Idaho-Nevada).	1. Pure quartzite, marble, schist; some volcanics, graywacke with or with-out "iron formation" and chert (Precambrian, Caledonides and Great Lakes region). 2. Quartz arenite inter-stratified with lavas (Jurassic, southeast California). 3. Quartz arenite, lime-stone, minor shale (Cape System, South Africa).

in both modern and ancient mobile belts, fringing islands and on broad shallow platforms.

Environment versus process. The most obvious and important single environmental factor is the relative position of sea level, geology's most fundamental interface. Yet sea level is a somewhat transitory and fickle phenomenon. Eustatic fluctuations may result from changes completely external to a mobile belt, (Dott, 1964a); therefore certainly no succession of deposits can be included or excluded as geosynclinal merely on the basis of contemporary position of sea level alone. It is contended, for example, that very thick, wholly nonmarine volcanic and sedimentary successions in the Cretaceous-Tertiary of central Chile (Munoz Cristi, 1956) are as truly geosynclinal as any others. Both north and south along the Andean belt, marine rocks were deposited simultaneously.

Clastic wedges. Tectonic lands have supplied the bulk of eugeosynclinal sediments. Large, high lands—tectonic borderlands—of complex composition were raised during major orogenies. These shed vast prisms or *clastic wedges* of material so rapidly as to exceed subsidence, thus building a succession of very large, subaerial alluvial plains grading distally to marine environments. These comprise most if not all of the deposits called *molasse.* Though molasse is generally considered "post-orogenic," it may as well be "syn-orogenic." The transition to molasse could, in fact be induced by externally caused eustatic changes.

At least two major, distinct types of clastic wedges exist, *red bed wedges* like the famous Old Red Sandstone of Britain and the Catskill deposits of the Appalachians, and *non-red coaly wedges* such as the Carboniferous of the Appalachians and Cretaceous of the Rocky Mountains. Both types contain thick prisms of clastic deposits coarsening toward their tectonic land sources within geosynclines. Sediment volume was so great that most examples spilled out of the geosyncline onto cratonic margins (Kay's exogeosynclines) where their strata intertongue with normal epeiric sea sediments. The differences between red and non-red types are largely environmental. Clastic wedges have accompanied practically every major orogeny, and commonly occur superposed in the stratigraphic record, as in the Appalachians, reflecting multiple pulses of orogeny.

Australia and New Guinea appear to provide a perfect modern analogue to the relations of ancient tectonic borderlands, alluvial plains, cratons and epeiric seas. New Guinea is a tectonic land within the Indonesian mobile belt bordering the Australian craton. From the youthful mountaions, a broad, low, swampy tropical alluvial plain spreads out to the south to the Arafura epeiric sea. The latter floods the north edge of the craton much as interior epeiric seas flooded central North America when clastic wedge deposits spread cratonward from Appalachian borderlands. The scale matches very closely the inferred dimensions of the Devonian borderland.

Cratonic sources. It is a long-acknowledged fact that, particularly in the early histories of North American geosynclines, much of the mature terrigenous clastic debris of miogeosynclines was craton derived (Kay, 1951). But cratonic-derived mature clastic detritus has even found its way into several eugeosynclinal belts. In north central Nevada and Idaho, very mature quartz arenite occurs in Ordovician eugeosynclinal slate-chert-volcanic successions (Kay and Crawford, 1964; Churkin, 1962), and similar sandstones occur among Jurassic lavas in southeast California (Grose, 1959). These sands apparently were cratonic derived ultimately and spread across the miogeosyncline into volcanic regions. Rather large volumes of feldspathic quartz sandstones of varied texture occur in the west Antarctic mobile belt (Halpern, 1963) and an ultimate source for the mature constituents may have been cratonic.

The Precambrian contains by far the greatest examples of very mature quartz sandstones in mobile belts, and their significance seems neglected. The Dalradian Series (late Precambrian-Cambrian) of Scotland takes the honors for being the earliest example studied. John Playfair in 1802, in establishing the abundance of detrital quartz sandstones in "Primitive" rocks hitherto supposed to be entirely chemical, cited exposures of Dalradian quartzites. Cross-stratified units occur chiefly in the middle Dalradian and are at least locally 15,000 feet thick (Knill, 1963). Extensive carbonate rocks are also characteristic, and in the upper part volcanic rocks and graywackes are important. According to Knill (1963) a landmass, presumably cratonic, lay to the northwest of the mobile belt.

In North America the classic Great Lakes middle Precambrian also contains well-known cross-stratified pure quartz arenites (now quartzites), carbonates and quartz wackes. Wherever paleocurrent analyses have been made, transport directions are shown to have been southerly or southeasterly (Pettijohn, 1957; Nilsen, 1964). There appears to be a clear transition from craton to geosynclinal accumulations that in many areas have important volcanic rocks, chiefly pillowed greenstones and some tuffs. The mobile belt fill rests upon an older granitic-metamorphic basement. Several masses over 1,000 feet thick of supermature, cross-stratified quartzites punctuate the succession. Undoubtedly the material was derived from a stable shelf area to the north, but was deposited in an embryonic mobile belt. Moreover, many of the less pure sandstones or graywackes of the region are quartz wackes and simply represent lesser textural maturity, thus lower energy environments with no significant difference in source.

It seems inescapable that these old mobile belts of Scotland and the Great Lakes region lay either within, or marginal to, large cratons stable for a long time in order to possess such staggering volumes of pure quartz sand. There is a good possibility that continents already approached their present sizes and were severed down their middles by mobile belts.

Mixed sources and temporal progressions of sediment types. An example from southwestern Oregon illustrates the possible mixed source and space-time complexities. Major Mesozoic orogenesis produced severe deformation, metamorphism and plutons in the Klamath region. A new Eocene eugeosyncline formed to the north. Submarine volcanism produced immense volumes of spilite and keratophyre in western Oregon and Washington while subaerial volcanism was rampant to the east. Flysch-like sedimentation of graywackes derived from contemporaneous volcanic and older metamorphic-sedimentary sources occurred in an oceanic embayment around the submarine volcanic pile. Shallow marine, in part deltaic, and even nonmarine coal-bearing molasse-like sediments accumulated simultaneously near the tectonic-volcanic land in the Klamath region (Dott, 1964b). Flysch and molasse accumulated side by side simultaneously and both were "post-orogenic" to Mesozoic diastrophism, but "pre-orogenic" in terms of the Cenozoic. By late Eocene, offshore volcanic islands had appeared to the northwest and provided sediment locally while the larger tectonic-volcanic land to the southeast continued to provide much more.

Conclusion

At present we understand very little of the gross tectonics and evolution of the crust. Perhaps the suggestions of Carey (1958) and others that the outer earth is laced with gigantic megashears crossing continents and ocean basins alike regardless of boundaries, are not so incredible as many at first thought. Some of the clear oceanic fracture zones appear onshore, notably in Guatemala, and magnetic data suggests (Fuller, 1964) that others may pass at deeper levels beneath continents. By comparison, the similar apparent disregard for continental boundaries shown by mobile belts is not so strange after all. And if continental crust did form early and was not so genetically tied to mobile belts, then these discordances seem more intelligible. Another puzzling facet of the crustal problem is the question of the meaning of mid-oceanic rises. And if a majority of geologists ever enthusiastically "allow" continents to drift, then a whole new world of possibilities will emerge. Ultimate big answers to these bigger questions, however, must lie buried in the mantle and core of the earth, inaccessible to conventional geologic methodology. But earth historians, too, must apply their own peculiar talents to these big questions in consort with geophysicists and geochemists.

References

CAREY, S. W. 1958. A tectonic approach to continental drift. *In* Continental Drift, A Symposium. S. W. Carey, Ed.: 177-355. Hobart Univ. Tasmania.

CHURKIN, M. 1962. Facies across paleozoic miogeosynclinal margin of central Idaho. Bull. Am. Assoc. Petrol. Geol. 46:569-591.

CRAM, I. H., Jr. 1962. Crustal structure of Texas coastal plain region. Bull. Am. Assoc. Petrol. Geol. 46: 1721-1727.

DALY, R. A. 1912. Geology of the North American Cordillera at the forty-ninth parallel. Geol. Surv. Can. Mem. 38 2: 547-857.

DICKINSON, W. R. 1962. Petrogenetic significance of geosynclinal andesitic volcanism along the Pacific margin of North America. Bull. Geol. Soc. Am. 73: 1241-1256.

DIETZ, R. S. 1963. Collapsing continental rises: An actualistic concept of geosynclines and mountain building. J. Geol. 71: 314-333.

DOTT, R. H., JR. 1964a. Superimposed rhythmic patterns in mobile belts. Symposium on cyclic sedimentation. Kansas Geol. Surv. (in press).

DOTT, R. H., JR. 1964b. Ancient deltaic sedimentation in eugeosynclinal belts. *In* Developments in Sedimentology. 1. Deltaic and shallow marine deposits. L. M. J. U. van Straaten, Ed.: 105-113. Elsevier. Amsterdam, Netherlands.

ENGEL, A. E. J. 1963. Geologic evolution of North America. Science 140: 143-152.

FULLER, M. D. 1964. Expression of E-W fractures in magnetic surveys in parts of the U.S.A. Geophys. 29: 602-622.

GASTIL, G. 1960. The distribution of mineral dates in time and space. Am. J. Sci. 258: 1-35.

GILLULY, J. 1963. The tectonic evolution of the western United States. Quart. J. Geol. Soc. (London) 119: 133-174.

GROSE, L. T. 1959. Structure and petrology of the northeast part of the Soda Mountains, San Bernardino County, California. Geol. Soc. Am. Bull. 70: 1509-1548.

HALPERN, M. 1963. Cretaceous sedimentation in Base O'Higgins area of Northwest Antarctic Peninsula. SCAR-IUGS Symp. Antarctic Geol. Capetown, Rep. South Africa. R. Adie, Ed. (in press).

HURLEY, P. J., H. HUGHES, G. FAURE, H. W. FAIRBAIRN & W. H. PINSON. 1962. Radiogenic strontium-87 model of continent formation. J. Geophys. Res. 67: 5315-5334.

JUDSON, S. & D. F. RITTER. 1964. Rates of regional denudation. J. Geophys. Res. 69: 3395-3401.

KAY, M. 1951. North American geosynclines. Geol. Soc. Am. Mem. 48: 143.

KAY, M. & J. P. CRAWFORD. 1964. Paleozoic facies from the miogeosynclinal to the eugeosynclinal belt in thrust slices, central Nevada. Geol. Soc. Am. Bull. 75: 425-454.

KNILL, J. L. 1963. A sedimentary history of the Dalradian Series. In The British Caledonides. M. R. W. Johnson & F. H. Stewart, Eds. : 99-116. Oliver & Boyd. Edinburgh, Scotland & London, England.

MACDONALD, G. J. F. 1964. The deep structure of continents. Science 143: 921-929.

MUNOZ CRISTI, J. 1956. Chile. In Handbook of South American Geology. W. F. Jenks, Ed. Geol. Soc. Am. Mem. 65: 187-214.

NILSEN, T. H. 1964. Geology of the Animikean Pine River (Breakwater) quartzite conglomerate and the Keyes Lake quartzite, Florence County, Wisconsin.: 100. Unpublished M. S. Thesis. Univ. Wisconsin. Madison, Wis.

PETTIJOHN, F. J. 1957. Paleocurrents of Lake Superior Precambrian quartzites. Geol. Soc. Am. Bull. 68: 469-480.

RUBEY, W. W. 1955. Development of the hydrosphere and atmosphere. In The Crust of the Earth. A. Poldervaart, Ed. Geol. Soc. Am. Special Paper 62: 631-650.

TATSUMOTO, M. & C. PATTERSON. 1964. Age studies of zircon and feldspar concentrates from the Franconia Sandstone. J. Geol. 72: 232-242.

TILTON, G. R. & S. R. HART. 1963. Geochronology. Science 140: 357-366.

ZARTMAN, R. E., J. J. NORTON & T. W. STERN. 1964. Ancient granite gneiss in the Black Hills, South Dakota. Science 145: 479-481.

ERRATA

The first word of line 3, page 135 should be "first order" rather than "initial." As it stands, "initial" is very contradictory with the general context.

Part VII

SEA-FLOOR SPREADING AND PLATE TECTONICS—EARLY FORMULATIONS

Editor's Comments
on Papers 20 Through 26

The transition from conventional fixist earth models to a model encompassing a mobilist plate tectonic Earth has had an uneven and curious history. Following the Dutch gravity surveys of the East Indian island arcs by Vening Meinesz, combined with stratigraphic and sedimentological studies by Umgrove and Kuenen, Haarmann (1926, p. 107) introduced the concept of the tectogene, an apparently viable device for producing geosynclines. Kuenen and his Dutch colleagues developed the idea with geophysical underpinnings. Griggs (Paper 14) demonstrated experimentally that such linear crustal downbuckles (and their surface expression, trenches) might be related to thermal convection currents in the mantle. The idea that Pacific-type island arc-trench systems might be reasonable analogues for ancient mobile belts was welcomed in Europe. Hess (1938) supported this concept by making a detailed comparison of igneous rocks in ancient alpine systems and modern arcs. Eardley would later extend this comparison to the overall stratigraphic succession in geosynclinal belts and ocean areas (Paper 16). Hess emphasized the significance of paired belts of serpentine in the Alps (and Appalachians), contending that they were mantle-derived intrusions plastically squeezed into the flanks of the tectogene during the early phase of downbuckling.

Paper 20 is included because in it Hess expands on his earlier themes. For Hess, the rocks of the Steinmann Trinity (E. B. Bailey's nickname for the spilite-serpentine-chert association, or *ophiolite suite*) were the diagnostic lithologies in alpine belts. Consequently, Hess was convinced that Hall's original geosynclinal concept could not be adequately applied to the alpine mountain systems, because Hall's (Appalachian) model lacked serpentines and did not resemble in any sense what for Hess were modern geosynclinal analogues, namely the modern arc-trench systems of the western Pacific. Hess believed that the thick sequence of dominantly clastic sediment, the main constituent of Hall's *geosyncline*, developed only after alpine mountain systems were produced by deformation of subsiding troughs parallel to rising mountain areas. For Hess as for Dana, although in quite different senses, the geosyncline was the *result* rather than the cause of mountain building. Island arcs and trenches, rather than broad, sediment-loaded basins (geosynclines), represented the early developmental stage for alpine mountain systems.

As others had already done, Hess pointed out that few modern geosynclines really existed as Hall defined them (for example, the Gulf Coast geosyncline). Conversely, most mobile belts (he cited the Laramide belt as a notable and curious exception) seemed more akin to Pacific island arc systems like the Philippines: parallel belts of serpentine, an axial zone of metamorphism, and a belt of andesitic volcanism. These modern island arcs, (geosynclines in a more actualistic sense) "seemed to form" out on the open ocean floor. Were most geosynclines in fact oceanic features in their developmental stage, rather than "creatures" of the continents? Speculating on the origin of the almost ubiquitous serpentine belts, Hess explained their possible origin as being due to a location directly above rising limbs of thermally-driven convection cells coinciding with the axis of the mid-ocean ridge system. Hess would soon return to this theme with dramatic results (Paper 23).

Soon after Hess published his paper on serpentines, Drake, Ewing, and Sutton (Paper 21) summarized much of the extensive geophysical work done along the eastern continental margin of North America from Cape Hatteras north in the 1930s, 1940s, and 1950s. They were particularly impressed with the remarkable shape, thickness, and overall (inferred) lithological make-up of the present-day shallow-water continental shelf and deeper-water continental slope and rise areas. They believed the continental shelf and rise matched the original Stille-Kay concept (as represented in the Appalachians) of an inner miogeosynclinal trough and an outer, volcanic-bearing eugeosynclinal trough. One could even detect a ridge separating the two modern troughs that resembled the Precambrian geanticlinal barrier documented by Kay (1951). Unfortunately, the modern trough contained no volcanics. Importantly, they showed that the floor of the geosyncline was trough-like, while the surface was a uniform slope.

Despite a certain similarity between this modern continental margin and some ancient geosynclinal basins, Drake and his co-authors could offer no plausible mechanism by which either the continental shelf or rise might develop initially. They were frankly skeptical that sediment loading could produce such troughs; nor could they suggest reasonable processes by which continental margins like those along eastern North America might eventually be deformed into an alpine orogenic belt. Particularly perplexing was the process by which the relatively thin, presumably oceanic crust beneath the outer continental slope and rise area (the *eugeosynclinal trough*) could be thickened and modified to produce a more continental-type crust. In a curious aside, the authors at the same time did recognize an overall similarity between alpine mountain systems and island arcs, emphasizing that some belts (the West Indies and Marianas) had no adjacent shallow-water *miogeosynclinal belts*, whereas other belts (southern Japan, the Ryukyu Islands, and easternmost Aleutians) included such areas of shallow marine sediment.

In 1960 and 1961, H. H. Hess discussed the essentials of his sea-floor spreading concept with numerous friends and colleagues. Preliminary drafts of what was to become the paper entitled "History of Ocean Basins" (Paper 23) were widely circulated. However, shortly before publication of this Hess manuscript, R. S. Dietz described an identical process, coining for it the name *sea-floor spreading* (Paper 22). (Dietz would later acknowledge that Hess formulated the concept, although both Hess and Dietz were indebted to Arthur Holmes.) In his short but classical piece, Dietz suggested that the concepts of the Moho, sialic crust and simatic crust were not genetically meaningful in terms for explaining the occurrence and tectonic consequences of mantle convection. With Holmes and Barrell, he assumed the existence of a rigid outer rind, the lithosphere, which overlies a plastic asthenosphere. A rising convection cell under the mid-ocean ridge created new simatic oceanic crust, separating the continental slabs like passengers on a conveyor belt. Hess's paper subsequently appeared in 1962. Papers 22 and 23 represent the first formal presentations of modern sea-floor spreading.

Curiously neither Hess nor Dietz used the term geosyncline at this time. Both recognized that ancient mobile belts would be fragmented by the opening of ocean basins and that the leading edges of converging (and descending) crustal segments, whether surficially oceanic or continental in character, would be subjected to downbuckling (producing trenches), compressional deformation, and metamorphism. Hess stated that sediment riding down the descending limb of a convection belt would probably he welded to the adjacent continental block. Dietz, following Suess (1885–1909) and Daly (1926), characterized continental margins as either tectonically stable (Atlantic type) or unstable (Pacific type).

In 1963, Dietz applied the sea-floor spreading idea directly to ancient geosynclinal belts in two important papers, included here as Papers 24 and 25. He proposed a truly actualistic model for geosynclines that differed dramatically from earlier actualistic models such as those of Eardley (Paper 16) and Hess (Paper 20). Dietz accepted the earlier suggestions of Drake, Ewing, and Sutton (Paper 21) that modern continental terrace (shelf) sedimentary prisms are equivalent to ancient miogeosynclines and the subjacent wedges of continental rise sediment are equivalent to ancient eugeosynclines. (The terms *miogeocline* and *eugeocline* were subsequently suggested as more appropriate by Dietz and Holden in 1966 because the thick sedimentary accumulations along such continental margins were deposited not in a linked pair of subsiding, synclinal troughs, but rather as wedge-shaped sequences of inclined sediment developing along foundering continental margins. The terms however, have not caught on.)

According to Dietz (Paper 25), orthogeosynclines such as those in North America (that is, situated around the periphery of a continent), originated as a result of normal subsidence along continental margins. He specifically excluded intracontinental, *collision* belts from his model, because to him such belts (Alps, Himalayas) were genetically different. Dietz believed that new

Atlantic-type continental margins were produced wherever sea-floor spreading rifted older continental blocks apart. During subsequent spreading, deposition of the thick wedges of shallow shelf (miogeosynclinal) and deep rise (eugeosynclinal) wedges of sediment produced the classic *geosynclinal prism*. Continued spreading eventually produced decoupling of the sial from the adjacent simatic ocean floor. The downsliding and slippage of dense sima toward and under the adjacent sial led to compression, folding, and thrusting beneath the continental rise and terrace deposits on the margin of the continental craton. Ultimately this underthrusting produced a deformed geosynclinal belt and continental growth by lateral accretion.

In point of fact, as Dott (Paper 19) would later point out, the Atlantic and Gulf Coast sediment prisms differ in many respects from both the conventional orthogeosynclinal belts (paired miogeosynclines and eugeosynclines) as originally defined by Stille and Kay, and most other early actualistic models. For example, little if any volcanism contemporaneous with the deposition of abyssal plain muds and turbidites characterizes modern Atlantic-type continental rise and slope areas. Similarly, the extensive tracts of uplifted local source areas (the *cordilleras*, or Kay's tectonic lands and volcanic arcs) that could be cannibalistically eroded to provide eugeosynclinal sediment have no modern analogues along the present Atlantic margin, yet their existence along the Appalachian belt was established. Kay had in fact already characterized the Atlantic and Gulf Coast sediment prisms as paraliageosynclines rather that orthogeosynclines. It seems curious that the authors of the Atlantic orthogeosynclinal concept (Drake, Ewing, and Sutton), although departmental colleagues of Kay, appeared to have little idea of the original definitions of miogeosynclines and eugeosynclines. Dott, in contrast, was a student of Kay.

In retrospect, it is obvious that Dietz was yet another victim of the rigid, fixist thinking which afflicted most geologists at that time. As sea-floor spreading and plate tectonics theory would soon demonstrate, ancient geosynclinal belts and modern continental margins both represent markedly different things in particular belts. Any orogenic belt or continental margin as it now exists may drastically differ in appearance from its past character. Despite such shortcomings, Dietz nevertheless focused the attention of the geological community on the new sea-floor spreading concept, on the possible relationship between ancient geosynclines and modern continental margins, and on the potential relationship between sea-floor spreading and geosynclinal evolution. Ancient conceptual models were being confronted by the new geological realities evident in the ocean basins.

J. Tuzo Wilson's landmark publication (Paper 26) introduced the essential elements of plate theory, or the new global tectonics. Mobile belts, the mid-ocean ridge system, ocean trenches, and island arcs could all be understood and interrelated if the earth's surface was regarded as being subdivided into a small number of shifting, internally rigid plates (Morgan, 1968; LePichon, 1968). Lateral shearing, separation, and collision of adjacent plates produced modern seismic and volcanic activity and through time would generate the major first-order features of the crust: continental blocks, ocean basins, and mobile belts (Isacks, Oliver, and Sykes, 1968; McKenzie and Parker, 1967). Wilson clearly had recognized the relationship between sea-floor spreading and continental drift. The latter term, in its Wegenerian sense, was inappropriate, because Wilson recognized that continents did not actually drift separately and distinctly from the sima to which they were firmly attached. Continents now drifted as passengers on (moving) lithospheric plates. The upper mantle was now seen as part of the lithosphere, while the asthenoshpere now took over the role that Wegener had wrongly attributed to the simatic crust. This revolution in the earth sciences would soon accelerate as geologists considered the implications of the new global tectonics for ancient mobile belts and continental margins. (Dott, 1978)

REFERENCES

Daly, R. G., 1926, *Our Mobile Earth*, C. Scribner's Sons, New York, 342 p.

Dietz, R. S., and J. C. Holden, 1966, Miogeoclines (Miogeosynclines) in Space and Time, *Jour. Geology* **74**:566–583.

Dott, R. H., Jr., 1978, Tectonics and Sedimentation a Century Later, *Earth-Sci. Rev.* **14**:35–63.

Haarmann, E., 1926, Tektogenese oder Gefugebildung statt Orogenese oder Gebirgsbildung, *Deutsch. Geol. Gesell. Zeitschr.* **78**:105–107.

Hess, H. H., 1938, Gravity Anomalies and Island Arc Structure with Particular Reference to the West Indies, *Am. Philos. Soc. Proc.* **79**:71–96.

Isacks, B., J. Oliver, and L. R. Sykes, 1968, Seismology and the New Global Tectonics, *Jour. Geophys. Research* **73**:5855–5899.

Kay, G. M., 1951, North American Geosynclines, *Geol. Soc. America Memoir* **48**:1–143.

LePichon, X., 1968, Sea Floor Spreading and Continental Drift, *Jour. Geophys. Research* **73**:3661–3697.

McKenzie, D. P., and R. L. Parker, 1967, The North Pacific: An Example of Tectonics on a Sphere, *Nature* **216**:1276–1280.

Morgan, W. J., 1968, Rises, Trenches, Great Faults, and Crustal Blocks, *Jour. Geophys. Research* **73**:1959–1982.

Suess, E., 1885–1909, *Das Anlitz der Erde*, 3 vols., Tempsky, Wien. (English translation: *The Face of the Earth,* 1904–1924, 5 vols., Clarendon Press, Oxford.)

20

Reprinted from pages 391–392, 400–401, 404–405, and 407 of *Geol. Soc. America Spec. Paper* **62**:391–407 (1955)

Serpentines, Orogeny, and Epeirogeny

H. H. Hess

ABSTRACT

Steinmann recognized the association of serpentinized peridotites, radiolarian cherts, and spilitic lavas 50 years ago in the Alps. Benson, 30 years ago, emphasized the world-wide nature of serpentines associated with alpine-type mountains. Nearly 20 years ago the present writer pointed out the relationship of serpentines to island arcs, adding weight to the hypothesis that island arcs represent an early stage in alpine-type mountain building.

Serpentinized peridotites are probably intruded during the first great deformation of a mountain belt and do not recur in subsequent deformations of the same belt. They are typically found in two belts about 120 miles apart, one on either side of the central axis of most intense deformation, but may also occur irregularly through this zone. This enables one to date orogenies by dating the serpentines and to follow the axis of an ancient orogenic belt in some cases for thousands of miles. On this basis it was pointed out in 1937 that the great orogeny in the Appalachians was in the Ordovician, not at the end of the Paleozoic, and the Caledonian Revolution in Scotland was of the same age, and not Silurian. These views are now widely accepted. The concept that geosynclines (long narrow troughs containing a thick section of clastic sediments, commonly of shallow-water origin) localize mountain building was challenged on the basis that such a feature is not present in island arcs before the first deformation, but normally develops later because of that deformation. This idea has met with strong resistance, but the writer maintains his original stand.

For most of this century a magnificent argument has gone on between field geologists who have worked on the peridotites of alpine mountains and laboratory investigators of their chemistry (particularly Bowen). The former stoutly maintains that the evidence indicates that they were intruded in a fluid state, as magmas; and the latter equally forcefully has proved to his own satisfaction that such magmas are not possible. The writer casts his vote with the field geologist and believes that the field evidence takes precedence. Probably there is some factor or constituent missing in the laboratory investigations. In recent years the idea that the serpentines were emplaced as solids has gained much favor. Some are unquestionably so emplaced, but many cannot be (*see* Hiessleitner on Balkan occurrences), so that this concept fails as a general explanation. At the other extreme we find Bailey and McCallien suggesting that the serpentines of Turkey are submarine lava flows.

In the last few years it has been demonstrated that peridotitic material occurs at shallow depth beneath the oceans (10–12 km). Placing it at such shallow depth beneath island arcs at the time of their initial deformation makes things appear much easier for the solid intrusionists. Were it not for the occurrence of concordant sills in the relatively little deformed flanks of the orogenic zone and the lack of an internal fabric suggesting solid flow, the hypothesis would be an attractive one. That peridotites occur on fault scarps on the Mid-Atlantic Ridge does not indicate that this feature is to be interpreted as an alpine type of mountain system. In this case the faults have merely brought the peridotitic substratum high enough to be exposed.

In the oldest rocks serpentinized peridotites do not occur in belts but are ubiquitous throughout the whole terrane. The best example of this is MacGregor's description of the Sebakwian of Southern Rhodesia, but it also seems to hold for the oldest rocks of the Canadian Shield. These rocks seem to represent something similar to the present oceanic crust strongly deformed.

Finally it is suggested that many features of suboceanic topography might be the result of uplift caused by serpentinization of peridotite below the Mohorovičić discontinuity brought about by water leaking from the interior of the earth. The reaction is reversible inasmuch as increase in temperature could cause deserpentinization. This uplift and subsequent subsidence could be accounted for by this reaction.

INTRODUCTION

This would have been a better paper had Columbia University had the foresight to have been born 20 years earlier or perhaps 10 years later. The facts can be summarized, and some conclusions drawn concerning the serpentines of alpine mountain systems. But satisfying general conclusions which would account in one all-embracing theory for their origin, mode of emplacement, and subsequent serpentinization are not forthcomng at this time.

The writer summarized the problem in a paper presented before the International Geological Congress in Russia in 1937. Today the ages of many of the serpentine belts shown in the illustrations need to be modified in light of further geological evidence. Particularly those shown as Eocene are all older, and most of them are Late Cretaceous. To make another comprehensive review of the world literature would take several years, so the complete task of revising the ages of the belts and adding newly discovered occurrences will not be undertaken here. Ideas on the nature of the crust under the oceans have been modified by recent seismic exploration at sea eliminating the granitic crust entirely and leaving only a few kilometers of basalt resting on the presumably peridotitic rocks below the Mohorovičić discontinuity. These new developments do not materially change the tectogene hypothesis derived from Vening Meinesz's earlier work. The structure remains the same, but the mass distributions have to be modified. The writer's hypothesis that the serpentines were emplaced as a hydrous peridotite magma is contradicted by laboratory experimental results of Bowen and Tuttle. Though several hypotheses consistent with the laboratory evidence have been advanced to account for emplacement of the ultramafics of alpine systems, none of these is really satisfactory when tested against the field evidence.

[*Editor's Note:* Material has been omitted at this point.]

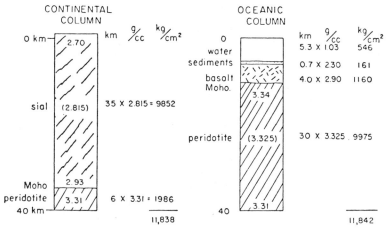

FIGURE 5.—*Crustal columns deduced from seismic evidence, petrological inference, and the assumption of isostatic equilibrium*

GEOSYNCLINES AND MOUNTAIN BUILDING

The original concept of a geosyncline (but not the name) dates back to an address by James Hall a century ago (1859). Hall had in mind the Appalachian geosyncline, a great elongate trough filled by a thick sequence of clastic sediments dominantly of shallow-water origin. It would be best to restrict the term to features close to the original concept. On this basis the Puerto Rico trench or Hudson's Bay for example are not today geosynclines. Hall also introduced the idea that mere existence of such a thick sequence of sediments in the geosyncline would inevitably cause it to become a folded mountain system. J. D. Dana (1873a; 1873b) was critical of this hypothesis and presented arguments against it. Nevertheless the concept was a hardy one and has persisted in one form or another over the generations to the present time. Thus one often sees statements that the present Gulf Coast geosyncline will be a future site of mountain building. I doubt if there is a real basis for such a supposition. The misconception can be correlated with a misconception as to the sequence of events in the Appalachian system. If one looks at these mountains with the false impression that the Appalachian Revolution at the end of the Paleozoic was the main event which brought them into being then the conclusion to which I am objecting follows. The great event in the Appalachians, however, was the Taconic Revolution in Mid-

Ordovician time. The trough which became the geosyncline came into being as a structure at this time, and the clastic sediments in huge amounts which subsequently filled it were derived from the metamorphic and igneous rocks of the present Piedmont belt which after the Taconic became an alpine mountain system. Most geosynclines form in troughs paralleling preceding alpine-type mountains and are filled by the debris from the same. It is certainly the fate of most such geosynclines to become folded mountains. But they were a consequence of preceding deformation, not themselves the agents which localized the initiation of the mountain building.

Geosynclines of the Gulf Coast type seem to be comparatively rare. If one looks back over the geologic record to locate one which subsequently became a mountain system, it is hard to find one. The great thickness of sediment in the Belt geosyncline may be a representative of the Gulf Coast type. Did it localize mountain building? Between the end of the Proterozoic and the end of the Cretaceous there were five or six episodes of strong deformation in North America, but none of them until the Laramide paid any attention to the Belt geosyncline. When finally this region was deformed in the Laramide it seems to have been pure chance that the orogenic movements transected the Belt.

If one agrees that island arcs are the initial stage of alpine mountain building, it is fairly obvious that no present-day arcs have developed over a geosyncline. Instead they seem to form on the ocean floor with little sedimentary cover though occasionally they may penetrate into continental slopes and shelves where a moderate thickness of sediment was present.

[*Editor's Note:* Material has been omitted at this point.]

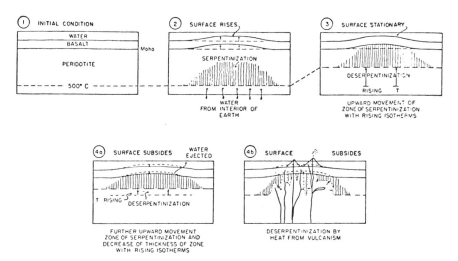

FIGURE 6.—*Consequences of the hypothesis of sub-Mohorovičić serpentinization by water rising from the mantle, deserpentinization by rising isotherms, and the consequent effects on the submarine topography*

APPLICATIONS TO OCEANIC FEATURES

Mid-Atlantic Ridge.—In general the Mid-Atlantic Ridge stands 2–3 km above the sea floor for about 10,000 km in length and an average width of perhaps 500 km. Until recently most geologists would have supposed that the Mid-Atlantic Ridge was either a folded mountain system with consequent thickening of the crust above the Mohorovičić discontinuity or alternatively that it was a thick section of volcanic material lying on normal oceanic crust or intruded into it. Now it seems more likely that it represents a welt of serpentine. Serpentinized peridotite was dredged from large fault scarps on the Ridge by Ewing *et al.* (Shand, 1949). The flanks of the Ridge once stood much higher than today as indicated by paired terraces along its east and west slopes. Presumably some deserpentinization has occurred since maximum serpentinization. The problem of why serpentinization was concentrated in the Atlantic along a median line can perhaps be explained in several ways. One hypothesis could be that convective circulation in the mantle occurs, and the ridge represents the trace of an upward limb of a cell. In this case water ejected from the top of the column might cause the serpentinization and heat which would move upward much more slowly by conduction might cause the later deserpentinization. Whatever hypothesis may be suggested to account for the localization of serpentinization along the Ridge is at present pure speculation. The idea that the topographic elevation of the Ridge may be due to serpentinization should be considered on its own merits apart from the above speculation. If the Ridge is so formed a huge amount of water would be involved. Serpentine necessary to sustain the Ridge would contain a volume of water perhaps 1 per cent as large as the oceans.

[*Editor's Note:* In the original, material follows this excerpt. Only the references cited in the preceding excerpt are reproduced here.]

REFERENCES CITED

Dana, J. D. (1873a), On the origin of mountains: Am. Jour. Sci., 3rd. ser., v. 5, p. 347–350.

———— (1873b), On some results of the Earth's contraction from cooling, including a discussion of the origin of mountains and the nature of the Earth's interior: Am. Jour. Sci., 3rd. ser., v. 5, p. 423–443.

Hall, J. (1859), Introduction, Natural History of New York: Geol. Surv. New York, pt. 6, Paleont., v. 3, p. 1–96.

Shand, S. J. (1949), Rocks of the Mid-Atlantic Ridge: Jour. Geol., v. 57, p. 89–92.

21

Reprinted from pages 110–111 and 176–184 of *Physics and Chemistry of the Earth*, vol. 3, Pergamon, London, 1959, pp. 110–198

CONTINENTAL MARGINS AND GEOSYNCLINES: THE EAST COAST OF NORTH AMERICA NORTH OF CAPE HATTERAS

By C. L. Drake, M. Ewing and G. H. Sutton

Abstract—Many geophysical measurements, including seismic refraction, gravity, magnetics, and echo soundings, have been made along the continental margin of eastern North America north of Cape Hatteras in the last twenty years. These have revealed the presence of two sedimentary troughs, one under the shelf, the other in deeper water under the continental slope and rise which are separated by a ridge in the basement near the edge of the shelf. The sediments in the shelf trough have been drilled to a depth of 10,000 ft and are of shallow water character. Cores of the upper part of the sediments of the outer trough have revealed features attributed to slumping, sliding, and turbidity current action, and, in part, sediments similar to the graywackes of PETTIJOHN's (1949) classification.

This sedimentary system is quite comparable to the Appalachian system as restored for early Paleozoic time. The sediments of the inner and outer troughs are similar to those of the Appalachian miogeosyncline and eugeosyncline (KAY, 1951), respectively, and the basement ridge resembles the Pre-Cambrian axis which separates these two troughs in the Appalachians. While there is no active volcanism in the outer (eugeosynclinal) trough at the present time, evidence of past volcanism is present in the form of partially buried seamounts with large magnetic anomalies. Conditions in the Appalachian eugeosyncline appear to have been similar with but limited volcanism prior to the beginnings of Taconic activity.

The gravity calculations reveal an abrupt change in depth of the Mohorovicic discontinuity near the edge of the shelf. There is some indication that the boundary between the crust and the mantle in this area is gradational rather than a sharp discontinuity.

The major process necessary to convert the present continental margin into a mountain system is the one which thickens the crust under the outer, or eugeosynclinal trough. Since the miogeosyncline is already based on a crust of continental proportions, its deformation requires only a means of folding and thrusting the surficial sediments. This is but a minor part of the overall tectonic activity.

INTRODUCTION

MUCH of the geologic nomenclature is inexact and a particular term may have different meaning or a different shading of meaning for different persons. This is due in part to original definitions which have proved to be too generalized, and in part to distortions of the original meaning. In order that difficulties may not arise from this source, it is worthwhile, at the outset, to state the manner in which some of these terms are used in this paper.

The term "geosyncline", as used here, refers to the orthogeosynclines of STILLE's (1936, 1941) classification. It implies a linear or arcuate belt marginal to a continental shield or craton.

"Orthogeosynclines were classified as eugeosynclines ('straight' or 'real' geosynclines) and miogeosynclines ('lesser geosynclines') by STILLE (1941, p. 15), who used the term quite informally, but clearly implied a volcanic basis of distinction. An eugeosyncline is a surface that has subsided deeply in a belt having active volcanism, a miogeosyncline in a belt lacking active volcanism." (KAY, 1951, p. 4.)

References to the Appalachian system include both the metamorphic and plutonic belt as well as the folded and thrust-faulted belt to the west. The former is termed the Appalachian eugeosyncline, the latter the Appalachian miogeosyncline following KAY (1951).

In identifying certain of the sediments as "graywackes", the definition given by PETTIJOHN (1949, p. 942–943) is used. The term has been defined in many ways and applied to sediments of different descriptions but the above definition is best suited for the discussions in this paper.

The terminology used for the seismically determined layers within the earth is explained in some detail in a later section. The sedimentary layers are in general discussed as a unit, since their relation to the stratigraphy is not constant. Crustal rocks are those above the Mohorovicic discontinuity and the region below this discontinuity is referred to as the mantle. This is the nomenclature usually used in seismic investigations although "crust" and "mantle" have been used differently in other work. The continental crustal rocks are referred to as "basement", in many cases, for the sake of brevity. This is not in conflict with the general geological use of this term.

The discussion of the seismic results uses the terminology of the other papers dealing with the emerged and submerged Atlantic coastal plain. Seismic *station* refers to the position of a recording vessel, seismic *profile* refers to the data along the line of shots and may include two *stations* if the profile is reversed, and seismic *section* refers to the geologic structure inferred from the data from a single profile or many profiles along a line.

References to "velocity" within any layer indicate compressional wave velocity. The discussion of velocities deals with the compressional velocities except in cases which state that shear wave velocities are being considered. The velocities indicated on the travel time curves are in kilometres per second while the abscissae represent direct water wave travel times in seconds.

Lastly, the descriptions of the basement rocks often have little petrological significance. "Granite" is an especially abused term and may refer to a variety of metamorphic rocks or even, in some cases, to the point at which drilling of a well becomes very difficult.

[*Editor's Note:* Material has been omitted at this point.]

STRUCTURE SECTIONS AND ISOPACHS

The depth and thickness data are presented in the form of structure sections (Fig. 35) and as an isopach map of total sediment thickness (Fig. 29). The sedimentary nomenclature is that outlined previously and may have no stratigraphic significance.

All of the structure sections show the same general features. Starting from the shore, the coastal plain sediments thicken to form a trough from 3 to 6 km in depth. This trough is bounded, near the edge of the continental shelf by a ridge in the basement which reduces the sediment thickness to between 1·5 to 2·5 km. On the ocean side of this is another trough containing up to 8 km of sediment. The sediment thins seaward from this trough until it reaches the average thickness of 1 km found in the ocean basin. The development of these features varies from section to section but the overall relationships are the same. The isopachs of Fig. 29 show these features in a very striking manner. These were not extended to the Newfoundland section due to lack of data between this section and that off Halifax.

Beneath the sediments the basement rocks thin and eventually pinch out under the ocean basin. There appears to be considerable variation in extent. Under the Newfoundland profile these rocks extend some 200 miles beyond the edge of the shelf, while they are limited to about 50 miles under the Matinicus–Georges Bank section. Near centres of submarine volcanism the picture becomes complicated since the volcanics have velocities in the same range as those of the basement rocks off the edge of the shelf.

There is an indication that the oceanic crustal layer rises somewhat near the edge of the continent before dipping down and possibly merging with the continental material. This rise is shown best on the Halifax and Portland sections and the four profiles between the Portland and Woods Hole sections and is suggested on some of the other sections. A similar rise has been observed off the coast of Brazil (Warren and Ewing, 1958) and Argentina (Ludwig, personal communication). In general, there is a thinning of the continental basement rocks where this rise occurs with the locus of the thinnest part near the axis of the outer trough mentioned earlier.

The depth to the Mohorovicic discontinuity was measured in a few places, but there are insufficient seismic data to allow conclusions to be drawn concerning its behaviour in the transition zone between oceanic and continental areas. It is well known that this discontinuity is at a depth of about 10 km beneath the surface of the oceans while under the continents it is three or four times as deep. The few measurements of its depth in this area show it to be dipping down toward the continent.

COMPARISON WITH APPALACHIAN SYSTEM

Kay (1951) divided the Appalachian orthogeosyncline into two major parts: an inner (shoreward) trough, termed a miogeosyncline, notable for its lack of volcanics, and an outer trough, the eugeosyncline, which is characterized by intermittent volcanic activity. The miogeosynclinal belt is typically folded and thrust-faulted while the major igneous and metamorphic activity took place in the eugeosynclinal belt. These two are separated by a geanticlinal barrier of some description which serves to divide them sharply and to confine the effects of igneous activity principally to the outer, eugeosynclinal, trough.

The configuration of the region between the Fall line and the continental rise shows a great resemblance to that of the Appalachian system as it was presumed to be during the early stages of its development (Fig. 30). Here, as in the Appalachians, there are two troughs separated by a basement ridge. The outer trough may be compared with the Appalachian eugeosyncline, the inner with the miogeosyncline. It should be noted that the total thickness of material of sedimentary character in the outer trough may be greater than indicated on the isopachous sections of Fig. 30. The velocities in the continental crustal layer of this area are in the range typical of consolidated sediments and all or part of this layer might be included in the total sediment thickness.

Eugeosyncline

If we are to identify the outer trough with eugeosynclines, we should find signs of volcanic activity. While there is no activity at the present time, the magnetic results have revealed evidence of volcanism in the past. The geologic information from deep wells south of the area under consideration supports the notion that the basement rocks under the shelf, slope, and rise may be composed to a great extent of volcanic and pyroclastic material. The seismic velocities determined for these rocks are in good agreement with those of the volcanic slates of the Piedmont and volcanic rocks in the vicinity of Bermuda and the Mid-Atlantic Ridge. Furthermore, the velocity contours parallel the structure so well that it is logical to expect that conditions are similar along the strike of the system.

It would appear, then, that while the sediments lie on a volcanic terrain, there is a possibility of only limited volcanism, represented by the offshore seamounts, during Mesozoic or Tertiary time. Although this seems at first to conflict with the evidence from the Appalachian region, a closer examination of the data does not reveal any basic inconsistencies.

KAY (1951, pp. 49–57) gives examples of volcanics in the Appalachian eugeosyncline which range from Cambrian through Pennsylvanian in age. The dating is often doubtful and the age limits are wide in many cases. The Cambrian examples are particularly questionable. Kay reports that he is:

". . . not familiar with any volcanic rock within a section having Cambrian fossils above and below it. Some of the volcanics are Lower Cambrian because they are within sedimentary sequences conformably underlying fossiliferous Cambrian; others are considered to be Cambrian because they lie below Ordovician, and still others are correlated on stratigraphic and structural grounds." (1951, p. 52.)

Recent investigations in Cape Breton, Nova Scotia (HUTCHINSON, 1952) have revealed the presence of Middle Cambrian volcanics with Cambrian fossils both above and below, but this appears to be an isolated case. The dating is complicated by uncertain structural relationships, lack of fossils, prejudice, and disagreement as to the location of the base of the Cambrian. It would seem, nevertheless, that there are few volcanic outcrops of unquestionable Cambrian age, and that most of these are limited to the Lower Cambrian. This leaves a long period of time, at least a good fraction of the Cambrian for which little positive evidence of active volcanism has been obtained.

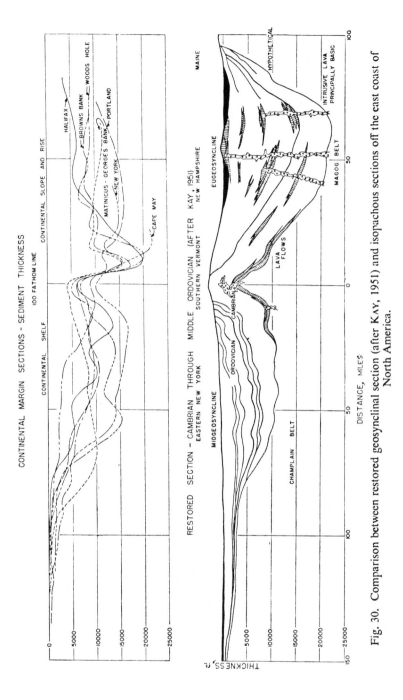

Fig. 30. Comparison between restored geosynclinal section (after Kay, 1951) and isopachous sections off the east coast of North America.

173

Activity was much more extensive in the Ordovician, particularly in the middle and upper parts (KAY, 1951, pp. 52–53), hence comparisons with the present continental margin must be restricted to the Cambrian or, perhaps, extended into the lower Ordovician. Within these limits, if we allow that some of the outcrops identified as Cambrian are, in reality, pre-Cambrian, there are no essential differences in quantity or distribution of volcanic activity between the two areas.

Placing the eugeosyncline in relatively deep water would serve to localize the effects of the volcanic activity. Although the miogeosyncline is immediately adjacent, it should receive no important amount of volcanic material if the activity is submarine. Furthermore, although the mechanism by which extrusive material is brought to the surface is poorly understood, it would logically seek the easiest path. The relatively abrupt transition zone between continent and ocean might be expected to be the region of maximum instability and hence favourable for igneous activity.

The association of greenstones and spilitic volcanics with the early geosynclinal sediments has been remarked by many observers (TURNER and VERHOOGEN, 1951, pp. 201–212; DEWEY and FLETT, 1911, pp. 202–209; BLOOMER and WERNER, 1955, pp. 589–594; KNOPF, 1948, pp. 649–670; PARK, 1946, pp. 305–323; TALIAFERO, 1943, pp. 144–147; and others) and the problems connected with the origin of spilites have been outlined by GILLULY (1935, pp. 234–244). The association of these rocks with radiolarian cherts and the frequently found pillow structure of the volcanics support a submarine origin. While the secondary origin of the albite is not universally accepted, the suggested deep-water environment of emplacement might favour such a secondary origin.

If eugeosynclinal sediments of older systems were deposited in an environment similar to that of the outer trough, there should be evidence of this in the geologic record. Most of the information regarding the sedimentary processes which affect the continental slope and rise is of recent origin. The high degree of metamorphism and the confusion caused by emplacement of igneous bodies and volcanics have made it difficult to determine the sedimentary character in eugeosynclinal belts except in more or less limited areas. The problem is made more complex in the Piedmont area of the southern Appalachians by the deep weathering.

In spite of the difficulties, it is becoming apparent that sediments from this type of environment are more widespread than has been assumed in the past. BAILEY (1930, pp. 86–88) recognized the deep water character of graywackes although it was not clear at that time how the sands were emplaced. PETTIJOHN (1943, pp. 925–972) noted that crossbedding and ripple marks were scarce in Archean graywackes and that graded bedding was common. A mechanism for the formation of these features was presented by KUENEN and MIGLIORINI (1950, pp. 91–127) in the form of turbidity currents. The characteristics to be expected of sediments subjected to these currents were outlined in some detail by KUENEN (1953, pp. 1045–1047). Studies of modern sediments in submarine cores from the continental slope and rise have shown that the graded bedding commonly found may be attributed to the transporting characteristics and the intermittent nature of turbidity currents (ERICSON et al., 1951, 1952; HEEZEN and EWING, 1952; EWING and HEEZEN, 1955; HEEZEN et al., 1954). In addition it has been shown that coarse, angular clastics can be deposited near the base of the continental slope, that shallow water fauna can be carried down without apparent damage, that breccias of soft clay and limestone can form, and that such minerals as glauconite, calcite, and feldspars may be present.

PETTIJOHN (1950) attributed the apparent lack of modern graywackes to the fact that they are being formed by turbidity currents in deep water and pointed out that the features attributed to turbidity current deposition were those found in graywackes from Archean to Tertiary in age. Similar features have been found by others. BAILEY (1936) noted the occurrence of graded bedding of many ages:

". . . . the younger Dalradian (probably late Pre-Cambrian), Cambrian, Ordovician, and Silurian of Scotland, Ireland, and Wales within the compass of the Caldonian geosyncline . . . it characterizes some of the Amorican geosyncline of the Devon Peninsula and . . . the Tertiary Flysch of the Alpine geosyncline."

The absence of ripple marks and cross bedding in the Franciscan (Jurassic) of California and the even bedding have been noted by DAVIS (1918, pp. 38–39). Sands are interbedded with radiolarian cherts, foraminiferal limestones, and shales (TALIAFERO, 1943, p. 132). KUENEN and MIGLIORINI (1950, pp. 109–113) believe that the Macigno and the brecciolas of the northern Apennines have been deposited by turbidity currents, the differences between the two being attributed to higher velocity currents in the latter case. KUENEN gives further examples from Wales and from California (1953, pp. 1060–1065), and from Germany (1956, pp. 649–671).

RICH (1950, pp. 717–741) has examined the Silurian rocks of Wales and finds evidence that they were deposited in a continental slope environment. The deposits are marked by a scarcity of fossils, evenness of bedding, flow markings, few ripple marks and only small-scale, vague cross-bedding. RICH (*op. cit.*, p. 739) has found that, within his experience, it is:

". . . difficult to find rocks consisting of thin evenly bedded alternations of siltstone and shale (or their calcareous equivalents) which do not show these flowage phenomena."

JONES (1955, p. 330) describes the Aberystwyth grit group near Cwmystwyth in Wales as a 4000 ft series of coarse and fine sediments alternating every foot or two. He further notes the presence of a slump structures of many types in the rocks of this area. In an earlier paper (1937) he mentions heavy shelled shelf fossils carried down by slumping.

Pettijohn (1949, pp. 254–255) notes that graywackes of his definition are the most typical sediments of some orogenic belts and do not occur in any important manner outside of this environment.

". . . The early pre-Cambrian of Ontario (excluding the Grenville), commonly called Timiskaming, is 90 per cent graywacke. Such apparently is also true of the Pre-Karelidic deposits of Finland. . . . Some parts of the Huronian also are marked by thick graywackes and associated slates. . . . The graywacke suite is typical of the Lower Paleozoic of Great Britain in the Caledonian geosyncline of Wales, the Lake District of England, and the Southern Highlands of Scotland, where it is associated with pillow lavas and cherts. Graywackes occur in the Devonian of south-western Cornwall, in the Amorican geosyncline, and in the Devonian and Lower Carboniferous of the Hartz Mountains of Germany. The Ordovician of the Green Mountain belt in Vermont and the Upper Devonian of Eastern Pennsylvania and New York (Rensselaer grit) contain important graywacke beds. The sandstones of the Stanley shale and the Jackfork (Carboniferous) of the Ouachitas of Oklahoma and Arkansas are graywackes."

It should be noted that the Rensselaer is regarded as Lower Cambrian and the equivalent of the Rowe schist to the east by PRINDLE and KNOPF (1932, p. 284).

ESPENSHADE (1937) described similar rocks in the Badger Bay (Ordovician) series of Newfoundland as

". . . graded bedding, fine conglomerate or coarse sandstone passing into finer-grained sandstone through an interval of several feet. This rhythmical or cyclical bedding occurs in the sandstones, shales, and graywackes throughout the Badger Bay series."

KING (1949, p. 633) notes in the southern Appalachians:

"The rocks of the lower part of the Ocoee series . . . are of graywacke facies and strongly resemble the graywackes of the Archean of the Canadian Shield as described by Pettijohn . . .

"In the graywackes of the Ocoee series, as in others, graded bedding is a striking feature, which suggests that the sediments were deposited rapidly, without sorting except that which was produced by settling in water. This is manifested by cyclical arrangement of the coarse and fine deposits, each cycle beginning with granular or pebbly deposits, followed by sandy and then silty or slaty deposits, the whole cycle being represented by beds 1 ft to more than 25 ft thick, and endlessly repeated through thousands of feet of section. In many places the slaty or silty beds at the tops of the layers are broken up and reincorporated as chips or slabs in the basal deposits of the succeeding unit. Ripple marks and cross-bedding are rare."

There are undoubtedly further examples of deposits of this sort to be found in the literature of orogenic belts. It is important to be selective since the term graywacke has various definitions. Graywacke is used in the sense defined by Pettijohn in this paper (see PETTIJOHN, 1943, pp. 942–943).

Further information regarding the depositional environment of eugeosynclinal belts has been presented by BUCHER (1953, p. 275–300). He listed occurrences of fossils in metamorphic rocks and reached the conclusion that:

". . . the total number of fossils recorded after 150 years of search is too disproportionately small to allow any other conclusion, but that the larger types of fossils were rare in the original sediment . . ."

and further that

". . . it pays to emphasize the conclusion that must be drawn from the habitual scarcity of benthonic fossils, and of fossils in general, in geosynclinal sediments. A careful analysis, for which this is not the place, leaves little doubt that, of all possible factors, considerable depth of water is the one that will reduce the benthonic fauna drastically, no matter what sediment accumulates on the sea floor."

Sediment cores taken off the edge of the continent reveal remarkably few fauna with the exception of microfossils such as foraminifera. Occasional megafossils have been reported but these are primarily of shallow-water origin and have been carried down by turbidity currents or sliding. All evidence from the bottom sediments in deep water off the east coast of North America supports Bucher's argument as to the nature of the environment.

Miogeosyncline

The sediments of the inner (miogeosynclinal) belt of the coastal plain prove to be similar in character to those of the folded Appalachians. Where exposed at the surface they are gently dipping, continental, clastic beds barren of glauconite, interbedded with glauconitic marine clastics and limestones. They are of shallow water marine or near-shore character and include lignite beds in parts. There is an increase in the calcareous content of the sediments to the south. This is also true of the sediments of the Appalachians.

There are a few deep wells in the coastal plain area which reach into the trough proper. The Hatteras Well No. 1 of the Standard Oil Company of New Jersey is one of these and shows a decrease in the percentage of non-marine sediment from that found in outcrops and from wells farther inland. Over basal clastics of possible Jurassic age are deposited Cretaceous to Recent limestones, shales, and calcareous sands with an abundant shallow-water fauna. Glauconite is common, especially in the upper part, and no lignite was noted (Swain, 1947, pp. 2054–2060). The log of the North Carolina Esso No. 2 shows similar lithologies. To the south, the increase in calcareous components becomes very great as illustrated by the wells described by Applin and Applin (1944, pp. 1673–1753) in Florida and adjacent states. It can be seen (*Ibid.* Figs. 22 and 23) that the clastics are more abundant to the north and to the west. It is also demonstrated that the clastic facies has been retreating and that it extended much farther on to the Florida peninsula in Cretaceous time than it does at the present time.

North of Cape Hatteras the deepest wells are west of the axis of the trough. The sediments have a larger non-marine fraction and contain a greater proportion of clastics. The Ohio Oil Company's Hammond No. 1 near Salisbury, Maryland, has lignite beds in both Miocene and Upper Cretaceous formations and a limited amount of limestone (Richards, 1945, p. 903). The major part of the section is made up of sands, clays, and gravels. Well No. 1927 at the United States Naval Receiving Station, Long Beach, Long Island, New York (Roberts and Brashears, 1946, p. 153), shows a section that is almost entirely clastic: sands, clays, and gravels, with occasional lignite beds and shells, but no limestone reported as such.

The early Paleozoic rocks of the Appalachian miogeosyncline are similar to the above. In north-west Newfoundland, Lower Cambrian arkosic and quartzose clastics are overlain by a great thickness of limestone and dolomite (King, 1951, p. 84; Schuchert and Dunbar, 1934). In the Champlain lowland of western Vermont, limestones and dolomites make up most of the sequence up to Middle Ordovician but there are thick quartzose clastics at the base and sandy beds higher up (Cady, 1945).

"Whatever their age, the basal Cambrian sediments are invariably terrigenous. . . . The earliest Cambrian of the west flank of the Green Mountains is coarse and conglomerate in the western belts of outcrop, finer to the east." (Kay, 1951, pp. 8–9).

In the southern Appalachians the sandstones and other clastics of the Chilhowee (Lower Cambrian) are the basal deposits. The higher Cambrian contains inter-bedded shales and sandstones but the dominating rocks are carbonates. The Upper Cambrian and Lower Ordovician dolomites and limestones of the Knox group reach 4000 ft in thickness (King, 1951, p. 121).

The principal difference between the early Paleozoic section of the Appalachian miogeosyncline and the sediments of the present trough in the coastal plain is the more widespread distribution of carbonate rocks in the former. This difference may, to some extent, only be apparent since well logs indicate a seaward increase in the percentage of limestone. Evidence seems to point to lower elevations along the coast in early Paleozoic time with a consequent decrease in the quantities of clastic material made available. The sediments of both areas have been identified as belonging to a shallow-water environment, faunas are plentiful, and the source of the sediments was to the west. The identification of the modern trough as a miogeosyncline seems most logical.

KAY (1951, pp. 82–83) terms the depositional areas offshore from the continents "paraliageosynclines" and does not believe that they should be considered miogeosynclinal.

". . . the paraliageosynclines are believed to differ in that they mark a new cycle of subsidence in a region that has passed through an orogenic and plutonic history after having been an eugeosynclinal belt of the same general trend. Moreover, the paraliageosynclines normally pass into ocean basins, whereas miogeosynclines normally are bordered by a belt of active deformation and volcanism. Paraliageosynclines are virtually miogeosyncline-like belts of subsidence having distinctive tectonic and chronologic positions."

The writers feel that the arguments in favour of an eugeosynclinal belt off the edge of the shelf answer the objections cited above in the case of the Atlantic coastal plain area.

Geanticlinal Barrier

The division between the two troughs of the Appalachian system is marked in most places by a ridge of Pre-Cambrian rocks. These rocks are exposed in many parts of this ridge in such areas as the Green Mountains, the Hudson Highlands, the Reading Prong, the Mine Ridge anticline, and many parts of the Blue Ridge Mountains.

As mentioned earlier in this paper, it is difficult to find Pre-Cambrian rocks outside of this rather limited region. The work of Cloos, Billings, Balk, King, and others has shown that many areas, formerly considered as Pre-Cambrian, are actually Paleozoic. The presence of additional areas of Pre-Cambrian rocks is suggested by high basement velocities in the Gulf of Maine and under parts of the continental shelf. These additional areas may represent the seaward margin of the Appalachian eugeosynclinal trough and possibly the flanks of additional troughs uplifted by post-taconic activity.

The sections of Figs. 35 and 30 show that there is a definite ridge in the basement which separates the two troughs of the continental margin of the eastern United States. This ridge is present on all of the sections but varies considerably in elevation and general configuration. A similar ridge is indicated by the seismic section north-east of Recife, Brazil, and by seismic observations made on the continental margin of Argentina. The topography of the continental shelf of eastern Labrador suggests the same feature. The central part of the shelf is depressed, reaching depths of 800 meters or greater, while the outer part is mostly less than 200 meters deep. The same type of topography is present off the coast of Greenland. While there is no assurance that the topography reflects basement structure, these coastlines are not of a type which may be expected to produce large quantities of sediment nor are marine organisms likely to produce large structures in such northern latitudes.

This ridge would be the major exposure of basement rocks following deformation, uplift, and erosion of the continental margin. There is a possibility of additional exposures on the outer margin of the continental rise trough (eugeosynclinal) due either to subsequent deformations or deeper erosion, but there should be none in or near the axis of this trough.

[*Editor's Note:* In the original, material follows this excerpt. Only the references cited in the preceding excerpt have been reproduced here.]

REFERENCES

Applin, P. L. and Applin, E. R. (1944) Regional subsurface stratigraphy and structure of Florida and southern Georgia. *Bull. Amer. Ass. Petrol. Geol.* **28**, 1673–1753.

Bailey, E. B. (1930) New light on sedimentation and tectonics. *Geol. Mag.* **67**, 77–92.

Bailey, E. B., (1936) Sedimentation in relation to tectonics. *Bull. Geol. Soc. Amer.* **47**, 1713–1726.

Bloomer, R. O. and Werner, H. J. (1955) Geology of the Blue Ridge region in central Virginia. *Bull. Geol. Soc. Amer.* **66**, 579–606.

Bucher, W. H. (1953) Fossils in metamorphic rocks: a review. *Bull. Geol. Soc. Amer.* **64**, 275–300.

Cady. W. M. (1945) Stratigraphy and structure of west-central Vermont. *Bull. Geol. Soc. Amer.* **56**, 515–558.

Dewey, H. and Flett, J. S. (1911) Some British pillow lavas and the rocks associated with them. *Geol. Mag.* **8**, 202–209.

Espenshade, G. H. (1937) Geology and mineral deposits of the Pilleys Island area (Newfoundland). *Newfoundland Dept. Natl. Res. Geol. Sec. Bull. 6*, 56pp.

Ewing, M. and Heezen, B. C. (1955) Puerto Rico topographic and geophysical data, pp. 255–268, *Geol. Soc. Amer. Special Paper 62*

Gilluly, J. (1935) Keratophyres of eastern Oregon and the spilite problem. *Amer. J. Sci.* **29**, 225–252, 336–352.

Heezen, B. C. and Ewing, M. (1952) Turbidity currents and submarine slumps, and the 1929 Grand Banks earthquake. *Amer. J. Sci.* **250**, 849–873.

Heezen, B. C., Ericson, D. B. and Ewing, M. (1954) Further evidence for a turbidity current following the 1929 Grand Banks earthquake. *Deep-Sea Res.* **1**, 192–202.

Hutchinson, R. D. (1952) The stratigraphy and trilobite fauna of the Cambrian sedimentary rocks of Cape Breton Island, Nova Scotia. *Geol. Surv. Can. Memoir 263*, 124pp.

Jones, O. T. (1937) On the sliding or slumping of submarine sediments in Denbighshire, North Wlaes, during the Ludlow period. *Quart. J. Geol. Soc. Lond.* **93**, 241–283d.

Jones, O. T. (1955) The geological evolution of Wales and the adjacent regions. *Quart. J. Geol. Soc. Lond.* **91**, 323–352.

Kay, G. M. (1951) North American geosynclines. *Geol. Soc. Amer. Memoir 48*, 143pp.

King, P. B. (1949) The base of the Cambrian in the southern Appalachians. *Amer. J. Sci.* **247**, 513–530.

King, P. B. (1951) *The Tectonics of Middle North America—Middle North America East of the Cordilleran System.* Princeton University Press, Princeton, 203pp.

Knopf, A. (1948) The geosynclinal theory, *Bull. Geol. Soc. Amer.* **57**, 649–670.

Kuenen, Ph. H. (1953) Significant features of graded bedding. *Bull. Amer. Ass. Petrol. Geol.* **37**, 1044–1066.

Kuenen, Ph. H. and Migliorini, C. I. (1950) Turbidity currents as a cause of graded bedding. *J. Geol.* **58**, 91–126.

Kuenen, Ph. H. and Sanders, J. E. (1956) Sedimentation phenomena in Kulm and Flozleeres graywackes, Sauerland and Oberharz, Germany. *Amer. J. Sci.* **254**, 649–671.

Park, C. F. (1946) The spilite and manganese problem of the Olympic Peninsula, Washington. *Amer. J. Sci.* **244**, 305–323.

Pettijohn, F. J. (1943) Archaean sedimentation. *Bull Geol. Soc. Amer.* **54**, 925–972.

Pettijohn, F. J. (1949) *Sedimentary Rocks*, Harper and Brothers, New York.

Pettijohn, F. J. (1950) Turbidity currents and graywackes—a discussion. *J. Geol.* **58**, 169–171.

Prindle, L. M. and Knopf, E. B. (1932) Geology of the Taconic quadrangle. *Amer. J. Sci. ser. 5*, **24**, 257–302.

Rich, J. L. (1950) Flowmarkings, groovings, and intra-stratal crumplings as criteria for recognition of slope deposits with illustrations from Silurian rocks of Wales. *Bull. Amer. Ass. Petrol. Geol.* **34**, 717–741.

Richards, H. G. (1945) Subsurface stratigraphy of Atlantic Coastal Plain between New Jersey and Georgia. *Am. Ass. Petrol. Geol. Bull.* **29**, 885–995.

Roberts, C. M. and Brashears, M. L. (1946) Record of wells in Suffolk County, New York. Suppl. No. 1 *Bull. St. N. Y. Dept. Conserv. Water Power Con. Comm.*, Albany GW-10.

Schuchert, C. and Dunbar, C. O. (1934) Stratigraphy of western Newfoundland. *Geol. Soc. Amer. Memoir 1.*

Stille, H. (1936) Wege und Ergbnisse des geologisch-tectonschen Forschung, *Festschr. Kaiser-Wilhelm Gesellsch. Förd, Wiss.*, Bd. 2.

Stille, H. (1941) *Einführung in den Bau Amerikas*, Borntraeger, Berlin.

Swain, F. M. (1947) Two recent wells in coastal plain of North Carolina. *Amer. Ass. Petrol. Geol. Bull.* **31**, 2054–2060.

Taliafero, N. L. (1943) Franciscan Knoxville problem. *Amer. Ass. Petrol. Geol. Bull.* **27**, 109–219.

Turner, F. J. and Verhoogen, J. (1951) *Igneous and Metamorphic Petrology* (1st Edition). McGraw-Hill, New York, 602pp.

Warren, D. and Ewing, M. (1958) Seismic refraction measurements in the Atlantic Ocean: Part VIII. Northeast of Recife, Brazil. In preparation.

22

CONTINENT AND OCEAN BASIN EVOLUTION BY SPREADING OF THE SEA FLOOR

By ROBERT S. DIETZ,

U.S. Navy Electronics Laboratory, San Diego 52, California

ANY concept of crustal evolution must be based on an Earth model involving assumptions not fully established regarding the nature of the Earth's outer shells and mantle processes. The concept proposed here, which can be termed the 'spreading sea-floor theory', is largely intuitive, having been derived through an attempt to interpret sea-floor bathymetry. Although no entirely new proposals need be postulated regarding crustal structure, the concept requires the acceptance of a specific crustal model, in some ways at variance with the present consensus of opinion. Since the model follows from the concept, no attempt is made to defend it. The assumed model is as follows:

(1) Large-scale thermal convection cells, fuelled by the decay of radioactive minerals, operate in the mantle. They must provide the primary diastrophic forces affecting the lithosphere.

(2) The sequence of crustal layers beneath the oceans is markedly different from that beneath the continents and is quite simple (Fig. 1). On an average 4·5 km. of water overlies 0·3 km. of unconsolidated sediments (layer 1). Underlying this is layer 2, consisting of about 2·0 km. of mixed volcanics and lithified sediments. Beneath this is the layer 3 (5 km. thick), commonly called the basalt layer and supposedly forming a world-encircling cap of effusive basic volcanics over the Earth's mantle from which it is separated by the Mohorovičić seismic discontinuity. Instead we must accept the growing opinion that the 'Moho' marks a change of phase rather than a chemical boundary, that is, layer 3 is chemically the same as the mantle rock but petrographically different with low-pressure phase minerals above the Moho and high-pressure minerals below. This change of phase may be either from eclogite to gabbro[1], or from peridotite to serpentine[2]; its exact nature is not vital to our concept, but we can tentatively accept the eclogite–gabbro transition as it has more adherents. Common usage requires that we reserve the term 'mantle' for the substance beneath the Moho, but in point of fact, the gabbro layer (as a change of phase) is also a part of the mantle—a sort of 'exo-mantle'. Except for a very thin veneer, then, the sea floor is the exposed mantle of the Earth in this larger sense.

(3) It is relevant to speak of the strength and rigidity of the Earth's outer shell. The term 'crust' has been effectively pre-empted from its classical meaning by seismological usage applying it to the layer above the Moho, that is, the sial in continental regions and the 'basaltic' layer under the oceans so that the continents have a thick crust and the ocean basins a thin crust. Used in this now accepted sense, any implications equating the crust with rigidity must be dismissed. For considerations of convective creep and tectonic yielding, we must refer to a lithosphere and an asthenosphere. Deviations from isostasy prove that approximately the outer 70 km. of the Earth (under the continents and ocean basins alike) is moderately strong and rigid even over time-spans of 100,000 years or more; this outer rind is the lithosphere. Beneath lies the asthenosphere separated from the lithosphere by the level of no strain or isopiestic level; it is a domain of rock plasticity and flowage where any stresses are quickly removed. No seismic discontinuity marks the isopiestic level and very likely it is actually a zone of uniform composition showing a gradual transition in strength as pressure and temperature rise; and in spite of the lithosphere's rigidity, to speak of it as a crust or shell greatly exaggerates its strength. Because of its grand dimensions, for model similitude we must think of it as weak[3]. If convection currents are operating 'sub-crustally', as is commonly written, they would be expected to shear below the lithosphere and not beneath the 'crust' as this term is now used.

(4) As gravity data have shown, the continents are low-density tabular masses of sial—a 'basement complex' of granitic rocks about 35 km. thick with a thin sedimentary veneer. Since they are buoyant and float high hydrostatically in the sima, they are analogous to icebergs in the ocean. This analogy

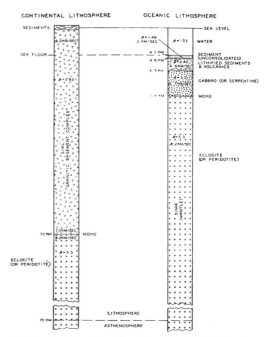

Fig. 1. Generalized crustal sections through the lithosphere beneath the continents and the ocean basins as presumed in this paper. Seismic velocities and densities are shown for the various layers

has additional merit in that convection of the sima cannot enter the sial. But the analogy gives the wrong impression of relative strength of sial and sima; the continental lithosphere is no stronger than the oceanic lithosphere, so it is mechanically impossible for the sial to 'sail through the sima' as Wegnerian continental drift proposes. The temperature and pressure are too high at the base of the sial to permit a gabbroic layer above the Moho; instead, there may be an abrupt transition from granite to eclogite.

Spreading Sea Floor Theory

Owing to the small strength of the lithosphere and the gradual transition in rigidity between it and the asthenosphere, the lithosphere is not a boundary to convection circulation, and neither is the Moho beneath the oceans because this is not a density boundary but simply a change of phase. Thus the oceanic 'crust' (the gabbroic layer) is almost wholly coupled with the convective overturn of the mantle creeping at a rate of a few cm./yr. Since the sea floor is covered by only a thin veneer of sediments with some mixed-in effusives, it is essentially the outcropping mantle. So the sea floor marks the tops of the convection cells and slowly spreads from zones of divergence to those of convergence. These cells have dimensions of several thousands of kilometres; some cells are quite active now while others are dead or dormant. They have changed position with geological time causing new tectonic patterns.

The gross structures of the sea floor are direct expressions of this convection. The median rises[4,5] mark the up-welling sites or divergences; the trenches are associated with the convergences or down-welling sites; and the fracture zones[6] mark shears between regions of slow and fast creep. The high heat-flow under the rises[7] is indicative of the ascending convection currents as also are the groups of volcanic seamounts which dot the backs of these rises.

Much of the minor sea-floor topography may be even directly ascribable to spreading of the sea floor. Great expanses of rough topography skirt both sides of the Mid-Atlantic Rift; similarly there are extensive regions of abyssal hills in the Pacific. The roughness is suggestive of youth, so it has commonly been assumed to be simply volcanic topography because the larger seamounts are volcanic. But this interpretation is not at all convincing, and no one has given this view formality by publishing a definitive study. Actually, the topography resembles neither volcanic flows nor incipient volcanoes. Can it not be that these expanses of abyssal hills are a 'chaos topography' developed as strips of juvenile sea-floor (by a process which can be visualized only as mixed intrusion and extrusion) and then placed under rupturing stresses as the sea floor moves outward?

The median position of the rises cannot be a matter of chance, so it might be supposed that the continents in some manner control the convection pattern. But the reverse is considered true: conditions deep within the mantle control the convective pattern without regard for continent positions. By vi cous drag, the continents initially are moved along with the sima until they attain a position of dynamic balance overlying a convergence. There the continents come to rest, but the sima continues to shear under and descend beneath them; so the continents generally cover the down-welling sites. If new upwells do happen to rise under a continental mass, it tends to be rifted. Thus, the entire North and South Atlantic Ocean marks an ancient rift which separated North and South America from Europe and Africa. Another such rift has opened up the Mediterranean. The axis of the East Pacific Rise now seems to be invading the North American continent, underlying the Gulf of California and California[8]. Similarly, the Indian Ocean Rise may extend into the African Rift Valleys, tending to fragment that continent.

The sialic continents, floating on the sima, provide a density barrier to convection circulation—unlike the Moho, which involves merely a change of phase. The convection circulation thus shears beneath the continents so that the sial is only partially coupled through drag forces. Since the continents are normally resting over convergences, so that convective spreading is moving toward them from opposite sides, the continents are placed consequently under compression. They tend to buckle, which accounts for alpine folding, thrust faulting, and similar compressional effects so characteristic of the continents. In contrast, the ocean basins are simultaneously domains of tension. If the continental block is drifted along with the sima, the margin is tectonically stable (Atlantic type). But if the sima is slipping under the sialic block, marginal mountains tend to form (Pacific type) owing to drag forces.

Implications of the Concept

Ad hoc hypotheses are likely to be wrong. On the other hand, one which is consonant with our broader understanding of the history of the Earth may have merit. While the thought of a highly mobile sea floor may seem alarming at first, it does little violence to geological history.

Volumetric changes of the Earth. Geologists have traditionally recognized that compression of the continents (and they assumed of the ocean floors as well) was the principal tectonic problem. It was supposed that the Earth was cooling and shrinking. But recently, geologists have been impressed by tensional structures, especially on the ocean floor. To account for sea floor rifting, Heezen[10], for example, has advocated an expanding Earth, a doubling of the diameter. Carey's[11] tectonic analysis has resulted in the need for a twenty-fold increase in volume of the Earth. Spreading of the sea floor offers the less-radical answer that the Earth's volume has remained constant. By creep from median upwellings, the ocean basins are mostly under tension, while the continents, normally balanced against sima creepage from opposite sides, are under compression.

The geological record is replete with transgressions and regressions of the sea, but these have been shallow and not catastrophic; fluctuations in sea-level as severe as those of the Pleistocene are abnormal. The spreading concept does no violence to this order of things, unlike dilation or contraction of the Earth. The volumetric capacity of the oceans is fully conserved.

Continental Drift. The spreading concept envisages limited continental drifting, with the sial blocks initially being rafted to down-welling sites and then being stablized in a balanced field of opposing drag forces. The sea floor is held to be more mobile and to migrate freely even after the continents come to rest. The sial moves largely *en bloc*, but the sea floor spreads more differentially.

Former scepticism about continental drift is rapidly vanishing, especially due to the palæomagnetic findings and new tectonic analyses. A principal objection to Wegener's continental drift hypothesis was that it was physically impossible for a continent to 'sail like a ship' through the sima; and nowhere is there any sea floor deformation ascribable to an on-coming continent. Sea floor spreading obviates this difficulty: continents never move through the sima—they either move along with it or stand still while the sima shears beneath them. The buoyancy of the continents, rather than their being stronger than the sima, accounts for this. Drag associated with the shearing could account for alpine folding and related compressional tectonic structures on the continents.

Persistent freeboard of the continents. A satisfactory theory of crustal evolution must explain why the continents have stood high throughout geological time in spite of constant erosional de-levelling. Many geologists believe that new buoyancy is added to continents through the gravitative differentiation from the mantle. Spreading of the sea floor provides a mechanism whereby the continents are placed over the down-wells where new sial would tend to collect, even though the convection is entirely a mantle process and the role of the continents is passive. It also follows that the clastic detritus swept into the deep sea from the continents is not permanently lost. Rather, it is carried slowly towards, and then beneath, the continents, where it is granitized and added anew to the sialic blocks.

Youth of the ocean floor. It follows paradoxically from the spreading concept that, although the ocean basins are old, the sea floor is young—much younger than the rocks of the continents. Marine sediments, seamounts, and other structures slowly impinge against the sialic blocks and are destroyed by underriding them. Pre-Cambrian and perhaps even most Palæozoic rocks should prove absent from the ocean floors; and Mohole drilling should not reveal the great missing sequence of the Lipalian interval

(Pre-Cambrian to Cambrian) as hoped for by some. All this may seem surprising, but marine geological evidence supports the concept.

On his discovery of the guyots of the Pacific, Hess[12] supposed these were Pre-Cambrian features protected from erosion by the cover of the sea. But Hamilton[13] proved the guyots of the Mid-Pacific Mountains were Cretaceous, and these seem to be among the oldest of the seamount groups. In an analysis of the various seamount groups of the western Pacific, I was forced to conclude that none of them was older than mid-Mesozoic. The young age of the seamounts has been puzzling; certainly they can neither erode away nor subside completely. Also, there seem to be too few volcanic seamounts, if the present population represents the entire number built over the past hundred million years or more. The puzzle dissolves if sea floor spreading has operated. Modern examples of impinging groups of seamounts may be the western end of the Caroline Islands, the Wake–Marcus Seamounts, and the Magellan Seamounts[14]. All may be moving into the western Pacific trenches. Seamount GA-1 south of Alaska may be moving into the Aleutian Trench[15].

The sedimentary layers under the sea also appear to be young. No fossiliferous rocks older than Cretaceous have yet been dredged from any ocean basin. Radioactive dating of a basalt from the Mid-Atlantic ridge gave a Tertiary age[16]. Kuenen[17] estimated that the ocean basins should contain on an average about 3·0 km. of sedimentary rocks assuming the basins are 200 million years old. But seismic reflexions indicate an average of only 0·3 km. of the unconsolidated sediments. Hamilton[18], however, believes that much of layer 2 may be lithified sediments. If *all* layer 2 is lithified sediments, Hamilton finds that the ocean basins may be Palæozoic or late Pre-Cambrian in age—but not Archæan. But very likely layer 2 includes much effusive material and sedimentary products of sea floor weathering. In summing up, the evidence from the sediments, although still fragmentary, suggests that the sea floors may be not older than Palæozoic or even Mesozoic.

Spreading and magnetic anomalies. Vacquier, V., *et al.* (in the press) recently have completed excellent sea-floor magnetic surveys off the west coast of North America. A striking north–south lineation shows up which seems to reveal a stress pattern (Mason, R. G.,

and Raff, A. D., in the press). Such interpretation would fit into spreading concept with the lineations being developed normal to the direction of convection creep. The lineation is interrupted by Menard's[6] three fracture-zones, and anomalies indicate shearing offsets of as much as 640 nautical miles in the case of the remarkable Mendocinco Escarpment[19]. Great mobility of the sea floor is thus suggested. The offsets have no significant expression after they strike the continental block; so apparently they may slip under the continent without any strong coupling. Another aspect is that the anomalies smooth out and virtually disappear under the continental shelf; so the sea floor may dive under the sial and lose magnetism by being heated above the Curie point.

By considering an Earth crustal model only slightly at variance with that commonly accepted, a novel concept of the evolution of continents and ocean basins has been suggested which seems to fit the 'facts' of marine geology. If this concept were correct, it would be most useful to apply the term 'crust', which now has a confusion of meanings, only to any layer which overlies and caps the convective circulation of the mantle. The sialic continental blocks do this, so they form the true crust. The ocean floor seemingly does not, so the ocean basin is 'crustless'.

I wish to express my appreciation to E. L. Hamilton, F. P. Shepard, H. W. Menard, V. Vacquier, R. Von Herzen and A. D. Raff for critical discussions.

[1] Kennedy, G. C., *Amer. Sci.*, **47**, 4, 491 (1959).
[2] Hess, H. H., *Abst. Bull. Geol. Soc. Amer.*, **71**, Pt. 2, 12, 2097 (1960).
[3] Griggs, D. A., *Amer. J. Sci.*, **237**, 611 (1939).
[4] Ewing, M., and Heezen, B. C., *Amer. Geophys. Union Geophys. Mon. No.* 1, 75 (1956).
[5] Menard, H. W., *Bull. Geol. Soc. Amer.*, **69**, 9, 1179 (1958).
[6] Menard, H. W., *Bull. Geol. Soc. Amer.*, **66**, 1149 (1955).
[7] Von Herzen, R. P., *Nature*, **183**, 882 (1959).
[8] Menard, H. W., *Science*, **132**, 1737 (1960).
[9] Heezen, B. C., *Sci. Amer.*, Oct. 2, 14 (1960).
[10] Heezen, B. C., Preprints, First Intern. Ocean. Cong., 26 (1959).
[11] Carey, W. S., *The Tectonic Approach to Continental Drift: in Continental Drift—A Symposium*, 177 (Univ. Tasmania, 1958.)
[12] Hess, H. H., *Amer. J. Sci.*, **244**, 772 (1946).
[13] Hamilton, E. L., *Geol. Soc. Amer. Mem.*, **64**, 97 (1956).
[14] Dietz, R. S., *Bull. Geol. Soc. Amer.*, **65**, 1199 (1954).
[15] Menard, H. W., and Dietz, R. S., *Bull. Geol. Soc. Amer.*, **62**, 1263 (1951).
[16] Carr, D., and Kulp, J., *Bull. Geol. Soc. Amer.*, **64**, 2, 263 (1953).
[17] Kuenen, Ph., *Marine Geology* (John Wiley and Sons, New York. 1950).
[18] Hamilton, E. L., *Bull. Geol. Soc. Amer.*, **70**, 1399 (1959); *J. Sed. Petrol.*, **30**, 3, 370 (1960).
[19] Menard, H. W., and Dietz, R. S., *J. Geol.*, **60**, 3 (1952).

23

History of Ocean Basins

H. H. Hess

Princeton University, Princeton, N. J.

ABSTRACT

For purposes of discussion certain simplifying assumptions are made as to initial conditions on the Earth soon after its formation. It is postulated that it had little in the way of an atmosphere or oceans and that the constituents for these were derived by leakage from the interior of the Earth in the course of geologic time. Heating by short-lived radio nuclides caused partial melting and a single-cell convective overturn within the Earth which segregated an iron core, produced the primordial continents, and gave the Earth its bilateral asymmetry.

Mid-ocean ridges have high heat flow, and many of them have median rifts and show lower seismic velocities than do the common oceanic areas. They are interpreted as representing the rising limbs of mantle-convection cells. The topographic elevation is related to thermal expansion, and the lower seismic velocities both to higher than normal temperatures and microfracturing. Convective flow comes right through to the surface, and the oceanic crust is formed by hydration of mantle material starting at a level 5 km below the sea floor. The water to produce serpentine of the oceanic crust comes from the mantle at a rate consistent with a gradual evolution of ocean water over 4 aeons.

Ocean ridges are ephemeral features as are the convection cells that produce them. An ancient trans-Pacific ridge from the Marianas Islands to Chile started to disappear 100 million years ago. Its trace is now evident only in a belt of atolls and guyots which have subsided 1–2 km. No indications of older generations of oceanic ridges are found. This, coupled with the small thickness of sediments on the ocean floor and comparatively small number of volcanic seamounts, suggests an age for all the ocean floor of not more than several times 10^8 years.

The Mid-Atlantic Ridge is truly median because each side of the convecting cell is moving away from the crest at the same velocity, *ca.* 1 cm/yr. A more acceptable mechanism is derived for continental drift whereby continents ride passively on convecting mantle instead of having to plow through oceanic crust.

Finally, the depth of the M discontinuity under continents is related to the depth of the oceans. Early in the Earth's history, when it is assumed there was much less sea water, the continental plates must have been much thinner.

INTRODUCTION

The birth of the oceans is a matter of conjecture, the subsequent history is obscure, and the present structure is just beginning to be understood. Fascinating speculation on these subjects has been plentiful, but not much of it pre-dating the last decade holds water. Little of Umbgrove's (1947) brilliant summary remains pertinent when confronted by the relatively small but crucial amount of factual information collected in the intervening years. Like Umbgrove, I shall consider this paper an essay in geopoetry. In order not to travel any further into the realm of fantasy than is absolutely necessary I shall hold

185

as closely as possible to a uniformitarian approach; even so, at least one great catastrophe will be required early in the Earth's history.

PREMISES ON INITIAL CONDITIONS

Assuming that the ages obtained from radioactive disintegrations in samples of meteorites approximate the age of the solar system, then the age of the Earth is close to 4.5 aeons.[1] The Earth, it is further assumed, was formed by accumulation of particles (of here unspecified character) which initially had solar composition. If this is true, then before condensation to a solid planet the Earth lost, during a great evaporation, a hundred times as much matter as it now contains. Most of this loss was hydrogen. An unknown but much smaller amount of heavier elements was lost to space as well. The deficiency of the atmosphere in the inert gases points clearly to their loss. Urey (1957) suggests loss of nitrogen, carbon, and water, and perhaps a considerable proportion of original silicate material. He also points out that the lack of concentration of certain very volatile substances at the Earth's surface indicates that it never had a high surface temperature. This low temperature more or less precluded escape of large amounts of material after the Earth condensed and suggests that the loss occurred when the material forming the Earth was very much more dispersed so that the escape velocity from its outer portion was comparatively low. The condensation was rapid, and some light elements and volatile compounds were trapped within the accumulated solid material of the primordial Earth. I will assume for convenience and without too much justification that at this stage the Earth had no oceans and perhaps very little atmosphere. It is postulated that volatile constituents trapped within its interior have during the past and are today leaking to the surface, and that by such means the present oceans and atmosphere have evolved.

THE GREAT CATASTROPHE

Immediately after formation of the solid Earth, it may have contained within it many short-lived radioactive elements; how many and how much depends on the time interval between nuclear genesis and condensation. The bricketted particles from which it was made might be expected to have a low thermal conductivity at least near its surface as suggested by Kuiper (1954). The temperature rose, lowering the strength and perhaps starting partial fusion. The stage was thus set for the *great catastrophe* which it is assumed happened forthwith. A single-cell (toroidal) convective overturn took place (Fig. 1) (Vening Meinesz, 1952), resulting in the formation of a nickel-iron core, and at the same time the low-melting silicates were extruded over the rising limbs of the current to form the primordial single continent (Fig. 1). The single-cell overturn also converted gravitational energy into thermal energy (Urey, 1953). It is postulated that this heat and a probably much larger amount of heat resulting from the energy involved in the accumulation of the Earth were not sufficient to produce a molten Earth. The great quantitative uncertainties in this assumption can be gauged from MacDonald's analysis (1959).

---

[1] Aeon = 10[9] years (H. C. Urey).

The proposed single-cell overturn brought about the bilateral asymmetry of the Earth, now possibly much modified but still evident in its land and water hemispheres. After this event, which segregated the core from the mantle, single-cell convection was no longer possible in the Earth as a whole (Chandrasekhar, 1953).

The critical question now facing us is what percentage of the continental crustal material and of the water of the oceans reached the surface in the *great catastrophe*. On the basis that continental material is still coming to the surface of the Earth from the mantle at the rate of 1 km³/year*, accepting Sapper's (1927, p. 424) figure on the contribution of volcanoes over the past 4 centuries, and assuming uniformitarianism, this means 4×10^9 km³ in 4 aeons or approximately 50 per cent of the continents. So we shall assume that the other half was extruded during the catastrophe. The percentage of water is much harder to estimate. Rapid convective overturn might be much less efficient in freeing the water as compared to the low-melting silicates. The water might be expected to be present as a monomolecular film on grain surfaces. The low-melting silicate droplets could coagulate into sizable masses as a result of strong shearing during the overturn. On the other hand, shearing that would break down solid crystals to smaller size might increase their surface areas and actually inhibit freeing of water films. The best guess that I can make is that up to one-third of the oceans appeared on the surface at this time.

It may be noted that a molten Earth hypothesis would tend toward the initial formation of a thin continental or sialic layer uniformly over the Earth with a very thin uniform world-encircling water layer above it. Later it would require breaking up of this continental layer to form the observed bilateral asymmetry. With the present set of postulates this seems to be a superfluous step. Bilateral asymmetry was attained at the start, and it would be impossible ever to attain it once a core had formed, unless George H. Darwin's hypothesis that the moon came out of the Earth were accepted.

We have now set the stage to proceed with the subject at hand. Dozens of assumptions and hypotheses have been introduced in the paragraphs above to establish a framework for consideration of the problem. I have attempted to chose reasonably among a myriad of possible alternatives, but no competent reader with an ounce of imagination is likely to be willing to accept all of the choices made. Unless some such set of confining assumptions is made, however, speculation spreads out into limitless variations, and the resulting geopoetry has neither rhyme nor reason.

TOPOGRAPHY AND CRUSTAL COLUMNS

If the water were removed from the Earth, two distinct topographic levels would be apparent: (1) the deep-sea floor about 5 km below sea level, and (2) the continental surface a few hundred meters above sea level. In other words, the continents stand up abruptly as plateaus or mesas above the general level of the sea floor. Seismic evidence shows that the so-called crustal thickness—depth to

---

* This figure includes felsic volcanic material probably derived from partial melting within the continental crust but does not include magmas that formed intrusions which did not reach the surface.

the M discontinuity—is 6 km under oceans and 34 km under continents on the average. Gravity data prove that these two types of crustal columns have the same mass—the pressure at some arbitrary level beneath them, such as 40 km, would be the same. They are in hydrostatic equilibrium. It is evident that one cannot consider the gross features of ocean basins independent of the continental plateaus; the two are truly complementary.

Whereas 29 per cent of the Earth's surface is land, it would be more appropriate here to include the continental shelves and the slopes to the 1000-m isobath with the continents, leaving the remainder as oceanic. This results in 40 per cent continental and 60 per cent oceanic crust. In 1955 I discussed the nature of the two crustal columns, which is here modified slightly to adjust the layer thicknesses to the more recent seismic work at sea (Raitt, 1956; Ewing and Ewing 1959) (Fig. 2). A drastic change, however, has been made in layer 3 of the oceanic column, substituting partially serpentinized peridotite for the basalt of the main crustal layer under the oceans as proposed elsewhere (Hess, 1959a). Let us look briefly into the facts that seemed to necessitate this change.

That the mantle material is peridotitic is a fairly common assumption (Harris and Rowell, 1960; Ross, Foster, and Myers, 1954; Hess, 1955). In looking at the now-numerous seismic profiles at sea the uniformity in thickness of layer 3 is striking. More than 80 per cent of the profiles show it to be 4.7 ± 0.7 km thick.

Considering the probable error in the seismic data to be about ± 0.5 km, the uniformity may be even greater than the figures indicate. It is inconceivable

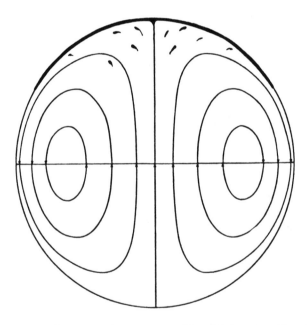

Figure 1. Single-cell (toroidal) convective overturn of Earth's interior. After Vening Meinesz. Continental material extruded over rising limb but would divide and move to descending limb if convection continued beyond a half cycle

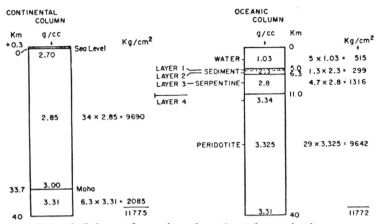

Figure 2. Balance of oceanic and continental crustal columns

that basalt flows poured out on the ocean floor could be so uniform in thickness. Rather, one would expect them to be thick near the fissures or vents from which they were erupted and thin or absent at great distance from the vents. The only likely manner in which a layer of uniform thickness could be formed would be if its bottom represented a present or past isotherm, at which temperature and pressure a reaction occurred. Two such reactions can be suggested: (1) the basalt to eclogite inversion (Sumner, 1954; Kennedy, 1959), and (2) the hydration of olivine to serpentine at about 500°C (Hess, 1954). The common occurrence of peridotitic inclusions in oceanic basaltic volcanic rocks (Ross, Foster, and Myers, 1954) and absence of eclogite inclusions lead the writer to accept postulate (2). Furthermore, the dredging of serpentinized peridotites from fault scarps in the oceans (Shand, 1949)[2], where the displacement on the faults may have been sufficient to expose layer 3, adds credence to this supposition. This choice of postulates is made here and will control much of the subsequent reasoning. The seismic velocity of layer 3 is highly variable; it ranges from 6.0 to 6.9 km/sec and averages near 6.7 km/sec, which would represent peridotite 70 per cent serpentinized (Fig. 3).

MID-OCEAN RIDGES

The Mid-Ocean Ridges are the largest topographic features on the surface of the Earth. Menard (1958) has shown that their crests closely correspond to median lines in the oceans and suggests (1959) that they may be ephemeral features. Bullard, Maxwell, and Revelle (1956) and Von Herzen (1959) show that they have unusually high heat flow along their crests. Heezen (1960) has demonstrated that a median graben exists along the crests of the Atlantic, Arctic, and Indian Ocean ridges and that shallow-depth earthquake foci are concentrated under the graben. This leads him to postulate extension of the crust at right angles to the trend of the ridges. Hess (1959b) also emphasizes the ephemeral

---

[2] J. B. Hersey reports dredging serpentinized peridotite from the northern slope of the Puerto Rico Trench (Personal communication, 1961)

character of the ridges and points to a trans-Pacific ridge that has almost disappeared since middle Cretaceous time, leaving a belt of atolls and guyots that has subsided 1–2 km. Its width is 3000 km and its length about 14,000 km (Fig. 4). The present active mid-ocean ridges have an average width of 1300 km, crest height of about 2½ km, and total length of perhaps 25,000 km.

The most significant information on the structural and petrologic character of the ridges comes from refraction seismic information of Ewing and Ewing (1959) (Fig. 5) on the Mid-Atlantic Ridge, and Raitt's (1956) refraction profiles on the East Pacific Rise. The sediment cover on the Mid-Atlantic Ridge appears to be thin and perhaps restricted to material ponded in depressions of the topog-

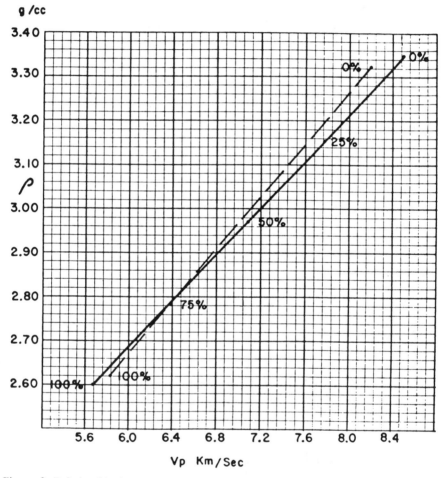

Figure 3. Relationship between seismic velocity, density, and per cent serpentinization. Solid curve for room temperature and pressure. Dashed curve estimated for T and P at 15 km depth. Curves based on measurements in laboratory by J. Green at the California Research Laboratory, La Habra, with variable temperatures up to 200° C and pressures up to 1 kilobar. The 100 per cent serpentinized sample measured by F. Birch at Harvard at pressures from 0 to 10 kilobars at room temperature (Hess, 1959a).

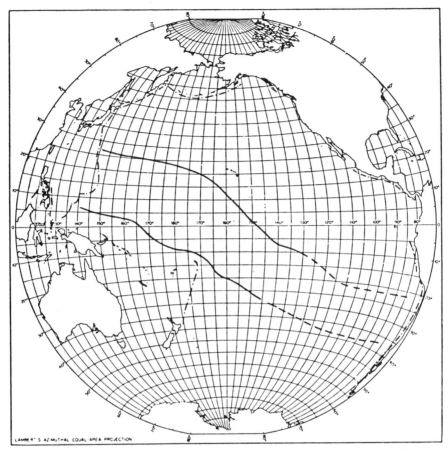

Figure 4. Former location of a Mid-Pacific Mesozoic ridge

raphy. On the ridge crest, layer 3 has a seismic velocity of from 4 to 5.5 km/sec instead of the normal 6 to 6.9 km/sec. The M discontinuity is not found or is represented by a transition from layer 3 to velocities near 7.4 km/sec. Normal velocities and layer thicknesses, however, appear on the flanks of ridges.

Earlier I (1955, 1959b) attributed the lower velocities (ca. 7.4 km/sec) in what should be mantle material to serpentinization, caused by olivine reacting with water released from below. The elevation of the ridge itself was thought to result from the change in density (olivine 3.3 g/cc to serpentine 2.6 g/cc). A 2-km rise of the ridge would require 8 km of complete serpentinization below, but a velocity of 7.4 km/sec is equivalent to only 40 per cent of the rock serpentinized. This serpentinization would have to extend to 20-km depth to produce the required elevation of the ridge. This reaction, however, cannot take place at a temperature much above 500° C, which, considering the heat flow, probably exists at the bottom of layer 3, about 5 km below the sea floor, and cannot reasonably be 20 km deep. Layer 3 is thought to be peridotite 70 per cent serpentinized. It would appear that the highest elevation that the 500° C isotherm can reach is approxi-

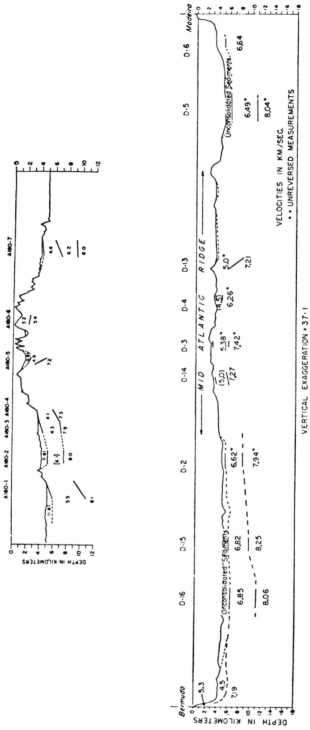

Figure 5. Seismic profiles on the Mid-Atlantic Ridge, by Ewing and Ewing (1959)

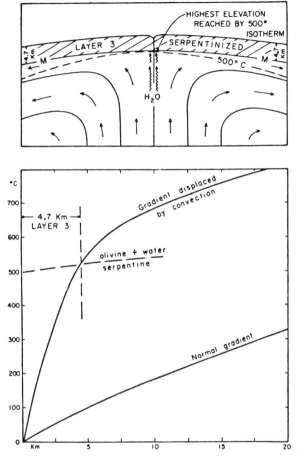

Figure 6. Diagram to portray highest elevation that 500° C isotherm can reach over the rising limb of a mantle convection cell, and expulsion of water from mantle which produces serpentinization above the 500° C isotherm

mately 5 km below the sea floor, and this supplies the reason for uniform thickness of layer 3 (Fig. 6).

CONVECTION CURRENTS IN THE MANTLE AND MID-OCEAN RIDGES

Long ago Holmes suggested convection currents in the mantle to account for deformation of the Earth's crust (Vening Meinesz, 1952; Griggs, 1939; 1954; Verhoogen, 1954; and many others). Nevertheless, mantle convection is considered a radical hypothesis not widely accepted by geologists and geophysicists. If it were accepted, a rather reasonable story could be constructed to describe the evolution of ocean basins and the waters within them. Whole realms of previously unrelated facts fall into a regular pattern, which suggests that close approach to satisfactory theory is being attained.

As mentioned earlier a single-cell convective overturn of the material within the Earth could have produced its bilateral asymmetry, segregating the iron core and primordial continents in the process. Since this event only multicell convection in the mantle has been possible. Vening Meinesz (1959) analyzed the spherical harmonics of the Earth's topography up to the thirty-first order. The peak shown in the values from the third to fifth harmonic would correlate very nicely with mantle-size convection currents; cells would have the approximate diameter of 3000 to 6000 km in cross section (the other horizontal dimension might be 10,000–20,000 km, giving them a banana-like shape).

The lower-order spherical harmonics of the topography show quite unexpected regularities. This means that the topography of a size smaller than continents and ocean basins has a greater regularity in distribution than previously recognized.

Paleomagnetic data presented by Runcorn (1959), Irving (1959), and others strongly suggest that the continents have moved by large amounts in geologically comparatively recent times. One may quibble over the details, but the general picture on paleomagnetism is sufficiently compelling that it is much more reasonable to accept it than to disregard it. The reasoning is that the Earth has always had a dipole magnetic field and that the magnetic poles have always been close to the axis of the Earth's rotation, which necessarily must remain fixed in space. Remanent magnetism of old rocks shows that position of the magnetic poles has changed in a rather regular manner with time, but this migration of the poles as measured in Europe, North America, Australia, India, etc., has not been the same for each of these land masses. This strongly indicates independent movement in direction and amount of large portions of the Earth's surface with respect to the rotational axis. This could be most easily accomplished by a convecting mantle system which involves actual movement of the Earth's surface passively riding on the upper part of the convecting cell. In this case at any given time continents over one cell would not move in the same direction as continents on another cell. The rate of motion suggested by paleomagnetic measurements lies between a fraction of a cm/yr to as much as 10 cm/yr. If one were to accept the old evidence, which was the strongest argument for continental drift, namely the separation of South America from Africa since the end of the Paleozoic, and apply uniformitarianism, a rate of 1 cm/yr results. This rate will be accepted in subsequent discussion. Heezen (1960) mentions a fracture zone crossing Iceland on the extension of the Mid-Atlantic rift zone which has been widening at a rate of 3.5 m/1000 yrs/km of width.

The unexpected regularities in the spherical harmonics of the Earth's topography might be attributed to a dynamic situation in the present Earth whereby the continents move to positions dictated by a fairly regular system of convection cells in the mantle. Menard's theorem that mid-ocean ridge crests correspond to median lines now takes on new meaning. The mid-ocean ridges could represent the traces of the rising limbs of convection cells, while the circum-Pacific belt of deformation and volcanism represents descending limbs. The Mid-Atlantic Ridge is median because the continental areas on each side of it have moved away from it at the same rate—1 cm/yr. This is not exactly the same as continental drift. The continents do not plow through oceanic crust impelled by unknown forces; rather they ride passively on mantle material as it comes to the surface at the crest of the ridge and then moves laterally away from it. On this basis the crest of the ridge should have only recent sediments on it, and recent and Tertiary sediments on its flanks; the whole Atlantic Ocean and possibly all of the oceans should have little sediment older than Mesozoic (Fig. 7). Let us look a bit further at the picture with regard to oceanic sediments.

Looking over the reported data on rates of sedimentation in the deep sea, rates somewhere between 2 cm and 5 mm/1000 yrs seem to be indicated. Writers in the last few years have tried hard to accept the lowest possible rate consistent with the data in order to make the thickness jibe with the comparatively thin cover of sediment on the ocean floor indicated by Seismic data. Schott's figures for the Atlantic and Indian oceans as corrected by Kuenen (1946) and further corrected by decreasing the number of years since the Pleistocene from 20,000 years to 11,000 years indicate a rate of 2 cm/1000 yrs. Hamilton's (1960) figures suggest 5 mm/1000 yrs. A rate of 1 cm/1000 yrs would yield 40 km in 4 aeons or 17 km after compaction, using Hamilton's compaction figures. A 5-mm rate would still give 8.5 km compacted thickness instead of 1.3 km as derived from seismic data. This 1 order of magnitude discrepancy had led some to suggest that the water of the oceans may be very young, that oceans came into existence largely since the Paleozoic. This violates uniformitarianism to which the writer is dedicated and also can hardly be reconciled with Rubey's (1951) analysis of the origin of sea water. On the system here suggested any sediment upon the sea floor ultimately gets incorporated in the continents. New mantle material with no sedimentary cover on it rises and moves outward from the ridge. The cover of young sediments it acquires in the course of time will move to the axis of a downward-moving limb of a convection current, be metamorphosed, and probably eventually be welded onto a continent.

Assuming a rate of 1 cm/1000 yrs one might ask how long, on the average, the present sea floor has been exposed to deposition if the present thickness of sediment is 1.3 km. The upper 0.2 km would not yet have been compacted and would represent 20 million years of deposition. The remaining 1.1 km now compacted would represent 240 million years of accumulation or in total an average age of the sea floor of 260 million years. Note that a clear distinction must be made between the age of the ocean floor and the age of the water in the oceans.

In order to explain the discrepancy between present rate of sedimentation in the deep sea and the relatively small thickness of sediment on the floor of the oceans, many have suggested that Pleistocene glaciation has greatly increased the rate of sedimentation. The writer is skeptical of this interpretation, as was Kuenen in his analysis (1946)[3]. Another discrepancy of the same type, the small number of volcanoes on the sea floor, also indicates the apparent youth of the floor. Menard estimates there are in all 10,000 volcanic seamounts in the oceans. If this represented 4 aeons of volcanism, and volcanoes appeared at a uniform

---

[3] The Mohole test drilling off Guadalupe Island in 1961 suggests a rate of sedimentation in the Miocene of 1 cm/1000 yrs or a little more.

rate, this would mean only one new volcano on the sea floor per 400,000 years. One new volcano in 10,000 years or less would seem like a better figure. This would suggest an average age of the floor of the ocean of perhaps 100 to 200 million years. It would account also for the fact that nothing older than late Cretaceous has ever been obtained from the deep sea or from oceanic islands.

Still another line of evidence pointing to the same conclusion relates to the ephemeral character of mid-ocean ridges and to the fact that evidence of only one old major ridge still remains on the ocean floor. The crest of this one began to subside about 100 million years ago. The question may be asked: Where are the Paleozoic and Precambrian mid-ocean ridges, or did the development of such features begin rather recently in the Earth's history?

Egyed (1957) introduced the concept of a great expansion in size of the Earth to account for apparent facts of continental drift. More recently Heezen (1960) tentatively advanced the same idea to explain paleomagnetic results coupled with an extension hypothesis for mid-ocean ridges. S. W. Carey (1958) developed an expansion hypothesis to account for many of the observed relationships of the Earth's topography and coupled this with an overall theory of the tectonics of the Earth's crust. Both Heezen and Carey require an expansion of the Earth since late Paleozoic time (ca. 2×10^8 years) such that the surface area has doubled. Both postulate that this expansion is largely confined to the ocean floor rather than to the continents. This means that the ocean basins have increased in area by more than 6 times and that the continents until the late Paleozoic occupied almost 80 per cent of the Earth's surface. With this greatly expanded ocean floor one could account for the present apparent deficiency of sediments, volcanoes, and old mid-ocean ridges upon it. While this would remove three of my most serious difficulties in dealing with the evolution of ocean basins, I hesitate to accept this easy way out. First of all, it is philosophically rather unsatisfying, in much the same way as were the older hypotheses of continental drift, in that there is no apparent mechanism within the Earth to cause a sudden (and exponential according to Carey) increase in the radius of the Earth. Second, it requires the addition of an enormous amount of water to the sea in just the right amount to maintain the axiomatic relationship between sea level-land surface and depth to the M discontinuity under continents, which is discussed later.

MESOZOIC MID-PACIFIC RIDGE

In the area between Hawaii, the Marshall Islands, and the Marianas scores of guyots were found during World War II. It was supposed that large numbers of them would be found elsewhere in the oceans. This was not the case. The Emperor seamounts running north-northwest from the west end of the Hawaiian chain are guyots, a single linear group of very large ones. An area of small guyots is known in the Gulf of Alaska (Gibson, 1960). There are a limited number in the Atlantic Ocean north of Bermuda on a line between Cape Cod and the Azores, and a few east of the Mid-Atlantic Ridge; other than these only rare isolated occurrences have been reported.

Excluding the areas of erratic uplift and depression represented by the island arcs, lines can be drawn in the mid-Pacific bounding the area of abundant guyots and atolls (Fig. 4), marking a broad band of subsidence 3000 km wide crossing the Pacific from the Marianas to Chile. The eastern end is poorly charted and complicated by the younger East Pacific Rise. The western end terminates with striking abruptness against the eastern margin of the island-arc structures. Not a single guyot is found in the Philippine Sea west of the Marianas trench and its extensions, although to the east they are abundant right up to the trenches.

Fossils are available to date the beginning of the subsidence, but only near the axis of the old ridge. Hamilton (1956) found middle Cretaceous shallow-water fossils on guyots of the Mid-Pacific mountains, and Ladd and Schlanger (1960) reported Eocene sediments above basalt at the bottom of the Eniwetok bore hole. It should also be noted that atolls of the Caroline, Marshall, Gilbert, and Ellice islands predominate on the southern side of the old ridge, whereas guyots greatly predominate on the northern side. Hess (1946) had difficulty in explaining why the guyots of the mid-Pacific mountain area did not become atolls as they subsided. He postulated a Precambrian age for their upper flat surfaces, moving the time back to an era before lime-secreting organisms appeared in the oceans. This became untenable after Hamilton found shallow-water Cretaceous fossils on them. Looking at the same problem today and considering that the North Pole in early Mesozoic time, as determined from paleomagnetic data from North America and Europe, was situated in southeastern Siberia, it seems likely that the Mid-Pacific mountain area was too far north for reef growth when it was subsiding. The boundary between reef growth and nonreef growth in late Mesozoic time is perhaps represented by the northern margins of the Marshall and Caroline islands, now a little north of 10° N, then perhaps 35° N. Paleomagnetic measurements from Mesozoic rocks, if they could be found within or close to this area, are needed to substantiate such a hypothesis.

The old Mesozoic band of subsidence is more than twice as wide as the topographic rise of present-day oceanic ridges. This has interesting implications regarding evolution of ridges which are worth considering here. Originally I attributed the rise of ridges to release of water above the upward-moving limb of a mantle convection cell and serpentinization of olivine when the water crossed the 500-degree C isotherm. As mentioned above, this hypothesis is no longer tenable because the high heat flow requires that the 500-degree C isotherm be at very shallow depth. The topographic rise of the ridge must be attributed to the fact that a rising column of a mantle convection cell is warmed and hence less dense than normal or descending columns. The geometry of a mantle convection cell (Fig. 8) fits rather nicely a 1300-km width assuming that the above effect causes the rise.

Looking now at the old Mesozoic Mid-Pacific Ridge with the above situation in mind, volcanoes truncated on the ridge crest move away from the ridge axis at a rate of 1 cm/yr. Eventually they move down the ridge flank and become guyots or atolls rising from the deep-sea floor. Those 1000 km from the axis, however, were truncated 100 million years before those now near the center of the

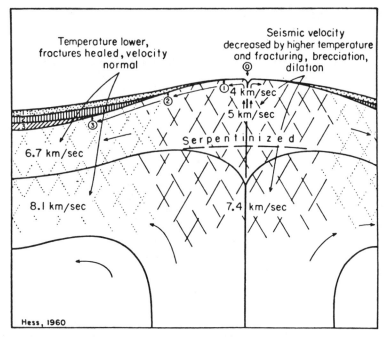

Figure 7. Diagram to represent (1) apparent progressive overlap of ocean sediments on a mid-ocean ridge which would actually be the effect of the mantle moving laterally away from ridge crest, and (2) the postulated fracturing where convective flow changes direction from vertical to horizontal. Fracturing and higher temperature could account for the lower seismic velocities on ridge crests, and cooling and healing of the fractures with time, the return to normal velocities on the flanks.

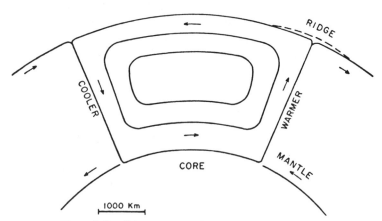

Figure 8. Possible geometry of a mantle convection cell

old ridge (Fig. 9). On this basis it would be very interesting to examine the fauna on guyots near the northern margin of the old ridge or to drill atolls near the southern margin to see if the truncated surfaces or bases have a Triassic or even Permian age. At any rate the greater width of the old ridge and its belt of sub-

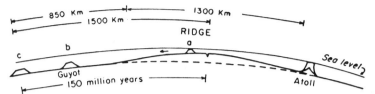

Figure 9. Diagram to show progressive migration of volcanic peaks, guyots, and atolls, from a ridge crest to the flanks, suggesting that the wave-cut surfaces of guyots or the bases of atolls may become older laterally away from the crest

sidence compared to present topographic ridges could be explained by the above reasoning.

Turning to a reconsideration of the Mid-Atlantic Ridge it appears that layer 3, with a thin and probably discontinuous cover of sediments, forms the sea floor. The dredging of serpentinized peridotite from fault scarps at three places on the ridge (Shand, 1949) points to such a conclusion. The abnormally low seismic velocity, if this is layer 3, might be attributed to intense fracturing and dilation where the convective flow changes direction from vertical to horizontal. The underlying material, which ordinarily would have a velocity of 8 km/sec or more, has a velocity approximately 7.4 km/sec partly for the same reason but also because of its abnormally high temperature (Fig. 7). The interface between layer 3 and the 7.4 km/sec material below is thus the M discontinuity. The increase in velocity of layer 3 to about 6.7 km/sec and of the sub-Moho material to 8 km/sec as one proceeds away from the ridge crest may be attributed to cooling and healing of the fractures by slight recrystallization or by deposition from solution in an interval of tens of millions of years.

DEVELOPMENT OF THE OCEANIC CRUST (LAYER 3) AND THE EVOLUTION OF SEA WATER

Assuming that layer 3 is serpentinized peridotite, that the water necessary to serpentinize it is derived by degassing of the rising column of a mantle convection cell, and that its uniform thickness (4.7 ± 0.5 km) is controlled by the highest level the 500° C isotherm can reach under these conditions, we have a set of reasonable hypotheses which can account for the observed facts (Fig. 6).

The present active ridge system in the oceans is about 25,000 km long. If the mantle is convecting with a velocity of 1 cm/yr a vertical layer 1 cm thick of layer 3 on each side of the ridge axis is being formed each year. The material formed is 70 per cent serpentinized, based on an average seismic velocity of 6.7 km/sec, and this serpentine contains 25 per cent water by volume. If we multiply these various quantities, the volume of water leaving the mantle each year can be estimated at 0.4 km³. Had this process operated at this rate for 4 aeons, 1.6×10^9 km³ of water would have been extracted from the mantle, and this less 0.3×10^9 km³ of water now in layer 3 equals 1.3×10^9 km³ or approximately the present volume of water in the oceans.[4]

---

[4] The estimate of how much of the present Mid-Ocean Ridge system is active is uncertain. That fraction of the sytem with a median rift was used in this estimate. The whole system is approximately 75,000 km long. The velocity of 1 cm/yr is also uncertain. If it were 0.35 cm/yr, as Heezen mentions for widening of the Iceland rift, this coupled with a 75,000 km ength of the ridge system would give the required amount of water for the sea in 4 aeons.

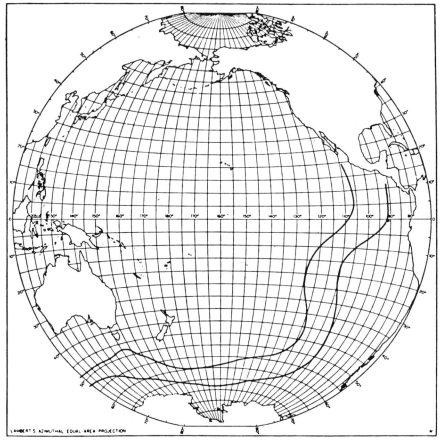

Figure 10. Approximate outline of East Pacific Rise, which possibly represents an oceanic
ridge so young that it has not yet developed a median rift zone and pre-Rise
sediments still cap most of its crest

The production of layer 3 by a convective system and serpentinization must be
reversed over the downward limbs of convection cells. That is, as layer 3 is depressed into the downward limb it will deserpentinize at 500° C and release its
water upward to the sea. Thus the rate of entry of juvenile water into the ocean
will equal the rate of acquisition of water from the mantle to form layer 3 over
the rising limbs of convection cells.

It is not at present possible to check against the record the assumption that the
process outlined went far back to the beginning of geologic history at a uniform
rate. If Africa and South America moved away from each other at the rate of 2
cm a year they would have been adjacent to each other about 200 million years
ago. Presumably this was the beginning of the convection cells under the present
ridge. The assumption of a rate of movement for convection of 1 cm/yr was
based on the above situation because the geologic record suggests splitting apart
near the end of the Paleozoic Era. The convection cells under the Mesozoic Mid-

Pacific Ridge ceased to function about 100 million ago inasmuch as the crest is known to have begun to subside at this time. It must have taken at least 150 million years at 1 cm/yr for the flanks of the ridge to spread to a width of 3000 km, and possibly the convection cells were in operation here for several times this long. The East Pacific Rise crosses the Mesozoic ridge at right angles and presumably did not come into existence until recent times, but certainly less than 100 million years ago. No evidence of older ridges is found in the oceans, suggesting that convection is effective in wiping the slate clean every 200 or 300 million years. This long and devious route leads to the conclusion that the present shapes and floors of ocean basins are comparatively young features.

RELATIONSHIP OF THICKNESS OF CONTINENTS TO DEPTH OF THE SEA

In Figure 2 the balance of oceanic and continental columns is portrayed. The layer thicknesses are derived from seismic profiles, and the densities are extrapolated from seismic velocities and petrologic deduction (Hess, 1955). Gravity measurements during the past half century have shown that the concept of isostasy is valid—in other words that a balance does exist. The oceanic column is simpler than the continental column and less subject to conjecture with regard to layer thicknesses or densities. The main uncertainty in the continental column is its mean density. Given the thickness of the crust, this value was derived by assuming that the pressure at 40 km below sea level under the continents equalled that for the same depth under the oceans, or 11,775 kg/cm². The mean density of the continental crust then becomes 2.85 g/cc. The latitude that one has for changing the numerical values in either of the two columns is small. The error in the pressure assumed for 40 km depth is probably less than 1 per cent.

The upper surface of the continent is adjusting to equilibrium with sea level by erosion. But as material is removed from its upper surface, ultimately to be deposited along its margins in the sea, the continent rises isostatically. If undisturbed by tectonic forces or thermal changes it will approach equilibrium at a rate estimated by Gilluly (1954) as 3.3×10^7 yrs half life. It is thus evident that, if the oceans were half as deep, the continents would be eroded to come to equilibrium with the new sea level, they would rise isostatically, and a new and much shallower depth to the M discontinuity under continents would gradually be established. A thinner continent but one of greater lateral extent would be formed inasmuch as volume would not be changed in this hypothetical procedure. The relationship between depth of the oceans, sea level, and the depth to the M discontinuity under continents is an axiomatic one and is a potent tool in reasoning about the past history of the Earth's surface and crust.

The oft-repeated statement that amount of water in the sea could not have changed appreciably since the beginning of the Paleozoic Era (or even much further back) because the sea has repeatedly lapped over and retreated from almost all continental areas during this time interval is invalid because the axiomatic relationship stated in the last paragraph would automatically require that this be so regardless of the amount of water in the sea.

One can compute the pressure at 40 km depth for an ocean with 1, 2, 3, or 4

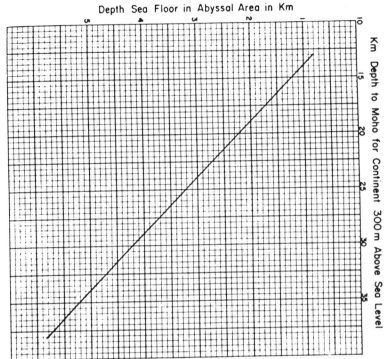

Figure 11. Graph portraying depth to the M discontinuity under continents *vs.* depth of abyssal areas in oceans, computed from balance of crustal columns

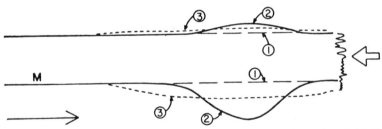

Figure 12. Diagram to illustrate thickening of a continent by deformation. Initially a mountain system and much larger root are formed, but both spread laterally with time and isostatic adjustment

km of water and equate this to continental columns for the same pressure at 40 km, distributing the amount of crustal material (density 2.85 g/cc) and mantle material (density 3.31 g/cc) in such proportion that balance is established. This computation is shown graphically in Figure 11. Assuming, as has been done in this chapter, that the oceans have grown gradually with time, one must suppose that the continents were much thinner in the early Precambrian. This could possibly be recognizable in the difference of tectonic pattern in very old terrains as compared to present continental structure.

If there is gradual increase of water in the sea one may ask why continents are not eventually flooded and why are there not continental-type areas now a kilo-meter or more below sea level. No extensive areas of this sort are found. Part of the answer might lie in the generation of new continental material at a rate equivalent to eruption of new water. An increase of depth of the sea by 1 km allows thickening of the continents by about 5 times this amount, which would be several times in excess of the estimated 1 km^3 per year extraction of magma from the mantle. Even if this were an underestimate there is no reason why continents might not extend laterally rather than grow thicker. The answer seems to be that there is more than enough energy in the crustal regime of the Earth to thicken the continents to an extent that they are maintained somewhat above the equilibrium level (Fig. 12). A continent will ride on convecting mantle until it reaches the downward-plunging limb of the cell. Because of its much lower density it cannot be forced down, so that its leading edge is strongly de-formed and thickened when this occurs. It might override the downward-flowing mantle current for a short distance, but thickening would be the result as before.

The Atlantic, Indian, and Arctic oceans are surrounded by the trailing edges of continents moving away from them, whereas the Pacific Ocean is faced by the leading edges of continents moving toward the island arcs and representing downward-flowing limbs of mantle convection cells or, as in the the case of the eastern Pacific margin, they have plunged into and in part overridden the zone of strong deformation over the downward-flowing limbs.

RECAPITULATION

The following assumptions were made, and the following conclusions reached:

(1)   The mantle is convecting at a rate of 1 cm/yr.

(2)   The convecting cells have rising limbs under the mid-ocean ridges.

(3)   The convecting cells account for the observed high heat flow and topo-graphic rise.

(4)   Mantle material comes to the surface on the crest of these ridges.

(5)   The oceanic crust is serpentinized peridotite, hydrated by release of water from the mantle over the rising limb of a current. In other words it is hydrated mantle material.

(6)   The uniform thickness of the oceanic crust results from the maximum height that the 500° C isotherm can reach under the mid-ocean ridge.

(7)   Seismic velocities under the crests of ridges are 10–20 per cent lower than normal for the various layers including the mantle, but become normal again on ridge flanks. This is attributed to higher temperature and intense fracturing with cooling and healing of the fractures away from the crest.

(8)   Mid-ocean ridges are ephemeral features having a life of 200 to 300 million years (the life of the convecting cell).

(9)   The Mid-Pacific Mesozoic Ridge is the only trace of a ridge of the last cycle of convecting cells.

(10)   The whole ocean is virtually swept clean (replaced by new mantle ma-terial) every 300 to 400 million years.

(11)   This accounts for the relatively thin veneer of sediments on the ocean

floor, the relatively small number of volcanic seamounts, and the present absence of evidence of rocks older than Cretaceous in the oceans.

(12) The oceanic column is in isostatic equilibrium with the continental column. The upper surface of continents approaches equilibrium with sea level by erosion. It is thus axiomatic that the thickness of continents is dependent on the depth of the oceans.

(13) Rising limbs coming up under continental areas move the fragmented parts away from one another at a uniform rate so a truly median ridge forms as in the Atlantic Ocean.

(14) The continents are carried passively on the mantle with convection and do not plow through oceanic crust.

(15) Their leading edges are strongly deformed when they impinge upon the downward moving limbs of convecting mantle.

(16) The oceanic crust, buckling down into the descending limb, is heated and loses its water to the ocean.

(17) The cover of oceanic sediments and the volcanic seamounts also ride down into the jaw crusher of the descending limb, are metamorphosed, and eventually probably are welded onto continents.

(18) The ocean basins are impermanent features, and the continents are permanent although they may be torn apart or welded together and their margins deformed.

(19) The Earth is a dynamic body with its surface constantly changing. The spherical harmonics of its topography show unexpected regularities, a reflection of the regularities of its mantle convection systems and their secondary effects.

In this chapter the writer has attempted to invent an evolution for ocean basins. It is hardly likely that all of the numerous assumptions made are correct. Nevertheless it appears to be a useful framework for testing various and sundry groups of hypotheses relating to the oceans. It is hoped that the framework with necessary patching and repair may eventually form the basis for a new and sounder structure.

ACKNOWLEDGMENTS

The writer's research on ocean basins has been supported by the Office of Naval Research. He is particularly indebted to Carl Bowin for critical evaluation of a number of the ideas discussed above. The writer is grateful for comments on the manuscript by W. W. Rubey, H. W. Menard, M. N. Bass, C. E. Helsley, A. E. J. Engel, C. Burk, and many others.

REFERENCES CITED

Bullard, E. C., Maxwell, A. E., Revelle, R., 1956, Heat flow through the deep sea floor: Advances in Geophysics, v. 3, p. 153–181

Carey, S. W., 1958, The tectonic approach to continental drift: Symposium, Univ. of Tasmania, Hobart, 1956, p. 177–358

Chandrasekhar, S., 1953, The onset of convection by thermal instability in spherical shells: Philos. Mag., ser. 7, v. 44, p. 233–241

Egyed, L., 1957, A new dynamic conception of the internal constitution of the Earth: Geol. Rundschau, v. 46, p. 101–121

Ewing, J., and Ewing, M., 1959, Seismic-refraction profiles in the Atlantic ocean basins, in the Mediterranean Sea, on the Mid-Atlantic Ridge and in the Norwegian Sea: Geol. Soc. America Bull., v. 70, p. 291–318

Gilluly, J., 1954, Geologic contrasts between continents and ocean basins, p. 7–18 in Poldervaart, Arie, Editor, Crust of the earth: Geol. Soc. America Spec. Paper 62, 762 p.

Gibson, W. M., 1960, Submarine topography in the Gulf of Alaska: Geol. Soc. America Bull., v. 71, p. 1087–1108

Griggs, D., 1939, A theory of mountain building: Am. Jour. Sci., v. 237, p. 611–650

———— 1954, Discussion, Verhoogen, 1954: Am. Geophys. Union Trans., v. 35, p. 93–96

Hamilton, E. L., 1956, Sunken islands of the Mid-Pacific mountains: Geol. Soc. America Mem. 64, 98 p.

———— 1960, Ocean basin ages and amounts of original sediments: Jour. Sediment. Petrology, v. 30, p. 370–379

Harris, P. G., and Rowell, J. A., 1960, Some geochemical aspects of the Mohorovicic discontinuity: Jour. Geophys. Research, v. 65, p. 2443–2460

Heezen, B. C., 1960, The rift in the ocean floor: Scient. American, v. 203, p. 98–110

Herzen, R. von, 1959, Heat flow values from the southern Pacific: Nature, v. 183, p. 882–883

Hess, H. H., 1946, Drowned ancient islands of the Pacific Basin: Am. Jour. Sci., v. 244, p. 772–791

———— 1954, Serpentines, orogeny and epeirogeny, p. 391–408 in Poldervaart, Arie, Editor, Crust of the earth: Geol. Soc. America Spec. Paper 62, 762 p.

———— 1955, The oceanic crust: Jour. Marine Research, v. 14, p. 423–439

———— 1959a, The AMSOC hole to the Earth's mantle: Am. Geophys. Union Trans., v. 40, p. 340–345; (1960, Am. Scientist, v. 47, p. 254–263)

———— 1959b, Nature of the great oceanic ridges: Internat. Ocean. Cong. Preprints, p. 33–34, AAAS, Washington, D. C.

Irving, E., 1959, Paleomagnetic pole positions: Roy. Astron. Soc. Geophys. Jour., v. 2, p. 51–77

Ladd, H. S., and Schlanger, S. O., 1960, Drilling operations on Eniwetok Atoll: U. S. Geol. Survey Prof. Paper 260Y, p. 863–903

Kennedy, G. C., 1959, The origin of continents, mountain ranges and ocean basins: Am. Scientist, v. 47, p. 491–504

Kuenen, Ph. H., 1946, Rate and mass of deep-sea sedimentation: Am. Jour. Sci., v. 244, p. 563–572

Kuiper, G., 1954, On the origin of the lunar surface features: Nat. Acad. Sci. Proc., v. 40, p. 1096–1112

MacDonald, G. J. F., 1959, Calculations on the thermal history of the Earth: Jour. Geophys. Research, v. 64, p. 1967–2000

Menard, H. W., 1958, Development of median elevations in the ocean basins: Geol. Soc. America, v. 69, p. 1179–1186

———— 1959, Geology of the Pacific sea floor: Experientia, v. XV/6, p. 205–213

Raitt, R. W., 1956, Seismic refraction studies of the Pacific Ocean Basin: Geol. Soc. America Bull., v. 67, p. 1623–1640

Ross, C. S., Foster, M. D., Myers, A. T., 1954, Origin of dunites and olivine rich inclusions in basaltic rocks: Am. Mineralogist, v. 39, p. 693–737

Rubey, W. W., 1951, Geologic history of sea water: Geol. Soc. America Bull., v. 62, p. 1111–1148

Runcorn, S. K., 1959, Rock magnetism: Science, v. 129, p. 1002–1011

Sapper, K., 1927, Vulkankunde, Stuttgart, Engelhorn, 358 p.

Shand, S. J., 1949, Rocks of the Mid-Atlantic Ridge: Jour. Geology, v. 57, p. 89–92

Sumner, J. S., 1954, Consequences of a polymorphic transition at the Mohorovicic discontinuity (Abstract): Am. Geophys. Union Trans., v. 35, p. 385

Umbgrove, J. H. F., 1947, The pulse of the Earth, 2nd ed.: The Hague, Martinus Nijhoff 359 p.

Urey, H. C., 1953, Comments on planetary convection as applied to the Earth: Philos. Mag., ser. 7, v. 44, p. 227–230

———— 1957, Boundary conditions for theories of the origin of the solar system: Progress in Physics and Chemistry of the Earth, v. 2, p. 46–76

Vening Meinesz, F. A., 1952, The origin of continents and oceans: Geol. en Mijnbouw, n. ser. v. 14, 373–384

———— 1959, The results of development of the Earth's topography in spherical harmonics up to the 31st order, provisional conclusions: Koninkl. Nederl. Akad. v. Wetenschappen Amsterdam Proc., ser. B, v. 62, p. 115–136

Verhoogen, J., 1954, Petrologic evidence on temperature distribution in the mantle of the Earth: Am. Geophys. Union Trans., v. 35, p. 50–59

24

ALPINE SERPENTINES AS OCEANIC RIND FRAGMENTS

ROBERT S. DIETZ

Abstract: Accepting that alpine serpentines are tectonically emplaced in the solid state, the writer suggests that they may be fragments of the sea floor derived from the Oceanic Layer. This concept is based on the writer's belief that eugeosynclinal graywackes may be equated with modern continental-rise turbidite prisms which are laid down on the deep-sea floor, abutting and overlapping, in part, the continental slope. When the sea floor thrusts toward the continent, the continental-rise prism is folded into a eugeosynclinal prism. The Franciscan graywackes prism of the California Coast Range would be an example. Pods of serpentine derived from the sea-floor sima underlying the eugeosyncline would be caught up in this folding, along with some deeper sub-M discontinuity ultramafic mantle rock. It is further supposed that the spilite-keratophyre suite characteristic of eugeosynclines is laid down in the deep ocean, Na metasomatism being caused by sea-water contact. On the floor of the open sea, away from the continental-rise turbidites, Layer 2 of sea-floor seismology probably is made up of spilite plus lithified eupelagic sediments altered to chert, ironstone, red argillite, and carbonate rock. This view has implications for the Mohole project.

Introduction

There are two major genetic types of ultramafic bodies, lopoliths or layered intrusions and so-called alpine serpentines that occur in the folded eugeosynclinal graywackes of orogenic belts. The writer offers a new suggestion concerning the origin of these curious and widely distributed ultramafic masses. This thesis is that eugeosynclines, the precursors of alpine mountains, are formed on the ocean floor rather than on the continents and that serpentines are fragments of the oceanic rind which are tectonically caught up in the folding when the orogenic pulse occurs.

Alpine-type serpentines occur in linear swarms of subparallel or en échelon masses along the axes of old and new belts of folded mountains, for example, along 1600 miles of the Appalachian axis from Nova Scotia to North Carolina; in North Carolina alone several hundred serpentine masses are known, many less than 100 m across. The pods of serpentine enclosed in the Franciscan graywackes of the Coast Ranges of California and Oregon are even more striking. Similar masses occur in Japan, the Philippines, New Caledonia, and New Zealand. Two more of the numerous examples occur in the folded strata along the eastern margin of Australia and in the Ural Mountains. Invariably these serpentines are in eugeosynclinal prisms of graywacke that have been orogenically folded into alpine mountains.

Previous Views

Several writers have recently reviewed the alpine-serpentine problem (Hess, 1955; Thayer, 1960; Turner and Verhoogen, 1960), so only brief comments are needed here. The lack of any strong metamorphic effects has led nearly all petrologists to reject emplacement by direct intrusion of a primary peridotite magma, as this would involve temperatures in excess of 1600°C. To account for the evidence of low temperatures and to allow for autoserpentinization Hess (1938) hypothesized a special peridotite magma with 25 per cent water, thus approximating serpentine in composition. Bowen and Tuttle (1949, p. 440), however, showed by laboratory experiments that ". . . there is no likelihood that any magma can exist that can be called a serpentine magma, and certainly no possibility of its existence below 1000°C." Their experiments further showed that serpentines can form only at temperatures below about 500°C, and if the temperature exceeds 500°C, serpentines revert to peridotite. Bowen's (1928) early view that alpine serpentines were injected in the quasi-liquid state as crystal mushes now seems acceptable to most petrologists. A temperature of 1100°–1200°C would still be needed, but if only small portions of the mass were magma, the heat of crystallization would be small. The emplacement of serpentine according to the usual tectogene concept is shown in Figure 1.

The author suggests an alternate explanation to the crystal-mush concept, which has several weaknesses. For example, an enormous amount of water would have to be extracted from the surrounding rocks to cause the complete serpentinization of the large masses of peridotite

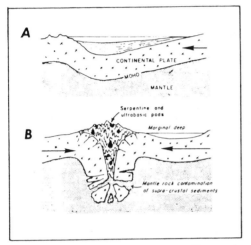

Figure 1.   Origin of serpentines, according to the tectogene concept. (A) Geosyncline is downbowed into the continental plate. (B) Ultramafic rocks from the mantle rise through the sial. These are serpentinized and intruded as pods into eugeosynclinal prisms of folded mountains of the so-called alpine type.

commonly observed. Bowen's thesis only partially avoids the problem of the general absence of heat effects around serpentine bodies. The layered character of some of these serpentines remains anomalous if the injection was quasi-liquid. Also, considerable expansion is involved with the conversion of peridotite to serpentine, unless large-scale extraction of ions is assumed. In brief, the crystal-mush concept is not an entirely satisfactory solution to this knotty problem in petrogenesis.

Serpentines as Oceanic Crust

The crystal-mush concept assumes the necessity of vertical migration through the sialic continental plate because such a long transfer would seem to require a fluid substance. But this also would involve a deep and hence hot

source. Certainly a simpler solution would be for the serpentine to be the basement rock upon which a eugeosynclinal prism is deposited. This would be true if eugeosynclines are formed on the sea floor marginal to continents and if the Oceanic Layer (i.e., the so-called Basaltic Layer or Oceanic Crust lying just above the M discontinuity) were serpentine. Regarding the latter "if," Hess (1954) has convincingly advanced the view that the mantle is feldspathic peridotite and that the Oceanic Layer is simply its hydrated equivalent, i.e., serpentine. Most geologists would accept this as a reasonable, if not a likely, possibility. On the other hand, the first "if" discards the classical and still almost universal belief that eugeosynclines, as with all other types of geosynclines, form on a sialic or continental basement.

Apart from the serpentine problem, the writer is convinced that eugeosynclinal turbidite prisms exist on the deep-sea floor (Dietz, 1963); others have given this view some support (King, 1959; Heezen and others, 1959; Drake and others, 1959). The writer's particular version is that eugeosynclines are fully equatable with modern continental-rise sedimentary prisms. Such rises apparently have been laid down by turbidity currents. The rise off the eastern United States may be taken as the type example. The graywacke suites of the compressed and uplifted eugeosynclinal orogens may be ancient continental rises.

In the writer's view the conventional geosyncline theory is not actualistic—modern examples are said not to exist (Kay, 1951). This deficiency is avoided if the continental-terrace sedimentary wedges are modern miogeosynclines and if the subjacent continental-rise prisms are eugeosynclines, together forming a couplet. This equating is reasonable if the wave-built-terrace concept of continental-terrace development is discarded (Dietz, 1952). More reasonably the terrace wedge (the miogeosyncline) is constructed by prograding paralic beds on an isostatically downwarping continental margin. Most sediment, however, is carried off the continental terrace and forms the continental rise (the eugeosynclinal prism), an apron of turbidity-current deposits of graywacke or Flysch facies. This development of the miogeosynclinal-eugeosynclinal couplet requires diastrophic quiescence, except for isostatic downbowing.

The orogenic or alpine-mountain-building cycle must be of such a nature as to convert a continental-rise prism into a folded mountain belt. This can be accomplished by continent-ward thrusting of the sea floor, possibly by the mechanism of sea-floor spreading which relates this thrusting to mantle thermal convection that invades the sea floor (Dietz, 1961). Thrusting, folding, and metamorphism would then account for alpine mountains. Batholithic injection might follow from fusible sialic components simultaneously carried to great depths. The inherent low density of the sedi-ments plus sialic plutonism would contribute buoyancy, so that the compressed prism would rise to continental elevations. With the onset of orogeny the serpentinized ultramafic rocks underlying the continental rise would be injected in the solid state and tectonically incorporated into the folding (Fig. 2). In this scheme, even if not correct in detail, the serpentines may have existed as hydrated ultra-mafic rocks before the sediments were laid

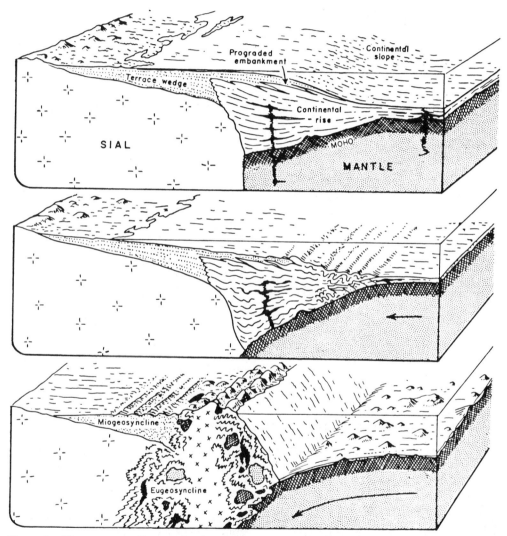

Figure 2. Time-sequence diagram showing writer's view on the origin of the spilite-keratophyre suite, serpentines, and ultramafic rocks in alpine-folded eugeosynclinal prisms. During tectonic quiescence a-long a continental margin, a continental-terrace wedge and a continental rise are laid down. Upon thrust-ing of the ocean floor toward the continents (by sea floor spreading?) the continental rise is converted into a eugeosyncline and the terrace wedge into a miogeosyncline. Bodies of spilite, serpentine, and ultramafic rock become incorporated as fragments of the sea floor.

down. Thus we can dispense with the problem of the origin of the water in serpentines; it would have been derived mostly from the sea plus some juvenile water from the mantle. The serpentines would have been intruded in a completely solid state and at a low temperature.

This suggestion is incomplete without some explanation of the other major components of the graywacke suites, the spilite-keratophyre association plus the radiolarian cherts, ironstones, and thin limestones. One may suggest that the graywackes form from hemipelagic sediments shed from the continental block, the spilite-kertophyre suite represents altered effusions of basaltic and andesitic fluids on the sea floor, and the associated sedimentary rocks are altered eupelagic sediments, red clay, and biogenous oozes. Turner and Verhoogen (1960) state that spilites apparently were derived from basalts uncontaminated by sialic components and that their albitization is a secondary replacement of their more calcic feldspars with Na. This Na metasomatism may be reasonably ascribed to a direct contact with sea water. The extensive development of pillow lavas and of zeolitization also suggests direct sea-water contact. Similarly the Na metasomatism and the thermal and pressure conditions implied by the development of glaucophane are consistent with this explanation. The keratophyres give evidence of sialic contamination, but this is not surprising in view of the marginal position of a continental rise with its turbidites being continental detritus.

The writer supposes that this association may be equated to Layer 2, a mixture of somewhat lithified sediments and effusive rocks overlying the serpentinized peridotite of the Oceanic Layer. When associated with the continental rise or eugeosyncline, these deposits would be mostly intercalated with the graywackes. However, they might also underlie and lie seaward of this prism so that some tectonic insertion might occur as with the serpentine.

The gross layering of the sea floor is doubtlessly complicated in detail—much more so than is suggested by their simple seismic designation as Layer 1, Layer 2, Oceanic Layer, and Mantle. But for purposes of generalization and following the rationale of this paper, the writer suggests in summary that Layer 1 (*ca.* 300 m thick) is unconsolidated sediment. Layer 2 (*ca.* 2 km thick) is a compacted or lithified mixture of spilites and keratophyres plus their usually associated rocks: cherts, limestones, dolomites, red argillites, and ironstones. The

Oceanic Layer (*ca.* 5 km thick) is partially or wholly serpentinized ultramafic rocks—especially peridotite and dunite in accordance with the contention of Hess (1955). New evidence of this has been recently provided by the dredging of rather fully serpentinized peridotite from a wall of the Puerto Rico Trench, where seismic data suggested an outcropping of the Oceanic Layer (Hersey, 1962). Of necessity, then, the upper mantle below the M discontinuity would be ultramafic. There is, of course, every reason to expect a considerable amount of gabbro in the Oceanic Layer, but it should be of secondary importance relative to serpentine (Fig. 3).

Discussion

The basement beneath eugeosynclines is only rarely unequivocally exposed; for example, no sialic basement can be identified for the Franciscan and Knoxville formations of California and Oregon. This supports the view that these sediments were laid down as a continental rise; thus the included serpentines and ultramafic rocks may be basement fragments. Even if a sialic basement can be demonstrated for parts of certain eugeosynclines, this does not disprove prove that eugeosynclines are laid down at the deep-ocean margin. This is expected as the sialic continental slope underlies nearly half of the sedimentary prism.

The present hypothesis suggests that extensive samples of the oceanic crust and mantle are accessible to the geologist's pick. The Franciscan, for example, provides numerous outcrops of these ultramafic rocks in only slightly metamorphosed condition. The concept then has important bearing on the Mohole project which proposes to drill through the oceanic rind into the mantle. Whereas serpentine would permit a faster drilling rate than basalt or gabbro (commonly suggested as components of the Oceanic Layer), its tendency for solid flow would be detrimental.

It is widely believed that deep-sea sediments are virtually or entirely absent from the continental blocks (Gilluly, 1955). But if the eugeosynclinal deposits are former continental-rise prisms of turbidites, then the deep-sea deposits on continents are of enormous extent and volume. They probably exceed the shallow-water continental-platform sediments in overall bulk. The deeper fabric of the continents is largely composed of this graywacke facies, as revealed on the deeply denuded Precambrian shields. The metamorphosed sedimentary rocks of such terrains are composed almost entirely of the eugeosynclinal graywacke suite. Continental-rise turbidites are hemipelagic (terrigenous detritus) rather than eupelagic sediments (red clay and biogenic oozes), and they are presumed here to have been incorporated into the continents by orogeny rather than "elevator tectonics" (localized high-amplitude block uplift). These factors may account for the geological difficulty of recognizing them as deep-sea sediments.

Hess (1955) also has stressed that serpentine

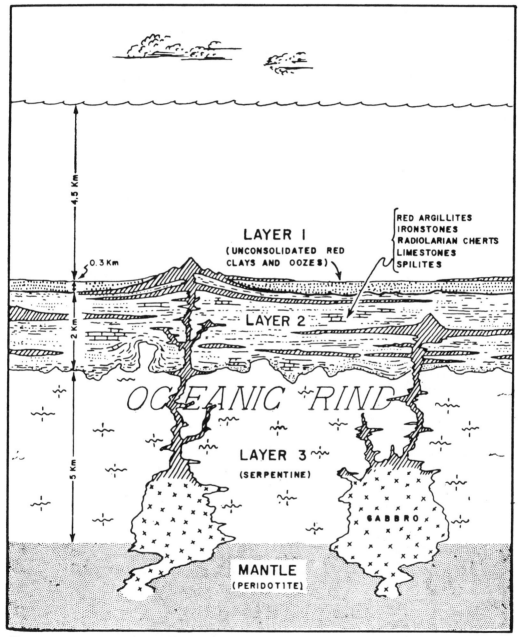

Figure 3. Generalized drawing of the subocean layers through which a Mohole would penetrate if the serpentines, ultramafic rocks, and spilites in eugeosynclinal prisms are derived from the sea floor.

intrusions may be used to mark the beginning of an orogeny. He surmised for example, that the initial and principal compression of the Appalachian revolution happened toward the end of the Ordovician rather than at the end of the Paleozoic. The writer's hypothesis is in agreement with this. If alpine-serpentine intrusions are ever radiometrically dated, they should prove older than the beds intruded. Hess's further contention that serpentines can be used to mark the sites of ancient trenches and associated island arcs does not follow; it involves the tectogene concept of a compressionally downfolded crust (Fig. 1). Instead the continental-rise prism is here considered as the locus of serpentine injections. It should be emphasized that the injection of serpentine would mark the beginning of the formation of geologic mountains (folded rocks) but not necessarily of geographic mountains above sea level. The Franciscan folds of Jurassic and Cretaceous age were not unroofed as exposed mountains until the Miocene.

Wilson (1949) has advanced the idea of marginal accretionary growth of continents following successive mountain-building orogenies. This concept has attracted many adherents, as it explains the successively younger mountain belts around the oldest rocks of shield nuclei. Continental contraction would necessarily occur if eugeosynclines are forever being formed upon the old continental platform; therefore, continental accretion is reasonable only if eugeosynclines are developed on the oceanic crust marginal to the continents.

References Cited

Bowen, N. L., 1928, The evolution of the igneous rocks: Princeton Univ. Press, 332 p.

Bowen, N. L., and Tuttle, O. F., 1949, The system $MgO-SiO_2-H_2O$: Geol. Soc. America Bull., v. 60, p. 439–460

Dietz, R. S., 1952, Geomorphic evolution of continental terrace (continental shelf and slope): Am. Assoc. Petroleum Geologists Bull., v. 36, p. 1802–1819

—— 1961, Continent and ocean basin evolution by spreading of the sea floor: Nature, v. 190, no. 4779, p. 854–857

—— 1963, Collapsing continental rises: An actualistic concept of geosynclines and mountain building: Jour. Geology, v. 71, no. 3, p. 314–333

Drake, C., Ewing, M., and Sutton, G., 1959, Continental margins and geosynclines—the east coast of North America north of Cape Hatteras, p. 110–198 in Ahrens, L. H., Press, Frank, Rankama, K., and Runcorn, S. K., Editors, Physics and chemistry of the earth, Volume 3: London, Pergamon Press, 464 p.

Gilluly, J., 1955, Geologic contrasts between continents and ocean basins, p. 7–18 in Poldervaart, Arie, Editor, Crust of the earth: Geol. Soc. America Special Paper 62, 762 p.

Heezen, B., Tharp, M., and Ewing, M., 1959, The floors of the oceans, Part 1, The North Atlantic: Geol. Soc. America Special Paper 65, 122 p.

Hersey, J. B., 1962, Findings made during the June 1961 cruise of CHAIN to the Puerto Rico Trench and Caryn Sea Mount: Jour. Geophys. Research, v. 67, no. 3, p. 1109–1116

Hess, H. H., 1938, A primary peridotite magma: Am. Jour. Sci., 5th ser., v. 35, no. 209, p. 321–344

—— 1954 [1955], Geological hypotheses and the earth's crust under the oceans, p. 341–348 in Bullard, E. C., Discussion Leader, A discussion on the floor of the Atlantic Ocean: Royal Soc. London Proc., ser. A, v. 222, p. 287–563

—— 1955, Serpentine, orogeny, and epeirogeny, p. 391–407 in Poldervaart, Arie, Editor, Crust of the earth: Geol. Soc. America Special Paper 62, 762 p.

Kay, M., 1951, North American geosynclines: Geol. Soc. America Mem. 48, 143 p.

King, P. B., 1959, The evolution of North America: Princeton Univ. Press, 189 p.

Thayer, T. P., 1960, Some critical differences between alpine-type and stratiform peridotite-gabbro complexes: 21st Internat. Geol. Cong. Rept., Copenhagen, 1960, pt. 13, p. 247–259

Turner, F. J., and Verhoogen, J., 1960, Igneous and metamorphic petrology (2nd ed.): New York, McGraw-Hill Book Co., Inc., 694 p.

Wilson, J. T., 1949, The origin of continents and Precambrian history: Royal Soc. Canada Trans., 3d ser., v. 43, sec. 4, p. 157–184

25

COLLAPSING CONTINENTAL RISES: AN ACTUALISTIC CONCEPT OF GEOSYNCLINES AND MOUNTAIN BUILDING[1]

ROBERT S. DIETZ

U.S. Navy Electronics Laboratory, San Diego 52, California

ABSTRACT

Conventional geosynclinal theory is not actualistic—modern examples are said not to exist. This deficiency is obviated if the continental terrace sedimentary wedges are modern miogeosynclines and if the subjacent continental-rise sedimentary prisms are modern eugeosynclines, together forming a couplet. Such equating is reasonable once the wave-built terrace concept of continental terrace development is discarded. Instead, the writer assumes construction of the terrace wedge (a miogeosyncline) by prograded wedges of paralic sediments on an isostatically downwarping continental margin. Most sediment, however, is assumed to be carried off the continental terrace to form the continental rise (a eugeosynclinal prism), an apron of turbidity current deposits of graywacke or flysch facies. The accumulation of the mio-eugeosynclinal couplet requires diastrophic quiescence except for isostatic depression.

The orogenic or alpine mountain-building cycle is initiated by thrusting probably related to decoupling of a spreading sea floor so that the sima tends to slip beneath the sialic continental raft. The continental rise is then thrust, folded, and metamorphosed, forming a folded eugeosyncline; ultramafics are intruded and are tectonically incorporated in the folding. Since it lies inside and upon buoyant sial, the continental terrace wedge, or miogeosyncline, is folded and thrust to a lesser extent. Intrusion of plutons is post primary folding.

Some implications of this actualistic concept of geosynclines are discussed. The eugeosyncline, for example, is oceanic and is composed of abyssal hemipelagic continental detritus. Orogeny generates new continental crust and causes continental accretion. The geosynclinal troughs are not formed by crustal compression. Island arcs, borderlands, trenches, and tectogenes play no part in this actualistic version.

I. INTRODUCTION

The geosynclinal theory is a pervasive and unifying geologic concept. But a wide divergence of views still exists, so the concept remains uncertain and unsatisfactory. The theory has been particularly unsuccessful actualistically. General agreement is lacking whether or not any particular modern structure on the continents or the ocean floor is, or is not, a geosyncline. According to the perceptive review of geosynclinal theory by Glaessner and Teichert (1947), Stille and others were inclined toward the view that the continents already have passed through their geosynclinal stage of development and are wholly consolidated. In an excellent memoir, Kay (1951) shows that the geologic development of North America has been dominated by a series of collapsing miogeosynclines and eugeosynclines. He states, however, that no examples of these types of geosynclines are being formed today. The modern Atlantic and Gulf Coast geosynclines are described as an entirely new type—a paraliageosyncline. Thus the most thoroughly probed modern geosynclines are said not to fit the normal geosynclinal pattern. For example, their sedimentary fill is coming from the distant continental interior rather than from either a borderland (an Appalachia) or a tectonic land (an island arc).

Some definitions are necessary since geosynclinal terminology is extensive and complex. The miogeosyncline and eugeosyncline are the two basic primary "true" or orthosynclines in the terminology of Stille. A eugeosyncline is a linear geosyncline parallel to the continental shield, or craton, and contains graywackes, volcanic rocks, cherts, etc. The Middle Ordovician, central New England to Newfoundland belt, is a good example; the Caledonian geosyncline of Wales is another (Jones, 1937).

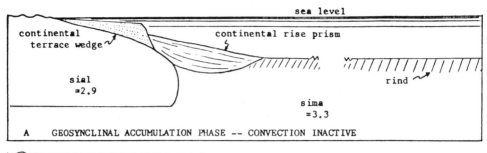

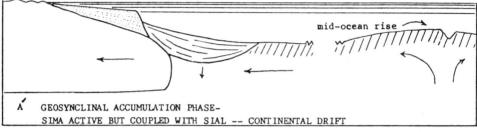

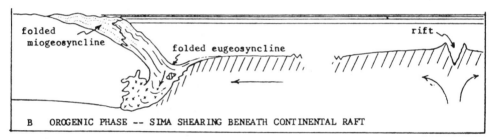

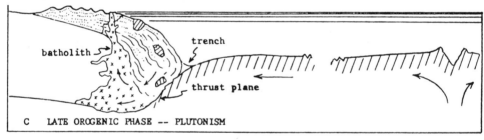

Fig. 1.—*A*, Schematic time-sequence diagram showing actualistic concept of geosynclines. Thermal convection circulation of mantle is assumed to be the fundamental diastrophic force. The ocean floor is considered to be crustless in that only the sea floor itself is an upper boundary to convection circulation. The oceanic rind is thought to be mostly serpentinized periodotite. *A'*, This is an alternate possibility (and the preferred view). Geosynclinal accumulation takes place during tectonic quiescence of the continental margin but is maintained in spite of sea-floor spreading of the ultrabasic sima. The continental raft is coupled so that continental drift occurs. *B*, Upon decoupling, the sima shears beneath continent, initiating an orogenic cycle. The continental rise is compressed, folded, and thrust against the continental raft, being added marginally to it. The miogeosyncline is also folded but to a lesser extent; it is only slightly metamorphosed. Black masses are symbolic of ultramafic serpentine masses that are broken-off pieces of the ocean floor incorporated in the eugeosyncline. *C*, The late orogenic or plutonic period is shown. A trench has developed, marking a thrust plane along which the sea floor pushed beneath the continent. The surficial sea-floor sedimentary layers and new detritus carried into the trench is thrust under the continents and sialized, and batholiths form. Water, provided by deserpentinization of the oceanic rind, helps to flux this reaction. Once the shearing ceases and the mountains erode, the conditions are appropriate for the commencement of phase *A* again.

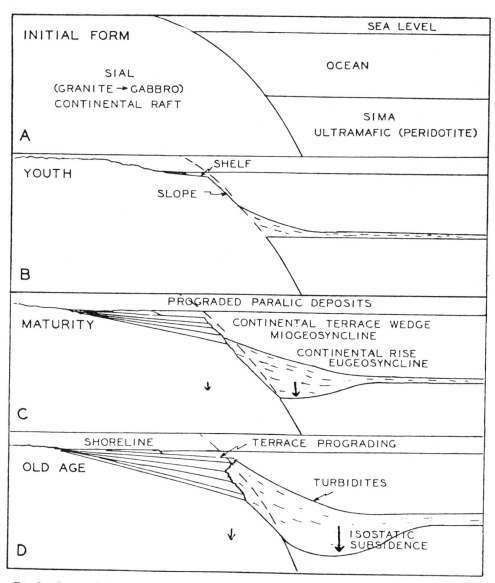

Fig. 2.—Geomorphic evolution of the continental terrace. *A, Initial Form.*—Immediately following a major orogeny, showing an unmodified continental margin. *B, Youth.*—A wave-cut terrace forms and a small continental rise occurs. The continental slope is eroded back slightly. A small prograded bed has been formed. *C, Maturity.*—A large continental rise composed of turbidites has been formed, causing isostatic depression of the sea floor with accompanying isostatic depression of the continental margin. This margin downwarping has allowed the preservation of numerous prograded wedges of paralic sediments laid down by transgression and regression of the shoreline. As each paralic wedge has a rather steep front, these wedges are capable of building up a terrace wedge on top of the continental slope. *D, Old Age.*—Eventually the continental-rise beds lap up to the top of the continental slope. This now permits the continental margin to grow seaward. It is supposed that this situation now exists off the Texas coast except that the slope has been rejuvenated by the formation of the Sigsbee fault at the base of the continental slope, so that it is now failing by mass gravity sliding (after Dietz, 1952).

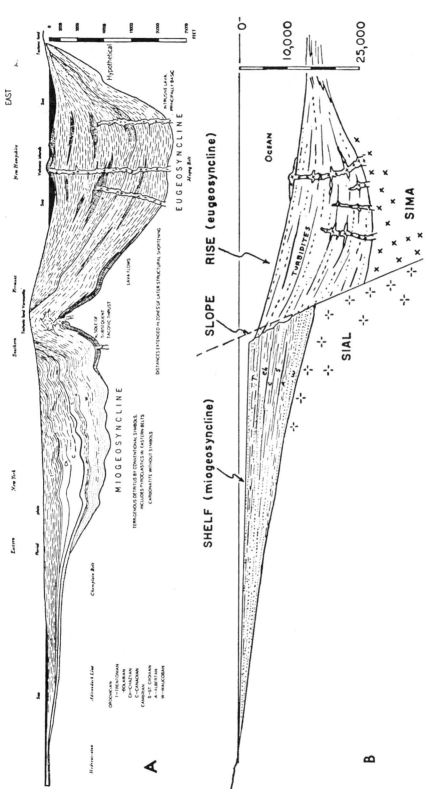

Fig. 3.—*A*, reconstruction according to Kay (1951) of the typical miogeosyncline and eugeosynclinal couplet just prior to orogenic compression. This is the New York–Maine Cambrian through middle Ordovician sedimentary prism shown in the immediate post-Trenton time. Kay envisions shallow sea deposition for the eugeosyncline. A tectonic highland is located between the two geosynclines and a volcanic-tectonic highland oceanward of the eugeosyncline. *B*, the author's suggested reconstruction of the same miogeosynclinal couplet at the end of Trenton time. The reconstruction is drawn to be analogous to our presumed existing situation off the eastern United States. The eugeosyncline is the continental-rise prism of sediments shed from the continental block and deposited in the deep ocean at the base of the continental slope largely by turbidity currents laying down turbidites (graywackes). The miogeosynclinal wedge is composed mostly of prograded paralic deposits. The continental rise is shown invaded by plutons and volcanics. Since these rise from suboceanic sima, they would be ultramafic, which conforms to the characteristic presence in eugeosynclines of pillow lavas, spilites, ophiolites, keratophyres, and melanophyres (Kay, 1951, p. 69). The continental-rise prism is shown as already having undergone considerable isostatic subsidence. The sediments, being somewhat lighter than the sima, have caused a shallowing of the ocean, reducing somewhat the original topographic contrast between the continental platform and the oceanic basin.

216

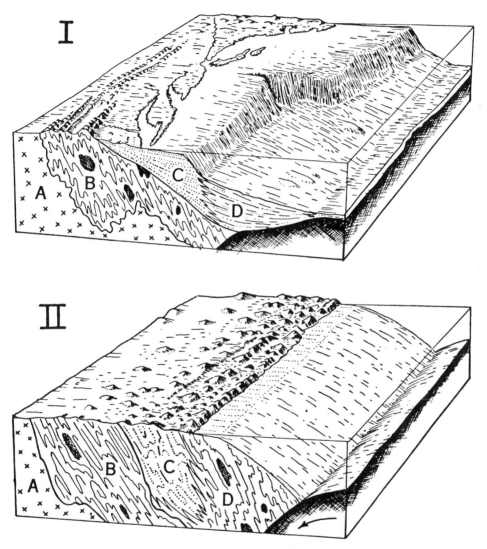

FIG. 5.—Block diagrams show the relationship of the continental terrace wedge and the continental rise prism to a miogeosynclinal and eugeosynclinal couplet. *I*, sedimentary accumulation during a long period of continental-margin quiescence. *II*, orogeny has occurred, causing the ocean floor to underthrust the continental block. The continental rise, being a buoyant mass, is folded against the nose of the continent. The flank of this orogeny creates a new continental slope. Diagrams are drawn to roughly approximate conditions along the eastern United States today (*I*) and at some future time (*II*). For simplicity, volcanism and plutonism are omitted from these diagrams. Cross-hatched masses are serpentine pods derived from the oceanic rind. *A*, *B*, *C*, and *D* refer to corresponding rock masses in *I* and *II*.

[*Editor's Note:* Only part of the introduction, figures 1 through 5, and the references cited therein are reproduced here.]

REFERENCES

Dietz, R., 1952, Geomorphic evolution of continental terrace (continental shelf and slope): *Am. Assoc. Petrol. Geologists Bull.*, v. 36, no. 9, p. 1802–1819.

Glaessner, M., and Teichert, C., 1947, Geosynclines: A fundamental concept in geology: *Am. Jour. Sci.*, v. 245, no. 8, p. 465–482, 571–591.

Jones, O., 1937, On the evolution of a geosyncline: *Geol. Soc. London Quart. Jour.*, v. 93, p. 241–283.

Kay, M., 1951, North American geosynclines: *Geol. Soc. America Memoir 48*, 143 p.

26

Copyright © 1965 by Macmillan Journals Ltd

Reprinted from *Nature* **207**:343–347 (1965)

A NEW CLASS OF FAULTS AND THEIR BEARING ON CONTINENTAL DRIFT

J. T. Wilson

*Institute of Earth Sciences, University of
Toronto*

*T*RANSFORMS *and half-shears.* Many geologists[1] have maintained that movements of the Earth's crust are concentrated in mobile belts, which may take the form of mountains, mid-ocean ridges or major faults with large horizontal movements. These features and the seismic activity along them often appear to end abruptly, which is puzzling. The problem has been difficult to investigate because most terminations lie in ocean basins.

This article suggests that these features are not isolated, that few come to dead ends, but that they are connected into a continuous network of mobile belts about the Earth which divide the surface into several large rigid plates (Fig. 1). Any feature at its apparent termination may be transformed into another feature of one of the other two types. For example, a fault may be transformed into a mid-ocean ridge as illustrated in Fig. 2a. At the point of transformation the horizontal shear motion along the fault ends abruptly by being changed into an expanding tensional motion across the ridge or rift with a change in seismicity.

A junction where one feature changes into another is here called a transform. This type and two others illustrated in Figs. 2b and c may also be termed half-shears (a name suggested in conversation by Prof. J. D. Bernal). Twice as many types of half-shears involve mountains as ridges, because mountains are asymmetrical whereas ridges have bilateral symmetry. This way of abruptly ending large horizontal shear motions is offered as an explanation of what has long been recognized as a puzzling feature of large faults like the San Andreas.

Another type of transform whereby a mountain is transformed into a mid-ocean ridge was suggested by S. W. Carey[2] when he proposed that the Pyrenees Mountains were compressed because of the rifting open of the Bay of Biscay (presumably by the formation of a mid-ocean ridge along its axis). The types illustrated are all dextral, but equivalent sinistral types exist.

In this article the term 'ridge' will be used to mean mid-ocean ridge and also rise (where that term has been used meaning mid-ocean ridge, as by Menard[3] in the Pacific basin). The terms mountains and mountain system may include island arcs. An arc is described as being convex or concave depending on which face is first reached when proceeding in the direction indicated by an arrow depicting relative motion (Figs. 2 and 3). The word fault may mean a system of several closely related faults.

Transform faults. Faults in which the displacement suddenly stops or changes form and direction are not true transcurrent faults. It is proposed that a separate class of horizontal shear faults exists which terminate abruptly at both ends, but which nevertheless may show great displacements. Each may be thought of as a pair of half-shears joined end to end. Any combination of pairs of the three dextral half-shears may be joined giving rise to the six types illustrated in Fig. 3. Another six sinistral forms can also exist. The name transform fault is proposed for the class, and members may be described in terms of the features which they connect (for example, dextral transform fault, ridge–convex arc type).

The distinctions between types might appear trivial until the variation in the habits of growth of the different types is considered as is shown in Fig. 4. These distinctions are that ridges expand to produce new crust, thus leaving residual inactive traces in the topography of their former positions. On the other hand oceanic crust moves down under island arcs absorbing old crust so that they leave no traces of past positions. The convex sides of arcs thus advance. For these reasons transform faults of types a, b and d in Fig. 4 grow in total width, type f diminishes and the behaviour of types c and e is indeterminate. It is significant that the direction of motion on transform faults of the type shown in Fig. 3a is the reverse of that required to offset the ridge. This is a fundamental difference between transform and transcurrent faulting.

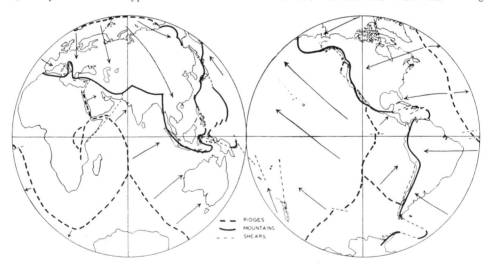

Fig 1. Sketch map illustrating the present network of mobile belts, comprising the active primary mountains and island arcs in compression (solid lines), active transform faults in horizontal shear (light dashed lines) and active mid-ocean ridges in tension (heavy dashed lines)

Many examples of these faults have been reported and their properties are known and will be shown to fit those required by the constructions above. If the class as a whole has not heretofore been recognized and defined, it is because all discussions of faulting, such as those of E. M. Anderson, have tacitly assumed that the faulted medium is continuous and conserved. If continents drift this assumption is not true. Large areas of crust must be swallowed up in front of an advancing continent and re-created in its wake. Transform faults cannot exist unless there is crustal displacement, and their existence would provide a powerful argument in favour of continental drift and a guide to the nature of the displacements involved. These proposals owe much to the ideas of S. W. Carey, but differ in that I suggest that the plates between mobile belts are not readily deformed except at their edges.

The data on which the ensuing accounts are based have largely been taken from papers in two recent symposia[4,5] and in several recent books[3,6,7] in which many additional references may be found.

North Atlantic ridge termination. If Europe and North America have moved apart, an explanation is required of how so large a rift as the Atlantic Ocean can come to a relatively abrupt and complete end in the cul-de-sac of the Arctic Sea. Fig. 5 illustrates one possible explanation.

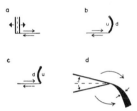

Fig. 2. Diagram illustrating the four possible right-hand transforms. *a*, Ridge to dextral half-shear; *b*, dextral half-shear to concave arc; *c*, dextral half-shear to convex arc; *d*, ridge to right-hand arc

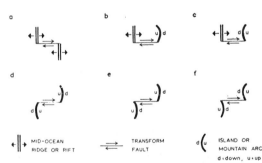

Fig. 3. Diagram illustrating the six possible types of dextral transform faults. *a*, Ridge to ridge type; *b*, ridge to concave arc; *c*, ridge to convex arc; *d*, concave arc to concave arc; *e*, concave arc to convex arc; *f*, convex arc to convex arc. Note that the direction of motion in *a* is the reverse of that required to offset the ridge

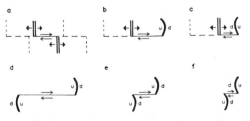

Fig. 4. Diagram illustrating the appearance of the six types of dextral transform faults shown in Fig. 3 after a period of growth. Traces of former positions now inactive, but still expressed in the topography, are shown by dashed lines

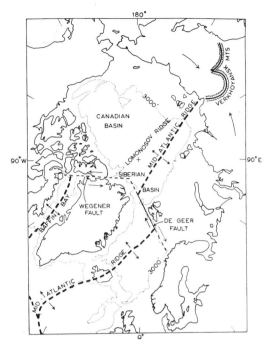

Fig. 5. Sketch map of the termination of the Mid-Atlantic ridge by two large transform faults (Wegener and De Geer faults) and by transformation into the Verkhoyansk Mountains

Wegener[8] suggested that the strait between Greenland and Ellesmere Island was formed by a fault, here postulated to be a sinistral transform fault (ridge–ridge type). Wegmann[9] named another between Norway, Spitsbergen and Greenland, the De Geer line, which is here regarded as a dextral transform fault (ridge–ridge type). The extension of the Mid-Atlantic ridge across the Siberian basin was traced by Heezen and Ewing[10], while Wilson[11] proposed its transform into the Verkhoyansk Mountains by rotation about a fulcrum in the New Siberian Islands. In accordance with the expectations from Fig. 4*a* earthquakes have been reported along the full line of the De Geer fault in Fig. 5, but not along the dashed older traces between Norway and Bear Island and to the north of Greenland. The Baffin Bay ridge and Wegener fault are at present quiescent. W. B. Harland[10] and Canadian geologists have commented on the similarities of Spitsbergen and Ellesmere Island.

Equatorial Atlantic fracture zones. If a continent in which there exist faults or lines of weakness splits into two parts (Fig. 6), the new tension fractures may trail and be affected by the existing faults.

The dextral transform faults (ridge–ridge type) such as *AA'* which would result from such a period of rifting can be seen to have peculiar features. The parts *AB* and *B'A'* are older than the rifting. *DD'* is young and is the only part now active. The offset of the ridge which it represents is not an ordinary faulted displacement such as a transcurrent fault would produce. It is independent of the distance through which the continents have moved. It is confusing, but true, that the direction of motion along *DD'* is in the reverse direction to that required to produce the apparent offset. The offset is merely a reflexion of the shape of the initial break between the continental blocks. The sections *BD* and *D'B'* of the fault are not now active, but are intermediate in age and are represented by fracture zones showing the path of former faulting.

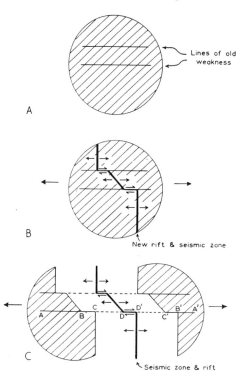

Fig. 6. Diagram illustrating three stages in the rifting of a continent into two parts (for example, South America and Africa). There will be seismic activity along the heavy lines only

Fig. 7 shows that the Mid-Atlantic ridge and the facture zones in the equatorial Atlantic may well be a more complex example of this kind. If so the apparent offsets on the ridge are not faulted offsets, but inherited from the shape of the break that first formed between the coasts of Africa and the Americas. Fig. 7 is traced from Heezen, Bunce, Hersey and Tharp[12] with additions to the north from Krause[13]. The fracture zones are here held to be right-hand transform faults and not left-hand transcurrent faults as previously stated. If the fracture zones can be traced across the Atlantic and are of the type postulated, then the points where they intersect the opposite coasts are conjugate points which would have been together before rifting.

It seems possible that the old fault in Pennsylvania and the offset of the Atlantic Coast described by Drake and Woodward[14] are of the same nature, although it is suggested that it is not usual for a fracture zone to follow a line of seamounts, and that the fracture zone may extend eastward, not south-east.

A possible explanation of the termination of the Carlsberg ridge. Another type of transform fault is found in the Indian Ocean (Fig. 8). If the Indian Ocean and Arabian Gulf opened during the Mesozoic and Cenozoic eras by the northward movement of India, new ocean floor must have been generated by spreading of the Carlsberg ridge. This ends abruptly in a transcurrent fault postulated by Gregory[15] off the east coast of Africa. A parallel fault has been found by Matthews[16] as an offset across the Carlsberg ridge and traced by him to the coast immediately west of Karachi. Here it joins the Ornach-Nal and other faults[17] which extend into Afghanistan and, according to such descriptions as I can find, probably merge with the western end of the Hindu Kush. This whole fault is thus an example of a sinistral transform fault (ridge-convex arc type).

At a later date, probably about Oligocene time according to papers quoted by Drake and Girdler[18], the ridge was extended up the Red Sea and again terminated in a sinistral transform fault (ridge-convex arc type) that forms the Jordan Valley[19] and terminates by joining a large thrust fault in south-eastern Turkey (Z. Ternek, private communication). The East African rift valleys are a still later extension formed in Upper Miocene time according to B. H. Baker (private communication).

The many offsets in the Gulf of Aden described by Laughton[20] provide another example of transform faults adjusting a rift to the shape of the adjacent coasts.

Possible relationships between active faults off the west coast of North America. This tendency of mid-ocean ridges to be offset parallel to adjacent coasts is thought to be evident again in the termination of the East Pacific ridge illustrated in Fig. 9. The San Andreas fault is here postulated to be a dextral transform fault (ridge-ridge type) and not a transcurrent fault. It connects the termination of the East Pacific ridge proper with another short length of ridge for which Menard[3] has found evidence off Vancouver Island. His explanation of the connexion—that the mid-ocean ridge connects across western United States—does not seem to be compatible with the view that the African rift valleys are also incipient mid-ocean ridges. The other end of the ridge off Vancouver Island appears to end in a second great submarine fault off British Columbia described by Benioff[7] as having dextral horizontal motion.

In Alaska are several large faults described by St. Amand[21]. Of the relations between them and those off

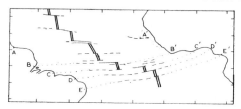

Fig. 7. Sketch (after Krause and Heezen *et al.*) showing how the Mid-Atlantic ridge is offset to the left by active transform faults which have dextral motions if the rift is expanding (see Fig. 4a). ||, Mid-ocean ridge; —, active fault; – – –, inactive fault trace; . . . , hypothetical extension of fault

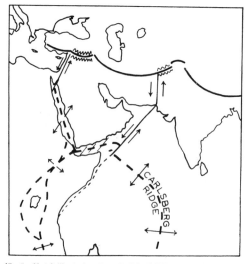

Fig. 8. Sketch illustrating the end of the Carlsberg mid-ocean ridge by a large transform fault (ridge-convex arc type) extending to the Hindu Kush, the end of the rift up the Red Sea by a similar transform fault extending into Turkey and the still younger East African rifts

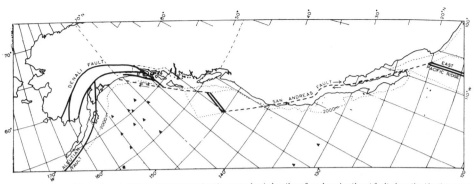

Fig. 9. Sketch map of the west coast of North America showing the approximate location of a submarine thrust fault along the Aleutian trench, the Denali faults (after St. Amand), the San Andreas and another large transform fault (after Benioff) and part of the East Pacific ridge and another mid-ocean ridge (after Menard)

the coast he writes: "If the two systems represent one consistent system, some interesting possibilities arise. One that the San Andreas and Alaska Complex is a gigantic tear fault, along which the Pacific Basin is being slid, relatively speaking under the Alaska Mainland, and the Bering Sea. On the other hand, if the whole system is a strike-slip fault having consistent right-lateral offset, then the whole of the western north Pacific Basin must be undergoing rotation".

St. Amand was uncertain, but preferred the latter alternative, whereas this interpretation would favour the former one. Thus the Denali system is considered to be predominantly a thrust, while the fault off British Columbia is a dextral transform fault.

At a first glance at Fig. 9 it might be held that the transform fault off British Columbia was of ridge–concave arc type and that it connects with the Denali system of thrust faults, but if the Pacific floor is sliding under Alaska, the submarine fault along the Aleutian arc that extends to Anchorage is more significant. In that case the Denali faults are part of a secondary arc system and the main fault is of ridge–convex arc type.

Further examples from the Eastern Pacific. If the examples given from the North and Equatorial Atlantic Ocean, Arabian Sea, Gulf of Aden and North-west Pacific are any guide, offsets of mid-ocean ridges along fracture zones are not faulted displacements, but are an inheritance from the shape of the original fracture. The fracture zones that cross the East Pacific ridge[22] are similar in that their seismicity is confined to the offset parts between ridges. An extension of this suggests that the offsets in the magnetic displacements observed in the aseismic facture zones off California may not be fault displacements as has usually been supposed, but that they reflect the shape of a contemporary rift in the Pacific Ocean. More complex variants of the kind postulated here seem to offer a better chance of explaining the different offsets noted by Vacquier[7] along different lengths of the Murray fracture zone than does transcurrent faulting. If the California fracture zones are of this character and are related to the Darwin rise as postulated by Hess, then the Darwin rise should be offset in a similar pattern.

The southern Andes appear to provide an example of compression combined with shearing. The compressional features are obvious. The existence of dextral shearing is also well known[23]. It is suggested that the latter may be due to the transformation of the West Chile ridge into a dextral transform fault (ridge–convex arc type) along the Andes which terminates at the northern end by thrusting under the Peruvian Andes (Fig. 10).

The observation that there is little seismicity and hence little movement south of the point where the West Chile ridge intersects the Andes can be explained if it is realized that the ridge system forms an almost complete ring about

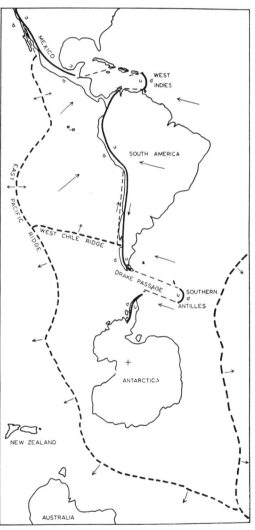

Fig. 10. Sketch map of Mexico, South America, Antarctica and part of the mid-ocean ridge system (heavy dashed lines) illustrating that the great loop of the ridge about Antarctica can only grow by increasing in diameter. Transform faults are shown by light dashed lines

Antarctica, from which expansion must everywhere be directed northwards. This may explain the absence of an isthmus across Drake Passage.

It would also appear that the faults at the two ends of the South Antilles and West Indies arcs are examples of dextral and sinistral pairs of transform faults (concave–concave arc types). According to Fig. 4 both these arcs should be advancing into the Atlantic and inactive east–west faults should not be found beyond the arcs.

This article began by suggesting that some aspects of faulting well known to be anomalous according to traditional concepts of transcurrent faults could be explained by defining a new class of transform faults of which twelve varieties were shown to be possible.

The demonstration by a few examples that at least six of the twelve types do appear to exist with the properties predicted justifies investigating the validity of this concept further.

It is particularly important to do this because transform faults can only exist if there is crustal displacement and proof of their existence would go far towards establishing the reality of continental drift and showing the nature of the displacements involved.

I thank the Departments of Geodesy and Geophysics and of Geology and Churchill College, University of Cambridge, for the opportunity to write this article, those whose data I have used, and colleagues—especially Sir Edward Bullard, W. B. Harland, H. H. Hess, D. H. Matthews and F. J. Vine—for advice; and Sue Chappell and Sue Vine for assistance. This is a contribution to the Vela Uniform programme and to the Canadian Upper Mantle Project.

[1] Bucher, W. H., *The Reformation of the Earth's Crust* (Princeton Univ. Press, 1933).

[2] Carey, S. W., *Proc. Roy. Soc. Tasmania*, **89**, 255 (1955).

[3] Menard, H. W., *Marine Geology of the Pacific* (McGraw-Hill Book Co., 1964).

[4] *Symp. on Continental Drift*, *Phil. Trans. Roy. Soc.*, edit. by Blackett, P. M. S., Bullard, E. C., and Runcorn, S. K. (in the press).

[5] *Symp. Earth Sciences, Mass. Inst. Tech.*, edit. by Townes, C. H. (in the press).

[6] *The Sea*, edit. by Hill, M. N., **3** (Interscience Pubs., New York and London, 1963).

[7] In *Continental Drift*, edit. by Runcorn, S. K. (Academic Press, New York and London, 1963).

[8] Wegener, A., *The Origin of Continents and Oceans* (E. P. Dutton and Co., New York, 1924).

[9] Wegmann, C. E., *Med. om Gronland*, **144**, No. 7 (1948).

[10] In *Geology of the Arctic*, edit. by Raasch, O., **1** (University of Toronto Press, 1961).

[11] Wilson, J. Tuzo, *Nature*, **198**, 925 (1963).

[12] Heezen, B. C., Bunce, E. T., Hersey, J. B., and Tharp, M., *Deep-Sea Res.*, **11**, 11 (1964).

[13] Krause, D. C., *Science*, **146**, 57 (1964).

[14] Drake, C. L., and Woodward, H. P., *Trans. N.Y. Acad. Sci.*, Ser. II, **26**, 48 (1963).

[15] Gregory, J. W., *Geog. J.*, **56**, 13 (1920).

[16] Matthews, D. H., *Nature*, **198**, 950 (1963).

[17] Hunting Survey Corp., *Recon. Geol. of Part of West Pakistan*, 365 (Toronto, 1960).

[18] Drake, C. L., and Girdler, R. W., *Geophys. J.*, **8**, 473 (1964).

[19] Quesnell, A. M., *Quart. J. Geol. Soc. Lond.*, **114**, 1 (1958).

[20] Laughton, A. S., *Proc. Roy. Soc.* (in the press).

[21] St. Amand, P., *Bull. Geol. Soc. Amer.*, **68**, 1343 (1957).

[22] Sykes, L. B., *J. Geophys. Res.*, **68**, 5999 (1963).

[23] St. Amand, P., *Los Terremotos de Mayo, Chile 1960* (Michelson Lab., U.S. Naval Ordnance Test Stations *NOTS* TP2701, China Lake, California, 1961).

Part VIII

MECHANISMS OF PLATE TECTONICS APPLIED TO ANCIENT GEOSYNCLINAL MOBILE BELTS— SOME EARLY ATTEMPTS

Editor's Comments
on Papers 27 Through 34

One of the first clear postulations that ancient geosynclinal belts could be produced by the opening and closing of ocean basins via sea-floor spreading and continental drift was that of J. Tuzo Wilson (Paper 27). He simply assumed a uniformitarian earth and concluded that the present tectonic processes must also have occurred in the past. Wilson contended that in Lower Paleozoic times, a proto-Atlantic ocean had gradually opened and separated the Old and New Worlds. The subsiding Atlantic-type continental margin that developed along the foundering coasts was the Appalachian geosyncline as it existed at the early stage of development, not a pair of elongate troughs, but simply subjacent continental shelf and rise areas as Drake, Ewing, and Sutton, as well as Dietz, had suggested (Papers 21, 24, and 25). The large, continent-sized borderland areas proposed originally by Barrell and Schuchert no longer had to mysteriously disappear; they survived as West Africa and Europe. A later episode of ocean closure, produced by the downward return of simatic lithosphere to the earth's interior, generated the bathymetric changes, island arc volcanicity, tectonic land uplift, and episodic metamorphism and orogeny dictated by the observable characteristics of the Appalachian eugeosyncline. Consequently, a different modern analogue existed for the Appalachian geosyncline, corresponding to a later stage of development and akin to the one suggested by Eardley (Paper 16) among others. Geosynclines, like the continental blocks and ocean basins, were not static phenomena. Individual geosynclines could collectively include the entire range of continental margin-ocean basin types visible at individual stages of development around the globe (the *Wilson cycle*, a label later proposed by Dewey and Burke, 1974).

Paper 28, also by Wilson, is included because it is a historically keyed, concise, clear statement of plate tectonics theory. This paper dramatically demonstrates how this new, globally unifying theory had rendered coherent such apparently diverse phenomena as seismic activity, global volcanicity, submarine physiography, faunal diversity, and apparent polar wandering. Wilson emphasized his earlier conclusion that if continental drift and sea-floor spreading were such obvious ongoing phenomena, they must have occurred widely in the geologic past and in belts other than the Appalachians. He elaborated on his earlier concept of a generalized life-cycle for ocean basins, defining a series of stages through which ocean basins (and by implication, geosynclines) evolved via initial continental rifting, ocean opening and subsequent closure, and final continental collision. The modern day analogues existing for each stage provided a reference standard with which to compare ancient mobile belts at specific points in space and time. Ancient mobile belts were simply the result of the vertical and horizontal juxtaposition of these various types of continental margin-ocean basin complexes.

Earlier attempts to understand ancient geosynclinal belts in comparison with modern continental margins admittedly had met with only mixed success. To some extent, sea-floor spreading and plate tectonics complicated the situation, because their existence meant that continental margins occupied unfixed positions and were subject to changes in their essential character and geosynclinal classification through time. Paper 29 by Mitchell and Reading is included here because it represents an early attempt to resolve this paradoxical situation. Mitchell and Reading carefully drew the important distinctions that exist among present-day "geosynclinal belts" (that is, the various kinds of modern continental margins). They are: Atlantic, Andean, Mediterranean, and island arc-inland sea (Western Pacific) types. Mitchell and Reading, like Wilson, were quite receptive to the concept that one continental margin type could readily undergo a transformation into another continental margin type. Paper 29 underscored the fact that the key to understanding ancient mobile belts was in an actualistic sense, with emphasis on flexibility of continental margins through time and space. Accepting such flexibility and assuming that modern continental margins could be accurately characterized, one could indeed "fingerprint" the various evolutionary stages of ancient geosynclinal belts. An integrated geotectonic-sedimentological methodology rendered this completely feasible as shown by Fairbridge (1958).

Papers 30, 31, 32, and 33 describe other early attempts to apply the concepts of the new global tectonics to specific mobile belts and to global geology in general. Dewey (Paper 30) documented the remarkable similarity in the structural and lithological framework of the northern (Newfoundland) Appalachians and the British Caledonides. His elegantly drawn, schematic cross-sections were convincing in their comprehensiveness. They incorporated spreading centers, flipping subduction zones, and sutures, and all developed as a proto-Atlantic Ocean opened and closed producing transformations from one type of continental margin to another. This model was strongly rooted in the earlier studies of Wilson (Paper 27) and Dietz (Papers 24 and 25). The importance of another milestone paper by Rodgers (1968), in which the modern Atlantic continental slope and Bahama Bank escarpment were used as analogues for the Appalachian miogeosynclinal-eugeosynclinal margin, cannot be overemphasized.

Dewey and Horsefield (Paper 31) considered the role of geosynclinal development via plate tectonics mechanisms in constructing continental blocks via lateral accretion. The authors almost indiscriminately applied plate theory to mountain belts across the globe.

Paper 32 by Bird and Dewey is included in excerpt form. This paper and with another paper by Dewey and Bird (1970); see also, Dennis, 1982) are probably two of the most widely read geological publications in this century. Profusely and beautifully illustrated, they awakened within the geological profession an awareness of how drastically differently mountain belts could be interpreted if one accepted continental drift, sea-floor spreading, and plate tectonics. Paper 33 by Dickinson expanded upon this theme of reinterpretation, revitalizing the analysis of ancient geosynclinal belts by incorporating into that analysis the perspectives of plate tectonics. The wild scramble aboard the new global tectonics bandwagon quickly followed.

However, Coney (Paper 34) questioned the fundamental premises of geosynclinal theory, raising the issue as to whether or not the theory ought to be completely rejected, rather than modified to fit the new schemes. It was obvious to Coney that new ground rules existed in tectonic theory, and equally obvious that new analytical tools must be developed to reinterpret mobile belt histories. Plate tectonics dictated that each mountain belt need not show the same sequence of developmental stages, thus there could be no common thread of dogma, no rigid regularity in sequence of rocks or deformational theory as geosynclinal theory had implied. From the first, geologists had insisted that geosynclines ultimately and irrevocably lead to the production of a mountain system. Plate tectonics implied no such certainty.

REFERENCES

Dennis, J., ed., in press, *Orogeny*, Hutchinson Ross Publishing Company, Stroudsburg, Pa.

Dewey, J. F., and J. M. Bird, 1970, Mountain Belts and the New Global Tectonics, *Jour. Geophys. Research* **75**:2625–2647.

Dewey, J. F., and K. Burke, 1974, Hot Spots and Continental Break-up: Implications for Collisional Orogeny, *Geology* **2**:57–60.

Fairbridge, R. W., 1958, What is a Consanguinous Association?, *Jour. Geology* **66**:319–324.

Rodgers, J., 1968, The Eastern Edge of the North American Continent During the Cambrian and Early Ordovician, in E.-An Zen, W. S. White, J. B. Hadley, and J. B. Thompson, Jr., eds., *Studies of Appalachian Geology: Northern and Maritime*, Wiley, New York, pp. 141–149.

27

Reprinted from *Nature* **211**:676–681 (1966)

DID THE ATLANTIC CLOSE AND THEN RE-OPEN?

By Prof. J. TUZO WILSON

Institute of Earth Sciences, University of Toronto

For more than a century it has been recognized that an unusual feature of the shallow water marine faunas of Lower Palaeozoic time is their division into two clearly marked geographic regions, which are commonly referred to as faunal realms. "The faunal assemblages are amazingly uniform throughout each realm so that correlation of any Cambrian section with another in the same realm is usually easy; on the other hand, the difference between the faunas in the two separate realms is so great as to make correlation between them very difficult"[1].

Two aspects of the distribution of these realms are remarkable. For one thing, some regions of similar faunas are separated by the whole width of the Atlantic Ocean; then, on the other hand, some regions of dissimilar faunas lie adjacent to one another. This is illustrated by Fig. 1, which is based on work by Cowie[2], Grabau[3] and Hutchinson[4].

Grabau showed that, if Europe and North America had become separated by continental drift, a simple reconstruction could explain the first anomaly in the distribution of the faunal realms in that, before the opening of the Atlantic Ocean, each realm would have been continuous, with no large gaps between outcrops of similar facies (Fig. 2).

It is the object of this article to show that drift can also explain the second anomaly. It is proposed that, in Lower Palaeozoic time, a proto-Atlantic Ocean existed so as to form the boundary between the two realms, and that during Middle and Upper Palaeozoic time the ocean closed by stages, so bringing dissimilar facies together

(Fig. 3). The supposed closing of the Tethys Sea by northward movement of India into contact with the rest of Asia, and the partial closing of the Mediterranean by northward movement of Africa, can be regarded as a similar but more recent event. The figures are based on a reconstruction by Bullard, Everett and Smith[5], but because those authors pointed out that no allowance had been made for the construction of post-Jurassic shelves, the continents have been brought more closely together.

Four lines of evidence suggest that this proposal is reasonable. (Unfortunately, so far as I can ascertain, palaeomagnetic evidence which might bear on this problem does not exist.)

First, this reconstruction of geological history is held to provide a unified explanation of the changes in rock types, fossils, mountain building episodes and palaeoclimates represented by the rocks of the Atlantic region.

Second, wherever the junction between contiguous parts of different realms is exposed, it is marked by extensive faulting, thrusting and crushing.

Third, there is evidence that the junction is everywhere along the eastern side of a series of ancient island arcs (Fig. 3).

Fourth, the fit appears to meet the geometric requirement that during a single cycle of closing and

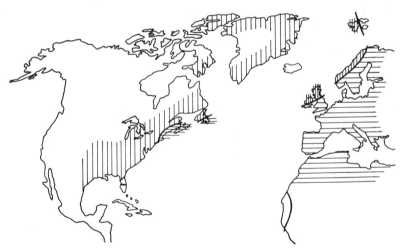

Fig. 1. The North Atlantic region showing the present distributions of the 'Atlantic' faunal realm (horizontal shading) and the 'Pacific' faunal realm (vertical shading). (After J. W. Cowie, A. W. Grabau and R. D. Hutchinson)

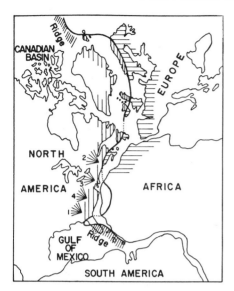

Fig. 2. The North Atlantic region in Upper Palaeozoic and Lower Mesozoic time showing that of the present Atlantic Ocean only the Canadian Basin and the Gulf of Mexico then existed. Four fans are shown which were formed: (1) during Middle Ordovician; (2) during Upper Ordovician; (3) during Upper Devonian; (4) during Pennsylvanian. The heavy line separates 'Pacific' and 'Atlantic' faunal realms. The two ridges are considered to have formed when the modern Atlantic started to open

reopening of an ocean, and in any latitudinal belt of the ocean, only one of the pair of opposing coasts can change sides (Fig. 4).

Recurrent Drift in the North Atlantic

The history proposed for the North Atlantic region can be stated very briefly as follows: (*a*) From the Late Pre-Cambrian to the close of Middle Ordovician time an open ocean existed in approximately, but not precisely, the same location as the present North Atlantic (Fig. 3). (*b*) From the Upper Ordovician to Carboniferous time this ocean closed by stages. (*c*) From Permian to Jurassic time there was no deep ocean in the North Atlantic region. The only marine deposits of that time are those connected with the Tethys Sea, with a shallow Jurassic invasion of Europe and with deeper Jurassic seas in the Gulf of Mexico and in the western Arctic Basin (Fig. 2). (*d*) Since the beginning of the Cretaceous period the present Atlantic Ocean has been opening, but this reopening did not follow the precise line of junction formed by the closing of the early Palaeozoic Atlantic Ocean; the result is that some coastal regions have been transposed (Fig. 1).

The Lower Palaeozoic continents may have first touched each other at the end of Middle Ordovician time, for thereafter the distinction between 'Atlantic' and 'Pacific' faunal realms ceases to be marked, but the complete closing of the ancient Atlantic may have required several periods.

For each continent, union meant replacing the open ocean by the other continent. This is offered as an explanation of the borderlands of J. Barrell and C. Schuchert for which there is no clear evidence until Upper Ordovician time. As Kay has suggested[6] concerning Eastern North America: "There has been little discussion of the evidence for borderlands in earlier Paleozoic time, though some have expressed scepticism". Kay's own support for island arcs is muted after Lower Palaeozoic time and he

accepts the view that the sediments of the "Late Devonian and Early Mississippian came from the land of Appalachia" —a borderland.

This view that extensive upland source areas lay to the east of the Appalachian geosyncline in the sites of the present coastal plain or ocean has been fully supported by recent work[7-9]. Tens of thousands of cubic miles of quartz-rich sediments, derived from the east, were deposited in shallow marine to sub-aerial deltas.

When the continents were pushed together, they would have touched first at one promontory and then at another. It can be expected that high mountains would have been formed locally and that they would have produced alluvial fans on both continents. As the ocean diminished the climate would have become increasingly arid. Such drastic alterations in the physiography would explain the change from predominantly marine and island arc deposition in the Lower Palaeozoic to conspicuous fans of Queenston, Catskill, Old Red Sandstone, and other deltas of Middle and Upper Palaeozoic time (see refs. 10 and 11, and Fig. 2). It can also be expected that the collision of continents would have produced great local uplifts which, if one continent overrode the other, would have migrated inland, perhaps pushing the Taconic and northern Newfoundland klippen[12] before them.

It would seem that by Permian and Triassic times the Atlantic Ocean was completely closed, because only continental beds, such as the Dunkard and Newark series, are found in North America. In Great Britain the New Red Sandstone is also continental as is the Permian of the Oslo district.

No Jurassic beds are known in eastern North America except in the Gulf of Mexico. Those of Europe were formed by a shallow marine invasion of the continent and are said by Hallam[13] to have fossils that "include many neritic forms that could not have crossed a deep ocean. The paleogeography for the Scottish Jurassic gives no hint of increasingly marine conditions to the west" (personal communication). Most of the available geological evidence suggests that the present Atlantic Ocean started to open at the beginning of Cretaceous time[14,15]. Although objections to this view are still being raised, they seem to be minor in comparison with the other evidence, and it is possible that they can be explained in other ways.

Faulted Contact between Faunal Realms

Starting our considerations in the north, the island of West Spitsbergen is underlain by a thick eugeosynclinal section of Lower Palaeozoic rocks named the Hecla Hoek succession. These strata rest on no known basement and were deformed, metamorphosed and intruded by granites during the Caledonian orogeny[16,17]. They contain fossils of the 'Pacific' fauna similar to those of Scotland and North America[18].

In Nordaustlandet, the adjacent, eastern island of the Spitsbergen group, Kulling[19] and Sandford[20] have mapped a thin section of unmetamorphosed and gently folded strata which a few fossils indicate to be of about the same age. These beds lie uncomformably on a basement which is regarded as part of the Baltic Shield. These strata do not thicken to the west as the much thicker section of West Spitsbergen is approached, nor do the few fossils necessarily belong to the 'Pacific' faunal realm.

Despite the considerable number of attempts to compare the sections in the adjacent islands, the correlation is not good. Changes in thickness, facies, and degree of metamorphism, basement and type of intrusives are all abrupt and striking. Orvin[21] and others have mapped faults in Hinlopen Strait between the islands. Klitin[22] summarizes the situation thus: "Of particular interest is the junction zone of the alleged Caledonian platform and Caledonian fold system. The transition to typical caledonids takes place in a zone not over 15 to 20 km wide, in the Hinlopen

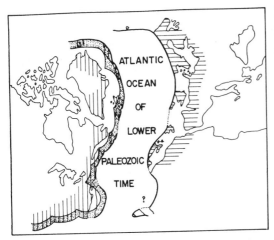

Fig. 3. The North Atlantic region in Lower Palaeozoic time. The proto-Atlantic Ocean would have formed a complete barrier between two faunal realms (shaded). Island arcs (dotted) probably lay along the North American coast. The floor of this ocean could have been absorbed in the trenches associated with these arcs as the ocean closed

Strait area, where the Hecla Hoek section abruptly increases in thickness, by a factor of four, with the appearance of extrusives in its metamorphosed and linearly folded beds. Such radical changes in thickness suggest a fault junction between an ancient platform and the caledonids." Following this interpretation I suggest that the Lower Palaeozoic ocean once separated the two islands and that, whereas Nordaustland formed part of the Baltic shield and shelf, West Spitsbergen was the site of a North American island arc.

In Scandinavia it is well established that the peninsula is divided longitudinally into two different provinces separated by a great zone of nappes and faults overthrust towards the east. According to the descriptions of Holtedahl[23], south-eastern Norway and most of Sweden are underlain by an extension of the Baltic Platform on which lie nearly flat, unfossiliferous rocks and "the eastern facies of the Cambro-Silurian (which) can be classed as miogeosynclinal in the terminology of Stille. The thickness is rather large and there is much terrigeneous material. Caledonian volcanic and intrusive igneous rocks are not found in the deposits of this type. The rocks are unmetamorphosed in the east and are of low metamorphic grade farther to the north-west. The deposits occur in an autochthonous or parautochthonous position above the original Archean basement."

The fossils are repeatedly referred to as being similar to those of the Baltic region, England and Wales. On the other hand, in the Trondheim area of western Norway, a thick succession of pillow lavas, shale and limestone with serpentinites in the lower part of the sequence contains a fauna "of American affinities . . . the limestone of Smøla is similar to the Durness in Scotland and to more or less contemporaneous limestone in Newfoundland, Bear Island and Spitsbergen. The limestone in Smøla thus seems to represent an American-Arctic facies of the Ordovician." In these "mainly pelitic sediments . . . we thus have a eugeosynclinal facies characteristic of the central parts of a geosyncline . . . probably all rocks of the present facies occur in allochthonous positions." An important event in central southern Norway was the close of marine deposition "in Late Ordovician or Early Silurian time brought about by the thrusting of nappes and deposition of the Valdres sparagmite (arkose), which was considered as a deposit of flysch type by Goldschmidt".

The zone of nappes is, therefore, held to be the boundary between faunal provinces and the line of closure of the Lower Palaeozoic ocean.

The geological relationship between Spitsbergen, Scandinavia and the British Isles has been discussed by Bailey and Holtedahl[24]. Of the Caledonian structures they state: " . . . in the present land area of Scandinavia we find the eastern part of the orogenetic belt only, while in Great Britain the whole of the orogenetic zone is represented. The Spitsbergen Group seems to lie rather centrally in the zone of deformation." Following Wegener, they consider that Greenland may have been formerly connected with Europe and suggest that, if so, it would have completed the western side of the mountain belt opposite Norway. This relationship has found support in recent years from several authors[14] including Umbgrove[25].

That the boundary between 'Atlantic' and 'Pacific' faunal realms crosses northern England between Scotland and Wales is supported by Walton, who writes of the Scottish occurrences: "The close affinity of the Durness and North America rocks was recognised long ago . . . . The long-recognised affinity of the Girvan Caradocian and Appalachian Mohawkian faunas was re-emphasized by Williams . . . . By contrast the faunas are only remotely connected with the Welsh Ordovician rocks"[26]. George[27] describes the relations thus: "The Salopian geosyncline may thus have extended unbrokenly from conjectural north-western limits in the Highlands to a south-eastern margin or marginal shelf in the English Midlands". It does not require much change to regard this geosyncline as a former ocean.

In Newfoundland the geology of the north-eastern coast has recently been described by Williams[28]. He suggests that in Cambrian time a deep basin or ocean, not underlain by continental crust, crossed the central part of the island and separated two shelves. The north-western shelf underlying the Long Peninsula has a basement of Grenville age overlain by strata with 'Pacific' faunas like those of Scotland, while the south-eastern shelf, which forms the Avalon peninsula, has a younger Pre-Cambrian basement overlain by strata with 'Atlantic' faunas like those of Wales. During Ordovician and subsequent time the intervening sea became filled with eugeosynclinal and volcanic sedimentary rocks, probably representing a former island arc and mountain belt.

Anderson[29] believes that faulting in Hermitage Bay on the south coast of Newfoundland not only separates the rocks of the south-eastern shelf from the central geosyncline, but may also completely divide the two faunas. On the north coast the corresponding fault zone may be in Freshwater Bay, but this point has not been settled as yet[30].

Among those who have recently correlated the Newfoundland and British sections are Dewey and Church[31]. Although Church does not favour continental drift, both he and Dewey make the same correlation as that already given here and extend the Caledonian mobile belt from northern England and central Ireland to central Newfoundland.

South of Newfoundland, the Gulf of St. Lawrence and younger rocks cover the key areas of Lower Palaeozoic formations almost as far as the Maine border. Little can be said except that the faunas of Cape Breton Island and St. John, New Brunswick, have European affinities while those of Gaspé and the Eastern Township of Quebec are typical of Scotland and most of North America[4].

In northern and eastern Maine the older literature is sparse and generalized. Recently a combined group of government and university geologists, including W. B. N. Berry, A. J. Boucot, E. Mencher, R. S. Naylor and L. Pavlides, have discovered new fossil localities there with both European and North American affinities, important Caledonian uplifts and large pre-Silurian faults[32]. Much of the state has been remapped, but for the reason that little of this work has yet been published (R. G. Doyle, personal communication), and because the structure is clearly much more complex than shown on early maps, this is not an opportune time to consider the area in detail.

From the southern part of Maine south across New Hampshire, Massachusetts and Connecticut a major zone can be traced from published accounts. Novotny[33] has described this "major fault zone" where it crosses the New Hampshire–Massachusetts boundary (Fig. 5). To the north it connects with several faults and silicified zones shown on the map of New Hampshire[34,35]. These lie along the line which separates those formations which underlie the greater part of New Hampshire from a suite of completely different formations, underlying south-eastern New Hampshire. This line may be continued northward into south-eastern Maine by the faults which Katz[36] has suggested bound the Berwich gneiss, itself crumpled, "closely folded and overturned". In Massachusetts, Novotny's fault may be exposed in the abandoned Worcester "coal" mine. The description suggests that much of the rock may be carbonaceous mylonite[37]. South of Worcester this fault connects with a major change in formations evident on the geological map of Massachusetts[38].

In Connecticut this boundary appears to join the Honey Hill fault and its northern continuation which bisect the state and separate two major rock sequences. Where best described in the south it "has been mapped for 25 miles eastward from Chester nearly to Preston without apparent repetition of the stratigraphy on either side of the fault. The fault plane dips 10°–35° N parallel to the underlying metasedimentary rocks . . . . The fault is marked by a zone a mile wide of mylonitized and crushed rocks . . . . Displacement must have been many miles"[39]. As in Norway, the orogenic belt appears to have been thrust eastwards over the eastern platform. To the south this major fault seems to strike into the Atlantic Ocean and the Appalachian belt, then narrows conspicuously. I have recently learned from L. R. Page and J. W. Peoples that mapping (for the most unpublished) by the federal and state surveys has defined this fault zone more satisfactorily, and that published aeromagnetic maps show a change in the strike of anomalies across it. This evidence suggests that New England is divided by a major fault zone into two provinces underlain by quite different rock formations. Of the few occurrences of the Lower Palaeozoic faunas, all those with European affinities lie to the east of the fault zone; all those typical of North America to the west[40,41].

In the light of this evidence it seems reasonable to suggest that this fault zone marks the line of closure of the Lower Palaeozoic Atlantic Ocean. It may seem strange to propose that a former position of the Atlantic Ocean lies through New England and that its full significance has not been realized, but it must be remembered that throughout the area outcrops are poor and that most of the mapping is old. Surface mapping reveals few faults, but new tunnels have shown that faults abound under the drift-filled valleys and that some of these are major (J. W. Skehan and A. Quinn, personal communications).

Most North American geologists have not accepted the idea of continental drift. Instead they have sought to explain the changes in faunas in terms of different environmental conditions. This has certainly been a factor and no doubt was responsible for the differences between the Durness, Girvan and Moffatt facies in Scotland, although those facies belong to the same faunal realm. In another example, G. Theokritoff (personal communication) has directed our attention to a possible mixing of Atlantic, Pacific and endemic faunas in the Taconic sequence of New York[42]. Christina Lochman[43] has emphasized the difficulties incurred in accepting such an interpretation. She states that the areas of mixed faunas

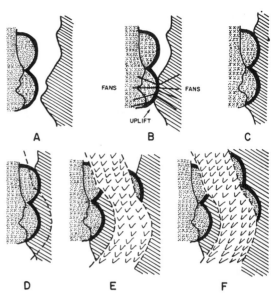

Fig. 4. *A*, A closing ocean, with island arcs on one coast, separating two different faunal realms. *B*, First contact between two opposite sides of a closing ocean. *C*, The ocean closed by overlap of the opposite coasts. *D*, A possible line (dashed) along which a younger ocean could reopen. *E*, A new ocean (checked) opening in an old continent. *F*, A geometrically impossible way for a younger ocean to open. (Note how the arcs overlap.)

lie in deeply down-warped basins between two shallow shelf deposits containing respectively Atlantic and Pacific faunas. These basins, she remarks, had connexions with the Atlantic Ocean and had a benthonic environment "similar to that of the open ocean . . . . Few normal benthonic species of the coastal shelf could establish themselves in such an alien environment, although, because of the geographic proximity of the two areas, sporadically drifted individuals might be found." She also refers to deep basins separating different faunas and to the evidence for a "biofacies regime ordinarily found on the floors of the continental shelf beyond the inner islands of the volcanic archipelago". These views would seem to admit the possibility of interpreting the palaeogeography as has been done in this article. It is suggested that one of the deep basins, instead of lying on a shelf, might have been an open ocean in Lower Palaeozoic time and that the mixed fauna may have come from the extreme edge of one continent.

The reconstruction (Fig. 2) then leads to Africa. Sougy[44] has described a large plate of metamorphic rocks east of Dakar thrust eastwards over Upper Devonian strata. Farther north in Spanish Morocco there is a folded belt. A large Cretaceous overlap along the coast and lack of diagnostic fossils in the overthrust block make correlation difficult, but Sougy concludes that: "In the future, when geologists study the relationships between America and Africa for evidence bearing on permanence of continents and oceans *versus* continental drift, they will have to consider that the western rim of Africa is made, from Guinea to Morocco, not of a Pre-Cambrian basement but of a mainly Hercynian orogenic belt, in some respects symmetrical to the Appalachian belt." His mention of drift suggests, as do most reconstructions[5], that this West African belt was formerly part of the Appalachians. This view is supported by P. A. Mohr, who writes: "the Cambrian manganese ores of Newfoundland, Wales and Morocco . . . I think were formed in a common geosynclinal trough" (personal communication). The view that these regions were

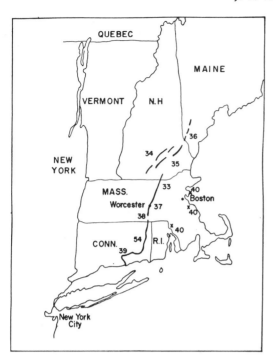

Fig. 5. Sketch map of New England showing location of some major faults and three fossil localities of the 'Atlantic' faunal realm (X). The numbers refer to papers in the list of references from which information was obtained

together until Cretaceous time is supported by the close similarity of Cretaceous fossils from these two regions now on opposite sides of the ocean[45]. Only further investigation, aided by extensive seismic investigations or drilling through the Cretaceous of both continental coasts, can show whether it is possible or likely that West Africa ever fitted into the central east coast of the United States. If so, some rotation of Africa clockwise and a closer fit than that shown by Bullard, Everett and Smith[5] may be indicated.

Although the northern termination of the West African fold belt is covered, its termination to the south is definite; the Pre-Cambrian of the African Shield extends to the Atlantic near Conakry[46]. The reconstruction suggests that this contact might reach Florida. In Florida and Georgia a cross-section of the Appalachians resembles that of Newfoundland in that these are the only parts of the Appalachians with a platform on both sides of the mobile belt. The platform on the south-eastern side of the mobile belt which lies beneath northern Florida and adjacent states is only known from well cores. These consist of Early Ordovician to Middle Devonian sandstones and shales said to have been deposited in shallow water[47]. Published accounts do not correlate the strata or fossils with formations elsewhere. It would seem that the basement has not been penetrated. Thus it is only speculative to suggest that this region was formerly part of Africa.

Island Arcs of Lower Palaeozoic Atlantic

According to a widely adopted hypothesis, island arcs and mountains represent places where the lithosphere is being compressed, while mid-ocean ridges represent places where it is being pulled apart and where new crust is being created. Thus the present Atlantic Ocean is expanding from the Mid-Atlantic Ridge.

For a Lower Palaeozoic Atlantic Ocean to have closed, it must, according to this hypothesis, have been marked not by a central ridge, but by a continuous system of island arcs and mountains. Observation shows that such systems commonly lie at the sides and not in the centre of oceans. Kay's view, that a system of island arcs existed off the eastern and southern coasts of North America in Lower Palaeozoic time, is entirely compatible with this hypothesis. This would have allowed the former Atlantic Ocean to close, and thus to convert the offshore island arcs of Lower Palaeozoic time into the intercontinental Appalachian mountains of Middle and Upper Palaeozoic time. Following Keith, Kay and King, I have sketched the location of seven former arcs in the Appalachians and their south-western continuation past the Ouachita and Marathon Mountains[48]. I see no reason to change this view.

The situation in Scotland has been described by Walton: "It seems likely that both the north-west Highlands and the Appalachians formed part of a very wide stable shelf which also included Greenland and other 'Boreal' regions having very similar Cambrian rock types. ... it is probable that sedimentation in the area of the southern Highlands during the Cambrian period was mainly of a greywacke, geosynclinal type and contrasted strongly with that in the north-west Highlands ... it is probable that Cambrian sedimentation stretched unbroken to an unknown distance south of the Highland Boundary fault"[26]. Ordovician volcanism followed and these correlations and descriptions suggest that, during Lower Palaeozoic time, arcs associated with a western land extended across Scotland. That there may have been contemporaneous volcanism and islands along the eastern coast in what is now Wales, merely adds complexity without affecting the history in Scotland. In the Silurian period conditions changed and the marine conditions "gave way to mixed environments ... at least partly in fluvio-deltaic environments with periods of emergence"[26].

Enough has been described of the conditions in Norway and West Spitsbergen and of their correlations with Scotland to show that they too were the sites of Lower Palaeozoic island arcs associated with a western continent. Thus the Caledonian–Appalachian arcs seem to have formed a continuous system along the western side of the former ocean. It is suggested that it was in the trenches of these arcs that the floor of the Lower Palaeozoic Ocean was swallowed up as that ocean closed.

Geometrical Control of Transposition

If two bodies are brought together so that they unite, and if they are later pulled apart so that they break on a different line from the line of union, then one important geometrical consideration holds. It is that, along any one stretch of the junction, only one fragment can change sides: one cannot transpose pieces from each margin at the same place. This is illustrated in Fig. 4 and a comparison with the other figures shows that our reconstruction obeys this principle. This does not prove that the reconstruction is correct, but the neat fashion in which fragments from either side are alternately transposed meets the geometrical requirements.

Some Possible Extensions

When, as is believed, the present Atlantic Ocean started to open at the beginning of Cretaceous time, it did so by breaking open a continent which was then continuous from West Spitsbergen to Florida (Fig. 2). North of Spitsbergen the coasts of North America and Siberia at that time diverged and the opening ceased to lie wholly within a continent and to have continental blocks on both sides. Following the descriptions of B. C. Heezen and M. Ewing, and Ya. Ya. Hakkel and N. A. Ostenso, I have suggested that the fracture followed the coast of Siberia forming the Lomonosov Ridge on the other side of the opening. Thus the Lomonosov Ridge separates an older Canadian ocean basin from a younger Siberian ocean basin in the Arctic Sea[14].

In the south the situation appears to be similar, but to understand it one must clearly separate the platform of northern Florida from the different geology of southern Florida. South of central Florida a southern Florida–Bahamas Ridge separates a main Atlantic, apparently of Cretaceous age, from a Gulf of Mexico which seems to have been a deep ocean and evaporite basin during the Jurassic period.

Drake, Heirtzler and Hirschman[49] emphasize the importance of the Florida-Bahamas Ridge for sealing off the Gulf of Mexico from the open ocean so that the Jurassic salt deposits could form. They suggest that this ridge is an "extension of the Ouachita system" and that it forms the "foundation for the entire chain of islands and banks". This ignores the earlier interpretations of magnetic and gravity anomaly maps made by Miller and Ewing[50] and by Lee[51]. Both papers suggested that southern Florida and the Bahamas are coral banks built on deeply submerged volcanoes. The Palaeozoic platform-type of sedimentary rocks, described from drill cores in northern Florida, are not like the Ouachita folds. Drilling has revealed no Palaeozoic rocks beneath southern Florida and the Bahamas. The magnetic anomaly map shows a marked change in central Florida. Drake *et al.*, in their Fig. 5, show three trend lines connecting southern Florida with the Bahamas and only one connecting it with northern Florida. It is suggested that the Florida-Bahamas Ridge, like the Lomonosov Ridge, formed when the main Atlantic Ocean started to open at the beginning of Cretaceous time and separated an ocean basin of Cretaceous age from an older ocean basin.

Drake *et al.* hold that the Florida-Bahamas Ridge extends to Navidad Bank, north of Hispaniola. It thus ends at the major zone of faulting which, according to Hess and Maxwell[52], and others, extends from Central America to the northern end of the West Indies arc. The Caribbean Sea and West Indies arc have often been regarded as associated with the Pacific Ocean, and I have discussed elsewhere their possible origin as a tongue, thrust from the Pacific and bounded by transform faults[53].

A complete discussion would require consideration of the Hercynian orogeny and faulting and post-Triassic faulting. This seems feasible but will not be attempted here.

It has been suggested in this article that during Lower Palaeozoic time North America and Europe were approaching each other, that this motion stopped and that it later reversed. If this is true, the onset of the reverse motion and the start of reopening of the Atlantic Ocean must have been an event of very major significance in world geology. The evidence suggests that it occurred at about the close of the Jurassic and the beginning of the Cretaceous periods. It seems reasonable to link it with other major events of that age in the Americas.

McLearn[55] has pointed out that at that time the drainage of much of western North America reversed its direction so that rivers which had been flowing west and building a great shelf along the Pacific coast were interrupted by the rise of the Cordillera and began to follow their present directions. Gilluly[56] has pointed out that "In Cretaceous time plutons probably a thousand times larger than those of all the rest of the Phanerozoic were emplaced". The onset of a relative advance of the Americas over the Pacific Ocean floor might well have caused a crumpling of the shelves along that coast with the creation and rise of extensive batholiths. Gilluly has pointed this out and concluded that the likely cause was that "it is probable that the continent as a whole is moving away from a widening Atlantic".

I am happy to acknowledge the benefits of discussions with W. B. Harland, A. Hallam, J. Dewey, W. R. Church, F. D. Anderson, H. Williams, E. W. R. Neale, H. H. Hess, F. J. Vine, E. Irving, A. Quinn, T. Mutch, J. Sougy and others, and of correspondence with many. I also thank F. W. Beales, C. Harper and D. York for help with the final version of this article.

This work was supported by the National Research Council of Canada, Vela-Uniform Program and Unesco.

[1] Hutchinson, R. D., *Geol. Surv. Canada, Mem.*, **263**, 52 (1952).

[2] Cowie, J. W., *Intern. Geol. Cong., Sess.* 21, *Copenhagen*, Part 8, 57 (1960).

[3] Grabau, A. W., *Palaeozoic Formations in the Light of the Pulsation Theory*, 1 (University Press, National University of Peking, 1936).

[4] Hutchinson, R. D., *Intern. Geol. Cong., Sess.* 20, *Mexico, The Cambrian System Symposium*, 2, 290 (1956).

[5] Bullard, E. C., Everett, J. E., and Smith, A. G., *Phil. Trans. Roy. Soc.*, A, **258**, 41 (1965).

[6] Kay, M., *Geol. Soc. Amer. Mem.*, **48**, 31, 56 (1951).

[7] Pettijohn, F. J., *Bull. Amer. Assoc. Petrol. Geol.*, **46**, 1468 (1962).

[8] Yeakel, jun., L. S., *Geol. Soc. Amer. Bull.*, **73**, 1515 (1962).

[9] Naylor, R. S., and Boucot, A. J., *Amer. J. Sci.*, **263**, 153 (1965).

[10] King, P. B., *The Evolution of North America*, 61 (Princeton University Press, 1959).

[11] Clark, T. H., and Stearn, C. W., *The Geological Evolution of North America*, 104, 114 (Ronald Press, New York, 1960).

[12] Rodgers, J., and Neale, E. R. W., *Amer. J. Sci.*, **261**, 713 (1963).

[13] Hallam, A., in *The Geology of Scotland*, edit. by Craig, G. Y. (Oliver and Boyd, Edinburgh, 1955).

[14] Blackett, P. M. S., Bullard, E. C., and Runcorn, S. K., *Phil. Trans. Roy. Soc.*, A, 258 (1965).

[15] Furon, R., *The Geology of Africa*, 49 (Oliver and Boyd, Edinburgh, 1963).

[16] Odell, N. E., *Quart. J. Geol. Soc. London*, **83**, 147 (1927).

[17] Harland, W. B., *Quart. J. Geol. Soc., London*, **114**, 307 (1958).

[18] Gobbett, D. J., and Wilson, C. B., *Geol. Mag.*, **97**, 441 (1960).

[19] Kulling, O., *Geogr. Annaler.*, *1934*, 161 (1934).

[20] Sandford, K. S., *Quart. J. Geol. Soc. London*, **112**, 339 (1956).

[21] Orvin, A. K., *Skr. Svalb. og Ishavet*, **78**, 1 (1940).

[22] Klitin, K. A., *Izvestiya Acad. Sci., U.S.S.R., Geol. Ser.* (Engl. Trans., Amer. Geol. Inst.), *1960*, 50 (1960).

[23] Holtedahl, O. (ed.), *Norges Geol. Undersökelse, Nr.* 208, 128, 153, 157, 165 (1960).

[24] Bailey, E. B., and Holtedahl, O., *Regionale Geologie der Erde*, 2 (Abschn. 2), 1 (Akad. Verlags. m. b. H., Leipzig, 1938).

[25] Umbgrove, J. H. F., *The Pulse of the Earth*, second ed., 232 (M. Nijhoff, The Hague, 1947).

[26] Walton, E. K., in *The Geology of Scotland*, edit. by Craig, G. Y., 167, 177, 201 (Oliver and Boyd, Edinburgh, 1965).

[27] George, T. N., in *The British Caledonides*, edit. by Johnson, M. R. W., and Stewart, F. H., 12 (Oliver and Boyd, Edinburgh, 1963).

[28] Williams, H., *Amer. J. Sci.*, **262**, 1137 (1964).

[29] Anderson, F. D., *Geol. Surv. Canada, Map* 8-1965, (1965).

[30] Jenness, S. E., *Geol. Surv. Canada, Mem.* 327 (1963).

[31] Church, W. R., *Can. Min. Metal. Bull.*, **58**, 219 (1944).

[32] *U.S. Geol. Survey Prof. Papers*, 424-B, 65 (1961); 475-B, 117 (1963); 501-C, 28 (1964); 525-A, 74 (1965).

[33] Novotny, R. F., *U.S. Geol. Surv. Prof. Paper*, 424-D, 48 (1961).

[34] Billings, M. P., *Geological Map of New Hampshire* (U.S. Geol. Surv., Washington, 1955).

[35] Freedman, J., *Geol. Soc. Amer. Bull.*, **61**, 449 (1950).

[36] Katz, F. J., *U.S. Geol. Surv. Prof. Paper*, **108**, 165 (1917).

[37] Zartman, R., Snyder, G., Stern, T. W., Marvin, R. F., and Buckman, R. C., *U.S. Geol. Surv. Prof. Paper*, 575-D, 1 (1965).

[38] Emerson, B. K., *U.S. Geol. Surv. Bull.*, 597 (1917).

[39] Lundgren, jun., L., Goldsmitt, R., and Snyder, G. L., *Geol. Soc. Amer. Bull.*, **69**, 1606 (1958).

[40] Billings, M. P., *The Geology of New Hampshire, Pt. II*, 105 (New Hampshire State Planning and Devel. Comm., Concord, 1956).

[41] Howell, B. F., *Intern. Geol. Cong. Sess.* 20, *Mexico, The Cambrian System Symposium*, 2, 315 (1956).

[42] Bird, J. M., and Theokritoff, G., *Geol. Soc. Amer. Bull.*, **77**, 13 (1966).

[43] Lochman, C., *Geol. Soc. Amer. Bull.*, **67**, 1331 (1956).

[44] Sougy, J., *Geol. Soc. Amer. Bull.*, **73**, 871 (1962).

[45] Reyment, R. A., *Nature*, **207**, 1384 (1965).

[46] Bureau Res. Geol. Min., *Carte Géol. Afrique Occid.*, Feuille No. 1 (1960).

[47] Carroll, D., *U.S. Geol. Surv. Prof. Paper*, 454-A, 1 (1963).

[48] Wilson, J. T., in *The Earth as a Planet*, edit. by Kuiper, G. P. (Univ. of Chicago Press, 1954).

[49] Drake, C. L., Heirtzler, J., and Hirschman, J., *J. Geophys. Res.*, **68**, 5289 (1963).

[50] Miller, E. T., and Ewing, M., *Geophysics*, **21**, 406 (1956).

[51] Lee, C. S., *Inst. Petroleum J.*, **37**, 633 (1951).

[52] Hess, H. H., and Maxwell, J. C., *Bull. Geol. Soc. Amer.*, **64**, 1 (1953).

[53] Wilson, J. T., *Earth and Planetary Science Letters* (in the press).

[54] Rodgers, J., Gates, R. M., and Rosenfeld, J. L., *Connecticut Geol. Nat. Hist. Surv. Bull.*, **84**, (1959).

[55] McLearn, F. H., *Geol. Surv. Canada, Mem.* (to be published).

[56] Gilluly, J., *Quart. J. Geol. Soc. London*, **119**, 133 (1963).

28

Reprinted from *Am. Philos. Soc. Proc.* **112**:309–320 (1968)

STATIC OR MOBILE EARTH: THE CURRENT SCIENTIFIC REVOLUTION

J. TUZO WILSON

Principal, Erindale College, University of Toronto

(Read April 19, 1968, in the Symposium on "Gondwanaland Revisited: New Evidence for Continental Drift")

I. INTRODUCTION

THE FIRST STUDENTS of the solid earth were miners. Early they founded the sciences of mineralogy and metallurgy, but their studies were local. In spite of the broader insights of such men as Pliny the Elder, Leonardo, Steno, Agricola, Werner, Cuvier, and Lamarck, it was not until the close of the eighteenth and the early nineteenth centuries that Hutton, Smith, and Lyell demolished semi-religious beliefs in a cataclysmic origin of the earth and established the principles of geology (Adams, 1938).

As a result, since 1830, the earth has been regarded as a rigid, stable body whose surface features have evolved slowly by processes which we see in action today. In this belief and with steadily improving techniques geologists have spent the past 150 years mapping the rocks exposed on land. Preoccupation with the beauty and intricacy of these discoveries and confinement to the land surface has prevented mankind from appreciating how little of the whole earth is visible. The inaccessibility of most of the earth is a limitation which has prevented geologists from developing general theories. One cannot discover everything about an egg by examining only one-third of its shell.

The first precise and general theories of the earth were formulated in the seventeenth century by William Gilbert, Newton and Halley from their investigations of the earth's magnetic and gravitational fields. Unfortunately their elegant theories were not matched by the development of good instruments until a few years ago. Whereas early geologists rapidly accumulated good data, but lacked theory, early geophysicists had precise theories, but could not make enough observations; the two groups could not correlate their ideas. Both long remained in ignorance of those greater parts of the earth, its interior and its ocean floors. Only during this century have seismologists discovered how to use earthquake waves, like giant X-rays, to illuminate the dark interior; only

since World War II have adequate ships, instruments and expenditures been available to explore the ocean basins.

Suddenly within the past two or three years, the wraps are off and we glimpse for the first time the full beauty of the naked earth. The vision is not what any of us had expected from our limited peeps at the constituent parts. The earth, instead of appearing as an inert statue, is a living, mobile thing. The vision is exciting. It is a major scientific revolution in our own time, but before expatiating upon its nature, let us examine the evidence.

II. THE CONCENTRIC SHELLS OF THE EARTH

The interior of the earth in its simplest terms consists in the main of a white-hot, solid mantle of dense rock. There is a central liquid core and a cool and rigid outer rind on which we live. This rind, which Daly (1940) called the lithosphere owes its separate existence and properties to its lower temperature. At a depth of about 50 km. the mantle becomes so hot, that like a red-hot poker it ceases to be rigid and becomes deformable. This layer Daly called the asthenosphere (see fig. 1).

The relationship of the lithosphere to the asthenosphere is like that of ice to water in a frozen lake, except that the asthenosphere is vastly more viscous than water. If one further imagines that some rafts of logs have been frozen into the ice and protrude from its surface, then one has a good analogy to the crustal blocks that constitute the continents. They are made of lighter rocks than the mantle and rise above the denser ocean floors. It should be noted that the contact of the crust and mantle at the Mohorovičić discontinuity is probably a change in composition. It is not the same as the deeper interface between rigid lithosphere and deformable asthenosphere which is a function of temperature.

It was geodesists, including Bouguer, Airy,

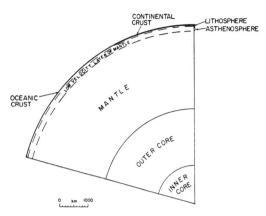

FIG. 1. The shells of the earth. The asthenosphere is thought to be deformable, which allows the lithosphere to break and its pieces to move about.

Hayford and Bowie (Heiskanen and Vening Meinesz, 1958) who first realized that the earth's lithosphere was floating in equilibrium upon a more mobile interior. Recently seismologists and geochemists have been able to refine and explain their ideas. Important contributors have included Miss Lehmann (1967), Gutenberg, Verhoogen, Dorman, Press, Brune, Ewing, McConnell, Ringwood, and Clark and Anderson (1966).

III. THE PLATES OF THE LITHOSPHERE

It has long been known that severe earthquakes are confined to a few belts about the earth. As the methods of seismology have improved and the location of foci has become more accurate, these belts have been narrowed so that they now appear to be cracks or fractures dividing the

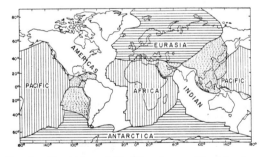

FIG. 2. The plates of the lithosphere outlined by fractures. Plates grow and move apart along mid-ocean ridges, but overlap and are absorbed along mountains and trenches. Six major plates are named and several minor ones are stippled. (After Le Pichon.)

lithosphere into half a dozen large plates and several smaller ones as shown in figure 2. The main plates are so large that each, except the Pacific plate, contains one or two continents and much surrounding ocean floor. The relative motion of the plates causes earthquakes, as they jostle one another. Morgan (1968), Le Pichon (1968) and Mackenzie and Parker (1967) have led in recognizing their nature. The discovery is important because, by showing that all significant motions are concentrated along a network of fractures, it explains why the plates themselves have remained rigid and why scientists working in the central parts of continents have often seen no evidence for motion and have long denied the existence of continental drift. They have been like sailors so occupied in examining the decks

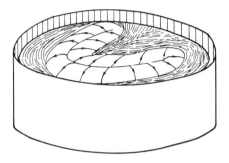

TWO PHASE SURFACE OF CONVECTING LIQUID

FIG. 3. The two-phase surface of a convecting liquid. The unshaded surface is like that of boiling soup and flows in the directions indicated. The shaded surface is like froth, which collects with a different pattern. On earth, ocean floors and continents behave similarly.

of their ships that they have failed to look further and thus notice that their ships were under way.

The causes of motion of the plates cannot be investigated directly but theoretical studies, particularly by Vening Meinesz, Orowan, Tozer, Knopoff and Elsasser (1966) suggest that it is probably due to heat generated by radioactive elements which sets up currents. These rise pushing some plates apart, forcing them to override others which are pushed down and becoming heated revert to the interior. The process is analogous to that in a cauldron of boiling soup, except that on earth the movements are so slow that the lithosphere has time to cool to a rigid brittle skin on the surface. The continents can be thought of as froth of a different composition which floats and manages to stay on top regardless of the currents beneath (see fig. 3).

IV. THE JUNCTIONS OF PLATES: THREE TYPES OF MOBILE BELT

The junctions between plates are of three types depending upon whether the plates are moving apart, sliding past one another or coming together.

Holmes (1931) and Hills (1947) with the imperfect evidence available to them suggested that upwelling occurs along mid-ocean ridges, but it was not until twelve years ago that Ewing and Heezen (1956) first recognized the full extent and continuity of these ridges, which are now known to be the greatest mountain range on earth. As the mantle wells up it cools and adds to the rigid plates. Hess (1962 and 1965) clarified these ideas and pointed out that the layer on the ocean floor below a veneer of sediment and lavas might be mantle rock altered by hydration to serpentinite. This view has been supported by discovery of serpentine exposures on the sea floor (Cann and Funnell, 1967). Dietz (1961) has likened the ridges to the junctions of pairs of conveyor belts which are rising, back to back, and carrying away the ocean floor in either direction.

Vine and Matthews (1963) and also Morley and Larochelle (1964) proposed that reversals in the earth's magnetic field might be imprinting the layer of basalt on the ocean floor. As Vine will explain in his paper in this symposium apparent imprinting has been discovered and seems to enable the rate of spreading along the mid-ocean ridges to be precisely measured. The maximum apparent rate of spreading is twelve centimeters a year. It seems that mid-ocean ridges tend to open in pure tension without shearing.

The second type of motion between two plates is horizontal shearing. Fracture zones are huge, linear features first discovered on the floor of the Pacific Ocean by the United States Coast and Geodetic Survey. They appear to be faults with large horizontal motions which offset mid-ocean ridges at right angles (Menard, 1964, and Heezen et al., 1964). They combine huge apparent displacements with abrupt terminations against continental margins, which was puzzling until it was suggested and demonstrated that they are a new class of fault with the opposite direction of motion to that which might have been expected (Wilson, 1965, and Sykes, 1967). A feature of this class of fault is that they can only form where lithosphere is being generated or reabsorbed. Their

existence on the ocean floors and the demonstration that earthquakes are distributed and move as predicted by this theory lends strong support to the idea of ocean floor spreading.

Euler long ago showed that any relative motion between two plates on a sphere takes the form of rotation about some axis. Morgan and Le Pichon have used the evidence that mid-ocean ridges are purely tensional and that fracture zones are purely shears to locate the axes for each pair of plates rather precisely. Magnetic imprinting of the ocean floor gives the rates of rotation. By examining successive pairs of plates, Le Pichon found the relative motions between all six major plates.

The third type of seismic belt underlies mountains and island arcs where plates come together. If, as seems likely, the earth is not expanding or is only changing in size by a small and negligible amount (Birch, 1964; Hospers and van Andel, 1967), as fast as lithosphere is created in some places it must be reabsorbed elsewhere. This causes the compression long recognized in mountains by geologists, but now that the relative movements of all major plates can be determined the rate and direction of closure of mountains can be stated. For example, Le Pichon has calculated that off the center of Honshu Island, Japan, the floor of the Pacific is approaching Asia at a rate of 8.8 cm. per year in a direction N 89° W, and that near Sicily Africa is approaching Europe at a rate of 2.4 cm. per year in a direction N 11° W.

Although plates separate in directions of pure tension and slide past one another in pure shears, they are irregular in shape and can seldom meet in purely compressional directions. As a consequence most mountain systems contain large longitudinal shear faults. The San Andreas fault, Great Glen fault, and Andes provide examples.

Closure is possible because, as Oliver and Isacks (1967) have demonstrated from seismic studies in the southwestern Pacific, the plates of the lithosphere override one another and the lower plate is forced down into the asthenosphere. The surface expression of this overlap takes the form of deep ocean trenches, overthrusts, and folded mountains. Four cases can be recognized. First, lithosphere containing oceanic crust may override another similar plate. Where shearing is slight this produces island arcs and trenches like those of the West Indies (Officer et al., 1959, and Wilson, 1966a), Alaska (Coats, 1962) and East Asia

(Matsumoto, 1967). Where shearing predominates, overriding forms straight features like the Tonga-Kermadec Islands and trench. Secondly, lithosphere with continental crust may override lithosphere with oceanic crust as in the Andes. Thirdly, lithosphere with oceanic crust may override continental material as Gass and Masson-Smith (1962) have proposed for Cyprus. Fourthly, one continent may be forced under another as seems to be the case in the Himalayas (Gansser, 1966).

A rate of motion of ten centimeters a year if continued would close an ocean ten thousand kilometers wide in a hundred million years. This makes it obvious that such motions must change from time to time. In this connection not only does the study of sediments support the idea of spreading because sediments thicken away from mid-ocean ridges, but also as Ewing and Ewing (1967) have shown irregular changes in thickness suggest discontinuities in motion. So do paleomagnetic results (Briden, 1967). Two reasons for changing the patterns of motion have been proposed, one is the great uplift and hence energy required to force one continent under another, and the other is Oliver and Isacks' (1967) suggestion that plates of the lithosphere might remain cold and hence strong long enough to reach the bottom of the asthenosphere. If the first proposal is correct it suggests that the reason why continental drift was not accepted sooner is that the central parts of continents are not

often affected by drift which has its chief impact in mountains. These are complex and have been difficult to interpret.

There is a great deal of older evidence which has been interpreted to favor continental drift. Much of it dates back to Wegener (1966), du Toit (1937) and before. Paleomagnetism is another important new aspect (Irving, 1964). There have also been strong contrary views and many authors have cited single pieces of evidence or an apparent lack of mechanisms as arguments against drift. It seems unnecessary to review this again since it has been fully discussed in several recent compilations and symposia (Carey, 1958; Runcorn, 1962; Blackett *et al.,* 1965; Johnson and Gilliland, in press; Kay, in press and Wilson, in press, *b*).

V. THE LIFE-CYCLE OF OCEAN BASINS

If continental drift has been going on for an appreciable part of geological time, at such rapid rates as recent work suggests, it means that a succession of ocean basins may have been born, grown, diminished, and closed again. Since ocean basins are the largest features of the earth's surface and would dominate other features it seems useful to outline the stages in their life cycle in terms of present examples. This makes it apparent that each stage has its own characteristic rock types and structures as outlined in table 1. This has been discussed in more detail elsewhere (Wilson, in press, *a*).

TABLE 1

Stages in the Life-cycle of Ocean Basins and Their Properties

Stage	Example	Motions	Sediments	Igneous rocks
1. Embryonic	East African rift valleys	Uplift	Negligible	Tholeiitic flood basalts, alkalic basalt centres
2. Young	Red Sea and Gulf of Aden	Uplift and spreading	Small shelves, evaporites	Tholeiitic seafloor, basaltic islands
3. Mature	Atlantic Ocean	Spreading	Great shelves (mio-geosynclinal type)	Tholeiitic ocean-floor, alkali basalt islands
4. Declining	Western Pacific Ocean	Compression	Island arcs (eugeo-synclinal type)	Andesitic volcanics, granodiorite-gneiss plutonics
5. Terminal	Mediterranean Sea	Compression and uplift	Evaporites, red beds, clastic wedges	Andesitic volcanics, granodiorite-gneiss plutonics
6. Relic scar (geosuture)	Indus line, Himalayas	Compression and uplift	Red beds	Negligible

VI. SOME PROPOSALS ABOUT MOUNTAIN BUILDING

As is well-known four mountain systems of varying Phanerozoic ages ring North America. By the beginning of this century the Appalachians were well studied and Barrell (1914) emphasized that, whereas the continent was the source for some of the sediments, at other times the sediments had come from marginal or off shore sources. No one has doubted the validity of this conclusion, which also applies to the sources of rocks of the other ranges. Three solutions have been proposed.

Barrell and Schuchert (1923) believed that extensive borderlands had existed in the ocean basins and that they had sunk, reducing the area of the continent (see fig. 4). It is now known that the ocean floor is not underlain by continental crust, so North American geologists have given up that idea, although some Soviet geologists support the concept of basification of continental crust to oceanic (Beloussov, 1966 and 1967).

The view now commonly held is that the sources were narrow ranges of mountains whose roots are now hidden by coastal plains. Some problems which have not been answered include why narrow zones should have been uplifted to the great heights necessary to produce such large volumes and whether the sediments are of the types expected under these conditions. It is also puzzling why in some cases there scarcely seems any room at the edge of the continent for these sources and why in other cases, like the Cretaceous orogeny in Nevada, the uplifts occurred in narrow zones far inland (Gilluly, 1963).

The third view which Wegener (1966), Grabau (1936), and others have advocated is that the sources were continents which later moved away horizontally. One can go further. If the concept of life-cycles of ocean basins is correct then one can visualize that throughout much of geological time continental blocks have been coming together along island arcs, piling up mountains when they collided, reversing directions and drifting apart again. This is the view here advocated.

It may be further observed that some former mountain ranges built by the closing of one ocean appear later to have been torn apart by the formation of another ocean on the same site. This seems to have been the case in the Caledonian and Appalachian systems which have fragments on both sides of the Atlantic. This suggests that

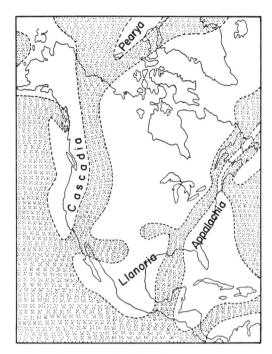

FIG. 4. Shows in simplified form the seaways and borderlands of North America. Land is unstippled and some borderlands are named. (After Schuchert.)

ocean basins tend to succeed one another in much the same places, a new ocean opening more or less along the suture on which a former ocean had closed. This concept has already been applied to the Atlantic (Wilson 1966a, 1966b, 1967).

While the writer certainly does not regard these suggestions as amounting to proof and admits that his proposals are certain to require modification, he believes it important to state these views because they are the sort of consequence that one must expect if drift has gone on at the rates indicated by magnetic imprinting of the ocean floor.

It does seem important to note that the evidence in the Atlantic meets the essential geometric requirement that in any stretch of ocean only one side can be transposed at a time (fig. 5).

VII. MAJOR EVENTS IN A POSSIBLE HISTORY OF THE SOUTHERN UNITED STATES

In a well-known map Kay (1951) showed the Lower Paleozoic island arcs of the Appalachians

A   B

FIG. 5. A. The lands about the closed proto-Atlantic Ocean of Permo-Triassic time. The ocean is considered to have closed along the heavy line which separates two different Cambrian faunal realms (after Grabau). B. An impossible case where the geometric requirement concerning transposition is not met.

continuing through the Ouachita and Marathon Mountains north of the Gulf of Mexico. If the Appalachian arcs were an indication that the proto-Atlantic was closing, it is reasonable to suppose that the other mountains suggest that a forerunner of the Gulf of Mexico was also closing. In Upper Ordovician and Silurian time some clastics spread northwards from the Gulf Coast, but it was not until Upper Mississippian and Pennsylvanian time that the main clastic wedges poured northwards, presumably marking Stage 5 (see table 1) of the closing of this southern ocean (Briggs and Cline, 1967). The main uplift followed in the Upper Pennsylvanian. During the ensuing Permian Period the seas which covered Texas were progressively cut off and evaporated giving rise to thick evaporite basins and important petroliferous reefs on top of the North American continent. This could be an evaporite deposit related to Stage 5. By Triassic times the seas had gone and all deposits became continental.

In Upper Triassic and Jurassic time seas again began to invade across the west coast of Mexico and farther to the south. Halbouty (1967) suggests that at that time the present Gulf of Mexico started to form. The Atlantic Ocean had not then· opened and as South America withdrew a basin formed in which the extensive Louann salt deposit formed on oceanic crust and around the

margins. This is the source of the salt domes in the Gulf Coast region and would be a deposit formed in Stage 2 of the new ocean. Thereafter the great shelf of Gulf Coast deposits developed, derived from within the continent.

This cursory discussion suggests that it is reasonable to believe that an earlier Gulf of Mexico closed and reopened again and the geology of northern South America suggests that this view is tenable (Harrington, 1962 and Bürgle, 1967).

VIII. MAJOR EVENTS IN A POSSIBLE HISTORY OF THE ARCTIC BASIN

It is well established by both North American and Soviet scientists that the Arctic basin is divided into two main parts separated by the Lomonosov Ridge. According to the views of Heezen and Ewing (1961), Harland (1965), Sykes (1965), Wilson (1963) and others the basin next to Siberia is a northern extension of the Atlantic Ocean which has opened since the start of the Cretaceous Period.

King et al. (1966), Deminitskaya and Karasik (1966) and others point out that a marked change occurs across the Lomonosov Ridge and that the basin on the Canadian side is shallower, has much thicker crust or sediments and seems to be older. In their study of the adjacent Canadian Arctic Islands Tozer and Thorsteinsson (1964) point out that in Ordovician and Silurian time a shallow water shelf lay across the islands with a deep trough along its northwestern side. This situation, which could be interpreted as a coast with an offshore system of island arcs changed drastically in Devonian time when, as Tozer and Thorsteinsson state, a "great belt of Devonian clastic rocks that extends throughout the Franklinian miogeosyncline and the adjacent Lowland was derived from an anomalous belt of 'tectonic lands' to the northwest."

In Mississippian times there was uplift and no rocks were deposited, but in the Pennsylvanian the direction of sedimentation again reversed and the Sverdrup Basin began to develop with sediments derived from the interior as a marginal shelf that has persisted to the present. It is significant that the oldest deposits in the Sverdrup basin are evaporites which today rise as domes analogous to those of the Gulf Coast.

Is it not easy to believe that a system of arcs of lower Paleozoic age led to closing of an early ocean and its reopening in the Pennsylvanian

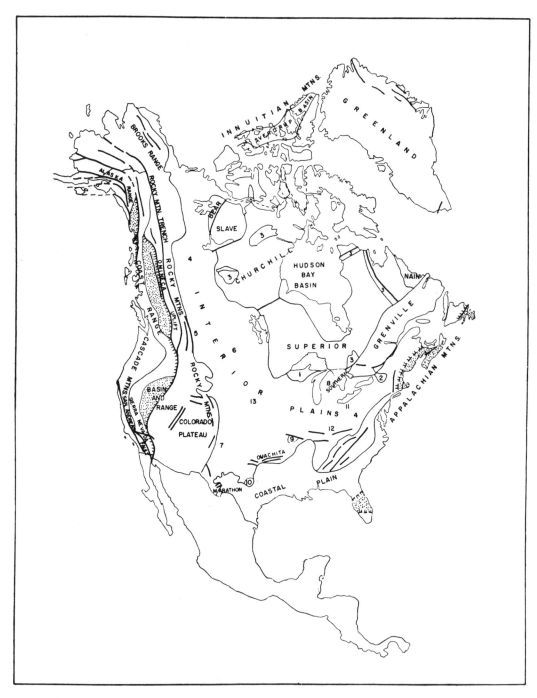

FIG. 6. Sketch map of North America showing junctions of formerly separate continents. Fragments in Florida, between Newfoundland and New England and from Nevada to Central British Columbia were once parts of Africa, Europe, and Asia, respectively.

Period after a period of impact and mountain development during the Devonian and Mississippian?

IX. MAJOR EVENTS IN A POSSIBLE HISTORY OF THE CORDILLERA

It seems foolhardy to attempt a brief summary of so complex a region as the Cordillera, but recent work and such summaries as those of Clark and Stearn (1960), Eardley (1962), Gilluly (1963 and 1965), Kay (1966), King (1959), Roddick *et al.* (1967), Wheeler (1967), White (1966), Ziegler (1967) among others have greatly clarified the picture.

It seems that there have been at least two island arc systems separated by plateaus. A shelf and offshore island arcs existed in the eastern Cordillera and Rocky Mountain region from Late Precambrian to Jurassic time. A western land seems first to have impinged on this coast in Upper Devonian to Mississippian time producing the Antler and equivalent orogenies. It also seems that these lands were not finally driven against and under North America until mid-Cretaceous time when Gilluly (1963) reports a second tremendous uplift in Nevada and Bally *et al.* (1966) discuss great thrusts in Alberta. Perhaps five million cubic kilometers of sediments were quickly eroded and poured eastward over the United States and Canadian plains.

The western lands were either small to begin with or broke off and were left behind as small continental blocks. It is suggested that one fragment now underlies western Nevada and adjacent areas and another underlies the Interior Plateau of British Columbia. Both developed a second shelf and arc system on their western sides which failed, as Bateman and Eaton (1967) have pointed out, to become the Sierra Nevada and Coastal Cordillera farther north (see fig. 6). The reason for suggesting or perhaps two more fragments is the increasing evidence that there are large differences and great dislocation between the Cordilleras of Canada and those of the United States. For example the Rocky Mountain Trench, which seems to be an old strike-slip fault of great displacement (Roddick, 1967), ends in Montana at a great transverse discontinuity recently described by Yates (1968) and Hamilton and Myers (1967).

The idea that one part of Asia collided with the Arctic coast of North America and another part with the Pacific Coast is in agreement with some modern Soviet views that Eurasia is a composite continent (Krapotkin, in press; Khramov *et al.,* 1966; and also Wilson, 1963).

X. SOME BIOLOGICAL IMPLICATIONS

All these tectonic outlines are tentative and will need to be studied and modified by experts. They have interesting biological implications. While the distribution of biota cannot prove or disprove continental drift, it would certainly be affected by it and should show this. It is therefore of great interest to note how many distributions fit the concept that Europe and most of North America were once together, and separated from the west coast of Canada and the United States.

Turning first to botany, Melville (1966) has written

During the Palaeozoic and the early part of the Mesozoic, four great floral regions can be distinguished. . . . In the Northern hemisphere, the coal flora occupied western Europe and eastern North America, which together formed a continent centered over the North Atlantic, which may be called Atlantica. A second distinctive flora, the *Gigantopteris* or Cathaysian flora occupied a large part of China and South-east Asia and also the western part of North America. These regions formed a continent centered over the North Pacific, which may be called "Pacifica."

At the same time Omodeo (1963) has pointed out that earthworms are conservative, slowly evolving creatures which cannot be transported alive across oceans and yet that European earthworms are very similar to those of Eastern North America as far west as the Rocky Mountain region and in the south even across the mountains to the Pacific coast of Mexico. On the other hand the west coast farther north in the United States and Canada has "five endemic species of the genus *Plutellus* of which genus the other endemic species are found principally in Australia, Tasmania and New Caledonia."

Again Waldén (1963), writing of land Gastropoda, divides North America in two parts with an eastern group somewhat related to European forms and

the other, West American group, comprises about three hundred endemic species, localized on the Pacific coast and in the Rocky Mountains. This group is dominated by two families of very old, autochtonous development, viz. the *Helminthoglyptidea* and the *Amaeridal*. The later is represented by other genera in Australia and East Africa.

These and many other peculiarities in the distribution of floras, faunas and their parasites have been discussed by many biologists including Darlington (1965), Brundin (1966), and Cameron (1950) and disposed them to favor drift.

Similar associations and barriers are well known to paleontologists. For example, Ross (1967) writing of early Permian fusulinids states that,

In California, and perhaps western Nevada and southern British Columbia, the distinctive *Chalaroschwagerina*-fusulinid association . . . was partially isolated from fusulinid associations to the north in the Cordilleran Geosyncline, to the east in the central interior and southwestern United States, and to the south in the Andean Geosyncline.

The boundary to which several of these biological associations point could coincide with the Antler orogeny which geologists have placed through northwestern Nevada and north into eastern British Columbia. It is also the locus of geophysical changes in the age of basement rocks according to compilations by Engel (1963) and by Kanasewich (1966) and of changes in the structure of the crust and mantle detected by Pakiser (1965) and by Caner *et al.* (1967).

XI. THE CURRENT REVOLUTION IN THE EARTH SCIENCES

In a series of compilations and symposia which have recently discussed continental drift (Carey, 1958; Runcorn, 1962; Blackett *et al.*, 1965; Johnson and Gilliland, in press; Kay, in press; Wilson, in press, *b*) opposition to drift was not strongly expressed and has declined. This cannot be entirely due to bias, so that it now seems as reasonable to assume that continents have been moving after the manner discussed in this paper as to adopt the older view, which is equally an assumption, that they are fixed.

What a major step this is appears not to be appreciated. The idea of drift has been around so long that most scientists are familiar with it and texts frequently mention it. This is not at all the same thing as accepting drift as a frame of reference for studying the earth.

T. S. Kuhn (1966) in a brilliant analysis of scientific methods has pointed out that in every age all the workers in the main stream of any branch of science accept a single frame of reference or paradigm within which all work and thought proceeds. Thus all early astronomers accepted Ptolemy's view that the earth was the center of the universe. To change that belief required a major scientific revolution and took fifty years after Copernicus' death to effect. It

is important to be clear that the essence of the revolution was not any improvement in techniques, not more or better data, not an advance in mathematics; it was a change in ideas. Copernicus was not a great observer or mathematician. He is remembered because he more than others persuaded men to accept the cogent idea advanced by Aristarchus that the earth is not fixed but moves.

Kuhn makes the point that a change in belief has been the essence of all the great scientific revolutions like those from phlogistics to modern chemistry, from caloric to modern thermodynamics, or from special creation to evolution. In our day it would appear that what earth science needs more than fresh data, better instrumentation, or new techniques is a simple change from our present belief that the structure of earth is static to the new concept that it has long been mobile. This is parallel and similar to the Copernican revolution and should perhaps be called the Wegenerian revolution from its chief advocate.

One can illustrate how important this change in belief is by an analogy. Imagine a school of scientists who studied whirlpools and also held the fixed doctrine that the water in whirlpools did not move. It would not matter how detailed their studies of the topography of the surface of the water, of its chemistry or of its physical properties; they would never understand whirlpools until they changed their doctrine and admitted that the water was moving. If, for the past century and a half, geologists and geophysicists alike have been trying to fit their observations into a wholly erroneous concept of the earth, this could explain why earth science has not progressed as it should. It could explain why geology has deteriorated from its original intention of being the study of the earth to being a study instead of rocks, minerals, and fossils and why geophysics has never been integrated but remains fragmented with studies of earthquakes by one group, of geomagnetism by another, and so on.

Another simple analogy which may help to explain the situation is to consider children playing with a Meccano set. At first each child is fascinated by some separate piece. It is not until they gain experience or find a plan that they can put the pieces together.

Earth scientists still seem to be in the stage of studying separate pieces. Each study is complex, beautiful, and satisfying in itself, but for real progress we need to assemble them. I suggest that our failure to do this has arisen from using the wrong plan.

It is easy enough to look back and see the logic of past revolutions, but difficult to change the belief of a lifetime and adopt a new revolution oneself. Although distressing, this is necessary. It is particularly hard for geologists to admit that it was a mistake to build their house of ideas on a firm rock foundation and to agree that a shifting base would have been better. Nevertheless, it seems likely that this gross misconception about the nature of the earth has prevented all earth scientists, no matter how good their observations in geology, geophysics, or geochemistry, from grasping the true nature of the earth.

It would seem to be time to abandon this doctrine and these divisive terms steeped for a century in the misconcept of a static earth-structure. We should start afresh and combine all investigations into the study of a mobile earth-structure under the name of one new science—geonomy.

BIBLIOGRAPHY

ADAMS, F. D. 1938. *The Birth and Development of the Geological Sciences* (Baltimore).

ANDERSON, D. L. 1966. "Earth's Viscosity." *Science* 151: pp. 321–322.

BALLY, A. E., R. L. GORDY and G. A. STEWART. 1966. "Structure, Seismic Data and Orogenic Evolution of Southern Canadian Rocky Mountains." *Bull. Canadian Petrol. Geol.* 14: pp. 337–381.

BARRELL, J. 1914. "The Upper Devonian Delta of the Appalachian Geosyncline." *Amer. Jour. Sci.* 4th ser., 37: pp. 87–109, 225–253.

BATEMAN, P. C., and J. P. EATON. 1967. "Sierra Nevada Batholith." *Science* 158: pp. 1407–1417.

BELOUSSOV, V. V. 1966. "Modern Concepts of the Structure and Development of the Earth's Crust and Upper Mantle of Continents." *Geol. Soc. London Quart. Jour.* 122: pp. 293–314.

——. 1967. "Against Continental Drift." *Sci. Jour.*, Jan. 1967: pp. 56–61.

BIRCH, F. 1964. "Density and Composition of Mantle and Core." *Jour. Geophys. Res.* 69: pp. 4377–4388.

BLACKETT, P. M. S., E. C. BULLARD and S. K. RUNCORN (eds.). 1965. "A Symposium on Continental Drift." *Phil. Trans. Roy. Soc. London*, ser. A, 258.

BRIDEN, J. C. 1967. "Recurrent Continental Drift of Gondwanaland." *Nature* 215: pp. 1334–1339.

BRIGGS, G., and L. M. CLINE. 1967. "Paleocurrents and Some Areas of Late Palaozoic Sediments of the Ouachita Mountains, Southeastern Oklahoma." *Jour. Sed. Petrol.* 37: pp. 985–1000.

BRUNDIN, L. 1966. "Transantarctic Relationships and Their Significance as Evidenced by Chironomid Midges." *Kungl. Svenska Vetenskap. Hand.* new ser. 11: no. 1.

BÜRGLE, H. 1967. "The Orogenesis in the Andean System of Columbia." *Tectonophysics* 4: pp. 429–444.

CAMERON, T. W. M. 1950. "Parasitology and Evolution." *Trans. Roy. Soc. Canada,* 3rd ser., 44: pp. 1–20.

CANER, B., W. M. CANNON, and C. E. LIVINGSTON. 1967. "Geomagnetic Depth Sounding and Upper Mantle Structure in the Cordillera Region of Western North America." *Jour. Geophys. Res.* 72: pp. 6335–6351.

CANN, J. R., and B. M. FUNNELL. 1967. "Palmer Ridge: A Section Through the Upper Part of the Ocean Crust." *Nature* 213: pp. 661–664.

CAREY, S. W. (ed.). 1958. *Continental Drift Symposium* (Dept. of Geology, University of Tasmania).

CLARK, T. H., and C. W. STEARN. 1960. *The Geological Evolution of North America* (New York).

COATS, R. R. 1962. "Magma Type and Crustal Structure in the Aleutian Arc." *Amer. Geophys. Union Monog.* 6: pp. 92–109.

DALY, R. A. 1940. *Strength and Structure of the Earth* (New York).

DARLINGTON, P. J., JR. 1965. *Biogeography at the Southern End of the World* (Cambridge, Mass, Harvard Univ. Press).

DEMINITSKAYA, R. M., and A. M. KARISIK. 1966. "Magnetic Data Confirm That the Nansen-Amundsen Basin Is of Normal Oceanic Type." *Geol. Surv. Canada Paper* 66–15: pp. 191–196.

DIETZ, R. S. 1961. "Continent and Ocean Basin Evolution by Spreading of the Sea Floor." *Nature* 190: pp. 854–857.

DU TOIT, A. L. 1937. *Our Wandering Continents* (Edinburgh).

EARDLEY, A. J. 1962. *Structural Geology of North America* (2nd ed., New York).

ELSASSER, W. M. 1966. "Thermal Structure of the Upper Mantle and Convection." in P. M. Hurley, ed., *Advances in Earth Science* (Cambridge, Mass.).

ENGEL, A. E. J. 1963. "Geologic Evolution of North America." *Science* 140: pp. 143–152.

EWING, J., and M. EWING. 1967. "Sediment Distribution on the Mid-ocean Ridge." *Science* 156: pp. 1590–1592.

EWING, M., and B. C. HEEZEN. 1956. "Some Problems of Antarctic Submarine Geology." *Amer. Geophys. Union Monog.* 1: pp. 75–81.

GANSSER, A. 1966. "The Indian Ocean and the Himalayas—A Geological Interpretation." *Eclog. Geol. Helvetiae* 59: pp. 831–848.

GASS, I. G., and D. MASSON-SMITH. 1962. "The Geology and Gravity Anomalies of the Troodos Massif, Cyprus." *Phil. Trans. Roy. Soc. London*, ser. A, 255: pp. 417–467.

GILLULY, J. 1963. "The Tectonic Evolution of the Western United States." *Geol. Soc. London Quart. Jour.* 119: pp. 133–174.

——. 1965. "Volcanism, Tectonism and Plutonism in the Western United States." *Geol. Soc. Amer. Spec. Paper* 80.

GRABAU, A. W. 1936. *Paleozoic Formations in the Light of the Pulsation Theory* 1 (Peking, Nat. Univ. Press).

HALBOUTY, M. T. 1967. *Salt Domes: Gulf Coast United States and Mexico* (Houston, Texas).

HAMILTON, W., and W. B. MYERS. 1967. "Cenozoic Tectonics of the Western United States." *Geol. Surv. Canada Paper* 66–14: pp. 291–306.

HARLAND, W. B. 1965. "Tectonic Evolution of the Arctic-North Atlantic Region." *Phil. Trans. Roy. Soc. London,* ser. A, 258: pp. 59–76.

HARRINGTON, H. J. 1962. "Paleogeographic Development of South America." *Amer. Assoc. Petrol. Geol. Bull.* 46: pp. 1773–1814.

HEEZEN, B. C., and M. EWING. 1961. "The Mid-Ocean Ridge and Its Extension Through the Arctic Basin," in O. Raasch (ed.) *Geol. of Arctic* 1: pp. 622–642.

HEEZEN, B. C., E. T. BUNCE, J. B. HERSEY and M. THARP. 1964. "Chain and Romanche fracture zones." *Deep-sea Res.* 11: pp. 11–33.

HEISKANEN, W. A., and R. A. VENING MEINESZ. 1958. *The Earth and Its Gravity Field* (New York).

HESS, H. H. 1962. "History of Ocean Basins," in A. E. J. Engel (ed.), *Petrologic Studies* (New York, Geol. Soc. Amer.), pp. 599–620.

——. 1965. "Mid-oceanic Ridges and Tectonics of the Sea-Floor," in W. F. Whittard and R. Bradshaw (eds.), *Submarine Geology and Geophysics,* (London, Butterworths), pp. 313–334.

HILLS, G. F. S. 1947. *The Formation of the Continents by Convection* (London, Arnold).

HOLMES, A. 1931. "Radioactivity and Earth Movements." *Trans. Geol. Soc. Glasgow* 18 (1928–1929): pp. 559–584.

HOSPERS, J., and S. I. VAN ANDEL. 1967. "Palaeomagnetism and the Hypothesis of an Expanding Earth." *Tectonophysics* 5: pp. 5–24.

IRVING, E. 1964. *Paleomagnetism* (New York, Wiley).

JOHNSON, H., and W. GILLILAND. 1968. *Symposium: What's Happening on Earth?* (New Brunswick, N. J., Rutgers Univ. Press). In press.

KAY, M. 1951. "North American Geosynclines." *Geol. Soc. Amer. Mem.* 48.

——. 1966. "Comparison of Lower Paleozoic Volcanic and Non-volcanic Geosynclinal Belts in Nevada and Newfoundland." *Bull. Canadian Petrol. Geol.* 14: pp. 579–595.

——. 1968. *Procs. Conf. on Continental Drift, Gander, Newfoundland.* In press.

KANASEWICH, E. R. 1966. "Seismicity and Other Properties of Geological Provinces." *Nature* 208: pp. 1275–1278.

KHRAMOV, A. N., V. P. RODINOV and R. A. KOMISSARAVA. 1966. "New Data on the Paleozoic History of the Geomagnetic Field in the U.S.S.R." *Canada Def. Res. Bd. Transl.* T.460R.

KING, E. R., I. ZIETZ, and L. R. ALLDREDGE. 1966. "Magnetic Data on the Structure of the Central Arctic Region." *Bull. Geol. Soc. Amer.* 77: pp. 619–646.

KING, P. B. 1959. *The Evolution of North America* (Princeton Univ. Press).

KRAPOTKIN, P. 1968. "Eurasia as a Composite Continent." *Proc. UNESCO-IUGS Symposium on Continental Drift, Montevideo.* In press.

KUHN, T. S. 1966. *The Structure of Scientific Revolutions* (Chicago, Phoenix Books).

LEHMANN, I. 1967. "Low-Velocity Layers," in T. F. Gaskell, ed., *The Earth's Mantle* (London and New York), pp. 41–62.

LE PICHON, X. 1968. "Sea-floor Spreading and Continental Drift." *Jour. Geophys. Res.* 73: pp. 3661–3698.

MACKENZIE, D. F., and R. L. PARKER. 1967. "The North Pacific; an Example of Tectonics on a Sphere." *Nature* 216: pp. 1276–1280.

MATSUMOTO, T. (ed.). 1967. "Age and Nature of the Circum-Pacific Orogenies." *Tectonophysics* 4, 4–6: pp. 317–613.

MELVILLE, R. 1966. "Continental Drift, Mesozoic Continents and the Migrations of the Angiosperms." *Nature* 211: pp. 116–120.

MENARD, H. W. 1964. *Marine Geology of the Pacific* (New York).

MORGAN, W. J. 1968. "Rises, Trenches, Great Faults and Crustal Blocks." *Jour. Geophys. Res.* 73: pp. 1959–1982.

MORLEY, L. W., and A. LAROCHELLE. 1964. "Paleomagnetism as a Means of Dating Geological Events." *Roy. Soc. Canada Trans.* 9: pp. 40–51.

OFFICER, C. B., J. I. EWING, J. F. HENNION, D. G. HARKRIDER and D. E. MILLAR. 1959. "Geophysical Investigations in the Eastern Caribbean, Summary of 1955 and 1956 Cruises." *Phys. Chem. of the Earth* 3: pp. 17–109.

OLIVER, J., and B. ISACKS. 1967. "Deep Earthquake Zones, Anomalous Structures in the Upper Mantle and the Lithosphere." *Jour. Geophys. Res.* 72: pp. 4259–4276.

OMODEO, P. 1963. "Distribution of the Terricolous Oligochaetes on the Two Shores of the Atlantic," in A. Löve and D. Löve (eds.), *North Atlantic Biota* (Pergamon, Oxford), pp. 127–152.

PAKISER, L. C. 1965. "The Basalt-eclogite Transformation and Crustal Structure in the Western United States." *U. S. Geol. Surv. Prof. Paper,* 525B.

RODDICK, J. A. 1967. "Tintina Trench," *Jour. Geol.* 75: pp. 23–33.

RODDICK, J. A., J. O. WHEELER, H. GABRIELSE and J. G. SOUTHER. 1967. "Age and Nature of the Canadian Part of the Circum-Pacific Orogenic Belt." *Tectonophysics* 4: pp. 319–338.

ROSS, C. A. 1967. "Development of Fusilinid (Foreminiferida) Faunal Realms." *Jour. Paleontol.* 41: pp. 1341–1354.

RUNCORN, S. K. (ed.). 1962. *Continental Drift* (London and New York, Academic Press).

SCHUCHERT, C. 1923. "Sites and Nature of the North American Geosyncline." *Bull. Geol. Soc. Amer.* 34: pp. 151–229.

SYKES, L. R. 1965. "The Seismicity of the Arctic." *Bull. Seismol. Soc. Amer.* 55: pp. 519–536.

——. 1967. "Mechanism of Earthquakes and Nature of Faulting on the Mid-ocean Ridges." *Jour. Geophys. Res.* 72: pp. 2131–2153.

TOZER, E. T., and R. THORSTEINSSON. 1964. "Western Queen Elizabeth Islands, Arctic Archipelago." *Geol. Surv. Canada, Mem.* 332: pp. 205, 207–217.

VINE, F. J., and D. H. MATTHEWS. 1963. "Magnetic Anomalies over Oceanic Ridges." *Nature* 199: pp. 947–949.

WALDÉN, H. W. 1963. "Historical and Taxonomical Aspects of the Land Gastropoda in the North Atlantic Region," in A. Löve and D. Löve (eds.), *North Atlantic Biota* (Pergamon, Oxford).

WEGENER, A. 1966. *The Origin of Continents and Oceans* (Transl. 4th ed., New York, Dover).

WHEELER, J. O. 1967. "Tectonics." *Geol. Surv. Canada Paper* 67–41: pp. 3–59.

WHITE, W. H. (ed.). 1966. *A Symposium on the Tectonic History and Mineral Deposits of the Western Cordillera* (Cdn. Inst. Min. Metal. Montreal).

WILSON, J. T. 1963. "Hypothesis of Earth's Behaviour." *Nature* **198**: pp. 925–929.

——. 1965. "A New Class of Faults and Their Bearing on Continental Drift." *Nature* **207**: pp. 343–347.

——. 1966a. "Are the Structures of the Caribbean and Scotia Arc Regions Analogous to Ice Rafting?" *Earth and Planet. Sci. Let.* **1**: pp. 335–338.

——. 1966b. "Did the Atlantic Close and Then Reopen?" *Nature* **211**: pp. 676–681.

——. 1967. "Some Implications of New Ideas on Ocean-floor Spreading upon the Geology of the Appalachians." *Roy. Soc. Canada Spec. Pub.* **10**: pp. 94–99.

——. In press, a. "A revolution in Earth Science: Life-Cycle of Ocean Basins." *Science.*

——. In press, b. *Proc. UNESCO-IUGS Symposium on Continental Drift, Montevideo.*

YATES, R. G. 1968. "The Trans-Idaho Discontinuity, a Major Tectonic Feature in the Northwestern United States." *Internat. Geol. Cong. 23rd Sess. Prague, Abstract* p. 38.

ZIEGLER, P. A. 1967. *Internat-Symposium on the Devonian System, Guidebook for Canadian Cordilleran Field Trip* (Alberta, Canada, Alta. Soc. Petrol. Geol., Calgary).

29

Reprinted with permission from *Jour. Geology* 77:629–646 (1969)

CONTINENTAL MARGINS, GEOSYNCLINES, AND OCEAN FLOOR SPREADING[1]

ANDREW H. MITCHELL[2] AND HAROLD G. READING

Department of Geology and Mineralogy, University of Oxford

ABSTRACT

On Atlantic-type continental margins there is no differential movement between ocean floor and continent. On Andean-type margins spreading ocean floor descends beneath a submarine trench and continental mountain arc, and on island arc–type margins ocean floor descends beneath a trench and island arc bordering a small ocean basin of Japan Sea type. The corresponding modern geosynclines of Atlantic, Andean, island arc, and Japan Sea type each have distinct associations of sediments and volcanics. Geosynclines of Mediterranean type occur in small ocean basins between continents. Orogeny of Andean and island arc type occurs during development of the respective geosynclines. Himalayan-type orogeny results from collision of migrating continents. Development of the types of continental margin and associated geosynclines, and changes in type of geosyncline, are related to the oscillation of continents between ocean rises. The sequence pre-flysch→ flysch→ molasse in many ancient geosynclines can be interpreted in terms of modern geosynclinal deposits and continental margins.

INTRODUCTION

For more than 50 years after Dana (1873) applied the term *geosynclinal* to Hall's (1857) concept of linear belts of thick sediments deposited on subsiding crust, most arguments about geosynclines concerned the relationship of the sediments to later mountain building, and whether the sediments were deposited within or beside continents (e.g., Schuchert 1923). Since Stille (1940) divided "true" or "ortho-" geosynclines into paired belts consisting of a miogeosyncline lacking volcanics near a continent, and a eugeosyncline with volcanics on the oceanic side of the miogeosyncline, attempts have been made to interpret an-

[1] Manuscript received February 14, 1969; revised June 23, 1969.

[2] Present address: Institute of Geological Sciences, 5 Princés Gate, London, England.

cient geosynclines in terms of present processes and of continental margins.

Kay (1951) suggested that many eugeosynclines, and specifically the North American Cordilleran and Appalachian eugeosynclines, were analogous to present island arcs, and that sediments were derived from active volcanic island arcs within the sedimentary basin.

Drake et al. (1959) described two thick belts of sediment off the east coast of the United States, one under the continental shelf and the other beneath the continental rise, and compared these with rocks of the early Paleozoic Appalachian geosyncline. They attempted to relate the present sedimentary belts to Kay's classification of geosynclines, equating the continental shelf deposits with those of the Appalachian geosyncline, and the sediments beneath the continental rise with those of the Appalachian eugeosyncline.

Dietz (1963) showed how eugeosynclinal continental rises may build out over oceanic crust, and his actualistic model of continental accretion by modern geosynclines was further developed by Dietz and Holden (1966) when they suggested that many ancient miogeosynclines formed by the growth of miogeoclines on continental margins. This view has been challenged by Hsu (1965), who used, among other arguments, the evidence that much of the fill of North American geosynclines was derived from "borderlands" situated on the oceanic side of the continental margins.

Most European geologists have considered that geosynclines developed either upon continents or in oceanic areas between continents, and the Alpine geosyncline has been compared, in its structure and development, to the Indonesian arcs (e.g., Argand 1916; Kuenen 1967; Sylvester-Bradley, 1968). Aubouin (1965), in particular, equated the sequence of events in the Mediterranean part of the Alpine geosyncline with the events in the Sunda arc, and suggested that a complex system of troughs and ridges similar to those of the Sunda arc were present in the Alpine geosyncline throughout much of its history.

The theory of continental drift and the recent hypotheses of ocean floor spreading and underthrusting of the lithosphere have demonstrated both the mobility of oceans and continents and the impermanent nature of some continental margins. It is the aim of this paper to consider the development of geosynclinal successions and continental margins in terms of ocean floor spreading and lithospheric underthrusting.

CONTINENTAL MARGINS AND
SPREADING OCEAN FLOOR

Since the work of Vine and Matthews (1963) on magnetic anomalies over ocean ridges, further evidence in favor of the ocean floor spreading of Dietz (1961) and Hess (1962) has been provided by a large volume of geophysical data and by the dating of ocean floor sediments and lavas. The recent spreading hypothesis of Morgan (1968) and of Le Pichon (1968) has been strongly supported by the seismic data of Isacks et al. (1968).

According to Le Pichon's hypothesis, six rigid lithospheric plates cover the earth's surface, the area of which remains constant. These plates spread from a world rift system which includes ocean rises, and descend either beneath submarine trenches or beneath seismically active mountain arcs on continents (fig. 1).

Although some submarine trenches, such as the Tonga-Kermadec trench, are remote from a continent, most are located either on, or within about 1,000 km of, a continental margin. Where spreading lithosphere descends beneath a trench on or immediately adjacent to a continent (e.g., Peru-Chile trench), a mountain range forms on the continental margin. Where the trench lies more than 200 km from a continent (e.g., Aleutians and Izu-Bonin trenches), an island arc develops on the concave side of the trench and is separated from the continent by a small ocean basin. A trench is absent on margins near which there is no differential movement between the continent and the ocean floor.

Modern continental margins can thus be divided into three types (fig. 1): Atlantic-type margins, lacking a trench; Andean-type margins, bordered by a trench; and island arc–type margins, separated by a small ocean basin from the trench and island arc.

Major belts of stratigraphically thick sediment in which deposition is continuing today may be termed modern geosynclines, and most of these are located on or near the borders of continents. As the lithology and composition of the rocks present in these geosynclines are determined largely by the type of margin on which they occur, different types of geosynclines can be recognized and related to the three types of continental margin (fig. 1): Atlantic-type geosynclines lie on and beside Atlantic-type margins; Andean-type geosynclines occur on and beside Andean-type margins; island arc-type geosynclines are located on and around active island arcs. A fourth type is the small ocean basin (Menard 1967), which is found on the concave side of many arcs and which is called here a Japan Sea–type geosyncline. Geosynclines of Mediterranean type occur in small ocean basins lying between continents.

ATLANTIC-TYPE GEOSYNCLINES

Continental margins of Atlantic type are not bordered by a submarine trench and are

The main features of Atlantic-type geosynclines are shown in table 1.

Shallow-water clastic sediments and in some cases constructional carbonate reefs accumulate on the continental margin, forming the continental shelf, and with subsidence a miogeosyncline or miogeocline develops (Dietz and Holden 1966).

Many major deltas, such as those of the Niger, Amazon, and Plate rivers, occur on Atlantic-type continental margins. The site of the deltas is probably due largely to the tectonic pattern on continents, as shown by

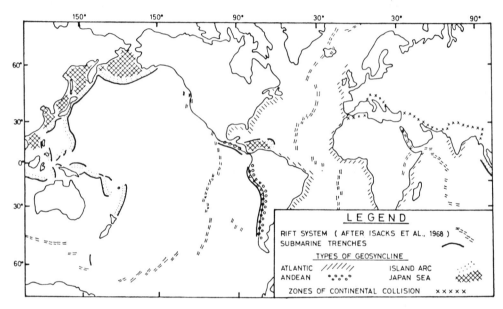

Fig. 1.—Position of modern geosynclines in relation to world system of rifts and submarine trenches

characterized by the absence of volcanic or seismic activity. On the continental shelf and rise seaward of these margins, thick sedimentary successions are accumulating, forming Atlantic-type geosynclines. The best-known example is the sedimentary belt off the east coast of North America (Dietz 1963; Drake 1966), and it has been suggested (Drake et al. 1959) that similar belts of miogeosynclinal and eugeosynclinal sediments are present off the coasts of Brazil and Argentina (fig. 1). Other examples are the east and west coasts of Africa and the coast of India.

Hospers (1965) in the case of the Niger delta; thus, in ancient geosynclines, deltas with an apparently random distribution may occur in place of or overlying miogeosynclinal sediments. Although most deltaic deposits are neither linear nor arcuate in shape, they form major sedimentary wedges on continental margins and hence cannot be excluded from a discussion of geosynclinal deposits.

Bordering the miogeosyncline, sediments consisting largely of material transported by geostrophic currents (Heezen et al. 1966), and in some cases interbedded with tur-

TABLE 1

SEDIMENTARY AND IGNEOUS ROCKS CHARACTERISTIC OF GEOSYNCLINES

Characteristic rock types:	ATLANTIC-TYPE GEOSYNCLINES		ANDEAN-TYPE GEOSYNCLINES		ISLAND ARC-TYPE GEOSYNCLINES		JAPAN SEA-TYPE GEOSYNCLINES
	Miogeosyncline Continental Shelf and Coastal Plain	Eugeosyncline Continental Rise, Abyssal Plains, Oceanic Rise	Arcuate Mountain Chain on Continental Margin	Trench	Volcanic Island Chain	Trench	Continental Margin in Restricted Basin
Shallow marine and coastal plain clastics	Abundant		Rare				Abundant (3)*
Carbonates	Abundant		Rare		Locally abundant		Locally common
Interbedded pelagics, tholeiitic lavas, and ultrabasics		Common preflysch (1)*		Abundant preflysch (1)*	Rare preflysch (1)*	Abundant preflysch (1)*	Present if basin floor is oceanic (1)*
Tholeiitic volcanic turbidites		Rare (1)*					
Compositionally mature turbidites		Abundant flysch (2)*		Rare to common flysch (2)*			Abundant flysch (2)*
Calc-alkaline volcanics and minor intrusions			Rare to abundant (2)*		Abundant (2)*		Tuffs
Calc-alkaline volcanic turbidites				Common to rare (2)*	Abundant flysch (2)*	Common flysch (2)*	Rare
Continent-derived coarse clastics			Abundant molasse (3)*				
Intermediate or acidic plutons			Common		Common		
Nature of underlying crust	Continental	Oceanic	Continental	Oceanic	Intermediate	Oceanic	Intermediate; modified oceanic and/or subsided continent

* (1),(2),(3) = upward stratigraphic sequence at any one locality in each geosyncline.

251

bidites and pelagics (Ericson et al. 1961), accumulate in deep water, forming the continental rise. The sediments are derived entirely from or across the miogeosyncline and continent and are thus compositionally mature. Subsidence, possibly due to isostatic adjustment, results in a thick succession (Dietz 1963).

Seaward of the continental rise are the continent-derived turbidites of the abyssal plains, interbedded with pelagics. These sediments lap against the flanks of the mid-ocean rise, where pelagic sediments alone accumulate. Nearer the crest of the rise tholeiitic volcanics, associated with ultrabasics, emplaced at the rise crest, are exposed. Turbidites derived from volcanic islands are in some places interbedded with the volcanics and may be cut by tholeiitic intrusions feeding volcanoes on the flanks of the rise.

Seaward prograding of the continental rise results in an upward vertical sequence from ultrabasics and tholeiitic volcanics through pelagic sediments into turbidites and geostrophic current deposits of the rise itself. Because volcanics occur at the base of this succession, it may be termed an Atlantic-type eugeosyncline.

ANDEAN-TYPE GEOSYNCLINES

The western border of South America forms an example of a margin on which a mountain belt is bordered by a submarine trench, and this and similar margins may thus be termed Andean-type. The southern part of the western margin of North America is broadly similar to that of the Andes, but much of the North American Pacific border is complicated by the underlying East Pacific rise (e.g., Vine 1966). Features typical of the geosynclinal deposits of an Andean-type margin are shown in table 1.

Little data is available on sediments accumulating in and near the Andean and Peru-Chile trench. However, Fisher and Raitt (1962) and Scholl, Huene, and Ridlon (1968) have shown that thick sequences of turbidites are accumulating locally both in part of the submarine trench and on its shoreward flank. Although little is known about their mineralogy, these turbidites are almost certainly derived from the Andean chain (Fisher and Raitt 1962), and their composition must reflect the detritus available at the source area and subsequent modification during transport. Much of the Andes consist of rocks of Mesozoic and Paleozoic age (Harrington 1962), but intense volcanism (Rutland and Guest 1965) and intrusive activity (Giletti and Day 1968) occurred during uplift of the chain in late Miocene and Pliocene time. Sediments derived from such sources must therefore include considerable calc-alkaline igneous detritus. The turbidites must also be mineralogically rather immature, as the narrow offshore shelf (Fisher and Raitt 1962) probably results in little reworking.

Sediments deposited on the continent, in and near the Andean-type mountain range, will contrast with those of other types of geosyncline. Among the most abundant late Cenozoic and Quaternary sediments in and around the Andes are coarse conglomeratic deposits (Harrington 1962), broadly similar in lithology to the Alpine molasse, often deposited in fault graben trending parallel to the mountain range (e.g., Galli-Olivier 1967). These sediments are mineralogically immature, consisting of calc-alkali volcanic detritus and older igneous or recycled material. Pyroclastics and lava flows may also accumulate with the conglomerates.

As the Andean margin of South America is tectonically mobile and may be underlain by a zone of descending lithosphere (Le Pichon 1968), it is possible that deformation of the trench and continental margin deposits is related to differential ocean floor-continent movement.

There is little evidence of isoclinal folding or low-angle thrusts indicative of intense crustal shortening in the Andes, and it seems probable that elevation of the mountain chain results largely from block faulting. In the Andes, orogeny, accompanied by volcanism and the emplacement of batholiths, is related to and contemporaneous with the development of the geosyncline.

Seyfert (1969) has attempted to explain the absence of thick sedimentary successions in much of the Peru-Chile trench by suggesting that they are thrust onto the continental margin as a result of the descent of the lithosphere. Thus deformation of the trench deposits may be a continuous process contemporaneous with sedimentation. Possibly some "mélange" deposits such as those of those of the Franciscan and Knoxville formations in California (Hsu 1968) were deformed on the margins of ancient trenches.

Development of an Andean-type margin might have a profound effect on sedimentation on the opposite side of the continent. The mountain chain on an Andean-type margin results in major river systems draining to the opposite coast; where this margin is of Atlantic or island arc type, geosynclines with deltas develop.

ISLAND ARC– AND JAPAN SEA–TYPE GEOSYNCLINES

Two main belts of sediment are associated with continental margins bordered by island arcs—one belt around the island arc itself and one belt within the small basin separating the arc from the continent. The sediments in these two belts differ from one another in lithology and composition and thus form two distinct types of geosynclinal deposit (table 1).

Island arc–type geosynclines.—Although many island arcs lie near a continental border, some, such as those in the southwest Pacific, are remote from a continent (fig. 1). However, rock types in and around island arcs are broadly similar regardless of the arc's location.

Most of the youngest islands in active island arcs consist of constructional volcanic cones. Coastal erosion of such islands is rapid, and steep subaerial slopes are maintained by volcanic upbuilding. In the absence of broad submarine shelves, coarse-grained epiclastic volcanic detritus is carried rapidly into deep water by turbidity currents and submarine lahars, where it accumulates with pyroclastic material, sub-

marine lavas, thin pelagic sediments, and in some cases reef-derived carbonate.

Migration or change in location of eruptive centers results in new islands overlying the clastic sediments, which are cut by minor intrusions. A belt of stratigraphically thick volcanics and compositionally immature volcanic sediments, with a predominantly calc-alkaline composition (Gorshkov 1962), thus results. As arcs are often contiguous (e.g., New Hebrides-Solomon Islands arcs), this belt may be elongate but not necessarily linear in shape; because pyroclastics, lava flows, and minor intrusions are present, it may be termed an island arc–type eugeosyncline. The lower Miocene rocks of the western belt of the New Hebrides island arc form an example of such deposits, features typical of which are shown in table 1.

The location of tectonically active island arcs in zones beneath which spreading lithosphere may descend suggests that orogeny in island arcs may be related to the movement of the lithosphere. The onset of tectonic activity could depend on the rate of descent of lithosphere, the depth to which it extends, or intermittent spreading.

Stratigraphic evidence (Mitchell and Warden, unpublished results) indicates that in many island arcs the present volcanic phase started in late Pliocene or early Pleistocene time. An earlier lower Miocene phase of volcanic and intrusive activity was accompanied or immediately followed by intense faulting and uplift. Data on oceanic sediments (Ewing and Ewing 1967) and magnetic anomalies (Moore and Buffington 1968) suggest that the present phase of ocean floor spreading began in the late Pliocene. It is thus possible that magmatic phases in island arcs are contemporaneous with spreading phases, and that orogeny in island arcs may accompany or follow the volcanism.

In many island arcs such as the northeast Honshu arc (Minato et al. 1965) and the New Hebrides, orogenic movements during the Tertiary have consisted largely of block faulting and uplift. In Japan the trend of the early Miocene arcs cuts older structural trends, and in the New Hebrides the present arc cuts the trend of the early Miocene arc. Thus ancient island arc–type geosynclines, which may have had a duration of many tens of millions of years, may consist of block-faulted deposits of successive island arcs with differing structural trends.

Miyashiro (1961) suggested that now uplifted belts of ultrabasic rocks with or without glaucophane schists in some arcs may have underlain ancient trenches or trench margins. In many arcs such as New Caledonia (Routhier 1953), the ultrabasics are similar in age to parallel volcanic arcs and have remained relatively stable since soon after the cessation of volcanism. Thus deformation and uplift of the trench or trench border deposits may accompany deformation of the volcanic arc.

It has been suggested (e.g. Oxburgh and Turcotte 1969) that ocean floor sediments may be "welded on" to island arcs, at the arc side of the submarine trenches, as a result of the descent of spreading lithosphere. This process presumably results in intense deformation of the sediments, similar to that which may occur in trenches bordering Andean-type margins.

Japan Sea–type geosynclines.—Small ocean basins lying between a continent and an island arc, such as the Japan Sea, Sea of Okhotsk, Andaman Sea, and Aleutian Basin, form geosynclines of Japan Sea type. Sedimentation processes in some of these basins have been discussed by Menard (1967), and Scholl, Buffington, and Hopkins (1968) have described continental margin deposits in the Aleutian Basin. On the continental margin shallow-water shelf or deltaic sediments accumulate; in basins where a distinct continental slope is present, these deposits may pass seaward into a wedge of turbidites and other mass-flow or slump deposits near the foot of the slope. Abyssal plain deposits consist of pelagics which are probably interbedded with turbidites and with tuffs erupted from the island arc. The sediments derived from the continent are compositionally mature. Be-

yond the abyssal plains, compositionally immature turbidites, derived from the island arc bordering the basin, may be interbedded with the continent-derived turbidites and pelagics.

Menard (1967) has shown that a very large volume of sediment is present in basins of Japan Sea type, due partly to the major rivers which feed some basins and partly to the rapid accumulation of sediment trapped behind the arc. If Japan Sea–type basins were as common in the past as they are today, it is probable that many ancient geosynclinal successions may have accumulated in similar basins.

Distinction between the deposits of Atlantic- and Japan Sea–type geosynclines will depend on both sedimentological criteria and on the nature of the rock beneath the abyssal plain sediments. Because Japan Sea–type basins are restricted basins, partially enclosed by island arcs, geostrophic contour current deposits will be absent, and tidal action may be limited. Volcanic tuffs will be present, and because the depth of the basins is generally less than the calcite compensation depth, pelagic sediments on the abyssal plains will be thicker than those of Atlantic-type geosynclines. The presence of marginal fault scarps in many small ocean basins (Menard 1967) may lead to contemporary mass flow of sediments resulting in deposition of pebbly mudstones and slump deposits.

Little is known about the nature of the rocks immediately beneath the abyssal plain deposits. The crust beneath Japan Sea–type basins is intermediate between continental and oceanic crust, but there is uncertainty as to whether this represents foundered continent or oceanic crust becoming continental. If the crust proves to be oceanic, the pelagic sediments and turbidites will presumably be underlain by ophiolites indistinguishable from those of Atlantic-type geosynclines. If, however, the crust beneath the basins represents subsided continent, the lower part of the succession in a Japan Sea–type geosyncline will consist of

TABLE 2

RELATIONSHIP OF GEOSYNCLINES AND OROGENY TO OCEAN FLOOR SPREADING

	Atlantic-Type Geosynclines	Andean-Type Geosynclines	Island Arc–Type Geosynclines	Japan Sea–Type Geosynclines
Location	Atlantic-type continental margin	On continental border with Andean-type margin	In front of small ocean basin; isolated arcs in ocean	Small ocean basins between continent and island arc
Relationship to spreading	No differential movement between continent and ocean floor	Differential movement between continent and ocean floor	Differential movement between lithospheric plates	Differential movement at associated island arc
Cause of Andean- or island arc–type orogeny	Start of differential movement between continent and ocean floor	Differential movement between continent and ocean floor	Intermittent differential movement between lithospheric plates	
Cause of Himalayan-type orogeny	Continental collision	Continental collision	Continental collision	Continental collision

any of the rock types found within continents.

MEDITERRANEAN-TYPE GEOSYNCLINES

Sedimentary deposits of geosynclinal dimensions are at present accumulating in some small marine basins situated between or within continents, such as the Mediterranean and Black Sea. These basins may represent oceanic areas remaining between continents which have collided, as discussed below. The sediments will be broadly similar to those in Japan Sea–type geosynclines, but as the basins are not directly related to present continental margins, they will not be discussed in detail.

CHANGES IN TYPE OF GEOSYNCLINE AND ASSOCIATED OROGENY

Types of continental margin and associated geosynclines have been discussed so far, and it has been seen that where there is no differential movement between continent and ocean, Atlantic-type geosynclines develop. Where there is differential movement at the continental margin, Andean-type geosynclines develop. Where a downward moving lithospheric plate descends some distance from the continental margin, island arc–type geosynclines occur, with, in some cases, Japan Sea–type geosynclines behind the island arcs (table 2).

Orogeny or mountain building connected with geosynclinal development may be one of three types, Andean, island arc, or Himalayan. Andean-type orogeny, characterized primarily by vertical movements, uplift of older continental material, contemporaneous igneous activity, and the formation of granite batholiths, and island arc–type orogeny, characterized by vertical movements, abundant volcanic activity, and intrusion of granites or diorites, have been described above. The third type of orogeny, Himalayan, is caused by the collision of two continental masses and is characterized by large lateral displacements and major thrusting due to the overriding of one continent by another; igneous activity is inconspicuous. Himalayan-type orogeny results in the tectonic intermixing of different geosynclinal facies and sequences.

If spreading ceases, or is renewed, or if the position of the zone of descending lithosphere changes, the type of geosyncline will also change. In theory there are twelve possible changes. However, on the hypothesis of ocean floor spreading, Atlantic-type geosynclines should precede other types. Also, Japan Sea–type geosynclines occur only in association with island arcs, and an orogeny is normally considered to conclude geosynclinal development. Therefore only three possible changes will be considered.

CHANGE FROM ISLAND ARC– AND JAPAN SEA– TO ATLANTIC-TYPE GEOSYNCLINES

With the cessation of differential ocean floor to continent movement at an island arc–type margin, volcanic activity in the arc may decrease and stop. Erosion will then reduce the island arc to sea level. The type of sediment which accumulates in the old Japan Sea–type geosyncline behind the eroded arc will depend largely on whether the basin floor and island arc subside, and there is considerable evidence that such subsidence occurs.

The coastline of the Andaman Sea, which is bordered by the Andaman-Nicobar island arcs, shows evidence of subsidence, at least during the late Cenozoic. Subsidence of the floor of the Japan Sea since the early Mesozoic is also widely accepted (Minato et al. 1965). Menard (1967) has argued that the thickness of sedimentary successions in many small ocean basins of Japan Sea–type is so great that subsidence cannot be due to isostatic adjustment alone. The apparent termination of continental structural trends in small ocean basins off the east coast of Asia, described by Beloussov and Ruditch (1961) and by Goryatchev (1962), could perhaps have resulted from subsidence of continental margins behind island arcs and burial of the margins by sediments. Evidence that inactive island arcs, as well as the basins behind them, undergo subsidence, is provided by the presence of largely submarine arcs such as the Iwo Jima arc.

Rapid sedimentation in subsiding Japan Sea–type basins will result in a very thick succession, and if the sediments of an advancing Japan Sea–type geosyncline accumulate more rapidly than the basin subsides, the basin may be filled. Assuming that the sunken arc remains as a submarine rim to the basin, detritus will spill over the rim and be transported onto the ocean floor beyond, forming continental rise and abyssal plain deposits. Thus the sediments deposited in the largely infilled basin will be shallow-water miogeosynclinal, and those deposited seaward of the ancient buried arc will be Atlantic-type eugeosynclinal. Examples of this may be present off the east coast of North America. Seismic refraction data show that a buried ridge lies between the miogeosynclinal shelf deposits and the eugeosynclinal continental rise deposits off the east coast of North America (Drake et al., 1959; Drake 1966). Watkins and Geddes (1965) have suggested on the basis of magnetic anomalies and other geophysical data that the buried ridge represents an inactive island arc, although Burk (1968) has discussed alternative origins for these and similar ridges elsewhere. The presence of a buried arc or chain of arcs in this locality is a possible explanation of the apparent thickening out seaward of the miogeosynclinal wedge discussed by Dietz and Holden (1966). This wedge would then be bounded on the seaward margin by the submerged arc, and the steep continental slope would represent the oceanward flank of the island arc, now blanketed in sediments.

CHANGE FROM ATLANTIC- TO ISLAND ARC– AND JAPAN SEA-TYPE GEOSYNCLINES

If spreading lithosphere starts to descend on or oceanward of an Atlantic-type geosyncline, an island arc will develop either across the continental rise or on the abyssal plain beyond. Development of an arc will cause continental rise and abyssal plain sediments to be intruded by calc-alkaline volcanics and to be overlain by the deposits of an island arc–type geosyncline. A Japan Sea–type geosyncline may form behind the

256

island arc. The geosynclinal deposits may lie partly on oceanic crust if the arc forms on the abyssal plains; if the arc develops on the continental rise, and the continental shelf subsides too rapidly for sedimentation to maintain equilibrium, the geosyncline will form on continental crust.

An example of an island arc forming across sediments of an earlier Atlantic-type geosyncline is provided by the Aleutians. The Aleutian trench, initiated in mid-Tertiary times, has cut late Mesozoic and early Tertiary continent-derived turbidites of an older abyssal plain and continental rise (Hamilton 1967), and an island arc–type geosyncline has developed across the earlier Atlantic-type geosyncline. Behind the Aleutian arc the Aleutian basin has been isolated from the main Pacific ocean (Shor 1964) with subsidence estimated at between 1.5 and 3–4 km during Cenozoic time (Scholl, Buffington, and Hopkins 1968). Thus a Japan Sea–type geosyncline has developed though whether subsidence is due to "oceanization" of a continental block or depression of former Pacific ocean floor is uncertain.

CHANGE FROM ATLANTIC- TO ANDEAN-
TYPE GEOSYNCLINES

Where the start of differential ocean floor to continent movement results in the descent of spreading lithosphere beneath the continental margin, an arcuate mountain range of Andean type will form. A submarine trench will develop on the site of the earlier Atlantic-type geosyncline, part of which will undergo orogeny and may be largely destroyed.

An example of this may be found in the Central Highlands of New Guinea, the northern edge of which probably represented the margin of the Australian continent in early Miocene times (Thompson and Fisher 1965). Paleomagnetic data indicate that the Australian continent was moving north on spreading ocean floor during the Mesozoic and Tertiary (Briden 1967), and thus prior to Miocene times the northern edge of the continent may have been the site of an At-

lantic-type geosyncline. With continued movement the continental margin approached the zone of descending lithosphere which now extends from the Andaman arc to the New Hebrides. Descent of lithosphere on the continental margin resulted in the development of an Andean-type chain, convex to the northeast, and consisting of calc-alkaline volcanics and older uplifted rocks, which now form the highlands of New Guinea. The region is complicated, however, by the present active arc of northern New Guinea, which is convex to the south; possibly this younger arc developed near the earlier lower Miocene trench, as a result of further northward movement of the continent.

A second example of a developing mountain range of Andean type may have been the Himalayas during the middle Miocene when the lower and middle Miocene Murree formation, with its fine-grained clastics derived from a low-relief shield area to the south, passed up into the upper Miocene Siwaliks, which are coarser and derived from a rising mountain chain to the north (Gansser 1964, pp. 40–46).

THE HIMALAYAS

One example of continental collision is the Himalayas. During the Mesozoic the Indian subcontinent was separated from Laurasia to the north by the Tethys ocean. The Mesozoic and Cenozoic history of the northern margin of the Indian shield is well documented in the foothills of the Himalayas, where an essentially miogeosynclinal sequence of continentally derived sediments persisted until the middle Miocene. The succession is divided by unconformities indicating regression and transgression, but there is little folding. The northern margin of India was thus an Atlantic-type geosyncline. Presumably, during the Cenozoic, India drifted northward on a spreading lithospheric plate which included the Indian shield, an expanding Indian ocean to the south, and a closing Tethys to the north.

In Late Cretaceous and Eocene times thick flysch deposits, such as the Indus

Flysch, interbedded with ophiolites and containing exotic blocks of essentially pelagic and oceanic material, were deposited in the northern Himalayas in a linear belt along the line of the Indus suture. This belt seems to have been the line of descending lithosphere at that time, and the lateral passage into the very complex basaltic and andesitic Dras volcanics with their associated basic and acid intrusions (Gansser 1964, p. 76) suggests that it was an island arc-type geosyncline. The Indus suture may have remained the zone of downward-moving lithosphere throughout Cenozoic times, but there is no direct evidence for this as there is no outcrop of post-Eocene, pre-Pleistocene sediments north of the southern foothills.

The direction in which lithosphere descended, and hence the orientation of the island arc, is uncertain. The regional structure, southward-moving thrust masses, and the general acceptance that India underthrust Laurasia suggest that the direction was northward. Support for this comes from the present arcuate pattern of the Himalayas, which are convex to the south. Frank (1968) pointed out that the radius of curvature of the Himalayas is similar to that of many island arcs, and he showed theoretically that the curvature of such arcs could be related to lithosphere descending at the convex side of the arc and plunging beneath the segment of crust on its concave side.

On the other hand, if, as suggested above, the lower and middle Siwaliks are the result of an Andean-type margin developing on the site of the north Indian Atlantic-type margin, the direction of downward-moving lithosphere must have been to the south, at least in upper Miocene to Pliocene times, and the change to northward-dipping lithosphere did not take place until late Pliocene, when, according Gansser (1964), the main Himalayan orogenic phase started. This would be the time of continental collision and major uplift, together with rapid erosion resulting in deposition of the very coarse conglomeratic upper Siwa-

liks and the major thrusts along the Main Boundary Fault.

However, an argument against the development of an Andean-type margin is the absence of calc-alkali volcanics, even in the Siwalik detritus, and it is possible that continental collision started in the late Miocene when spreading, according to Le Pichon (1968), recommenced after a period, during the Miocene, of quiescence.

OSCILLATION OF CONTINENTS AND GEOSYNCLINAL DEVELOPMENT

Using Le Pichon's (1968) hypothesis of ocean floor spreading, development of the continental margins and geosynclines described above can be related to a simple model in which continents oscillate between ocean rises. In the model parallel ocean rises are separated by continents. The position of the zone beneath which lithosphere descends changes from time to time, and the continents move backward and forward between the rises on spreading lithospheric plates. In order to simplify the diagram, a fixed frame of reference is maintained by assuming that the rises remain fixed relative to one another.

Spreading lithosphere in some places descends beside a continent, resulting in an Andean-type mountain chain, and in others descends many hundreds of kilometers from a continental margin, giving rise to an island arc. There is no explanation for this difference, though possibly it is partly governed by the distance between spreading and descending lithosphere.

The margin of a lithospheric plate beneath which ocean floor descends and at which differential movement occurs is termed *active*. A continental margin near which differential movement does not occur is termed *passive*. This concept is essentially similar to Dietz's (1963) idea of uncoupled and coupled continental margins, but it is applied to lithospheric plates instead of to sialic and simatic crusts.

In figure 2A it is assumed that continent I moves westward with spreading ocean floor. If lithosphere descends under the ac-

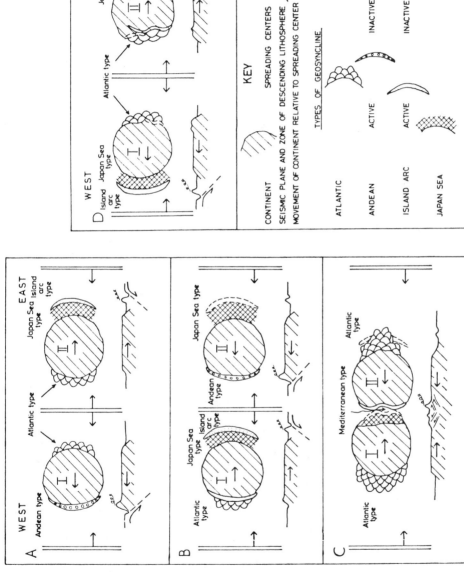
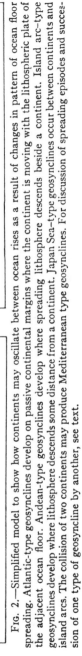

Fig. 2.—Simplified model to show how continents may oscillate between ocean rises as a result of changes in pattern of ocean floor spreading. Atlantic-type geosynclines develop on passive continental margins where the continent is moving with the lithospheric plate of the adjacent ocean floor. Andean-type geosynclines develop where spreading lithosphere descends beside a continent. Island arc–type geosynclines develop where lithosphere descends some distance from a continent. Japan Sea–type geosynclines occur between continents and island arcs. The collision of two continents may produce Mediterranean type geosynclines. For discussion of spreading episodes and succession of one type of geosyncline by another, see text.

259

tive western margin, an Andean-type margin and geosyncline result, while on the eastern passive margin an Atlantic-type geosyncline develops. If continent II moves east, and lithosphere descends oceanward of the eastern margin, an island arc–type geosyncline is formed, with a Japan Sea–type geosyncline behind the arc. On the passive western margin an Atlantic-type geosyncline develops.

If continent I, having approached the rise to the west, reverses its direction of movement and starts to move east (fig. 2B), the Andean-type geosyncline on the now passive western margin will be eroded, and eventually an Atlantic-type geosyncline will develop. Assuming that on the active eastern margin lithosphere descends on the abyssal plains at some distance from the continent, an island arc–type margin is formed. The former Atlantic-type geosyncline is now enclosed in a small ocean basin and becomes a geosyncline of Japan Sea–type.

Assuming that continent II in figure 2B starts to move west, and that lithosphere descends beneath the active western margin, an Andean-type margin develops across the earlier Atlantic-type geosynclinal deposits. On the passive eastern margin the now inactive island arc is eroded to sea level and may subside to form a submarine ridge. The basin on the concave side of the arc will hence remain essentially a Japan Sea–type geosyncline.

In figure 2C continued convergence of the two continents has resulted in their collision and the development of a Himalayan-type orogeny. This may require cessation of spreading at the central ocean rise. The earlier geosynclinal deposits on and oceanward of the continental margins are folded and thrust onto continents, and the original arcuate shape of the Andean- and island arc–type geosynclines may be destroyed. Small ocean basins of Mediterranean type may remain where the continents are not in contact.

On the western margin of continent I in figure 2C, complete erosion of the Andean mountain arc, accompanied and followed by sedimentation, results in continued development of an Atlantic-type margin.

Development of the eastern margin of continent II (fig. 2C) will depend largely on the rate of sedimentation relative to subsidence. Assuming that sediments build up faster than the basin floor subsides, the Japan Sea–type geosyncline becomes filled with sediment. This spills over the submerged island arc and accumulates on its oceanward flank and on the abyssal plains beyond, forming an Atlantic-type geosyncline.

In figure 2D, spreading along the line of the Himalayan-type orogenic belt is resumed, and the two continents move apart, carrying deformed, metamorphosed, and intruded geosynclinal deposits on their borders. Erosion of the orogenic belts will lead to the development of Atlantic-type geosynclines on the eastern margin of continent I and on the western margin of continent II. On the opposite margins of each continent descent of lithosphere will result in development of either island arc– or Andean-type margins.

STRATIGRAPHIC SEQUENCES
IN GEOSYNCLINES

SEQUENCES IN ANCIENT GEOSYNCLINES

Numerous authors have emphasized that vertical sequences of sedimentary facies occur in many geosynclines (Pettijohn 1957; Trümpy 1960; Aubouin 1965) as shown in table 3. Some (e.g., Tercier 1948; de Sitter 1964) have stressed the relationships of these facies to orogenic phases.

These sequences are broadly similar, although there are distinctions in terminology and emphasis and differences in interpretation. We shall now try to relate these facies, and the concepts implied by the sequences, to the types of geosyncline and their associated orogenies, and to explain the differences.

The orthoquartzite-carbonate suite consists of shallow-water sandstones and carbonates, without volcanics, and is shown by

Pettijohn (1957, fig. 165) as occurring beneath the euxinic suite in several geosynclinal sequences. Trümpy (1960) considers that miogeosynclinal shallow-water platform sediments may underlie the eugeosyncline. Nevertheless, both authors admit that it is more usual to find this miogeosynclinal facies alongside the euxinic and flysch facies. We believe that the present close geographical proximity of the euxinic and orthoquartzite suites in ancient geosynclines has often led geologists to suppose that the euxinic follows the orthoquartzite suite, although no stratigraphic contact between the two facies has been observed.

The preflysch (also euxinic, leptogeosynclinal, or bathyal lull of European Devonian

the Devonian as *becken* facies; in the lower Paleozoic it makes up a large proportion of the monotonous siltstone formations.

The flysch is principally composed of turbidites with interbedded pelagic mudstones. De Raaf (1968) suggested that those successions which accumulated in orthogeosynclinal basins during the advanced stage of orogenic development (flysch *sensu strictu*) should be separated from those which accumulated in an earlier or later stage of development (flyschoid formations). However, such a separation is difficult to apply in practice, and the term *flysch* is used in this paper to include all thick successions of "gravitational" and pelagic deposits. In some sequences flysch contains olistho-

TABLE 3

GEOSYNCLINAL SUCCESSIONS

Pettijohn (1957)	Aubouin (1965)	Trümpy (1960)
Molasse......................	Molasse	Molasse
Flysch.........................	Flysch	Flysch
Euxinic........................	Preflysch	Leptogeosynclinal/ Schistes lustrés
Orthoquartzite/carbonate.......	....	Miogeosynclinal (platform)

geologists) consists of pelagic sediments such as black shales and radiolarian cherts, and tholeiitic volcanics and ultrabasics. Although Pettijohn stresses the black shales and omits the volcanics, to the other authors ophiolites (serpentines, gabbros, and spilites) are an essential aspect of the preflysch.

Sedimentary facies of the preflysch can be divided into two subfacies. The first consists of very thin condensed deposits such as the leptogeosynclinal formations of Trümpy (1960), the *schwelle* of European Devonian geologists (e.g., Rabien 1956), or the condensed graptolitic shales of the lower Paleozoic. The second consists of thicker sequences of fine-grained turbidites and volcanics in addition to pelagic sediments. This subfacies is known in the Alps as *Schistes lustrés* or *Bündnerschiefer* and in

stromes or wildflysch, consisting of conglomerates and exotic blocks deposited as slide masses, the sedimentary or tectonic origin of which is often difficult to ascertain. In other sequences wildflysch is absent. In most flysch formations the sandstones are moderately mature in composition with a significant proportion of quartz. In a few cases there is little quartz, volcanic fragments predominate, and the flysch is associated with contemporary volcanic rocks.

Molasse deposits are largely continental and conglomeratic and are considered to be postorogenic, although Pettijohn equated his molasse suite with paralic, deltaic deposits. Molasse may overlie the flysch gradationally, with no break; or it may occur marginal to the flysch and be partly its time equivalent; or it may lie upon strongly folded flysch with angular unconformity.

In an Atlantic-type geosyncline preflysch deposits initially form on the mid-oceanic ridge. The early volcanic outpourings may possibly rest upon continental crust, as in the Red Sea, and therefore the preflysch may locally overlie earlier continental deposits. *Schwelle* subfacies will occur on rises and *becken* subfacies in small basins on the mid-ocean ridge.

Deposits of preflysch will come to form the abyssal plains and hills where more pelagic sediments and possibly tholeiitic volcanics will accumulate, but as early formed crust moves away from the ridge, volcanics will give way to pelagics. These will be followed and to some extent overlapped by a flysch which consists of continentally derived, compositionally mature turbidites. This extensive eugeosyncline will have a well-developed miogeosyncline on its flank. The continental rise will be formed partly by the relatively fine-grained deposits of geostrophic contour currents. Calc-alkaline volcanism will be absent, and there will be no angular unconformities within the flysch, as deformation will not take place until the geosyncline changes to a different type. The flysch, in fact, will be preorogenic, as suggested by Tercier (1948), and is in no way the result of any orogeny. Flysch deposition will continue indefinitely, merely building out the continental margin until there is a change to an island arc–type or Andean-type geosyncline. An orogenic stage will then follow the flysch, but there will be no causal relationship between flysch deposition and orogenesis. Therefore strictly the flysch should be called "flyschoid" in the sense of de Raaf (1968).

In an Andean-type geosyncline wild-flysch will be well represented. The flysch will be fairly coarse-grained, consisting mainly of continental and some calc-alkaline material. It will overlie preflysch of the usual oceanic type. There will be no clear differentiation between eugeosyncline and miogeosyncline, as the continental shelf will be virtually absent. Molasse deposition will be contemporaneous with and rather difficult to separate from flysch.

In an island arc–type geosyncline volcanic material, together possibly with limestone debris, will predominate in the flysch which will overlie preflysch. Quartz will be rare, and the flysch will be compositionally immature. Andesitic volcanics will be abundant. Although shallow water deposits may be found around volcanic islands, these will be locally derived volcanic clastics and clastic carbonates, and there will be no miogeosyncline. Unconformities within the flysch will be common. Molasse will be absent.

In Japan Sea–type geosynclines which have formed on subsiding continental crust, the preflysch may overlie an orthoquartzite-carbonate facies. It will resemble Pettijohn's euxinic facies without ophiolites, although tuffaceous material may be locally abundant. If the geosyncline has formed on oceanic crust, the preflysch will be normal oceanic material. The flysch will be compositionally mature and may include mass flow deposits if fault scarps develop. An absence of evidence for ocean bottom currents or geostrophic contour currents will indicate that the basin is restricted. It is even possible that restriction will lead to development of a fresh-water basin, indicated by a change to nonmarine faunas. Oceanward the flysch will pass into volcanics and volcanic turbidites derived from the island arc. Shallow marine and coastal plain deposits resembling fine-grained molasse will overlie the flysch conformably. Unless the geosyncline changes to an Andean type, or continental collision occurs, there will be little evidence of orogeny. Deformation will be slight and high-grade metamorphism and granitic intrusion will be absent.

CONCLUSIONS

1. Sedimentary and volcanic deposits of geosynclinal dimensions are accumulating on and oceanward of modern continental margins, around island arcs, and in some small ocean basins.

2. Geosynclines of Atlantic type occur on and oceanward of continental margins near

which there is no differential movement between lithospheric plates. The geosyncline includes an ocean rise, abyssal plains, continental rise, continental shelf, and coastal plain. Deposits show the broad upward sequence: ultrabasics → tholeiitic volcanics and pelagics → compositionally mature turbidites and pelagics → geostrophic contour current deposits → shallow water and paralic sediments.

3. Geosynclines of Andean type occur on and beside continental margins beneath which spreading lithosphere descends, and include a submarine trench and mountain arc. Trench deposits are pelagics, turbidites which are compositionally both mature and immature, and wildflysch. Coarse-grained molasse and calc-alkaline volcanics are typical of the mountain arc.

4. Geosynclines of island arc–type occur where spreading lithosphere descends some distance from a continent, and include a submarine trench and volcanic island arc. Deposits are compositionally immature turbidites and pelagics with calc-alkaline volcanics around the island arc, where there may also be derived shallow-water carbonates.

5. Japan Sea–type geosynclines occur in small ocean basins between continental margins and bordering island arcs. Deposits are compositionally mature turbidites, mass flow material, tuffs, and pelagics which pass up into shallow-water sediments. Depending on the nature of the basin floor, these deposits may overlie tholeiites, typical of Atlantic-type geosynclines, or they may overlie continental crust.

6. Mediterranean-type geosynclines occur in small ocean basins within or between continents. Deposits are similar to those of Japan Sea-type geosynclines.

7. Andean-type orogeny and island arc–type orogeny occur within, and are associated with the development of their respective geosynclines. An Atlantic-type geosyncline may change to a geosyncline of Andean or island arc–type, and the change will be accompanied by Andean or island arc–type orogeny, respectively. Himalayan-type orogeny results from the collision of continents and may affect all types of geosyncline.

8. Oscillation of continents between ocean rises is related to alterations in the positions of zones of descending lithosphere. Changes in type of geosyncline and orogeny of geosynclinal deposits are connected with the migration and intermittent collision of continents.

9. Continents grow by accretion at their margins. Decrease in the surface area of continents may accompany continental collision, and continental material may be lost as a result of subsidence in some small ocean basins of Japan Sea-type. Geosynclines form between continents, though sediment accumulation occurs mainly at the margins of continents.

10. The common occurrence in ancient geosynclinal deposits of the sequence preflysch (euxinic) → flysch → molasse can be related to type of geosyncline. Preflysch is the ocean floor accumulation of pelagics and tholeiitic volcanics, associated with ultrabasics. It may be present in all types of geosyncline. Euxinic facies, lacking volcanics, are confined to Japan Sea–type geosynclines. Flysch is different in each type of geosyncline. In the Atlantic type it is compositionally mature and lacks andesitic volcanics and wildflysch. It is preorogenic and has no causal relationship to orogeny. In the Andean type wildflysch predominates. In the island arc type flysch is compositionally immature and contains abundant andesitic volcanic detritus. In Japan Sea–type geosynclines flysch resembles that of Atlantic-type geosynclines, but tuffs and mass flow deposits are more abundant. Molasse occurs in Andean-type geosynclines, where it partly overlies the flysch and is partly its lateral equivalent. In Atlantic-, island arc– and Japan Sea–type geosynclines molasse does not occur unless the geosyncline changes to Andean type or continental collision takes place. The molasse will then be stratigraphically separate from and later than the flysch.

ACKNOWLEDGMENTS.—The authors are grateful for discussions with colleagues at the Department of Geology and Mineralogy, University of Oxford, especially Dr. E. R. Oxburgh, who criticized an early draft of the manuscript. Author Mitchell was supported, while at Oxford, by a grant provided by the Natural Environmental Research Council.

REFERENCES CITED

ARGAND, E., 1916, Sur l'arc des Alpes occidentales: Eclogae Geologicae Helvetiae, v. 16, p. 179–182.

AUBOUIN, J., 1965, Geosynclines: Amsterdam, Elsevier Publishing Co., 335 p.

BELOUSSOV, V. V., and RUDITCH, E. M., 1961, Island arcs in the development of the earth's structure (especially in the region of Japan and the Sea of Okhotsk): Jour. Geology, v. 69, p. 647–658.

BRIDEN, J. C., 1967, Recurrent continental drift of Gondwanaland: Nature, v. 215, no. 5108, p. 1334–1339.

BURK, C. A., 1968, Buried ridges within continental margins: New York Acad. Sci., Trans., ser. 3, v. 30, no. 3, p. 377–409.

DANA, J. D., 1873, On some results of the earth's contraction from cooling, including a discussion of the origin of mountains and the nature of the earth's interior: Am. Jour. Sci., ser. 3, v. 5, p. 423–443; v. 6, p. 6–14, 104–115, 161–172.

DIETZ, R., 1961, Continent and ocean basin evolution by spreading of the sea floor: Nature, v. 190, no. 4779, p. 854–857.

—— 1963, Collapsing continental rises: an actualistic concept of geosynclines and mountain building: Jour. Geology, v. 71, p. 314–333.

—— and HOLDEN, J. C., 1966, Miogeoclines in space and time: Jour. Geology, v. 74, p. 566–583.

DRAKE, C. L., 1966, Recent investigations on the continental margin of eastern United States, in POOLE, W. H. (ed.), Continental margins and island arcs: Geol. Survey Canada Paper 66-15, p. 33–47.

——; EWING, M.; and SUTTON, G. H., 1959, Continental margins and geosynclines: the east coast of North America north of Cape Hatteras, in Physics and chemistry of the earth: London, Pergamon Press, v. 3, p. 110–198.

ERICSON, D.; EWING, M.; WOOLIN, G; and HEEZEN, B., 1961, Atlantic deep-sea sediment cores: Geol. Soc. America Bull., v. 72, no. 2, p. 193–286.

EWING, J., and EWING, M. 1967, Sediment distribution on the mid-ocean ridges with respect to spreading of the sea floor: Science, v. 156, no. 3782, p. 1590–1592.

FISHER, R. L., and RAITT, R. W., 1962, Topography and structure of the Peru-Chile trench: Deep-Sea Research, v. 9, p. 423–443.

FRANK, F. C., 1968, Curvature of island arcs: Nature, v. 220, no. 5165, p. 363.

GALLI-OLIVIER, C., 1967, Pediplain in northern Chile and the Andean uplift: Science, v. 158, no. 3801, p. 653–655.

GANSSER, A., 1964, Geology of the Himalayas: London, Interscience Publishers, 289 p.

GILETTI, B. J., and DAY, H. W., 1968, Potassium-argon ages of igneous intrusive rocks in Peru: Nature, v. 220, no. 5167, p. 570–572.

GORSHKOV, G. S., 1962, Petrochemical features of vulcanism, in The crust of the Pacific Basin: Am. Geophys. Union, Geophys.Mon. 6, p. 110–115.

GORYATCHEV, A. V., 1962, On the relationship between geotectonic and geophysical phenomena of the Kuril-Kamchatka folding zone at the junction zone of the Asiatic continent with the Pacific Ocean, in The crust of the Pacific Basin: Am. Geophys. Union, Geophys. Mon. 6, p. 41–51.

HALL, J., 1857, Direction of the currents of deposition and source of the materials of the older Palaeozoic rocks: Canadian Naturalist and Geologist, v. 2, p. 284–286; Canadian Jour. Industry Sci. Art, n.s., v. 3, p. 88.

HAMILTON, E. L., 1967, Marine geology of abyssal plains in the Gulf of Alaska: Jour. Geophys. Research, v. 72, no. 16, p. 4189–4213.

HARRINGTON, H. J., 1962, Paleogeographic development of South America: Am. Assoc. Petroleum Geologists Bull., v. 46, p. 1773–1814.

HEEZEN, B. C.; HOLLISTER, D. C.; and RUDDIMAN W. F., 1966, Shaping of the continental rise by deep geostrophic contour currents: Science, v. 152, no. 3721, p. 502–508.

HESS, H. H., 1962, History of ocean basins, in Petrologic studies: a volume to honor A. F. Buddington: New York, Geol. Soc. America, p. 599–620.

HOSPERS, J., 1965, Gravity field and structure of the Niger delta, Nigeria, West Africa: Geol. Soc. America Bull., v. 76, p. 407–422.

HSU, K. J., 1965, Collapsing continental rises: an actualistic concept of geosynclines and mountain building: a discussion: Jour. Geology, v. 73, p. 897–900.

—— 1968, Principles of melanges and their bearing on the Franciscan-Knoxville paradox: Geol. Soc. America Bull., v. 79, p. 1063–1074.

ISACKS, B. I.; OLIVER, J.; and SYKES, L. R., 1968, Seismology and the new global tectonics: Jour. Geophys. Research, v. 73, no. 18, p. 5855–5899.

KAY, M., 1951, North American Geosynclines: Geol. Soc. America Mem. 48, 143 p.

KUENEN, PH. H., 1967, Geosynclinal sedimentation: Geologische Rundschau, v. 56, p. 1–19.

LE PICHON, X., 1968, Sea-floor spreading and continental drift: Jour. Geophys. Research, v. 73, no. 12, p. 3661–3697.

MENARD, H. W., 1967, Transitional types of crust under small ocean basins: Jour. Geophys. Research, v. 72, no. 12, p. 3061–3073.

MINATO, M.; GORAI, M.; and HUNAHASHI, M., eds., 1965, The geologic development of the Japanese islands: Tokyo, Tsukiji Shokan Co., 442 p.

MIYASHIRO, A., 1961, Evolution of metamorphic belts: Jour. Petrology, v. 2, p. 277–311.

MOORE, D. G., and BUFFINGTON, E. C., 1968, Transform faulting and growth of the Gulf of California since the late Pliocene: Science, v. 161, p. 1238–1241.

MORGAN, W. J., 1968, Rises, trenches, great faults, and crustal blocks: J. Geophys. Research, v. 73, p. 1959–1982.

OXBURGH, E. R., and TURCOTTE, D. L., 1969, The thermal structure of island arcs: Geol. Soc. America Mem. (in press).

PETTIJOHN, F. J., 1957, Sedimentary rocks (2d ed.): New York, Harper Bros., 718 p.

RAAF, J. F. M. DE, 1968, Turbidites et associations sédimentaires apparentées: Koninkl. Nederlandse Akad. Wetensch., Proc., v. 71. p. 1–23.

RABIEN, A., 1956, Zur Stratigraphie und Fazies des Ober-Devons in der Waldecker Hauptmulde: Abh. hess. landesamt Bodenforsch, v. 16, p. 1–83.

ROUTHIER, P., 1953, Etude géologique du versant occidental de la Nouvelle-Calédonie entre le Cola de Boghen et la Pointe D'Arana: Soc. géol. France Mem., v. 67, p. 1–271.

RUTLAND, R. W. R., and GUEST, J. E., 1965, Isotopic ages and Andean uplift: Nature, v. 208, no. 5011, p. 677–678.

SCHOLL, D. W.; BUFFINGTON, E. C.; and HOPKINS D. M., 1968, Geologic history of the continental margin of North America in the Bering Sea: Marine Geology, v. 6, no. 4, p. 297–330.

———; HUENE, R. VON, and RIDLON, J. B., 1968, Spreading of the ocean floor: undeformed sediments in the Peru-Chile trench: Science, v. 159, no. 3817, p. 869–871.

SCHUCHERT, C., 1923, Sites and natures of the North American geosynclines: Geol. Soc. America Bull., v. 34, p. 151–260.

SEYFERT, C. K., 1969. Undeformed sediments in oceanic trenches with sea floor spreading: Nature, v. 222, no. 5188, p. 70.

SHOR, G. G., JR., 1964, The structure of the Bering Sea and the Aleutian ridge: Marine Geology, v. 1, p. 213–219.

SITTER, L. U. DE, 1964, Structural geology: New York, McGraw-Hill Book Co. 551 p.

STILLE, H., 1940, Einführung in den Bau Nordamerikas: Berlin, Borntraeger Verlagsbuchhandlung, 717 p.

SYLVESTER-BRADLEY, P. C., 1968, Tethys, the lost ocean: Sci. Jour., v. 4, no. 9, p. 47–53.

TERCIER, J., 1948, Le Flysch dans la sédimentation Alpine: Eclogae Geologicae Helvetiae, v. 40, p. 164–198.

THOMPSON, J. E., and FISHER, N. H., 1965, Mineral deposits of New Guinea and Papua and their tectonic setting: Bureau Mineral Resources, Canberra, Preprint no. 129, 31 p.

TRÜMPY, R., 1960, Paleotectonic evolution of the Central and Western Alps: Geol. Soc. America Bull., v. 71, p. 843–907.

VINE, F. J., 1966, Spreading of the ocean floor: new evidence: Science, v. 154, p. 1405–1415.

——— and MATTHEWS, P.M., 1963, Magnetic anomalies over ocean ridges: Nature, v. 199, no. 4897, p. 947–949.

WATKINS, J. S., and GEDDES, W. H., 1965, Magnetic anomaly and possible orogenic significance of geological structure of the Atlantic shelf: Jour. Geophys. Research, v. 70, no. 6, p. 1357–1361.

30

Copyright © 1969 by Macmillan Journals Ltd

Reprinted from *Nature* **222**:124–129 (1969)

Evolution of the Appalachian/Caledonian Orogen

by

J. F. DEWEY

Department of Geology,
University of Cambridge,
Cambridge

The structure of the Appalachian/Caledonian orogen is described and a new model for its evolution proposed.

THE Appalachian and Caledonian orogens form a continuous belt[1] when the North Atlantic is reconstructed by fitting the 500 fathom line of the present continental margins[2]. On this reconstruction, the forelands consist of Pre-Cambrian basement complexes and the orogen appears to be ensialic, that is, cutting an older basement, like the Variscan orogen of the Urals. There is a remarkable correspondence, however, between the orogen and the subsequent axis of Atlantic rifting[3,4]. This led Wilson[5] to propose the hypothesis of a Proto-Atlantic Ocean, and recently (my unpublished results) a model for the evolution of orogenic belts based on an oceanic expansion-contraction cycle has been suggested, following earlier views[6,7] that orogens are sited on collapsed continental rises. In this article the structure of the Appalachian/Caledonian orogen is outlined and a possible model for its evolution presented.

Structural Zones of the Orogen

It has long been recognized that the British Caledonides consist of two greatly contrasting zones, previously referred to as the northern metamorphic and southern, non-metamorphic, Caledonides[8]. The northern zone (zone A, Fig. 1) includes the Moine and Dalradian Series and consists of a thick sedimentary sequence of late Pre-Cambrian, Cambrian and probably Lower Ordovician[9] age. Zone A is characterized by large scale recumbent folding and high-grade metamorphism. Various ages have been suggested for the age of climactic deformation and metamorphism in zone A, referred to here as the Grampian event (Fig. 1), including Pre-Cambrian[8], pre-Arenigian[10], Arenigian[11] and pre-Caradocian[12]. The details of these arguments will not be discussed here, but the greater part of the deformation and metamorphism occurred during early Ordovician times. In Norway, early Ordovician movements have long been known[13] and increasing recognition is being given to such movements in the northern Appalachians[14]. It is suggested that zone A continues throughout the north-west side of the Appalachians (Fig. 1) and includes the following: the Fleur-de-Lys and Grand Lake Schist Groups of Newfoundland[15]; the Shickshock and Macquereau complexes of the Gaspe Peninsula; the Sillery Belt of Quebec; the pre-Beauceville Group rocks of the Eastern Townships; the Grand Pitch Formation of Maine[16]; the pre-Silurian rocks of, and west of, the Oliverian Belt[14]; the New York City Group; the pre-Silurian rocks of Pennsylvania and Maryland; and probably most of the Piedmont[14]. In Scotland, the rocks of zone A are thrust north-westwards across a shelf sequence of Cambrian and Lower Ordovician[17] carbonates. In the Appalachians, relationships with the Laurentian Foreland are more complex. Sedimentation in the miogeocline[7], although interrupted by post-Beekmantown pre-Chazy non-sequence and disconformity, persisted until late Ordovician times when Taconian deformation[18] accompanied and followed[19] the emplacement of klippen.

The position and nature of the south-eastern margin of zone A are difficult to define, except in Newfoundland where the Luke Arm Fault[21] (Fig. 1) forms the southern boundary of a complex of spilites and tholeites (Lushs

Bight Complex). The ophiolite suite of the Ballantrae Complex in Scotland (Fig. 1) is characterized locally by glaucophane-schists and is unconformably overlain by Caradocian conglomerates. It is suggested that the Ballantrae Complex represents the southern margin of zone A, and continues westwards into the Tyrone Volcanic Complex in Ireland and thence south-westwards along a major magnetic anomaly at Strokestown[22] to the Dingle Peninsula (personal communication with N. Rast). In Scandinavia the south-eastern margin of zone A is probably represented by a belt of deformed basic volcanics (Fig. 1) including the Støren Group[13].

After the Ordovician climactic deformation in zone A, graben such as the South Mayo Trough (Fig. 1) developed, and marine transgression progressed generally northwards[23]. The molasse of the Ordovician deformation on the south-eastern margin of zone A is represented by the Caradocian conglomerates of the Girvan region and thick Ordovician argille scagliosa and flysch deposits south of the Luke Arm Fault in Newfoundland[21]. Zone A molasse on the north-western margin of the Appalachians is seen in the miogeoclinal Ordovician klippen, clastic wedges such as the greywackes of the Austin Glen Group and the Richmond–Juniata red-bed sequences, with a south-easterly provenance.

Pre-Appalachian/Caledonian basement is known in many areas of the north-western parts of zone A. In Scotland and Ireland[24], inliers of Lewisian rocks lie within the Moine outcrop (Fig. 1). Grenville rocks (Fig. 1) occur in the Great Northern Peninsula of Newfoundland, and in the Green Mountain (Berkshire Highlands–Hudson Highlands) Blue Ridge axis of the north-western Appalachians, where they form a basement to autochthonous Cambro–Ordovician miogeoclinal sequences. Grenville rocks also form a rejuvenated basement (Baltimore Gneiss of Maryland, Fordham Gneiss of New York) to the thicker, heavily deformed clastic sequences of the Piedmont.

The southern non-metamorphic Caledonides are divisible into a region (zone B) between the southern margin of zone A and the northern, fault-bounded margin of the Cambro–Ordovician Irish Sea Horst, and a second region (zone C) to the south of this line (Fig. 1). Zone B, consisting of Cambrian, Ordovician and Silurian strata, is also recognized in Newfoundland[25], New Brunswick[26] and New England[14]. Zone B was climactically deformed during the late Silurian or Devonian period. Fairly simple structures associated with steep slaty, and flat strain-slip, cleavages were produced with little or no regional metamorphism. In New England, however, more complex recumbent structures and high-grade regional metamorphism were developed. Vulcanism was important in zone B in which three Ordovician volcanic lineaments (Fig. 1) were developed. The oldest fossiliferous strata in zone B are Middle Cambrian[27] and the age of the base of the sequence and the nature of the underlying basement are unknown.

Zone C is characterized by widespread deformation just before the Cambrian period (Fig. 1). The rocks involved in this deformation are the Mona Complex and Cullenstown Group[28], the Uriconian, and the Long-

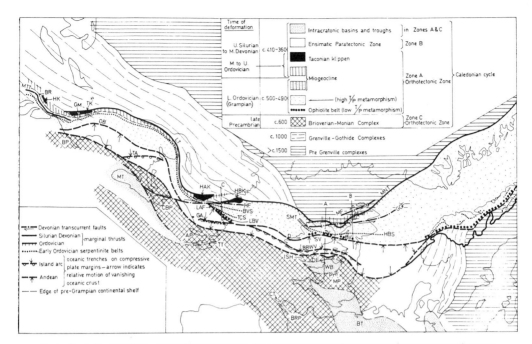

Fig. 1. Structural zones of the Appalachian/Caledonian orogen. AP, Avalon Platform; BBWV, Borrowdale, Balbriggan, Waterford Volcanic lineament; BP, Boston Platform; BR, Blue Ridge; BRP, Breton Platform; BT, Brabant Trough; BV, Ballantrae Volcanics; BVS, Baie Verte Serpentinite; CBP, Cape Breton Platform; CT, Crozon Trough; D, Dingle; DF, Dinorwic Fault; GA, Gander Volcanic lineament; GGF, Great Glen Fault; GM, Green Mts; HAK, Humber Arm Klippe; HBK, Hare Bay Klippe; HBS, Highland Boundary Serpentinite; HF, Hampden Fault; HK, Hamburg Klippe; ISH, Irish Sea Horst; LAF, Luke Arm Fault; LBV, Lushs Bight Volcanics; LT, Logans Thrust; MF, Minch Fault; MP, Midland Platform; MNT, Moine Thrust; MT, Meguma Trough; MTT, Martie Thrust; OB, Oliverian Belt; PHF, Pontesford Hill Fault; SMT, South Mayo Trough; SV, Strokestown Volcanics; TA, Tetagouche Volcanic lineament; TCS, Tilt Cove Serpentinite; TK, Taconic Klippe; TT, Trinity Trough; TV, Tyrone Volcanics; WB, Welsh Basin.

myndian in Britain, the Brioverian in Brittany, and late Hadrynian strata[27] in south-east Newfoundland, Cape Breton and New Brunswick. Zone C glaucophane-schists are known only in the Mona Complex of Anglesey.

During Lower Palaeozoic times, zone C was a platform for thin widespread deposition. There were, however, several regions of greater subsidence and vulcanism, such as the Meguma Trough, the Trinity Trough, the Welsh Basin and the Brabant Trough (Fig. 1). The Lower Palaeozoic sequences of these mobile areas were strongly deformed in late Silurian or Devonian times. Pre-Hadrynian basement is unknown in zone C in the Appalachians, but occurs locally in northern France (Pentevrian), England (Malvernian) and south-east Ireland (Rosslare Complex[28]).

The timing of the late Silurian/Devonian climactic deformation, dominant in zone B, but also important locally in zones A and C, is reflected in the age of Devonian molasse sediments[29]. The main directions of molasse transport were away from zone B to produce, for example, the fluviatile fans of the Anglo–Welsh Cuvette and the Catskill region of the Appalachians.

Evolution of the Orogen

It is suggested that the evolution of the Appalachian/ Caledonian orogen was related to a cycle of oceanic expansion and contraction schematically illustrated in Fig. 2. During the spreading–expanding phase (Fig. 2*A*, *B*) wedges of sediment (Torridonian and Longmyndian) accumulated on the continental terraces and continental rises (Moine–Dalradian and sediments of the Mona Complex) were built out onto the ocean floor. The development

of Benioff Zones associated with lithosphere descent, and frictional melting of basalt, in late Pre-Cambrian (Fig. 2*B*) and Upper Cambrian times (Fig. 2*C*) produced the Gwna Volcanics and Ballantrae–Tayvallich Volcanics respectively. The orthotectonic orogens of zones C and A were produced by progressive deformation of the continental rises (unpublished results). Subsequently, during Caradocian times (Fig. 2*D*), flysch wedges were built out from the oceanic side of zones C and A and thin pelagic sequences (for example, Moffat Shales in the Southern Uplands) were laid down on oceanic crust. During Silurian times, the flysch wedges completely swamped zone B (Fig. 2*E*). Finally, in late Silurian/Devonian times, zone B sequences were compressed and deformed during the last stages of Proto-Atlantic contraction (Fig. 2*F*). Trough and basin sequences which post-date the orthotectonic deformation in zones A and C were also deformed at this time, and in Scandinavia (Fig. 2*G*) intense contraction and suturing caused the overspilling of zone A and B sequences to produce great thrust sheets across the Baltic Foreland (Fig. 1). An outline section across the Himalayas[30] is shown in Fig. 2*H* to illustrate that continued contraction in a paratectonic orogen may produce continental underthrusting and the development of a sialic root.

One of the most remarkable and significant features of zone A is the long thermal history, evidenced by the emplacement of post-tectonic granite plutons in Scotland up to 100 m.y. after Ordovician metamorphism[31], the local development of a Devonian metamorphism[32] and the great spread of isotopic ages[33]. It has been argued (my unpublished results) that the long thermal histories,

J. F. Dewey

characterizing orthotectonic orogens, are related to the rise of magmas and volatiles from an underlying Benioff Zone progressively later on the continental side.

The age of volcanics is of great significance in establishing the time of initiation of the spreading–expanding and spreading–contracting phases of ocean development (my unpublished results). Throughout the Appalachians and Caledonides there is evidence of vulcanism at, or near, the base of the miogeoclinal sequences of zones A and C. In Appalachian zone A the Catotcin Volcanics, the Tibbit Hill Volcanics and the Labrador Basalts probably developed at the onset of Proto-Atlantic expansion. The Uriconian Volcanics in Caledonian zone C probably

bear a similar relationship. Lower Palaeozoic faunas are of great potential value for elucidating the timing of oceanic expansion and contraction. Trilobite[34] and graptolite[35] faunas show varying degrees of provinciality but were particularly different in zones A and C during Lower Ordovician times, supporting the contention that this was the approximate time of maximum separation.

Evolution of Zone A

Figs. 3 and 4 illustrate, schematically, the suggested sequence of events leading to the development of the zone A orthotectonic orogen in Britain and Newfoundland respectively. In both cases early oceanward palaeo-

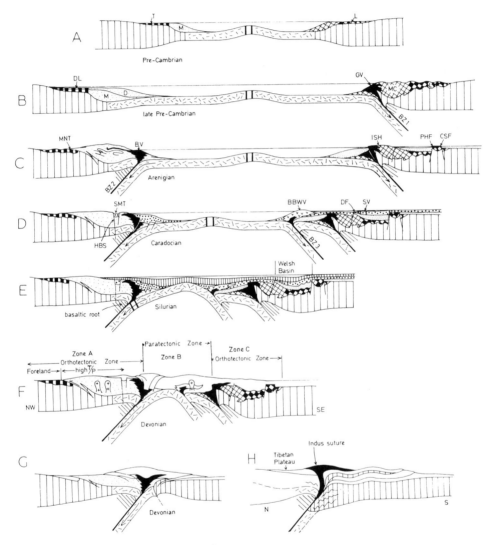

Fig. 2. Schematic sections illustrating proposed evolution (*A–F*) of the Caledonian orogen in Britain (line of section *A* in Fig. 1). Schematic section through Caledonides of east Greenland and Norway (*G*). Schematic section across the Himalayas (*F*). BZ1 . . . 3, Benioff zones in order of development; CSF, Church Stretton Fault; D, Dalradian; DL, Durness Limestone; GV, Gwna Volcanics; L, Longmyndian; MC, Mona Complex; SV, Snowdon Volcanics; T, Torridonian; other abbreviations as for Fig. 1.

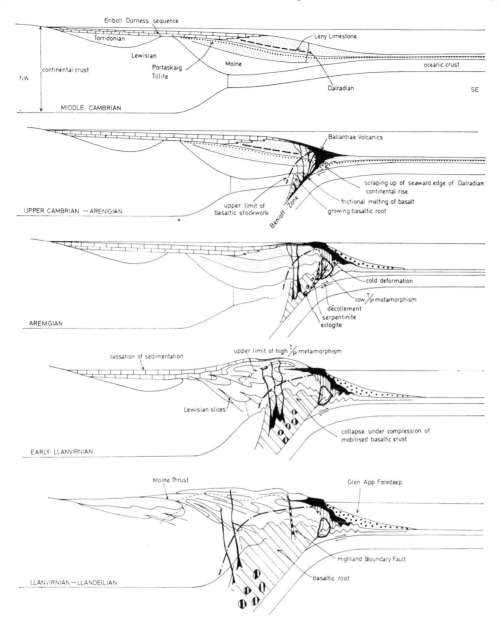

Fig. 3. Proposed development of zone A in the British Caledonides along section B of Fig. 1. Basic intrusions and volcanics shown in black.

slopes progressively changed to continentward slopes as deformation of the continental rise progressed above a growing basaltic root (my unpublished results). High gravity values characterize zone A in Newfoundland and the suggested crustal structure includes a Taconian basaltic root[36]. A basaltic substructure has been suggested for the Caledonides in Finmark[37], similar to the Alpine Ivrea Zone, and for the Caledonides of the Kola Peninsula[38]. In the Appalachians, the sedimentary and tectonic history of the miogeocline is longer than in the Caledonides, and Durness sedimentation persisted in Scotland into Llanvirnian times[17], while deformation was occurring in Zone A to the south-east (unpublished results of J. F. D. and J. M. Bird) and was finally obliterated by the emplacement of the Moine Nappe (Fig. 3). In Appalachian zone A of Newfoundland (Fig. 4), the

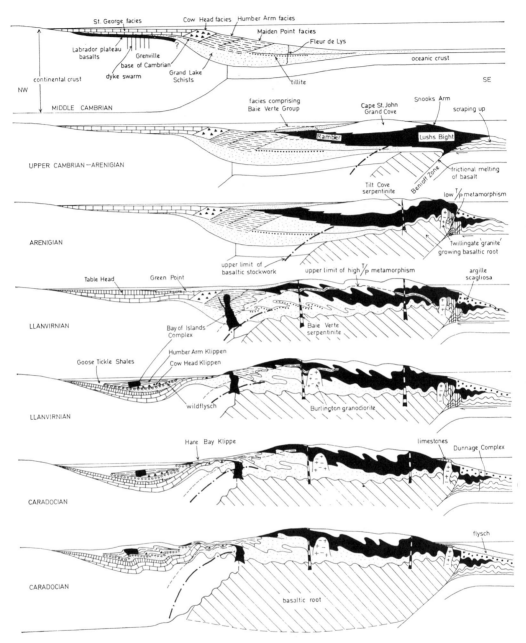

Fig. 4. Proposed development of zone A in the Appalachians of Newfoundland along section C of Fig. 1. Basic intrusions and volcanics shown in black.

Taconic region, and Pennsylvania there is a clear record of gradual relief inversion well into late Ordovician times. The pre-Taconian transition from continental shelf to continental rise can be traced almost continuously from Newfoundland to Pennsylvania and is represented by a rapid facies change from carbonates to clastics. A distinctive limestone–breccia facies (Cow Head of Newfoundland (Fig. 4), Conestoga Limestone of Pennsylvania) often marks the transition. It is suggested that relief inversion progressed from the oceanic side of zone A towards the continent. Taconian movements in late Ordovician times, finally obliterating the miogeocline, were the last stages of a progressive deformation begun with Penobscot movements[16] on the oceanic side.

It is thus suggested that the landmass of Appalachia[18] was developed by deformation of an Ordovician continental rise and does not imply the existence of a major continental mass on the site of the present-day Atlantic Ocean. Criticisms[39] of Wilson's[5] Proto-Atlantic theory are based on the assumption that the Appalachian/Caledonian orogen is founded on a Pre-Cambrian basement. An alternative view of North Atlantic history has been proposed[4] in which mobile zones were progressively stabilized, eventually leading to oceanic spreading and expansion in Mesozoic times. I believe this to be incorrect for the following reasons. First, it involves the former existence of linear ensialic zones of mobility in which no mechanism is provided for the large contractions evidenced by structural analysis[40] and by the faunal evidence. Second, the intermittent and distinctive nature of the thermal history is not considered. Vulcanicity occurred in zone A only during the initial and final stages of Pre-Cambrian/Cambrian/early Ordovician sedimentation, and the pattern of zone A sedimentation is asymmetrical and extremely similar to the present Atlantic margin of North America[6].

I thank Professor Marshall Kay and Professor Charles Drake for their help and encouragement. I also thank Dr P. F. Friend, Mr D. J. W. Piper and Dr A. Gilbert Smith for critically reading the manuscript.

Received December 17, 1968.

[1] Dewey, J. F., and Kay, M., in *History of the Crust of the Earth* (Princeton Univ. Press, in the press).

[2] Bullard, E. C., Everett, J. E., and Smith, A. Gilbert, *Roy. Soc. Lond. Phil. Trans.*, No. 1088, 41 (1965).

[3] Dearnley, R., *Nature*, **206**, 1083 (1965).

[4] Sutton, J., *Proc. Geol. Assoc.*, **79**, 275 (1968).

[5] Wilson, J. T., *Nature*, **211**, 676 (1966).

[6] Drake, C. L., Ewing, M., and Sutton, G. H., in *Physics and Chemistry of the Earth*, **3**, 110 (Pergamon, 1959).

[7] Dietz, R. S., and Holden, J. C., *J. Geol.*, **65**, 566 (1967).

[8] Read, H. H., *Liv. Manch. Geol. J.*, **2**, 563 (1961).

[9] Johnson, M. R. W., and Harris, A. L., *Scot. J. Geol.*, **3**, 1 (1967).

[10] Dewey, J. F., *Geol. Mag.*, **98**, 399 (1961).

[11] Skevington, D., and Sturt, B. A., *Nature*, **215**, 608 (1967).

[12] Kennedy, W. Q., *Trans. Geol. Soc. Glasgow*, **23**, 106 (1958).

[13] Sturt, B. A., Miller, J. A., and Fitch, F. J., *Norsk Geol. Tidsskr.*, **47**, 255 (1967).

[14] Rodgers, John, *Amer. J. Sci.*, **265**, 408 (1967).

[15] Church, W. R., *Canad. Min. Metall. Bull.*, **58**, 219 (1965).

[16] Neuman, R. B., *US Geol. Surv. Prof.*, Paper 524 (1967).

[17] Higgins, A. C., *Scot. J. Geol.*, **3**, 382 (1967).

[18] Cady, Marshall, and Colbert, E. H., in *Stratigraphy and Life History* (John Wiley, 1965).

[19] Cady, W. M., *Amer. J. Sci.*, **266**, 563 (1968).

[20] Zen, e-an, *Geol. Soc. Amer. Spec.*, Paper 97 (1967).

[21] Kay, Marshall, *Amer. Assoc. Petrol. Geol. Bull.*, **51**, 579 (1967).

[22] Murphy, T., *School of Cosmic Physics, Dublin*, Bull. 11 (1955).

[23] Piper, D. J. W., *Geol. Mag.*, **104**, 253 (1967).

[24] Sutton, S., and Max, M., *Geol. Mag.* (in the press).

[25] Williams, Harold, *Amer. J. Sci.*, **262**, 1137 (1964).

[26] Poole, W. H., *Geol. Assoc. Canad. Spec.*, Paper 4 (1967).

[27] Kay, Marshall, and Eldredge, N., *Geol. Mag.*, **105**, 372 (1968).

[28] Crimes, T. P., and Dhonau, N. B., *Geol. Mag.*, **104**, 213 (1967).

[29] Friend, P. F., *Amer. Assoc. Petrol. Geol. Bull.* (in the press).

[30] Gansser, A., in *Geology of the Himalayas* (Interscience, 1964).

[31] Brown, P. E., Miller, J. A., and Grasty, R. L., *Proc. Yorks. Geol. Soc.*, **36**, 251 (1968).

[32] Dearnley, R., *Scott. J. Geol.*, **3**, 449 (1967).

[33] Harper, C. T., *Earth and Planet. Sci. Letters*, **3**, 128 (1967).

[34] Whittington, H. B., *J. Palaeontol.*, **40**, 696 (1966).

[35] Berry, W. B. N., *Geol. Soc. Amer. Bull.*, **78**, 419 (1967).

[36] Sheridan, R. E., and Drake, C. L., *Canad. J. Earth Sci.*, **5**, 337 (1968).

[37] Brooks, M., *Proc. Geol. Soc. Lond.*, No. 1649, 139 (1968).

[38] Zhdarov, V. V., *Intern. Geol. Rev.*, **10**, 1138 (1968).

[39] Young, G. M., *Amer. J. Sci.*, **265**, 225 (1967).

[40] Dewey, J. F., *Amer. Assoc. Petrol. Geol. Bull.* (in the press).

31

Reprinted from *Nature* **225**:521–525 (1970)

Plate Tectonics, Orogeny and Continental Growth

by

J. F. DEWEY
Department of Geology,
University of Cambridge

BRENDA HORSFIELD
Further Education Television Department,
BBC, London

Ocean-driven plate mechanisms, the authors conclude, have been responsible for the growth and evolution of continents for at least 3×10^9 years. Some of the implications of this idea are discussed.

A GLOBAL tectonic mechanism in which essentially rigid lithosphere plates[1-4] are created at oceanic ridges, and consumed in oceanic trenches, provides a powerful framework for evaluating the development of orogenic belts and continents. Although continents are superficial passengers on moving plates, they place significant restraints[4] on plate evolution in that their buoyancy prevents their destruction by plate consumption. A variety of mountain, and continent, building mechanisms operate on leading plate margins as a result of the consumption of oceanic lithosphere. The plates of the present oceans probably contain no oceanic crust older than late Palaeozoic or early Mesozoic (personal communications from W. C. Pitman). It is therefore probable that the oceans suffer a complete turnover by plate accretion and consumption in about 250 m.y. The history of oceans before about 250 m.y. ago can only be elucidated by analysis of orogenic belts, in island arcs and continents, in which slices of oceanic crust (ophiolite suite) are preserved and mark the sites of former oceans. We take the view in this article that no kind of plate margin necessarily maintains a fixed position, that rates of plate production and consumption sum to approximately zero[5] on an Earth of essentially constant radius, and that lithosphere consumption only occurs on one side of a plate boundary. We consider below that the relationship of continents, island arcs and oceans has been determined by an ocean-based plate mechanism for at least 3×10^9 yr.

fold belt of Europe and the Middle East (unpublished results of J. Bonnin, J. F. Dewey, W. C. Pitman and W. B. F. Ryan). East of the Mediterranean further complexity arises from the growth of the Indian Ocean, whereby fragments of Gondwanaland such as India and Australia moved northwards to collide with Asia. The Indian Ocean thus consists of ocean developed from Cretaceous times with a much older Tethyan portion east of the Ninety East Ridge and west of Australia (Fig. 1).

In the Pacific, although lithosphere is being created at the East Pacific Rise and subsidiary ridges, it is being consumed in trenches at or near the Pacific Ocean margins. In western North America, the Mesozoic Cordilleran

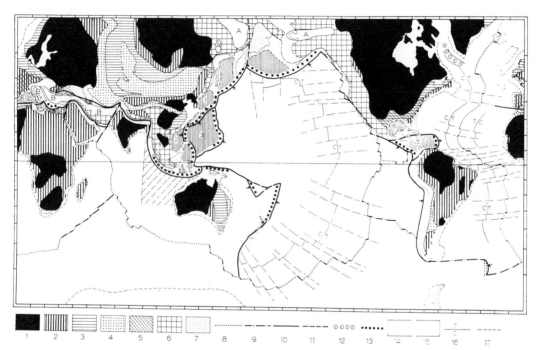

Fig. 1. Boundaries of present lithosphere plates and distribution of orogenic belts younger than 10^9 yr. Key: 1—Cratonic nuclei consolidated since 10^9 yr ago; 2—Late Pre-Cambrian (Baikalian) orogenic belts; 3—Lower Palaeozoic (Caledonian) orogenic belts; 4—Upper Palaeozoic (Uralian) orogenic belts; 5—Regions of Upper Palaeozoic regeneration (Alleghenian-Variscan); 6—Mesozoic (Cimmerian) orogens; 7—Tertiary-Quaternary (Alpine) orogens; 8—Accreting plate margins; 9—Transform plate margins; 10—Consuming plate margins (consuming side shown by small arrow); 11—Plate margins of indeterminate nature; 12—Extinct accreting plate margins; 13—Calc-alkaline vulcanicity; 14—Small ocean basins; 15—Remnants of Tethys; 16—Boundary of Cretaceous/Tertiary ocean floor; 17—Continental margins; A—Pre-Cambrian cratons of uncertain age; B—Internal microcontinents of the Alpine system of Europe and the Middle East. Latitude and longitude shown in 10° intervals.

Present Plate System

The Atlantic Ocean (Fig. 1) has been expanding in a varied and complex way since late Triassic times. The volume of sediments forming the continental rises depends on the rate of sediment discharge from adjacent continents from the time of continental separation. These continental margin sediments are largely non-volcanic and are thickest on the eastern margin of North America[6], reflecting accumulation since late Triassic times when Africa separated from North America. Elsewhere sediment volumes are smaller, reflecting separation of South America from Africa[7], Greenland and Europe from North America in Cretaceous times and Europe from Greenland in Palaeocene times. The result of the complex opening history of the Atlantic has been an exceedingly complex history of closure of Tethys now represented by remnant patches of ocean such as the Black, Caspian and Eastern Mediterranean Seas associated with the Alpine

Orogen is bounded to the west by a ridge transform system argued[8] to be the result of driving the East Pacific Rise into a trench complex. In western South America, the Peru–Chile Trench precisely delineates the edge of the Mesozoic–Tertiary Andean Orogen on the continental margin. In the Western Pacific, however, plate consumption is occurring in trenches in front of island arc volcanic/orogenic belts within the ocean (Fig. 1). The simplest situation holds where, as in Japan, a small ocean basin lies between continental margin and island arc and the trench consumes the Pacific Plate. Further south there are complications. The Philippine Sea is a small contracting plate bounded by island arcs, pushing a finger westwards into, and therefore consuming part of, the South China Sea. The extending Fiji transform fault joins the New Hebrides Trench (consuming the northern Tasman Sea) to the Tonga Kermadec Trench (consuming the Pacific Plate). A range of present plate/ocean/continent combinations is shown in Fig. 2.

Evolution of Oceans

Lithosphere is being created by plate accretion along oceanic ridges in all the major oceans. In the Atlantic lithosphere consumption occurs in only two small regions (the Lesser Antilles and the Scotia Arc) and the continental margins are not plate margins. Because the Atlantic Ocean is therefore expanding, it is axiomatic that another major ocean is contracting. This is probably the Pacific, which has therefore to consume lithosphere at a rate that is the sum of its own rate of production and the rate of expanding oceans. The marginal Pacific trenches must therefore retreat into the ocean in front of continental margins and island arcs.

Although a pattern of linear oceanic expansion and contraction is an over-simplification, the effect of continental disruption, by plate growth, and convergence, by plate consumption, is that any continental margin will be the site of thick extensive sedimentation during continental separation and subsequently the site of orogeny during continental convergence[9]. This is the basis for our model of continental growth discussed below. In a sense this is an oceanic cycle but does not imply that the Atlantic, Pacific, Indian and Tethyan Oceans have all suffered alternate phases of expansion and contraction. There is evidence[10-12], however, that the Appalachian/Caledonian Orogen developed by contraction of a Proto-Atlantic Ocean roughly along the axis of sub-

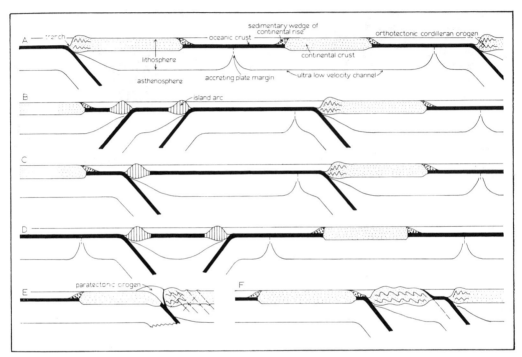

Fig. 2. Schematic relationships between lithosphere plates, continents, oceans, island arcs and orogenic belts. (A) Section from the Pacific Ocean, across South America, the Atlantic Ocean, Africa, and the Indian Ocean, to the Java Trench. (B) Section from South-East Asia, across the South China Sea, the Philippines, the Philippine Sea, the Marianas Arc, the Pacific Ocean, and South America, to the Atlantic Ocean. (C) Section from Australia, across the Tasman Sea, the New Hebrides Arc, the Pacific Ocean, and South America, to the Atlantic Ocean. (D) Section from the Pacific Ocean, across the Caribbean Sea, the Atlantic Ocean, and Africa, to the Indian Ocean. (E) Section from the Indian Ocean, across the Himalayas (site of lost Tethyan Ocean) to Tibet. (F) Section from the Atlantic Ocean, across Africa, the Eastern Mediterranean, Turkey, and the Black Sea, to the Caucasus.

A simplified view[9] of the present oceans is thus that they are either expanding (Atlantic) or contracting (Tethys). This is a grossly oversimplified way of analysing the behaviour of oceans because, although the Pacific is contracting, the oceans form a continuous system and cannot be considered separately. It is more accurate and instructive to consider the evolution of plates which may or may not carry continents as passengers. This emphasizes that the oceans and therefore plates are mutually interdependent and that continents may be disrupted and swept apart by plate motion, and may collide with island arcs or other continents. The great possible latitude and longitude changes involving continental separation and conjunction have fundamental implications for the distribution of animals and plants, particularly land-dwelling types. Faunal and floral province studies will contribute a great deal to our understanding of the relative positions of continents in the past.

sequent opening of the central and northern Atlantic. Plate growth may begin along old lines of weakness and the effect of this will be a close association of orogenic belts of widely varying ages[13,14].

Orogeny and Continental Accretion

The close spatial relationships of consuming plate margins, island arcs and the younger orogenic belts indicate with certainty that the energy source for mountain building is associated with plate consumption. Orogenic belts seem to be associated with both lithosphere consumption at continental margins (Andes) or in island arcs (Japan), and with continental collision (Himalayas). The conditions that determine the site and timing of plate consumption are so far obscure. It has been suggested[9,15] that an Atlantic-type continental margin with its thick pile of sediments on the margin of a wide old ocean, where the oceanic lithosphere is oldest and thickest, is eventually likely to become the site of lithosphere consumption and orogeny. Island arcs may be the result of the nucleation of trenches at microcontinents near continental margins. The onset of trench development may be related to the distance of a section of lithosphere from a ridge. If the lithosphere is regarded as a boundary conduction layer, steadily cooling and thickening away from the ridge, there may be a critical distance/thickness function at which time the plate sinks to form a trench. An alternative, or concurrent, possibility is that a major continental collision, after the closing of an ocean, resulting in cessation of plate consumption, leads to a situation where other active trenches cannot cope with lithosphere disposal and new trenches are forced to open, the likeliest positions being those already discussed.

There are at least four kinds of orogenic belt (Fig. 3) and the effectiveness of each type in accreting continents is discussed below.

convergent tholeite-alkali basalt and calc-alkalic andesite suite are erupted and thus an orogen characterized by a paired metamorphic zonation is developed. The Jurassic/Cretaceous (Nevadan to Laramide) cordilleran orogen is believed to have developed in this way. The amount of lateral continental accretion achieved by a cordilleran orogeny may be quite small because, although continent-derived sediments are spread on to oceanic crust, the strip of oceanic crust trapped against the continent by the trench may be narrow and furthermore both this strip and the continental basement behind it probably suffer considerable contractions. The total width of an orogen such as the Cordilleras, the Andes, or the early Caledonides[11] cannot be considered to be a prism accreted during the orogeny. Only those parts underlain by ophiolite complexes may be taken as the accreted portion and the paucity of exposed ophiolites militates against an estimate.

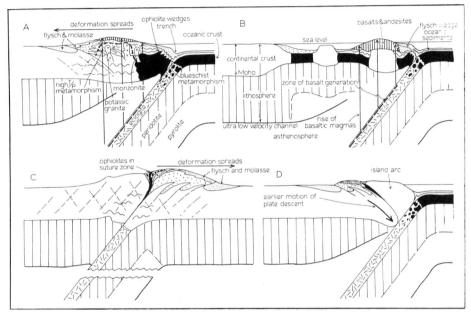

Fig. 3. Models for four basic types of orogenic belts developed at consuming plate margins. (*A*) Cordilleran-type orthotectonic orogen developed at the continental margin of a leading plate edge. The Mesozoic cordilleran orogen of the western United States is taken as the type example. (*B*) Island arc orthotectonic orogen developed on leading plate edge, perhaps nucleated on a microcontinent, within the ocean. (*C*) Himalayan-type paratectonic orogen developed by continental collision. (*D*) Paratectonic orogen developed by collision of a continent and island arc, followed by the development of a new consuming plate margin on the oceanic side of the arc. The Tertiary fold belt of New Guinea seems to have developed in this way. Southern New Guinea was part of the Australian continent and, after the contraction of an ocean (part of Tethys) between southern and northern New Guinea (Bismarck island arc), collided with the Bismarck Island arc in Miocene times. A new trench then formed north of New Guinea to consume the Pacific Plate.

(1) *Cordilleran-type.* These develop at the continental margin of a leading plate edge when lithosphere consumption begins near the foot of a continental rise[9,13]. They are characterized by a deformation pattern (migrating towards the continent) developed by the contraction of the leading plate edge (Fig. 3*A*) weakened by the high heat-flow associated with the region behind the trench. On the inner wall of the trench, wedges of oceanic crust (ophiolite suite consisting of ultrabasic rocks overlain by basic intrusives and extrusives) are thrust up. These and the mélange of sediment scraped from the descending plate are deformed and metamorphosed in the high pressure, low temperature, regime of the blueschist facies. As the deformation spreads away from the trench, the sediments of the old continental rise are thrust, above basement wedges, on to the continent and affected by a high temperature metamorphism. Volcanics of the

(2) *Island arcs.* We suggest that island arcs are built when lithosphere consumption begins at some distance from the continental margin (Fig. 3*B*). Upthrusting of ophiolite wedges and the development of paired metamorphic belts occur[17] as in cordilleran orogeny. The crust of island arcs, perhaps nucleated on microcontinents, is built by the scraping up of oceanic sediments (of which chert is a significant component) and the accumulation of a thick pile of basalts and andesites and their acid derivatives[18]. Island arc development is an extremely effective mode for continental accretion. Not only is it a direct mechanism of building continental crust within an ocean, but the small ocean basins behind arcs are major sediment traps. The accumulation of a thick sedimentary sequence may convert a small ocean basin into continental crust, particularly as the high geothermal gradients[16] are likely to result in metamorphism.

(3) *Himalayan-type.* These are the result of the collision of an aseismic continental margin carrying a sedimentary wedge, against an opposing continental margin in the final stages of oceanic contraction. They are characteristically asymmetric and result from the squeezing of sediments which are heaped onto the continent of the underthrust plate (Fig. 3*C*). Upthrust wedges of oceanic crust (ophiolites) occurring in suture zones (for example, Indus Suture of the high Himalayas; Ivrea Zone of the Alps) are of great significance because these ophiolite belts represent the join-lines of collided continents. Underthrusting of a continent cannot proceed to any great degree because of buoyancy considerations and crushing and shortening begins to occur over a wide zone possibly following the detachment and sinking of the descending plate[4]. Himalayan-type orogeny works against continental accretion in that a small amount of continent may be underthrust.

(4) *New Guinea-type.* This type (Fig. 3*D*) results from the consumption of oceanic lithosphere on the continental side of an island arc, for example the consumption of the Tasman Sea on the west side of the New Hebrides arc. Eventually the continent carrying a marginal sedimentary wedge collides with, and partially underthrusts, the arc. Continued underthrusting is prevented by the buoyancy of the continent[4] and at this stage the descending slab is detached and a new trench is formed to consume lithosphere from the oceanic side. An island arc is thus accreted on to a continent. This notion of flip-over[4] in the direction of lithosphere underthrusting may be a partial explanation of why trenches are commonly associated with orogenic belts on continental margins. The trenches and island arcs develop within the ocean perhaps to be subsequently driven towards, and accumulate at, the continental margin.

Pre-Mesozoic Oceans

Plate tectonics will probably be used as a rigorous geometrical tool in working out the relationship between continents, orogenic belts and oceans well back into Mesozoic times. It will, however, only be valuable as a qualitative notion in understanding the relationships between older orogenic belts and long-vanished pre-Mesozoic oceans, unless large segments of transform faults like the San Andreas Fault are preserved and if palaeomagnetic isoclines are systematically deduced. If pre-Mesozoic orogenic belts were related to the evolution of the oceans, and the view that geosynclines are simply oceans and continental margins is accepted, detailed synthesis of sedimentary, igneous and metamorphic facies-assemblages in the older orogenic belts should yield valuable information for the evolution of the early oceans. This has been attempted for the Appalachian/Caledonian Orogen[10-12] where the geological history is consistent with this approach. The precise site of lithosphere consumption within the older orogens is likely to be represented by blueschist belts and ophiolite complexes. The ophiolites of the median suture of the Urals are associated with blueschist metamorphism and almost certainly lie on the site of a contracted ocean. The late pre-Cambrian (Mozambique–Damaran belts of Africa, Bajkalian belt of Siberia, Fig. 1) orogens apparently present an obstacle to an oceanic origin in that large areas of these belts consist of regenerated earlier continental crust. The existence of earlier basement rocks beneath an orogenic belt, however, as argued above, does not imply a totally intra-continental origin, for during continental collision the ocean may be reduced to a narrow ophiolite bearing suture. A section across the Himalayas at a depth of 30 km might reveal a wide "regeneration" zone. Ophiolites occur along the edge of the Dahomey Belt in Ghana[19] but are rare in other African belts.

Within the oldest cratonic nuclei of Africa, such as the Rhodesian Craton, the mafic volcanic rocks and serpentine forming the lower parts (Sebakwian and Bulawayan) of the Schist Belt sequences[20] are overlain by more andesitic volcanics and coarse clastic sequences. It may be that the andesites and coarse clastic sediments represent island arcs built on a Sebakwian/Bulawayan ocean floor older than 3×10^9 yr. In the Yellowknife region of the Great Slave province of Canada tholeitic and calc-alkalic volcanics about $2 \cdot 6 \times 10^9$ yr old have been suggested to belong to an old island arc complex[21].

We submit, then, that ocean-driven plate mechanisms have been responsible for the growth and evolution of continents since at least 3×10^9 yr ago. During the early Pre-Cambrian, oceanic lithosphere plates perhaps considerably thinner and more numerous than at present controlled the growth of continental crust by the eruption of basalts and andesites and their acidic derivatives in island arc complexes. These were amalgamated by the moving plates until at some stage they formed small continental nuclei. We believe this process to be happening now, in a small scale way in the south-west Pacific, where the New Hebrides/Fiji/Tonga arc-transform system is building continental crust. Since the nuclei were formed, they were added to by the processes of accretion described here. During the early stages of continental growth island arc mechanisms must have been dominant, but as the nuclei enlarged they contributed increasing volumes of sediments to be spread on to the ocean floor as continental rises. Cordilleran accretion mechanisms have thus probably become progressively more important. This model of continental growth involving accretion on earlier nuclei does not imply a concentric accretion. The continuous evolution of oceans by plate growth and consumption leads to complex alternate phases of accretion on different margins of continents. As the continents grow and the oceans diminish in size, oceanic expansion–contraction cycles are likely to become more important. The cyclic opening and closing of oceans are inherent in the concept of plate tectonics because, if this were not the case, a single large continent would have been developed long ago. Continental accretion, however, may be exceedingly irregular, involving a continuous sequence of separation and amalgamation of continents, sometimes along old lines of earlier orogenic belts and sometimes across them to open new continent/ocean interfaces. Such terms as Gondwana, Laurasia and Pangea can therefore only properly be used to describe temporary conjunctions of continents.

We thank Professor Sir Edward Bullard for his encouragement and for critically reading the manuscript. J. F. D. acknowledges discussions with Professor C. L. Drake and Dr D. P. McKenzie.

Received December 10, 1969.

[1] McKenzie, D. P., and Parker, R. L., *Nature*, 216, 1276 (1967).
[2] Morgan, W. J., *J. Geophys. Res.*, 73, 1959 (1968).
[3] Isacks, B. L., Oliver, J., and Sykes, L. R., *J. Geophys. Res.*, 73, 5885 (1968).
[4] McKenzie, D. P., *Geophys. J. Roy. Astron. Soc.*, 18, 1 (1969).
[5] Le Pichon, X., *J. Geophys. Res.*, 73, 3661 (1968).
[6] Drake, C. L., Ewing, M., and Sutton, G. H., in *Physics and Chemistry of the Earth*, 3, 110 (Pergamon, Oxford, 1959).
[7] Dickson, G. O., Pitman, W. C., and Heirtzler, J. R., *J. Geophys. Res.*, 73, 2087 (1968).
[8] McKenzie, D. P., and Morgan, W. J., *Nature*, 224, 125 (1969).
[9] Dewey, J. F., *Earth Plan. Sci. Lett.*, 6, 189 (1969).
[10] Wilson, J. T., *Nature*, 211, 676 (1966).
[11] Dewey, J. F., *Nature*, 222, 124 (1969).
[12] Bird, J. M., and Dewey, J. F., *Geol. Soc. Amer. Bull.* (in the press).
[13] Sutton, J., *Proc. Geol. Assoc.*, 79, 275 (1968).
[14] Dearnley, R., *Nature*, 206, 1083 (1965).
[15] Dietz, R. S., and Holden, J. C., *J. Geol.*, 65, 566 (1967).
[16] McKenzie, D. P., and Sclater, J. G., *J. Geophys. Res.*, 73, 3173 (1968).
[17] Miyashiro, A., *Medd. Dansk Geol. Foren.*, 17, 390 (1967).
[18] Sugimura, A., Matsuda, T., Chinzei, K., and Nakamura, K., *Bull. Volc.*, 26, 125 (1963).
[19] Grant, N. K., *Geol. Soc. Amer. Bull.*, 80, 45 (1969).
[20] Amm, F. L., *Rhodesia Geol. Surv. Bull.*, 35, 307 (1940).
[21] Folinsbee, R. E., Baadsgaard, H., Cumming, G. L., and Green, D. C., *Amer. Geophys. Union Geophys. Mem.*, 13, 441 (1968).

32

Reprinted from pages 1031–1032, 1043–1046, and Figures 9A and 9B of Geol. Soc.
America Bull. **81**:1031–1060 (1970)

Lithosphere Plate-Continental Margin Tectonics and the Evolution of the Appalachian Orogen

JOHN M. BIRD   *Department of Geological Sciences, State University of New York at Albany, Albany, New York 12203*

JOHN F. DEWEY   *Department of Geology, Cambridge University, Cambridge, England*

INTRODUCTION

In recent years, geophysical studies of the world's oceans and seismic belts have led to a new appraisal of the role of sea-floor spreading in the evolution of oceans and continents. The corroboration of Hess's (1962) theory of sea-floor spreading by the discovery (Vacquier, 1962; Mason and Raff, 1961; Raff and Mason, 1961) of linear oceanic magnetic anomalies, and their interpretation as zones of normal and reversed polarity (Vine and Matthews, 1963), the work of Sykes (1967, 1968) on seismic first motions, and the discovery of the low velocity channel (Gutenburg, 1959, Dorman and others, 1960; Anderson, 1962) were significant advances toward the development of the theory of lithosphere plate tectonics (Morgan, 1968; McKenzie and Parker, 1967; Isacks and others, 1968). The essential tenets of the theory are as follows:

1. The lithosphere is segmented into rigid plates bounded by the major seismic zones. Dr. D. P. McKenzie (1969, personal commun.) suggests that oceanic plates, representing a boundary conduction layer, range from as little as 5 to 10 km thick along oceanic ridges to as much as 120 km thick near the trenches; shield areas lie on plates that may be as much as 150 km thick.

2. Plates grow by accretion along the tensional zones of the ocean rises by the injection of wedges of basic magma.

3. Plates are consumed in oceanic trenches and return into the asthenosphere as cold, seismically active slabs to depths of as much as 700 km.

4. Major displacements of the oceanic anomaly patterns occur along transform faults; these appear as lines of strike-slip displacement where lithosphere is neither created nor destroyed. Traces of these faults give, uniquely, the relative vector of plate displacement.

5. Global plate accretion and consumption vectors sum to approximately zero, and thus sea-floor spreading may be accommodated with the earth's radius remaining essentially constant. Any variations of earth radius are minor compared with plate displacements.

6. Lithosphere plates move as rigid slabs on the low velocity channel, and continents passively ride as superficial parts of plates. Plate boundaries may or may not coincide with continental margins. Continental drift is merely an effect of plate movement, although continents may place important restraints on plate

motion when collisions of continents against continents or island arcs occur. Buoyancy considerations suggest that those parts of plates carrying continents cannot be under-thrust to any great extent. A continent/continent collision (for example, peninsular India against the Tibetan Plateau) leads to the loss of a deep seismic zone and the development of a wide region of shallow seismicity. A continent/island arc collision may lead to "flipping" of the direction of plate descent.

It is difficult to challenge the concept of plate tectonics as applied to the present-day earth, since it is the only unifying global tectonic model that satisfactorily explains the seismic and magnetic data. Furthermore, plate tectonics is rapidly becoming a powerful tool in giving new insight into the evolution of recent continental margins, island arcs and orogenic belts. If the plate paradigm is viable for the modern earth and is the mechanism by which all modern major global tectonics are driven, it seems most unlikely that it is valid for only the last 100 m. y. or so of earth history, and that it has no relevance to the evolution of earlier orogenic belts. The authors take the view that lithosphere plate theory is the most fundamental paradigm change that has occurred in the earth sciences, and explains the tectonic evolution of the earth during at least the last 1000 m. y.

The purpose of this paper is to examine the Late Precambrian to Devonian history of two segments of the Appalachian orogen, New England and Newfoundland, in terms of plate theory. This is now possible because during the past decade, there has been a substantial increase in knowledge of the stratigraphy and structure of the Appalachian orogenic belt. Particularly significant gains have been made in New England, especially through the work of Professor Marland P. Billings, his students and associates, and in Newfoundland through the work of the Canadian Geological Survey and Professor Marshall Kay.

We propose an evolution model for the Appalachian-Caledonian "Atlantic" Ocean and its system of lithosphere plates and their role in the formation of the orogen. Emphasis will be given to the pre-Silurian evolution; Silurian/Devonian history will be given relatively brief treatment. Furthermore, particular attention will be paid to the northwestern margin of the orogen. Since this paper involves the synthesis of a large amount of stratigraphic and structural data, the authors

have concentrated on incorporating most of this information into Figure 5 and the Appendix with a minimum of description in the text. *Geographical directions are referred to the present; that is, when we refer to "north" for Ordovician times, we mean present compass north and imply a direction probably other than north for that time.*

OUTLINE OF NORTHERN APPALACHIAN STRUCTURAL FRAMEWORK AND HISTORY

Although the Appalachian orogen lies along almost the entire eastern continental margin of North America, it clearly formed part of a continuous Appalachian/Caledonian Orogen (Dewey and Kay, 1968; Dewey, 1969a) extending in Late Paleozoic times from at least Florida to Spitzbergen on the Bullard, Everett and Smith (1965) reconstruction of the North Atlantic continents. The orogen consists of a group of distinct mobile zones (Zones A, B and C; Fig. 1), particularly well-shown in New England and Newfoundland.

Zone A consists of a region dominated by Ordovician orogeny (Taconian/Humberian) but affected by Devonian orogeny (Acadian). Zone A may be divided into: (1) a northwestern strip, here termed Logan's Zone (E-an Zen, 1968, personal commun., has also proposed this term), which acted as a stable nonvolcanic, orthoquartzite-carbonate miogeocline (Dietz and Holden, 1967) before the Taconian/Humberian Orogeny, and then became a linear zone of exogeosynclinal subsidence and westward thrusting, receiving gigantic klippen and Taconian/Humberian flysch and molasse from the southeast, and (2) a southeastern strip, here termed the Piedmont, which accumulated a prism of Late Precambrian [often referred to as Cambrian(?) or Eocambrian] to Ordovician clastic sediments and volcanics and was deformed and metamorphosed by a diachronous Taconian/Humberian Orogeny in Ordovician times. The southeastern edge of Logan's Zone was a site of major facies change before the Taconian/Humberian Orogeny. A Cambrian/Early Ordovician carbonate shelf/miogeocline in Logan's Zone gave way southeastward through a bank edge, carbonate-slide-breccia facies to a predominantly shale sequence (Zen, 1967; Rodgers, 1968). The site of the bank edge (southeastern edge of Logan's Zone) approximately coincides with the western edge of deposition of Late Precambrian coarse clastics.

[*Editor's Note:* Material has been omitted at this point.]

PRE-TACONIAN/HUMBERIAN EVOLUTION

The authors contend that the pre-Taconian/Humberian stratigraphic framework of Zone A, depicted for New England in Figure 7A and Newfoundland in Figure 8A, represents the sedimentary evolution of the continental margin of eastern North America from Late Precambrian to Early Ordovician times. It is suggested that this framework was established during the spreading of a Proto-Atlantic ocean by plate-accretion (Dewey, 1969a). The palautochthonous carbonate sequence of Logan's Zone was deposited on the continental shelf, equivalent to a starved continental rise mud facies. Carbonate breccia facies such as the Cow Head and Rugg Brook formed on the upper continental slope, and thin carbonate breccias in the starved sequence (for example,

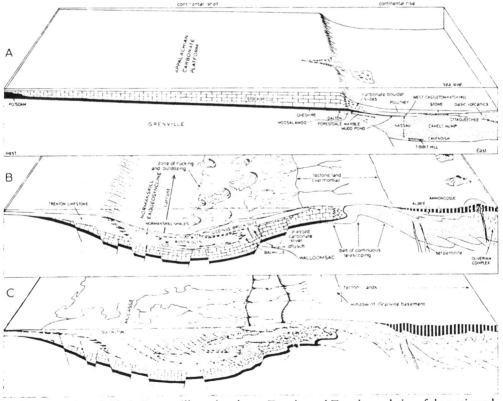

Figure 7. Schematic block diagrams illustrating the pre-Taconian and Taconian evolution of the continental margin of North America in western New England: A. pre-Taconian; B. early Taconian; C. late Taconian.

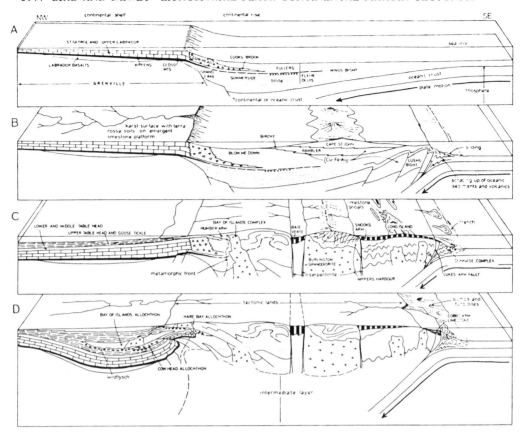

Figure 8. Schematic block diagrams illustrating the pre-Humberian and Humberian evolution of the continental margin of North America in Newfoundland: A. pre-Humberian; B. Initiation of trench and island arc; C. early Humberian; D. late Humberian.

West Castleton-Hatch Hill-Poultney) are their distal downslope equivalents. The carbonate palautochthon extends with an amazingly constant stratigraphy for at least 4000 mi on the western margin of the Appalachian/Caledonian Orogen and must have been located more or less parallel to an equatorial climatic belt within about 30° of the early Paleozoic equator. The absence of coarse terrigenous clastics in the starved sequence testifies not only to the effect of sediment blocking by the laterally extensive carbonate shelf but, more importantly, to the absence of clastic source areas to the east, and to the absence of any significant relief to the west in the craton. It is of the utmost importance to emphasize that pre-Taconian/Humberian coarse clastics (other than the palautochthonous basal orthoquartzite blanket) are restricted to the Zone A Piedmont where they are overlain by the Cambro-Ordovician starved sequence. The source for

these coarse clastics was mainly Grenville basement rocks; Grenville blue quartz occurs in the Bull-Nassau sequence. Thus, before the Cambrian marine transgression occurred westward across the miogeocline, the emergent basement of Logan's Zone was a source region for the pre-*Olenellus* biozone clastic sequences of the continental rise. The Late Precambrian/? Early Cambrian glaciation predates the orthoquartzite transgression of the shelf, and there is some evidence in Newfoundland of glacial scouring of the Grenville basement (Keene Swett, 1969, personal commun.). There are rare limestone clasts in the Rensselaer facies of the Bull-Nassau sequence and in the Maiden Point Formation. Some of these clasts are Grenville Marble, but others are marly limestones, and since at least most of the Rensselaer and Maiden Point facies are preglacial, they may indicate that pre-orthoquartzite shelf carbonate sequences existed which were later stripped from Logan's Zone by the glacial scouring. Basic volcanic rocks occur low in the pre-*Olenellus* sequence of the Piedmont as the Tibbit Hill Volcanic Member of the Pinnacle Formation and as greenstones in the Rensselaer facies of the Bull-Nassau sequence. In Newfoundland, the Labrador Basalts predate the basal orthoquartzite blanket of Logan's Zone and are probably related to the Long Range basic dike swarm. The pre-*Olenellus* biozone sediments of the Taconic Allochthon include large amounts of green, chloritic sediments postdating the Rensselaer facies. These may be a reflection of the stripping of large plateau basalt spreads from the site of the future continental shelf. These early volcanics are probably a reflection of the earliest stages of development of an Appalachian/Atlantic when the Laurasian continental crust was stretching, necking, and rifting. Tensional graben accumulated pre-*Olenellus* biozone clastics such as the Rensselaer (containing local basaltic flows; N. M. Ratcliffe and J. M. Bird, observations) during the earliest phases of rifting, and the Proto-Atlantic shelf/continental rise pair developed later on the continental margin. Although Grenville basement rocks occur below the graben/continental rise sequences of the Zone A Piedmont (Chester Dome), the authors suggest that a full thickness of continental basement ended near the shelf edge. This line was an important stratigraphic/tectonic break throughout the evolution of the pre-Taconian/ Humberian graben, shelf and continental rise and, later, became a significant tectonic abutment during the Taconian/Humberian Orogeny.

THE TACONIAN/HUMBERIAN OROGENY

Figures 7B and 7C, and 8B, 8C, and 8D illustrate the proposed sequence leading to the development of the Taconian/Humberian Orogen in New England and Newfoundland, respectively. After the development of the carbonate palautochthon of Logan's Zone, an unconformity developed by the emergence, warping, and block faulting of the shelf/miogeocline, followed by the Tablehead and Balmville transgressions. This unconformity is not recorded as a physical break in the allochthonous sequences, but it is noteworthy that correlative red lutites occur in the Indian River and Blow-Me-Down Formations. It is suggested that the red lutite was derived from *terra rosso* karst soils on the uplifted limestone platform. At several localities, green, iron-rich clays occur on the palautochthon/autochthon unconformity and may be reduced residual deposits. On the southeastern margin of the Zone A Piedmont in Newfoundland, the Lush's Bight basic volcanic sequence, here interpreted as oceanic crust (layer 2) of the Appalachian Atlantic, was thrust up as wedges behind a trench. The Lush's Bight was deformed, metamorphosed, and partially eroded before the deposition of the Early Ordovician Catchers Pond sequence and the eruption of the Long Island and Upper Wild Bight volcanics. In New England, deformation in the Bronson Hill Anticlinorium, and its northern continuation in Maine, occurred between Cambrian times and the deposition of the Middle Ordovician Shin Brook Formation and Ammonoosuc Volcanics. Before the deformation and metamorphism of the Piedmont sequence of the Burlington Peninsula in Newfoundland, basic followed by acidic volcanicity spread westward, and the deformation was postdated by the eruption of the Snooks Arm volcanics and the development of the Baie Verte grabensuture with its associated serpentinites. The authors suggest that this sequence (basic volcanics-deformation and metamorphism-mixed volcanics) represents the Early Ordovician establishment of a complex island arc system near the southern margin of the Piedmont on the distal edge of the pre-Taconian/ Humberian continental rise. Dewey (1969a) has suggested that this was a direct consequence of the initiation of underthrusting of Appalachian Atlantic lithosphere along a Benioff seismic zone whose surface expression was an oceanic trench in which an argille scagliose facies (Dunnage Complex) accumulated. The authors believe this mechanism to explain the following relationships: (1) The eruption of basic volcanics (Long Island, Rambler) spreading westward from the volcanic front as the lithosphere plate descended farther and farther under the continental rise. The origin of the basaltic magmas may be related either to a frictional melting origin (Oxburgh and Turcotte, 1968) or to partial melting of oceanic crust as the plate descends into lower and hotter regions of the asthenosphere (Dr. D. P. McKenzie, 1969, personal commun.). (2) The development of the argille scagliose facies, by the scraping up of oceanic sediments and volcanics along a décollement horizon above oceanic crust and by the sliding of sediments and gigantic volcanic blocks down the inner wall of the trench. (3) The sharp disappearance of the Zone A Humberian deformation along the Lukes Arm fault belt. The authors can think of no other mechanism which explains the fact that the Dunnage Complex passes conformably up into Upper Ordovician shales and graywackes of the Dildo sequence while a major intra-volcanic unconformity exists north of the Lukes Arm fault belt. The Dildo Sequence suffered its first major deformation, apart from soft-rock sliding and scraping deformation, in Acadian times.

Rodgers (1967) has suggested that the Lush's Bight/Penobscot/Oliverian deformation occurred within a developing island arc system and also made the illuminating observation that it may represent a "precocious" Taconian event. The present authors would expand this concept to the suggestion that the Penobscot movements were the initial Early Ordovician deformation of the continental rise that spread westward and culminated in later Ordovician times with the Taconian/Humberian deformation of Logan's Zone. The first clastics with an easterly provenance appear in the Moretown Formation (≡ Albee Formation) which rests unconformably upon the Pinnacle-?Stowe sequence in the Chester Dome and upon the Stowe Schist in the serpentinite belt. Slightly later, during the deposition of the Indian River and the Blow-Me-Down of the upper continental rise, fine clastic tongues had an easterly provenance. At least some of the Mt. Merino cherts may have been originally deposited from thin air-fall tuffs. Thus, the island-arc complex shed clastics and thin distal volcanics to the west. Deformation and metamorphism and the associated westward flow of clastics was spreading westward while the bank edge was still shedding carbonate slide-breccias onto the upper continental rise. As the deformation spread toward the upper continental rise, the carbonate bank edge retreated rapidly westward across Logan's Zone to the western margin of the then subsiding exogeosyncline. The basal Pawlet unconformity in the allochthonous sequences evidences the spreading of the flysch onto the then uplifted continental rise. The Austin Glen flysch facies spread into the exogeosyncline accompanied by the emplacement of the early Taconian detached gravity slides (Zen, 1967; Bird, in press).

In Newfoundland, the site of basic and ultrabasic magma intrusion (Bay of Islands Complex) and extrusion (Humber Arm Volcanics) had spread westward into the upper continental rise before the emplacement of the gravity slides. There is some suggestion in the Taconic region that basic volcanicity occurred on the upper continental rise before uplift and westward sliding. At Starks Knob, north of Albany, New York, a gigantic block in the wildflysch consists of pillow lavas, intricately interfilled with fine-grained marly limestone. The authors believe this volcanic/limestone assemblage to have originated in an atoll-like limestone shoal on a volcano east of, but close to, the main carbonate bank edge.

As the early allochthon slid westward, the inner regions of the Piedmont underwent strong stratal shortening which telescoped the collapsed upper continental rise and progressively led to the later hard-rock thrust emplacement of the higher allochthonous slices. The metamorphism (biotite/chloritoid/chlorite grade) spread westward, and partly affected the higher slices before they moved out into Logan's Zone. The resulting structural relations were altered, and, locally, entirely obscured by the later Acadian deformation.

[*Editor's Note:* In the original, material follows this excerpt. Only the references cited in the preceding excerpt are reproduced here.]

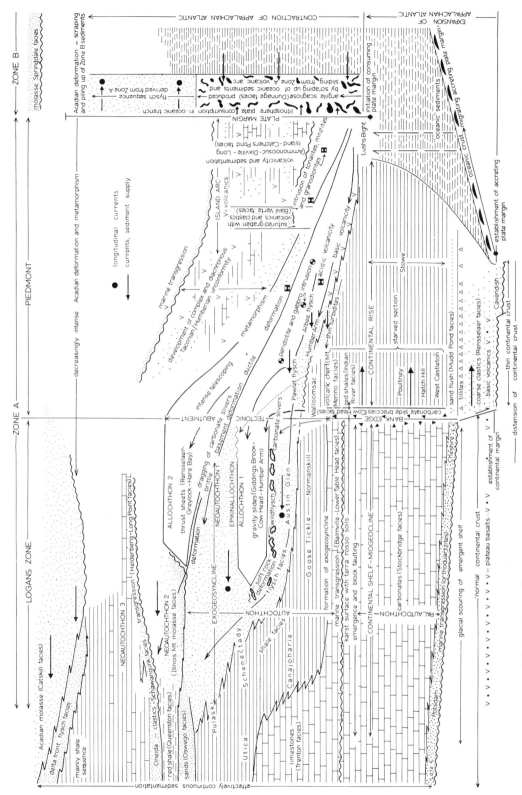

Figure 9A. Model for the evolution of the Appalachian orogen. Time/space framework of sedimentation, volcanicity, deformation, and metamorphism in Zones A and B.

283

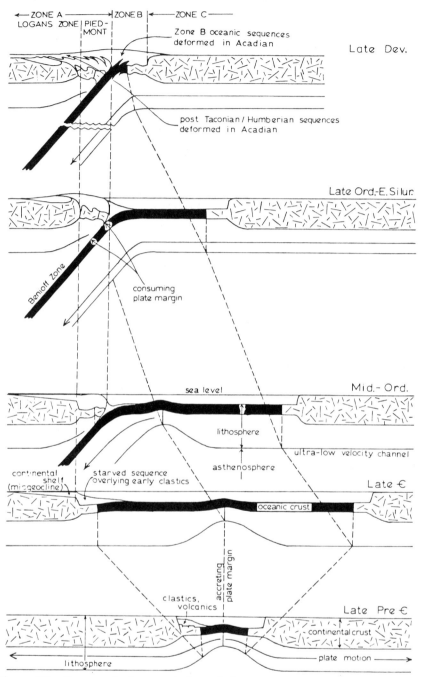

Figure 9B. Suggested plate evolution of the Appalachian Atlantic and relationships of plate consumption to the Taconian/Humberian and Acadian orogenies. The positions of plate margins are entirely relative. We do not imply a stable position for either accreting or consuming plate margins.

REFERENCES

Anderson, D. L., 1962, The plastic layer of the earth's mantle: *Sci. American*, July, 1962, p. 52–59.

Bird, J. M., in press, Middle Ordovician gravity sliding—Taconic region, in Kay, Marshall, Editor, *North Atlantic—Geology and continental drift:* Am. Assoc. Petroleum Geologists Mem. 12.

Bullard, E., Everett, J. E., and Smith, A. G., 1965, The fit of the continents around the Atlantic: *Royal Soc. London Philos. Trans.,* ser. A, v. 258, p. 41–51.

Dewey, J. F., 1969a, Evolution of the Appalachian/Caledonian orogen: *Nature,* v. 22, p. 124–129.

Dewey, J. F., and Kay, Marshall, 1968, Appalachian/Caledonian evidence for drift in the North Atlantic, in Phinney, R. A., Editor, *The history of the earth's crust, a symposium:* Princeton, New Jersey, Princeton Univ. Press, p. 161–167.

Dietz, R. S., and Holden, J. C., 1967, Miogeoclines in space and time: *Jour. Geology,* v. 65, p. 566–583.

Dorman, J., Ewing, M., and Oliver, J., 1960, Study of shear velocity distribution by mantle Rayleigh waves: *Seismol. Soc. America Bull.* v. 50, no. 1, p. 87–115.

Gutenberg, B., 1959, *Physics of the Earth's interior:* New York, Academic Press, 240 p.

Hess, H. H., 1962, History of the ocean basins, in Engel, A. E. J., and others, *Petrological studies: A volume in honor of A. F. Buddington:* Geol. Soc. America, p. 599.

Isacks, B., Oliver, J., and Sykes, L. R., 1968, Seismology and the new global tectonics: *Jour. Geophys. Research,* v. 73, p. 5855–5900.

Mason, R. G., and Raff, A. D., 1961, Magnetic survey of the west coast of North America, 32° N. latitude to 42° N. latitude: *Geol. Soc. America Bull.,* v. 72, p. 1259–1266.

McKenzie, D. P., and Parker, R. L., 1967, The north Pacific: An example of tectonics on a sphere: *Nature,* v. 216, p. 1276–1280.

Morgan, J., 1968, Rises, trenches, great fualts and crustal blocks: *Jour. Geophys. Research,* v. 73, p. 1959–1982.

Oxburgh, E. R., and Turcotte, D. L., 1968, Problem of high heat flow and volcanism associated with zones of descending mantle convective flow: *Nature,* v. 218, no. 5146, p. 1041–1043.

Raff, A. D., and Mason, R. G., 1961, Magnetic survey off the west coast of North America, 40° N.-52° N. latitude: *Geol. Soc. America Bull.* v. 72, p. 1267–1270.

Rodgers, John, 1967, Chronology of tectonic movements in the Appalachian region of eastern North America: *Am. Jour. Sci.,* v. 265, p. 408–427.

Rodgers, John, 1968, The eastern edge of the North American continent during the Cambrian and Early Ordovician, Chap. 10, in Zen, E-an, White, W. S., Hadley, J. B., and Thompson, J. B., Jr., Editors, *Studies of Appalachian geology: Northern and maritime:* New York, Interscience, 475 p.

Sykes, L. R., 1967, Mechanism of earthquakes and nature of faulting on the mid-oceanic ridges: *Jour. Geophys. Research,* v. 72, p. 2131–2153.

Sykes, L. R., 1968, Seismological evidence for transform faults, seafloor spreading, and continental drift, in Phinney, R. A., Editor, *Proceedings of NASA Symposium, History of the Earth's Crust:* Princeton, New Jersey, Princeton, Univ. Press, p. 120–150.

Vacquier, V., 1962, Magnetic evidence for horizontal displacements in the floor of the Pacific Ocean, in Runcorn, S. K., Editor, *Continental drift:* New York, Academic Press, p. 135.

Vine, F. J., and Matthews, D. H., 1963, Magnetic anomalies over oceanic ridges: *Nature,* v. 199, p. 947–949.

Zen, E-an, 1967, Time and space relationships of the Taconic allochthon and autochthon: *Geol. Soc. America Spec. Paper no. 97,* 107 p.

33

Reprinted from *Earth and Planetary Sci. Letters* **10**:165–174 (1971)

PLATE TECTONIC MODELS OF GEOSYNCLINES

William R. DICKINSON

Geology Department, Stanford University,
Stanford, California, USA

Received 17 October 1970

Identification of the sites of miogeosynclinal and eugeosynclinal assemblages of various kinds in terms of plate tectonic theory permits interpretation of previously studied orogenic belts in terms of the new global tectonics. The most important sites are elements of arc-trench systems at convergent plate junctures and miogeoclinal settings at stable continental margins in plate interiors. Orogenic events mainly include crustal subduction and crustal collision related to plate consumption. Successions of orogenic events can be understood within the logic of plate tectonic theory. Orogenic belts record a variety of events controlled by both plate divergence and convergence. The progression of events is likely to be dissimilar in different belts even though the rock assemblages present all belong to a limited catalogue of principal types.

1. Introduction

Plate tectonic theory of the so-called new global tectonics [1] permits a fresh and deductive analysis of the geologic history of mountain belts [2]. However, much of the past literature on mountain belts is couched in the terminology of the inductive geosynclinorial theory [3] of mountain-building. Briefly, this now outmoded view of mountain-building, or orogenesis, held that elongate, geographically fixed belts of deep subsidence and related thick sedimentation called geosynclines are the necessary precursors of later mountain ranges in which the exceptionally thick geosynclinal strate, mainly marine, are exposed by grand uplift following deformation, or tectogenesis, of the geosynclinal prism. To many, the theory implied that the existence of a geosyncline predestines later orogeny.

The current view of tectonics holds instead that deformation of crustal rocks by folding and faulting is mainly concentrated along the boundaries between semirigid plates of lithosphere in semiconstant relative motions with respect to one another. Broadly, three kinds of plate junctures can be identified: (a) divergent junctures where new lithosphere is formed in the widening gap between two plates moving away from one another; (b) shear junctures where two plates slip laterally past one another; and (c) convergent junctures where the margins of two plates moving toward one another are either both crumpled against the join, or else one is destroyed or consumed as one plate slides beneath the other. Major divergent junctures are represented by the midoceanic rises. The shear junctures are the transform faults [4] that bridge between the other two kinds of plate boundaries where lithosphere is either created or consumed. The convergent junctures are the loci of orogeny involving plate consumption, continental collision, stratal crumpling, thrusting, crustal thickening, and isostatic uplift, together with accompanying magmatic and metamorphic events, the whole tracing different orogenic patterns.in time and space as the circumstances of convergence vary [5]. Unlike the older geosynclinorial theory, the new synthesis does not require a regular sequence, or geotectonic cycle, of orogenic events to be recognized in the same order in all mountain belts [6]. Taking subduction zones as the controlling features for orogeny [30], alternate scenarios are possible and plausible.

2. Geosynclinal terminology

The transition from stabilist to mobilist concepts

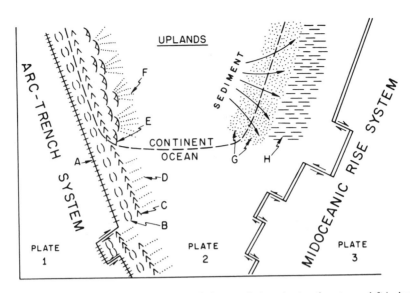

Fig. 1. Diagrammatic map illustrating potential sites of geosynclinal accumulations. Arc-trench system on left is along convergent juncture of plates 1 and 2. Midoceanic rise (double-line) on right is along divergent juncture of plates 2 and 3. Half-arrows indicate sense of relative shear motion along transform faults (single lines). Continent-ocean interface (dashed line) lies partly along arc-trench system and partly in the interior of plate 2. Elements of arc-trench system denoted as follows: A (crossed line) is trench-style subduction zone; B (parentheses) is arc-trench gap including sedimented troughs, slopes, and shelves; C (inverted V's) is volcano-plutonic arc; D (dotted lines) is clastic wedge in ocean basin behind arc; E (crescents with triangles) is intracontinental thrust belt behind arc; F (dotted fans) is clastic wedge deposits of exogeosynclinal character in foreland basins behind arc. Strata associated with stable continental margin, where arrows show overall sediment movement, denoted as follows: G (stippled) is miogeoclinal region of shelf deposits on continental crust and continental rise on oceanic crust; H (dashes) is offshore region of oceanic facies equivalents. See text for discussion.

about the rock masses in orogenic belts requires that the major tectonic elements recognized by past geosynclinal terminology be identified in terms of plate tectonic theory. This change in thinking requires only that observational data on orogenic rock masses be passed through a new theoretical filter. This approach is true to the traditional view that the present is the key to the past, for the plate tectonic theory was constructed from knowledge of the actual behavior of existing plates of lithosphere. Where the nature and setting of ancient orogenic rock masses can be specified adequately by analogy with modern tectonic and depositional features, no valid purpose can be served by continued usage of conceptual terms like eugeosynclinal and miogeosynclinal.

Loose usage of the term geosyncline for all basins with a thick sedimentary record, folded or not, led to the coining of the term orthogeosyncline, in the sense

of a proper geosyncline, for the sites of thick sedimentary accumulations later incorporated within orogenic belts [7]. Within the orogenic belts, two broadly contrasting kinds of stratified, orthogeosynclinal rock masses can be identified: (a) miogeosynclinal sequences lacking volcanic rocks, deposited wholly or partly in shallow water, and resting unconformably, in at least local exposures, on sialic basement rocks of the continental crust; and (b) eugeosynclinal sequences including voluminous volcanic strata, deposited wholly or partly in deep water, and lacking a visible underlying floor of continental crust over large regions. Re-evaluation of the geologic history of orogenic belts must rest mainly on the re-identification of miogeosynclinal and eugeosynclinal rock masses in terms of plate tectonic theory.

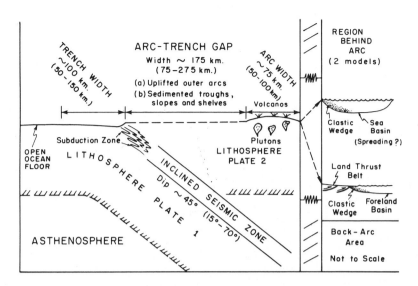

Fig. 2. Diagrammatic illustrating elements of arc-trench system to approximate true scale. Vertical distance from arc volcanos and plutons, shown schematically, to mean position of inclined seismic zone in mantle beneath varies from 75 to 275 km, hence median value of ∼175 km is shown. See text for discussion.

3. Loci of geosynclines

Suspected sites of accumulation of geosynclinal sequences can be restricted to those places where surficial materials, of sedimentary or volcanic origin, can be piled, depositionally or tectonically, to great thicknesses in linear belts. This is not a trivial conclusion, for the greater part of the earth's surface can be ruled out on one of two grounds. In the continental interiors, elevated tracts are assured of erosion and lowlands cannot subside greatly to receive thick sediment because of the isostatic balance of the light continental crust. In the ocean basins far from continents, oceanographic research has shown that the sediment cover is thin because there is no effective mechanism to deliver large volumes of sediment to such distal areas. Our attention is drawn primarily, then, to the margins of the oceans where sediment delivery is effective and to the margins of the continents where thinning of the continental crust is sufficient to allow large subsidence under sedimentary loading. Secondary attention is also drawn to the vicinity of intraoceanic island arcs where volcanogenic sediment is in abundant supply, and to intra-

continental regions of crustal extension where concomitant crustal thinning likewise permits large subsidence.

Our eye, then, is upon the continent-ocean interfaces, and upon places which mimic them in availability of sediment and the means to receive it, by reason either of deep water or the capacity to subside greatly (fig. 1). These places can be divided into those which are at or near plate junctures and those that lie within plate interiors. As the rafts of continental crust are mere passengers riding on the tops of the much thicker plates of lithosphere, continental margins are not, in general, coincident with plate boundaries.

4. Plate-juncture settings

The plate-juncture sequences of first importance are those that accumulate near convergent plate junctures. In this setting, the crumpling and consumption of plate margins insures that any thick sedimentary or volcanic sequences accumulated near the juncture will be affected variously by deformation and

289

metamorphism within a few millions or tens of millions of years after their formation. Any stratified pile in this setting will automatically be incorporated in what will later be called an orogenic belt.

At convergent plate junctures where at least one plate margin is capped by oceanic crust, the characteristic tectonic expression of the juncture is an arc-trench system (fig. 2). An oceanic plate-margin descends beneath the edge of the adjacent plate, and a trench forms on the descending plate along the curvilinear line of juncture. The thermal regime beneath the region of plate consumption leads to the construction of a magmatic arc near the edge of the over-riding plate. The magmatic arc is of the type common along the annular array of volcanic chains making up the ring of fire or andesite zone around the Pacific where plate consumption is now active. Where the over-riding plate margin is also capped by oceanic crust, the magmatic arc is an intra-oceanic island arc structure [8]; where it is capped by continental crust, the magmatic arc is a volcano-plutonic complex near the continental margin [9]. In rare instances an oceanic plate margin may ride over the lip of a continental plate margin with uncertain tectonic consequences [10]. In cases where both plate margins are continental, the two rafts of continental crust are juxtaposed along a complexly sutured join, or superimposed and telescoped by thrusting [11].

5. Arc-trench systems

Arc-trench systems contain as integral parts two kinds of sequences that have been called eugeo-synclinal in the past [12]. The magmatic arc is built as a sequence of generally andesitic rocks, varying from basaltic to dacitic in composition, intruded by comagmatic granitic plutons. Both extrusive and intrusive phases of the arc magmatism display a transverse petrologic asymmetry, such that the level of potash content increases in the direction away from the trench. The asymmetry may be used to infer the polarity of ancient arcs [13]. The eruptives are partly lavas but in large part are pyroclastic rocks. Volcanogenic materials are dispersed widely from the volcanic centers as volcaniclastic detritus, and are deposited as bedded sediments in a variety of marine and terrestrial environments within the geographic

confines of the arc, as well as to either side of it as discussed below. The arc structure thickens as older parts founder into the roots of the arc and younger volcanogenic materials are added to the top to build volcanic cones and fill blocky volcaniclastic basins. The net result is an arc-type eugeosynclinal sequence even though the site of accumulation is a broadly positive topographic feature throughout the constructional activity.

The trench is a negative topographic feature whose floor is continually shifting as the oceanic plate margin being consumed at the convergent plate juncture glides beneath the arc capping the edge of the over-riding plate. The motion causes ocean-floor basaltic lavas and pelagic sediments to be fed steadily into a subduction zone of crumpling and thrusting beneath the arc-side wall of the trench [14]. The lavas that ride thus into the trench region were probably erupted mainly at midoceanic rises where oceanic crust is formed at divergent plate junctures. Beneath them are comagmatic gabbro and dolerite intrusives of the lower oceanic crust and ultramafic rocks of the underlying mantle. Above them are thin layers of pelagic sediment added in transit between rise and trench. The pseudostratigraphic oceanic crustal sequence of partly accumulate but mainly residual ultramafites at the base, gabbro-dolerite intrusives at middle levels, and basaltic pillow lavas at the top [10], the whole capped by deep-water marine pelagites, is the assemblage long known as an ophiolite sequence where exposed in orogenic belts (table 1). Also led into the subduction zone are clastic turbidites added to the oceanic crustal rind as it approaches the arc-trench system. The clastic sediment is derived either from the nearby volcano-plutonic arc or from uplifts, discussed below, that may form to either side of the arc as results of plate convergence. Depending upon the interplay between the rate of subduction and the rate of sediment delivery, the turbidites may form only a thin scum on the trench floor fed by longitudinal turbidity currents, or may partly bury the trench beneath large subsea fan ramparts fed by transverse turbidity currents. In either case, the stratigraphic thickness of undisturbed turbidites preserved on the descending plate at any one time is probably no index to the ultimate volume or tectonic thickness scraped through time off the descending plate and stuffed with shreds

Table 1

Interpretation of typical elements of oceanic ophiolite sequences as displayed intact in overthrust oceanic slabs, as shreds in under-thrust melanges, and in varied disorder in suture zones formed by crustal collision. See text for discussion.

Rock types	Mode of origin
B. *Crust elements*	
7. Graywacke turbidites (not essential)	Trench fills or subsea fans added just before subduction at convergent plate juncture
6. Chert and argillite (may be thin scum)	Pelagic marine sediments added in transit from divergent to convergent plate juncture
5. Basaltic lavas and pillow lavas	Sea-floor eruptions mainly where oceanic crust formed at midoceanic rise but partly younger seamounts
4. Dolerite screens and sills	Feeders for rise eruptions where new lithosphere formed at divergent plate juncture.
3. Varied gabbro bodies	Supramantle magma chambers formed beneath midoceanic rise.
A. *Mantle elements*	
2. Cumulus peridotite	Remnants of magma chamber floors
1. Peridotite tectonite	Refractory rind of residual outer mantle.

of ophiolotes into the subduction zone. The characteristic regime of deformation in a trench-style subduction zone leads to the chaotic tectonic style of melanges [15], rock masses composed of a jumbled mixture of rock slices whose structural geometry is dominated by pervasive shearing. The net result of the structural telescoping and tectonic accumulation of disordered oceanic strata in such a subduction zone is a trench-type eugeosynclinal prism of great aggregate thickness.

Orogenic magmatism with associated metamorphism in the arc region, and orogenic deformation accompanied by thrusting and metamorphism in the trench region, occur while the arc-trench system is active. In addition, when plate convergence ceases for whatever reason, gravimetric considerations dictate that the subducted materials, depressed by plate descent and thickened by tectonic stacking, must rise isostatically to cause uplift, which may involve gravitational spreading and back-thrusting of melanges toward the arc.

6. Fore-arc and back-arc settings

In addition to the distinctive trench and arc assemblages of eugeosynclinal character, arc-trench systems commonly include two other kinds of

sedimentary sequences that have been regarded variously as eugeosynclinal or miogeosynclinal in the past. The two may be difficult or impossible to distinguish from internal characteristics alone, but one accumulates between the trench and the arc, whereas the other accumulates behind the arc. Both are commonly composed mainly of volcano-plutonic debris eroded from the arc as provenance.

In modern arc-trench systems, there are elongate geographic gaps, within which elongate sedimentary basins occur, between the volcanic fronts of the arcs and the slope breaks above the crests of acoustic basement ridges at the tops of the trenches' inner walls [16, 17]. The gaps are 75 to 275 km wide, the most characteristic widths being 125 ± 50 km for island arcs and 225 ± 50 km for continental margin arcs. Within the arc-trench gaps are local elongate uplifts, the well known outer sedimentary arcs and related features. Also present, but largely undetected until recently, are sedimented linear troughs, slopes, and shelves beneath which undisturbed marine sediments are commonly too thick to penetrate with standard continuous profiling techniques. In California, the late Mesozoic Great Valley sequence, commonly described as miogeosynclinal from its lack of volcanic rocks, was deposited beneath varying water depths in an arc-trench gap [18]. The coeval Franciscan assemblage, a trench-type eugeosynclinal prism of subducted melanges and tectonites, was thrust regionally from the west beneath the Great Valley sequence [19], whose clastic detritus was derived from the adjacent Sierran-Klamath arc terrane on the east [20] where magmatic activity was contemporaneous with sedimentation and thrusting to the west [21]. Tectonically similar sequences, little deformed and positioned between coeval volcano-plutonic arc terranes and their associated trench-style subduction zones, are probably common in the circumoceanic orogenic belts. Their sedimentary petrology reflects derivation in the main from the volcano-plutonic arcs, but supplementary sources in isostatically uplifted melange belts and other exposures in outer sedimentary arcs, or analogous features, may be dominant locally.

Clastic sequences similar in sedimentary petrology to those of the arc-trench gaps may accumulate in back of an arc, on the side away from the trench. Behind some modern island arcs are clastic wedges and marine basin fills fed backward from the volcanic spine of the arcs into the flanks of marginal seas, some of which may be subsiding regions formed in the wakes of migrating arcs [16]. The deposits of the successor basins in the Mesozoic of British Columbia [22] may be analogous to these accumulations, except that the receiving basins in this case are floored largely by oceanic materials accreted to the continental margin by previous subduction [30]. Where the marginal sea also receives detritus from a continental mainland on the flank away from the arc, then the arc-derived basin fill may grade laterally to deposits indistinguishable from the plate-interior accumulations at continental margins that are discussed below [23].

Despite their continental setting, magmatic arcs standing near or on the edges of continents may also build clastic wedges into foreland basins lying on the side of the arc away from the trench. Thick Tertiary sequences in this setting are known behind the Sunda arc in Indonesia [24] and behind the Andean arc in South America [25]. An ancient analogue is afforded by the broad late Mesozoic basin in the interior of North America between the Cordilleran fold belt on the west and the Paleozoic platform and Precambrian shield to the east [9]. The detritus in such a sequence behind an arc may be derived mainly from the arc itself, as in Indonesia, or from uplifted

and overthrust masses of pre-arc continental rocks exposed in highlands between the volcano-plutonic belt and the foreland basin, as is the case for the interior Cretaceous of North America. Sedimentation in such foreland basins is accompanied along the arc-side margins of the basins by contemporaneous folding and thrusting that carries rocks of the adjacent highlands over strata deposited along the edges of the basins. This contractional motion behind some continental arcs appears opposite to the extensional motion inferred behind some island arcs, but may be caused in part by gravitational spreading of the adjacent highlands into flanking isostatic moats which may form the foreland basins [26]. Clastic wedges in foreland basins behind volcano-plutonic arcs and their associated highlands have commonly been described as exogeosynclinal in that they encroach upon cratons from uplifted sources beyond the cratons. Clastic wedges in successor basins may be similar except that the encroachment is upon previously accreted oceanic materials rather than upon the craton.

7. Plate-interior settings

Sedimentation along inactive continental margins within the interior of lithosphere plates gives rise to the classic miogeosynclinal sequences that are more properly called miogeoclinal from the fact that they are open to the sea on one side [7]. Characteristic assemblages of this kind include near the base coarse clastic strata deposited rapidly on sialic basement rocks as a thick conformable sequence. These typically are overlain by thinner, or at least less rapidly accumulated, sections that include both carbonate and clastic strata, and are broken by disconformities. Both the basal coarse clastic beds and the succeeding strata are commonly deposited in shallow water in subsiding shelves. Facies equivalents deposited offshore on the continental rise and the adjacent ocean floor are commonly called eugeosynclinal when later incorporated into orogenic belts because they include turbidites and marine pelagites deposited on a pillow lava substrate. The provenance for the clastic detritus in all water depths may be uplands located anywhere within the adjacent continent.

In terms of plate tectonics, the development of a

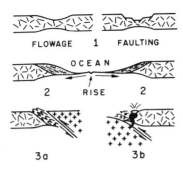

Fig. 3. Sketch of sequence of events leading to miogeoclinal sedimentation followed by orogeny: (1) crustal thinning during intracontinental rifting, (2) construction of miogeoclinal wedges as continental terraces on thinned trailing edges flanking a new and growing ocean basin, (3) deformation when miogeoclinal sequence encounters subduction zone at convergent plate juncture (a) or becomes site of volcanism and plutonism at convergent plate juncture newly initiated at previously stable continental margin (b). See text for discussion.

miogeoclinal pile and its oceanic equivalents offshore can be related to continental rifting and drifting with the formation of a new ocean basin between two continental fragments (fig. 3). Only along the raw edges of a rifted gash in a continental interior can pristine sialic basement of old age be exposed to the influence of much younger sedimentation along a continental margin. As the extensional gash expands to form a nascent ocean basin, deep-crustal flowage and extensional faulting will combine to thin the continental crust by necking along the trend of the rift. As new oceanic crust begins to form along the divergent plate juncture between the separating continental fragments, the finally ruptured continental crust will thin progressively from the interior of each of the fragments toward the rifted margins, which are able to subside as they receive sediment. The miogeoclinal clastic wedges, with their overlying scum of shelf strata and their associated oceanic equivalents, are then built as continent-flank terraces into the new ocean basin, but lap also on to the trailing edges of the two receding continental masses. The depositional areas of shallow-water miogeoclinal strata and proximal turbidites of newly constructed continental rises have curvilinear trends controlled by

the shapes of the new continental margins, but the pelagites and more distal turbidites deposited on the abyssal plains of the new ocean basin occupy a depositional area of indeterminate width. Large areas may be starved basin nearly free of clastic detritus.

In the scheme outlined, there is no requirement that the migeoclinal sequences be involved in orogeny. Unlike plate-juncture sequences, plate-interior sequences are born in a stable tectonic environment. The commonly observed incorporation of plate-interior sequences into orogenic belts is explained by the rapid tempo and shifting patterns of plate motions through time. Oceans open and close. Continents tear and collide. Miogeoclinal sequences on stable continental margins find themselves engaged in a random walk to a destiny which likely will include eventual mangling at a convergent plate juncture (fig. 3). Although miogeoclinal growth is fastest at the onset of development when the basal clastic wedge is deposited, deposition will continue at a reduced pace until sedimentation is terminated by deformational events associated with the close approach of the miogeoclinal pile to a convergent plate juncture, or initiation of convergence near the miogeoclinal region. Only in this passive sense does deposition of a miogeoclinal, or miogeosynclinal sequence, herald orogeny.

8. Collision events

Where lithosphere capped by oceanic crust covered only by thin pelagic sediment acts as the descending plate at a convergent plate juncture, consideration of modern tectonic events indicates that crustal consumption by underthrusting beneath a trench and construction of a nearby magmatic arc are the significant tectonic effects. Where thick turbidites also ride the descending oceanic plate, the scrape-off and isostatic pile-up of internally sheared melanges can be expected to complicate the picture. Where continental crust, or the semicontinental crust of a magmatic arc, forms the top tier of a descending plate of lithosphere, gravimetric considerations suggest that the possibility of bulk descent of an intact slab of lithosphere is severely limited. Such thick crust probably cannot be pushed or dragged down into the mantle. If continents or island arcs are brought together by convergence, we may speak of crustal collision as a major tectonic effect. The crustal collision will act to brake plate consumption, and may force plate convergence to cease at the site of collision, although we are not yet in a position to evaluate the dynamics of collision events in detail. If collision can stop convergence and force rearrangements of plate boundaries and motions, it may be that new plate junctures of convergence will form preferentially at formerly inactive continental margins where contrasts in the internal architecture of the lithosphere are greatest [28].

These considerations allow us to surmise the broad outlines of some of the orogenic events that may affect plate-interior sequences of generally miogeoclinal character with the passage of time. The onset of orogeny probably records one of two things. On the one hand, it may be a collision event caused by the arrival of a plate-interior sequence at a convergent plate juncture when the sequence is riding a descending plate of lithosphere. The short time interval represented by some orogenies may reflect the possibility that the duration of active collision is severely constrained by its braking effect on plate consumption. The consequent necessity to shift the site of plate convergence to the edge of some continental margin or dormant island arc that lay within the interior of some plate prior to the collision would terminate the orogeny. On the other hand, orogeny may represent the initiation of plate convergence and the onset of an arc-trench regime at a previously stable continental margin of miogeoclinal character. Events that would accompany either crustal collision or arc initiation could be expected to terminate miogeoclinal sedimentation, and to deform and metamorphose the miogeoclinal prism and its oceanic associates offshore. An additional likely event, of the kind recorded late in the history of many miogeosynclinal regions, would be the construction of a clastic wedge of exogeosynclinal character across the top of the miogeoclinal prism. The detritus would be fed toward the interior of the adjacent continent from marginal highlands formed either by crumpling and crustal overlapping during collision, or by initiation of a magmatic arc.

9. Complex orogenic belts

The facile concept of paired miogeosynclinal and eugeosynclinal belts of similar identity in all orogenic regions has crept into many minds, but is clearly too simple a view. Also to be rejected is the idea of a regular and fixed progression of orogenic events. Instead, we must accept the fact that past concepts of miogeosynclinal and eugeosynclinal sequences each embrace rock masses formed in a number of different tectonic and depositional settings. Major orogenic belts with a long history commonly include the results of a long string of diverse events. For western North America, for example, we are forced to contemplate the possibility of several episodes of rifting and continental separation to form raw continental edges; several collisions of the main North American craton with other continents, microcontinents, or island arcs; and the initiation of several arc-trench systems; as well as the complex unfolding of each major event with a special local flavor. Rock masses now adjacent to one another or now far apart may not have formed in their present relative positions. We should anticipate juxtaposed shreds of different continents, island arcs, and ocean basins brought together at fossil subduction zones representing ancient plate consumption, and at fossil suture zones representing ancient crustal collision [29, 30].

Nevertheless, the progression of supposed orogenic events cannot be entirely random, but ought to proceed from the logic of plate tectonic theory. Rock masses of different character must lie in patterns that can find explanation in a catalogue of possible kinds of plate behavior. Interpretive constraints are apparently firm, and lend a tight structure to what appears at first glance to be conceptual chaos. The most important guidelines seem clear and compact:

(a) Arc-type and trench-type eugeosynclinal successions occur in coeval association, and the polarity of arc-trench systems is given independently by the potash asymmetry within the igneous rocks of the arcs. Sequences of arc-trench gaps, and of marginal seas , successor basins, or foreland basins behind the arcs, occupy their proper geographic positions with respect to the arc and trench assemblages.

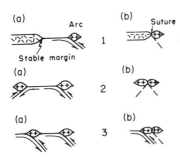

Fig. 4. Sketches of principal types of crustal collision: (1) continent-arc where arc may be intraoceanic or marginal to continent in which case collision is continent-continent, (2) arc-arc (front-to-front) where collision may be continent-arc if right-hand arc is marginal to continent. (3) arc-arc (front-to-back) where collision may be continent-arc if right-hand arc is marginal to continent. See text for discussion.

(b) Miogeosynclinal successions of the miogeoclinal type are continent-edge terrace assemblages masking rifted continental margins formed by separation at divergent plate junctures. Facies equivalents are deposited offshore in the deeper water of the new ocean basin formed by continental separation. Deposition of the miogeoclinal strata and adjacent turbidites of the nearby continental rise begins at the time of separation and is concluded by orogeny when the stable continental margin later drifts against an arc-trench system at a convergent plate juncture or becomes active as a nascent arc-trench system through newly initiated plate convergence localized at or near the continental margin.

(c) Crustal collision occurs only when plate consumption eliminates intervening ocean basins, hence at least one member of each collision pair is an island arc or active continental margin preserved as a volcano-plutonic orogen (fig. 4). Collision may thus juxtapose a plate-interior miogeosynclinal assemblage deposited on a stable continental margin against a eugeosynclinal assemblage of an arc-trench system. Two arcs may also collide, either face to face if both arc-trench systems consume parts of the intervening oceanic lithosphere, or face to back if one arc-trench system overtakes another having the same polarity by consuming the intervening oceanic lithosphere.

(d) Vanished ocean basins are marked by elongate suture zones containing melanges, ophiolites, and oceanic pelagites unless complex thrusting during crustal collision entirely masks them. Suture zones may lie between continents or continental fragments, between volcano-plutonic arcs or orogens, or between continents and arcs.

Vagaries of local behavior give flexibility to each of the guidelines. As arcs develop deep crustal roots, they may behave as microcontinents or accreted continental additions. In both cases, the polarity of the arcs may switch with time if plate consumption and crustal subduction occur on opposite flanks of the arcs at different times. Accreted arcs that are dormant may serve as parts of stable continental margins. In this case, they may be flanked or buried by pseudomiogeoclinal sequences whose characters mimic those of the miogeoclinal sequences deposited along rifted continental margins. Where narrow marginal seas lie between stable continental margins and offshore island arcs, miogeoclinal sequences of continental edges and eugeosynclinal assemblages of arc-trench systems may form side by side [23].

10. Conclusions

The depositonal, compositional, and deformational characteristics of classic miogeosynclinal and eugeosynclinal assemblages can be explained as a consequence of the motions of lithosphere plates and the events which take place at divergent and convergent plate junctures. The course of the dramatic events in orogenic belts can be interpreted plainly by deduction from plate tectonic theory. Although the tenets of classic geosynclinorial theory have played a valuable role as a means to classify tectonic elements, they are unnecessary impediments to clear thinking in the future.

In concentrating on certain facets of the problem of orogenic belts, other aspects, some equally important, have been slighted. A glaring example is the subject of metamorphism, which differs in the various settings discussed. In particular, blueschist belts may be key indicators of trench-style subduction zones. Also ignored are special aspects of sedimentation and deformation near the transform faults along shear junctures between lithosphere plates. In this category are the oceanic fracture zones and also continental faults like the San Andreas, which influenced late Cenozoic events in California. These omitted matters do not negate the main thesis of the paper.

Acknowledgements

The ideas presented here were fostered by discussions at the second Penrose Research Conference at Asilomar, Pacific Grove, California in December, 1969.

References

[1] B.Isacks, J.Oliver and L.R.Sykes, Seismology and the new global tectonics, J. Geophys. Res. 73 (1968) 5855.

[2] J.F.Dewey and J.M.Bird, Mountain belts and the new global tectonics, J. Geophys. Res. 75 (1970) 2625.

[3] A.Knopf, Analysis of some recent geosynclinal theory, Am. J. Sci. 258-A, Bradley Volume (1960) 126.

[4] J.T.Wilson, A new class of faults and their bearing on continental drift, Nature 207 (1965) 343.

[5] J.F.Dewey and B.Horsfield, Plate tectonics, orogeny, and continental growth, Nature 225 (1970) 521.

[6] P.J.Coney, Geotectonic cycle and the new global tectonics, Geol. Soc. Am. Bull. 81 (1970) 739.

[7] M.Kay, North American geosynclines, Geol. Soc. Am. Mem. 48 (1951) 1.

[8] A.H.Mitchell, Evolution of island arcs, J. Geology, in press.

[9] W.Hamilton, The volcanic central Andes – a modern model for the Cretaceous batholiths and tectonics of western North America, Ore Dept. Geol. Min. Ind. Bull. 65 (1969) 175.

[10] H.L.Davies, Papuan ultramafic belt, 23rd Internat. Geol. Cong. Rept. 1 (1968) 209.

[11] A.Gansser, The Indian Ocean and the Himalayas, a geological interpretation, Eclogae Geol. Helvetiae 59 (1967) 831.

[12] W.Hamilton, Origin of the volcanic rocks of eugeosynclines and island arcs, Geol. Surv. Canada Paper 66-15 (1966) 348.

[13] W.R.Dickinson, Relations of andesites, granites, and derivative sandstones to arc-trench tectonics, Rev. Geophysics 8 (1970) in press.

[14] R.L.Chase and E.T.Bunce, Underthrusting of the eastern margin of the Antilles by the floor of the western North Atlantic Ocean and origin of the Barbados Ridge, J. Geophys. Res. 74 (1969) 1413.

[15] K.J.Hsu, Principles of melanges and their bearing on the Franciscan-Knoxville paradox, Geol. Soc. Am. Bull. 79 (1968) 1063.

[16] D.E.Karig, Ridges and basins of the Tonga-Kermadec island arc system, J. Geophys. Res. 75 (1970) 239.

[17] W.R.Dickinson, Clastic sedimentary sequences of arc-trench gaps, Pacific Geol. 3 (1970) in press.

[18] W.R.Dickinson, Tectonic setting and sedimentary petrology of the Great Valley sequence, Geol. Soc. Am. Abstr. Prog. 2 (1970) 86.

[19] W.G.Ernst, Tectonic contact between the Franciscan melange and the Great Valley sequence – crustal expression of a late Mesozoic Benioff Zone, J. Geophys. Res. 75 (1970) 886.

[20] R.W.Ojakangas, Cretaceous sedimentation, Sacramento Valley, California, Geol. Soc. Am. Bull. 79 (1968) 973.

[21] J.F.Evernden and R.W.Kistler, Chronology of emplacement of Mesozoic batholithic complexes in California and western Nevada, U.S. Geol. Surv. Prof. Paper 623 (1970) 1.

[22] J.O.Wheeler, Introduction, Geol. Assoc. Can. Spec. Paper 6 (1970) 1.

[23] A.H.Mitchell and H.G.Reading, Continental margins, geosynclines, and ocean floor spreading, J. Geology 77 (1969) 629.

[24] W.Hamilton, personal communication, 1970.

[25] C.K.Ham and L.J.Herrerra, Jr., Role of Subandean fault system in tectonics of eastern Peru and Ecuador, Am. Assoc. Petroleum Geologists Mem. 2 (1963) 47.

[26] R.A.Price and E.W.Mountjoy, Geologic structure of the Canadian Rocky Mountains between Bow and Athabaska Rivers, Geol. Assoc. Can. Spec. Paper 6 (1970) 7.

[27] R.S.Dietz and J.C.Holden, Miogeoclines in space and time, J. Geology 65 (1967) 566.

[28] D.P.McKenzie, Speculations on the consequences and causes of plate motions, Geophys. J. 18 (1969) 1.

[29] W.Hamilton, Mesozoic California and the underflow of Pacific mantle, Geol. Soc. Am. Bull. 80 (1969) 2409.

[30] W.Hamilton, The Uralides and the motion of the Russian and Siberian platforms, Geol. Soc. Am. Bull. 81 (1970) 2553.

34

Reprinted from Geol. Soc. America Bull. **81**:739–747 (1970)

The Geotectonic Cycle
and the New Global Tectonics

PETER J. CONEY   *Department of Geology, Middlebury College, Middlebury, Vermont 05753*

INTRODUCTION

Advances in submarine geology and geophysics during the last decade have described and partially explained the geology of two-thirds of the surface of our planet, the ocean basins. Rejuvenation of the notion of continental drift (Runcorn, 1962) and the addition of two totally new notions—sea-floor spreading (Hess, 1962; Dietz, 1962) and transform faults (Wilson, 1965)—have revolutionized geotectonic thought and have been referred to as the "new global tectonics" (Isacks and others, 1968).

The particular concerns of this paper are the implications of the "new global tectonics" for our theories about mountain system evolution sometimes referred to as the "geotectonic cycle" (Dennis, 1967, p. 154). Long considered a prime process in continental evolution, the geotectonic cycle is briefly examined in light of recent developments. The conclusion seems inescapable that the explosion of oceanographic information and inference, coupled with steady, but seemingly less dramatic, advances on many fronts of "continental geology" demands a review of our models of mountain evolution and a re-examination of the significance of the geotectonic cycle.

THE GEOTECTONIC CYCLE

One of the fundamental abstractions in earth science has been the concept of a geotectonic cycle. It has gone hand-in-hand with various classifications of tectonic elements and is intricately interwoven in geosynclinal theory. The idea goes back at least to Hall (1859), who recognized the connection between thick accumulations of sediments and folded mountains.

Cady (1950) and Glaessner and Teichert (1947) have reviewed the history of geotectonic thought. By 1930, Bubnoff (1931) recognized time-space arrangements of tectonic elements, and by 1940, Stille (1913, 1941) and Kraus (1927, 1928) had worked out an evolutionary cycle which attempted to integrate sedimentary, igneous, and structural events into a "unified deterministic picture" (Glaessner and Teichert, 1947, p. 573–576) of geosynclinal evolution. By 1950, the concept of a geotectonic cycle was well established and Cady (1950) summarized geotectonic theory and nomenclature. Since 1950, refinements and adjustments have been made, including the works of DeSitter (1964), Rittmann (1962), Beloussov (1962), Badgley (1965), and Van Bemmelen (1966). The geotectonic cycle has provided a framework for various sub-cycles, such as the sedimentary-tectonic cycle (Krynine, 1941) and the igneous-tectonic cycle (Tyrrell, 1955).

The basic scheme which has emerged from these geotectonic syntheses is a progressive and somewhat deterministic sequence of events in space and time which begins with a thick, more or less linear, accumulation of sediments (geosynclinal phase), passes through a period of intense deformation and plutonism (tectogenic phase), and ends with differential uplift and collapse, volcanism, and final incorporation into the continental block (orogenic phase). All workers have recognized both quantitative and qualitative variations from mountain chain to chain, but the basic model has been generally recognized as valid.

The notion of a tectonic cycle has been in part dependent on a classification system of tectonic elements. Definition of a tectonic cycle has been, in fact, based on recognition of fairly definite litho-tectonic features following one another in a more or less orderly fashion in space and time. This has resulted in terminology applied to tectonic patterns *because of* position in a geotectonic cycle, with the result that classifications have been quite complex and the investigator often has difficulty deciding on appropriate terminology. For example, using current systems of tectonic regionalization, the Basin and Range province of western North America is considered as epieugeosynclinal, taphro-geosynclinal, a post-orogenic nuclear basin (Badgley, 1965, Fig. 11.17, p. 485), a Zwischengebirge (Moores and others, 1968), and a stable intrageanticline (Beloussov, 1962, p. 697–698).

Table 1 compares some organizations of the geotectonic cycle that have been proposed and Figure 1 is a series of diagrammatic structure-sections indicating evolutionary development of some tectonic elements. Such high-level syntheses form the foundations and landmarks of our understanding of the earth. As Marshall Kay has said (1967, p. 315) ". . . the very endeavor to classify has been rewarding."

THE GEOTECTONIC CYCLE IN LIGHT OF RECENT ADVANCES IN EARTH SCIENCE

The last decade has been a revolutionary one in earth science (Wilson, 1968). Advances have been made in many areas, several of which have been particularly significant in recent geotectonic thought. In this section, new developments in geosynclinal theory, studies in tectogenesis, and submarine geology and geophysics are examined; all of these have bearing on the geotectonic cycle.

Geosynclinal Theory

Recent geological work in mountain systems and geophysical work along continental margins has modified the image of a geosyncline. Drake and others (1958), Dietz (1963), and Worzel (1968) have argued similarity between Mesozoic-Cenozoic sedimentary prisms along the continental margin of eastern North America, making up the continental shelf, slope, and rise, and the classic mio-eugeosynclinal couple of historical geology. This has suggested to Dietz and Holden (1966) that the subsiding surface may not be synclinal and that new terms are needed, such as "miogeocline." Recent application of these concepts to the lower Paleozoic Appalachian "geosyncline" by Rodgers (1968) and Dewey (1969) has made Appalachian geology more comprehensible. A more drastic position has been taken by Gansser (1967). Based on geologic analysis of the Himalayan mountain system and bathymetric studies in the Indian Ocean, the theory of northward drift of India is accepted by Gansser; he concludes that the Himalayan chain had no preparatory geosyncline. Stocklin (1968, p. 1232–1233) working in Iran, has concluded similarly for other segments of the Tethys belt. This idea has been extended by Ahmad (1968) to all Tertiary mountain systems, the inference being that drifting continents rather than preparatory geosynclines produce mountains. Also, it has been suggested that island arcs may not necessarily represent embryonic mountain systems, as has been sometimes supposed, but may simply be an oceanic manifestation of some subcrustal process that builds mountains on continents (Hess, 1960, p. 237). Finally, those who have worked most closely with the problem over the years and have provided much of the inspiration for the models we use, caution us to emphasize differences between geosynclines. Further, they prefer to keep the tectonics of geosynclinal development separate from subsequent tectonic events (Kay, 1967, p. 314).

Studies in Tectogenesis

Of all the several phases of the geotectonic cycle, the "tectogenic phase" has been fundamental, because it has been viewed as a

Table 1. Examples of Geotectonic Cycles

	PRIMARY GEOSYNCLINE	PRIMARY MOUNTAIN BUILDING	SECONDARY MOUNTAIN BUILDING	FINAL DIFFERENTIAL UPLIFT AND SUBSIDENCE
CADY (1950)	Arcuate belts of mioeugeosynclinal couples, submarine volcanism	Nappes, syn-orogenic batholiths, metamorphism, and development of geanticlines, secondary geosynclines	Folds and thrusts in secondary geosynclines, granite plutons, and alkalic intrusives	Block faulting, uplift, plateau basalts
TYRRELL (1955)	GEOSYNCLINAL PHASE Ophiolite series	OROGENIC PHASE Foliated granites, augen gneiss, concordant sheets, and injection complexes, followed by discordant batholiths, ring complexes, ending with andesitic and rhyolitic lavas		POST-OROGENIC PHASE Basaltic lava fields and associated intrusions
RITTMANN (1962)	DEEP-SEA TRENCH / GEOSYNCLINAL PHASE Primary volcanicity — Marine transgressions, formation of geosyncline	TECTOGENESIS Regional metamorphism, anatexis of sial, nappes, basic effusions in marginal deeps	OROGENESIS Uplift, diapiric plutonism, orogenic volcanicity, down-sliding nappes, deposition of flysch, followed by erosion and deposition of molasse, followed by massive ignimbrite sheets, leading to erosion, cratonic conditions	
BELOUSSOV (1962)	GENERAL SUBSIDENCE Marine transgression, subsidence, great accumulation of sediment, minor folding and faulting	INVERSION Transition from subsidence to uplift, development of marginal deeps, main folding, batholiths, etc.	GENERAL UPLIFT General uplift, "mountain building," fracturing, leading to extrusive igneous activity	
BADGLEY (1965)	EUGEOSYNCLINAL PHASE Submarine volcanism, graywackes, cherts, etc.	EARLY TECTONIC PHASE Ultramafic intrusions — SYNTECTONIC PHASE Migmatites, gneiss domes, catazonal emplacement of batholiths, folding — MAIN-TECTONIC PHASE Meso-zonal emplacement of batholiths	LATE-TECTONIC PHASE Magmatic stoping, caldron subsidence, alkalic igneous activity	POST-TECTONIC PHASE Collapse, epeirogenesis, tholeiitic basalts, olivine basalts

299

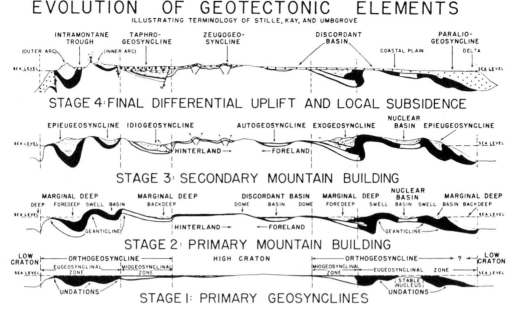

Figure 1. Evolutionary development of "orogens" during the geotectonic cycle (*from* Cady, 1950; with permission).

time of maximum deformation, metamorphism, and plutonism—as the event which welds an unstable belt into a continental block. It has often been assumed that successive welding of mountain systems has, indeed, constructed sialic crust by continental accretion (Wilson, 1964; Engel, 1963; Dietz, 1966).

Recent work, particularly the careful analysis of regional geologic data, studies in metamorphic petrology, and radiometric dating, suggests that a revision of classic theory is in order. It would seem that events normally associated with "tectogenesis" progress over enormous lengths of time, extending far down into the classic "geosynclinal phase" and far up into the "orogenic phase," so that boundaries or even transitions become difficult to isolate in space or time.

It has become increasingly clear that axial cores of mountain systems may record a prolonged episodic series of tectonic, metamorphic, and plutonic events distributed in irregular time-space patterns which may not conform to each other or to previous tectonic patterns. Sutton (1965, p. 29-31) has emphasized that the total life span of individual metamorphic belts may range from 30 to 900 m. y. For example, in the Appalachians of eastern North America, if we assume the "Appalachian cycle"

began some 700 to 800 m. y. ago and ended about 170 m. y. ago, we find tectogenic deformation and/or metamorphism and plutonism reported from 580 to about 180 m. y. ago (Rodgers, 1967, p. 421; Badgley, 1965, p. 360). Typical mio-eugeosynclinal deposition, however, extended at least to about 380 m. y. ago. The geosynclinal phase of the Cordillera of western North America began about 700 m. y. ago and extended at least to about 200 m. y. ago; compressive tectonism and/or plutonism and volcanism are recorded from about 520 m. y. ago to the present (Gilluly, 1967), and metamorphism is spread over a period from about 350 m. y. ago to Late Tertiary (Davis, 1968; Gilluly, 1967; Moores and others, 1968, p. 1722-1724).

We have lived through the Appalachian revolution and the Taconic disturbance, and the Taconic revolution and the Appalachian disturbance (Hess, 1955, p. 400-401). It would seem that we are not entirely clear what exactly the word "revolution" refers to, or what exactly a phase of the "geotectonic cycle" represents. One characteristic of confusion is use of words that have lost their meanings and dependence on models which no longer apply. I would suggest we need new terminology and a new model.

Submarine Geology and Geophysics

The most startling discovery of oceanic research has been gradual recognition of the oceanic rise as a first-order feature of our planet taking the form of a world rift system (Heezen, 1962; Girdler, 1964; Menard, 1964; Irvine, 1967). This feature, coupled with inferences based on paleomagnetic data (Cox and others, 1967; Vine, 1966) suggesting sea-floor spreading and transform faulting, has become the keystone of the "new global tectonics."

Of particular concern here is the notion that the oceanic rise system passes beneath continents at several points (Heezen, 1962, p. 259), such as the northwestern coast of Mexico and the northeastern coast of Africa. In the first case, the junction is with a major Circum-Pacific mountain system, and it has been suggested (Menard, 1960; Larson and others, 1968) that spreading along the rise has drifted the peninsula of Baja California away from the Mexican coast. This theory has been extended to the San Andreas and other strike-slip faults in California and Baja California, treating them as transform faults which relay the crest of the East Pacific rise up the western coast of North America beneath the continental margin (Wilson, 1965). Finally, the idea has been further extended to encompass and explain the entire pattern of Basin and Range faulting in the western United States as incipient continental fragmentation caused by spreading beneath the continental plate (Armstrong, 1968, p. 450; Cook, 1967). The important point is that this suggests the entire post-Eocene ("post-tectogenic"?) evolution of southwestern North America might be only indirectly related to any orderly geotectonic cycle in the deterministic sense, and due rather to accidental mergence of two distinct processes. Similarly, the White Mountain magma series of New Hampshire (Billings, 1956, p. 129–135), a complex of felsic ring dikes, volcanics, small plutons, and calderas of early Mesozoic(?) age, seem more related to Triassic rifting and opening of the Atlantic Ocean than to a "post-tectonic" (Eardley, 1962, p. 178; Chapman, 1967, p. 49) position in an "Appalachian geotectonic cycle."

A significant development in geotectonics has been recent evidence from seismology supporting the "new global tectonics" (Isacks and others, 1968). In essence, this work has reaffirmed the importance of concepts long ago elucidated, such as the role of a lithosphere and an asthenosphere (Barrell, 1914; Daly, 1940), and the deep seismic zones dipping beneath island arcs and some continental margins (Benioff, 1954). The new evidence has reaffirmed the dip-slip character of deformation under the arcs, and has isolated a slab of high velocity lithosphere dipping steeply below the arcs and penetrating the asthenosphere (Isacks and others, 1968, Fig. 7). Most important, it has integrated older concepts with newer ones, such as spreading of the sea floor, making for a remarkably simple system relating the oceanic rises to island arcs.

The implications for geotectonics are many, but one point is worthy of note here. These discoveries suggest island arcs are simply overprinted along some continental margins and may have only indirect relationship to much of the preceding tectonics that affected a continental margin. For example, if the western coast of South America is an arc-trench couple associated with spreading of the sea floor, say during the Neogene, this suggests Neogene "post tectonic" or "orogenic" tectonic activity of the "Andean cycle" may have only accidental or indirect relation to previous events in Andean "tectogenesis." Rather, these events may be a volcano-tectonic overprint on previous and independent tectonic patterns.

CONCLUSIONS

It would seem, in light of recent developments in earth science, that some questions can be raised as to the validity of the geotectonic cycle as a deterministic model of mountain system evolution. It can be argued that definition of each of the so-called "phases" of the cycle is obscure; the relationship of any one phase to another is in doubt; and there is uncertainty that the entire cycle can be attributed to a single internal mechanism. This is not to say that there has not been a series of events; mountain systems have obviously had a history. What is being said is that these histories are complex, probably different from one chain to another, and not due to a single internally deterministic cycle. Certainly, geologic histories of tectonic complexes are more comprehensible if they are simply considered the result of successive and interpenetrative tectonic responses of various sorts rather than due to any orderly "geotectonic cycle."

It is the thesis of the "new global tectonics" that lithosphere is being generated along the world rift system and destroyed along island arcs and active mountain systems. This focuses attention on boundaries of lithosphere plates—the axial rift of oceanic rises, island arcs, and certain transform and transcurrent faults. Clearly, some rift patterns can be related to past or present tensional spreading overprinted on various tectonic inheritances, including the ocean floor (Mid-Atlantic Rise), mountain systems (western United States), and a stable continental lithosphere (eastern Africa). The island arcs, however, lack the simplicity of the world rift system, and their precise relationship to such diverse features as continental margins, small ocean basins (Menard, 1967), and active mountain systems is obscure.

It seems likely that the island arc-trench couple represents a response to sea-floor spreading. Here oceanic lithosphere is being underthrust and the pattern can be conceptualized as marking a line of uncoupling. The question which now presents itself is how to relate island arcs to mountain systems where they appear to be extensions of one another, such as the arcs of the East Indies and the Alpine-Himalaya mountain system. A way out of this dilemma is to consider the Alpine-Himalaya system to be the result of intercontinental collision and

closing of the Tethys Sea (Wilson, 1968). Thus, where oceanic lithosphere is spreading into and below an adjacent plate, an arc-trench results. Where continental plates are colliding, or interacting in some way, certain types of mountain systems result.

It would appear that other types of mountain systems are positioned in such a way in space and time that no colliding continent was available to provide the mechanism for "tectogenic" structures, batholiths, and so on. The Cretaceous and earliest Tertiary Andes of Peru are an example, because examination of sea-floor spreading charts (Heirtzler, 1968, p. 65–66; Vine, 1968) seems to prohibit acceptance of the theory of colliding continents during times of "tectogenesis," say from 100 to 5 m. y. ago. Perhaps like granites, there are mountains and there are mountains: colliding types, like the Alpine Himalaya; uncoupled types, like the late Mesozoic and Tertiary Andes of Peru; combinations of the above, like the Paleozoic-Mesozoic Cordillera of North America (Wilson, 1968) and the Appalachian/Caledonian orogen (Dewey, 1969); and still others, such as the Barberton Mountain Land of South Africa, which may be truly intracontinental and due to some unique, but as yet

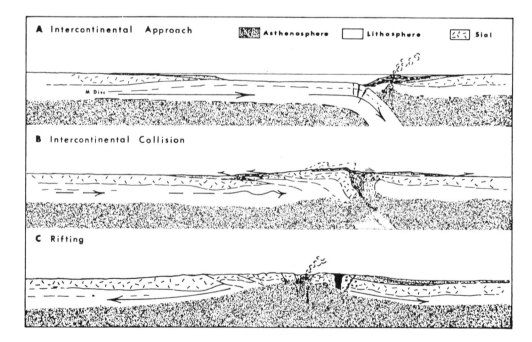

Figure 2. Tectonic history of a mountain complex.

obscure, process which has no relationship to classic geosynclinal theory or models of alpine tectonics (Anhaeusser and others, 1968). Periods of rifting would further complicate evolutionary histories. These possibilities inherent in coupled, uncoupled, rifting, and colliding plates, added to varying motions along transform boundaries and intra-plate transcurrent shifts, and irregularities in shapes of colliding margins, could explain why no single model fits all mountain systems. A succession of several of the above responses would produce what we might call a mountain complex, as diagrammed in Figure 2. It would seem one of the objectives now is to determine what the variants are in these complex histories as well as the invariants so that we may approach a clearer understanding of the tectonic evolution of mountain systems.

In any case, I suggest that we take seriously the recent data of the "new global tectonics" and abandon the geotectonic cycle as a deterministic model of mountain evolution. Similarly, the terminology of the geotectonic cycle, such as pre-, syn-, and post-tectonic, should be abandoned as meaningless words. Instead, the over-all interplay of global tectonic patterns should be recognized and a new

model sought for, based on the principles of sea-floor spreading, shifting lithosphere plates, transform faults, and colliding, coupled, and uncoupled continental margins.

Seen in this light, tectonic patterns become the result of interference of various tectonic responses, the order of which may be highly irregular. In this way, some linear sedimentary accumulations become features characteristic of continental margins. If ocean basins are opening and closing, these prisms can hardly escape collision or uncoupling. Saying geosynclinal prisms lead to mountain systems is a little like saying automobile fenders lead to accidents. Similarly, "post-tectogenic" rifting and other "orogenic" events become the result of fortunate position relative to other tectonic systems, such as oceanic rises.

Many fundamental problems remain. The "new global tectonics" has not yet adequately explained much of the complexity of continental geology, such as metamorphic imprints, nor the massive development of infrastructure patterns in axial cores of many mountain systems. Such features as small ocean basins have not been explained. We will either find the answers to these questions in the "new global tectonics" or move on to some

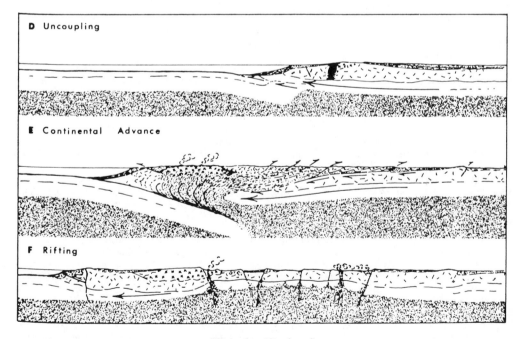

Figure 2. (Continued)

future model as a result of answering these questions.

ACKNOWLEDGMENTS

I am most grateful to W. M. Cady, J. C. Maxwell, D. W. Strangway, L. R. Sykes, and J. T. Wilson for helpful criticism and encouragement to proceed with the ideas expressed in this paper. Discussions with Brewster Baldwin, Wolfgang Elston, Marshall Kay, and Roger Laurent contributed much to the content.

REFERENCES CITED

Ahmad, F., 1968, Orogeny, geosynclines, and continental drift: Tectonophysics, v. 5, p. 177–189.

Anhaeusser, C. R., Mason, R., Viljoen, M. J., and Viljoen, R. P., 1968, A reappraisal of some aspects of Precambrian shield geology: Johannesburg, Univ. of the Witwatersrand, Econ. Geol. Research Unit Information Circ. 49, 30 p.

Armstrong, R. L., 1968, Sevier orogenic belt in Nevada and Utah: Geol. Soc. America Bull., v. 79, p. 429–458.

Badgley, P. C., 1965, Structural and tectonic principles: New York, Harper and Row, 521 p.

Barrell, J., 1914, The strength of the earth's crust: Jour. Geology, v. 22, p. 289–314, 441–468, 655–683.

Beloussov, V. V., 1962, Basic problems in geotectonics: New York, McGraw-Hill, 809 p.

Benioff, H., 1954, Orogenesis and deep crustal structure: Geol. Soc. America Bull., v. 65, p. 385–400.

Billings, M. P., 1956, The geology of New Hampshire, pt. II. Bedrock geology: New Hampshire Plan. and Devel. Comm., 204 p.

Bubnoff, S. von, 1931, Grundprobleme der geologie: Berlin, Borntraeger, 237 p. (not seen, fide Cady, 1950).

Cady, W. M., 1950, Classification of geotectonic elements: Am. Geophys. Union Trans., v. 31, p. 780–785.

Chapman, C. A., 1967, Magmatic central complexes and tectonic evolution of certain orogenic belts, in Etages tectoniques, Colloque de L'Institut de Géologie de L'Université de Neuchâtel, p. 41–51.

Cook, K. L., 1967, Rift system in the Basin and Range province, in Irvine, T. N., Editor, The world rift system: Canada Geol. Survey Paper 66-14, p. 246–279.

Cox, A., Dalrymple, G. B., and Doell, R. R., 1967, Reversals of the earth's magnetic field: Sci. American, v. 216, no. 2, p. 44–54.

Daly, R. A., 1940, Strength and structure of the earth's crust: New York, Prentice-Hall Inc., 434 p.

Davis, G. A., 1968, Westward thrust faulting in the south-central Klamath Mountains, California: Geol. Soc. America Bull., v. 79, p. 911–934.

Dennis, J. G., 1967, International tectonic dictionary: Am. Assoc. Petroleum Geologists Mem. 7, 196 p.

DeSitter, L. U., 1964, Structural geology: New York, McGraw-Hill, 551 p.

Dewey, J. F., 1969, Evolution of the Appalachian/Caledonian orogen: Nature, v. 222, p. 124–129.

Dietz, R. S., 1962, Ocean basin evolution by sea floor spreading, in Runcorn, S. K., Editor, Continental drift: Academic Press, p. 289–298.

—— 1963, Collapsing continental rises: An actualistic concept of geosynclines and mountain building: Jour. Geology, v. 71, p. 314–333.

—— 1966, Passive continents, spreading sea floors, and collapsing continental rises: Am. Jour. Sci., v. 264, p. 177–193.

Dietz, R. S., and Holden, J. C., 1966, Miogeoclines (miogeosynclines) in space and time: Jour. Geology, v. 74, no. 5, pt. 1, p. 566–583.

Drake, C. L., Ewing, M., Sutton, G. H., 1958, Continental margins and geosynclines: The east coast of North America north of Cape Hatteras, in Ahrens, L. H., Press, F., Rankama, K., and Runcorn, S. K., Editors, Physics and chemistry of the Earth, v. 3: London, Pergamon Press, Ltd., p. 110–193.

Eardley, A. J., 1962, Structural geology of North America (2nd ed.): New York, Harper and Row, 743 p.

Engel, A.E.J., 1963, Geologic evolution of North America: Science, v. 140, p. 143–152.

Gansser, A., 1967, The Indian Ocean and the Himalayas, a geological interpretation: Eclogae Geol. Helvetiae, v. 59, p. 831–848.

Gilluly, J., 1967, Chronology of tectonic movements in western United States: Am. Jour. Sci., v. 265, p. 305–331.

Girdler, R. W., 1964, Geophysical studies in rift valleys, in Ahrens, L. H., Press, F., and Runcorn, S. K., Editors, Physics and chemistry of the earth, v. 5: London, Pergamon Press, Ltd., p. 121–156.

Glaessner, M. F., and Teichert, C., 1947, Geosynclines, a fundamental concept in geology: Am. Jour. Sci., v. 245, p. 465–482, 571–591.

Hall, James, 1859, Natural history of New York: Paleontology, v. 3, pt. 1, p. 1–96.

Heezen, B. C., 1962, The deep-sea floor, in Runcorn, S. K., Editor, Continental drift: New York, Academic Press, p. 235–288.

Heirtzler, J. R., 1968, Sea-floor spreading: Sci. American, v. 219, no. 6, p. 60–70.

Hess, H. H., 1955, Serpentine, orogeny, and epeirogeny: Geol. Soc. America Spec. Paper 6, p. 391–408.

—— 1960, Caribbean research project, progress

report: Geol. Soc. America Bull., v. 71, p. 235–240.

Hess, H. H., 1962, History of ocean basins, *in* Engel, A. E. J., James, H. L., and Leonard, B. F., *Editors*, Petrologic studies: Buddington Volume, Geol. Soc. America Mem. 28, p. 599–620.

Irvine, T. N., *Editor*, 1967, The world rift system: Canada Geol. Survey Paper 66-14, 471 p.

Isacks, B., Oliver J., and Sykes, L. R., 1968, Seismology and the new global tectonics: Jour. Geophys. Research, v. 72, p. 5855–5900.

Kay, M., 1967, On geosynclinal nomenclature: Geol. Mag., v. 104, no. 4, p. 311–316.

Kraus, E., 1927, Der orogene zyklus und seine stadien: Centralbl. für Min., Abt. B., p. 216–233.

—— 1928, Das wachstun der kontinente nach der zyklustheorie: Geol. Rundschau, v. 19, p. 353–386, 481–493.

Krynine, P. D., 1941, Differentiation of sediments during the life history of a land mass (abs.): Geol. Soc. America Bull., v. 52, p. 1915.

Larson, R. L., Menard, H. W., and Smith, S. M., 1968, Gulf of California: A result of ocean floor spreading and transform faulting: Science, v. 161, p. 781–783.

Menard, H. W., 1960, The East Pacific Rise: Science, v. 132, p. 1737–1746.

—— 1964, Marine geology of the Pacific: New York, McGraw-Hill, 271 p.

—— 1967, Transitional types of crust under small ocean basins: Jour. Geophys. Research, v. 72, p. 3061–3073.

Moores, E. M., Scott, R. B., and Lumsden, W. W., 1968, Tertiary tectonics of the White Pine-Grant Range region, East central Nevada, and some regional implications: Geol. Soc. America Bull., v. 79, p. 1703–1726.

Rittmann, A., 1962, Volcanoes and their activity: New York, John Wiley and Sons, 305 p.

Rodgers, John, 1967, Chronology of tectonic movements in the Appalachian region of eastern North America: Am. Jour. Sci., v. 265, p. 408–427.

—— 1968, The eastern edge of the North American continent during the Cambrian and early Ordovician, *in* Zen, E-An, White, W. S., Hadley, J. B., and Thompson, J. B., *Editors*, Studies of Appalachian geology, northern and maritime: New York, Interscience, p. 141–149.

Runcorn, S. K., *Editor*, 1962, Continental drift: New York, Academic Press, 338 p.

Stille, H., 1913, Evolutionen und revolutionen in der erdgeschichte: Berlin (not seen, *fide* Glaessner and Teichert, 1947).

—— 1941, Einfuhrung in den bau Nordamerikas: Berlin, Borntraeger, 717 p. (not seen, *fide* Kay, 1967).

Stocklin, J., 1968, Structural history and tectonics of Iran: A review: Am. Assoc. Petroleum Geologists Bull., v. 52, p. 1229–1258.

Sutton, John, 1965, Some recent advances in our understanding of the controls of metamorphism, *in* Pitcher, W. S., and Flinn, G. W., *Editors*, Controls of metamorphism: New York, John Wiley & Sons, p. 22–45.

Tyrrell, G. W., 1955, Distribution of igneous rocks in space and time: Geol. Soc. America Bull., v. 66, p. 405–426.

Van Bemmelen, R. W., 1966, On mega-undations: A new model for the earth's evolution: Tectonophysics, v. 3, p. 83–127.

Vine, F. J., 1966, Spreading of the ocean floor: New evidence: Science, v. 154, p. 1405–1415.

—— 1968, Evidence from submarine geology, *in* Gondwanaland revisited: New evidence for continental drift: Am. Philos. Soc. Proc., v. 112, p. 325–334.

Wilson, J. T., 1964, The crust, *in* Bates, D. R., *Editor*, The planet earth (2nd ed.): New York, Pergamon Press, p. 52–78.

—— 1965, A new class of faults and their bearing on continental drift: Nature, v. 207, no. 4995, p. 343–347.

—— 1968, Static or mobile earth: The current scientific revolution, *in* Gondwanaland revisited: New evidence for continental drift: Am. Philos. Soc. Proc., v. 112, no. 5, p. 309–320.

Worzel, J. L., 1968, Advances in marine geophysical research on the continental margins: Canadian Jour. Earth Sci., v. 5, p. 963–983.

Part IX

ANALYTICAL KEYS FOR DECIPHERING ANCIENT GEOSYNCLINES PRODUCED BY PLATE TECTONIC MECHANISMS

Editor's Comments
on Papers 35 Through 42

The mixed success of the early papers that attempted to reinterpret ancient geosynclinal belts in terms of plate tectonics mechanisms underscored the need for pinpointing those aspects of the ancient rock record that clearly indicated a plate tectonic origin. In Paper 35, Dickinson underscored this point as he proposed a general outline of the petrogenetic criteria by which plate tectonics mechanisms could be inferred. It was obvious that many careful analyses were necessary to establish the structural, stratigraphic, and lithological characteristics of modern plate margins.

In general, over the course of the last decade, efforts by leading specialists in each of geology's subdisciplines have been directed towards this goal. Limited space precludes reproducing significant papers from all the essential fields of study. I have chosen representative articles from only one area, the field of stratigraphy and sedimentation.

However, it is appropriate to at least briefly review some of the milestone studies in some other fields. For example, in the area of igneous petrology, Gilluly (1971), Green (1971), Engel et al. (1974), Wood, Joron, and Treuil (1979), and Wyllie (1979) among others summarize efforts relating the origin and temporal and spatial distribution of igneous magmas to plate boundaries and configurations. These studies point out that such an approach brings fresh insights to such problems as the origin of ophiolites (Coleman, 1971; Cann, 1969; Ross et al., 1980; and Sengor, Yilmaz, and Ketin, 1980), tholeiitic basalt (Engel and Engel, 1964; Green, Hibberson, and Jacques, 1979; Bryan et al., 1976; and Presnall et al., 1979), rift zone alkalic volcanism (Baker and Wohlenberg, 1971), and andesites and granitic batholiths (Dickinson, 1970; Boettcher, 1973; Whitford, Compston, and Nicholls, 1977; Johnson, Mackenzie, and Smith, 1978; and Green, 1980).

The efforts of metamorphic petrologists are equally impressive. It was immediately obvious that because most metamorphism occurs in response to mechanical (tectonic) and chemical (pressure-temperature) processes that are in turn largely the result of lithospheric plate motions, individual metamorphic petrogenetic terrains exposed within geosynclinal belts can indicate the structural setting of geosynclines at various stages of development. Two companion Benchmark volumes, *Subduction Zone Metamorphism* (Ernst, 1975a) and *Metamorphism and Plate Tectonics* (Ernst, 1975b) historically document the progress in this field. Important papers illustrating how integrated models for metamorphosed geosynclinal belts can be developed include those of Ernst (1970, 1973), Hamilton (1969), and Miyashiro (1961, 1972a, 1972b, 1973).

No group of geologists has been more prolific at redefining their discipline within the premises and constraints of plate tectonics than the stratigraphers and sedimentologists. Two recent papers published by the Society of Economic Paleontologists and Mineralogists are devoted to this objective. Special Publication 19 (Dott and Shaver, 1974), *Modern and Ancient Geosynclinal*

Sedimentation, documents the lithological and structural characteristics of modern continental margins and their supposed ancient analogues. Special Publication 22, *Tectonics and Sedimentation,* evaluates "the interplay between tectonic events and sedimentation for which or from which plate tectonic implications can be drawn" (Dickinson, 1974, p. iii). Another noteworthy effort is *The Geology of Continental Margins* edited by Burk and Drake (1974). A particularly useful regional series on *The Ocean Basins and Continental Margins* is edited by Nairn and Stehli (1973, 1974, 1975); Nairn, Kanes, and Stehli (1977, 1978); and Nairn, Churkin, and Stehli (1981). The six volumes that have appeared so far cover the Atlantic Ocean, the Gulf of Mexico, the Caribbean Sea, the Arctic Ocean and the Mediterranean Sea.

An interesting multidisciplinary memoir published by the American Association of Petroleum Geologists is entitled *Plate Tectonics—Assessments and Reassessments* (Kahle, 1974). This volume contains many of the arguments appropriate to the growing pains of a new scientific revolution. Of considerable value are the last-ditch stands of the anti-drift spokesmen: the Meyerhoffs and Curt Teichert for the United States and V. V. Beloussov for the USSR (see also, Beloussov, 1970, 1979; and Beloussov, Ruditch, and Shapiro, 1979). As a curious sidelight on history, the American Association of Petroleum Geologists had forty-six years earlier published the symposium on the *Theory of Continental Drift* that had played a major role in suppressing the drift concept in the English-speaking world.

Several other studies, none of which are portions of the publications listed above, are reprinted here. An extract of Aubouin's study comparing frontal or prefatory mobile belts rimming continental margins (the Andes) with truly geosynclinal intracontinental belts such as the Alps (Paper 36) illustrates some of the difficulties stratigraphers encountered in attempting to reorient conventional approaches with the reality of global plate tectonics. Rather rigid mental straightjacketing had to be discarded and replaced with a not-so-easily attained flexibility.

As understanding of plate tectonics mechanisms improved, earlier plate tectonic basin evolution models were rejected or modified. For example, Moberly (Paper 37) demonstrated that the six lithospheric plates originally postulated represented an oversimplified picture of the real world. Ancient geosynclinal basin development could presumably be as diverse. A brief excerpt from Burk (Paper 38) illustrates how complex eugeosynclinal assemblages whose origin had heretofore been unfathomable could be profitably approached with the aid of the plate tectonics paradigm.

Papers 39 and 40 specifically relate the various geosynclinal sedimentary suites (particularly the sandstone component) to the many types of continental margin sedimentary associations dictated as their equivalent by plate tectonic theory. The contrasts between these studies and an earlier attempt by Mitchell and Reading (Paper 29) is striking. Paper 41, largely inspired by an earlier study by Crook (1974), postulates that specific continental margin sites are characterized by unique sandstone mineralogy and chemistry. I include this paper because it represents an early attempt to provide a quantitative fingerprint for identifying the plate setting for ancient mobile belts. Finally, Paper 42 attempts to discriminate among the various types of sedimentary basins existing within a plate tectonic framework (including geosynclines) by documenting systematic differences in yet another quantitative stratigraphic characteristic, the rate of accumulation of sediment fill.

REFERENCES

Baker, B. H., and J. Wohlenberg, 1971, Structure and Evolution of the Kenya Rift Valley, *Nature* **229**:538–542.

Beloussov, V. V., 1970, Against the Hypothesis of Ocean-floor Spreading. *Tectonophysics* **9**:498–511.

Beloussov, V. V., 1979, Why Do I Not Accept Plate Tectonics?, *Eos* **60**: 207–211.

Beloussov, V. V., E. M. Ruditch, and M. N. Shapiro, 1979, Intercontinental Structural Ties and Mobilistic Reconstructions, *Geol. Rundschau* **68**:393–427.

Boettcher, A. L., 1973, Volcanism and Orogenic Belts—the Origin of Andesites, *Tectonophysics* **17**:223–240.

Bryan, W. B., G. Thompson, F. A. Frey, and J. S. Dickey, 1976, Inferred Geological Settings and Differentiation in Basalts from the Deep Sea Drilling Project, *Jour. Geophys. Research* **81**:4285–4304.

Burk, C. A., and C. L. Drake, eds., 1974, *The Geology of Continental Margins*, Springer-Verlag, New York, 1009 p.

Cann, J. R., 1969, New Model for the Structure of the Ocean Crust, *Nature* **226**:928–930.

Coleman, R. G., 1971, Plate Tectonic Emplacement of Upper Mantle Peridotites Along Continental Edges, *Jour. Geophys. Research* **76**: 1212–1222.

Crook, K. A., 1974, Lithogenesis and Geotectonics: The Significance of Compositional Variation in Flysch Arenites (Graywackes), in R. H. Dott, Jr., and R. H. Shaver, eds., *Modern and Ancient Geosynclinal Sedimentation*, Society of Economic Paleontologists and Mineralogists Special Publication 19, pp. 304–310.

Dickinson, W. R., 1970, Relation of Andesites, Granites, and Derivative Sandstones to Arc-trench Tectonics, *Rev. Geophys. Space Sci.* **8**: 813–860.

Dickinson, W. R., 1974, Tectonics and Sedimentation, *Soc. Econ. Paleontologists and Mineralogists Spec. Pub.* 22, 204 p.

Dott, R. H., Jr., and R. H. Shaver, eds., 1974, Modern and Ancient Geosynclinal Sedimentation, *Soc. Econ. Paleontologists and Mineralogists Spec. Pub.* 19, 380 p.

Engle, A. E. J., and C. G. Engel, 1964, Composition of Basalts from the Mid-Atlantic Ridge, *Science* **144**:1330-1333.

Engle, A. E. J., S. P. Itson, C. G. Engel, and D. M. Stickney, 1974, Crustal Evolution and Global Tectonics: A Petrographic View, *Geol. Soc. America Bull.* **85**:843-858.

Ernst, W. G., 1970, Tectonic Contact Between the Franciscan Melange and the Great Valley Sequence, Crustal Expression of a Late Mesozoic Benioff Zone, *Jour. Geophys. Research* **76**:886-902.

Ernst, W. G., 1973, Blueschist Metamorphism and P-T reigmes in Active Subduction Zones, *Tectonophysics* **17**:255-272.

Ernst, W. G., ed., 1975a, *Metamorphism and Plate Tectonics Regimes*, Dowden, Hutchinson & Ross, Stroudsburg, Pa., 440 p.

Ernst, W. G., ed., 1975b, *Subduction Zone Metamorphism*, Dowden Hutchinson & Ross, Stroudsburg, Pa., 445 p.

Gilluly, J., 1971, Plate Tectonics and Magmatic Evolution, *Geol. Soc. America Bull.* **82**:2382-2396.

Green, D. H., 1971, Composition of Baslatic Magmas as Indicators of Origin with Application to Oceanic Volcanism, *Royal Soc. London Philos. Trans.* **268A:**707-725.

Green, D. H., 1980, Island Arc and Continent-building Magmatism—A Review of Petrogenic Models Based on Experimental Petrology and Geochemistry, *Tectonophysics* **63**:367-385.

Green, D. H., W. D. Hibberson, and A. L. Jaques, 1979, Petrogenesis of Mid-ocean Ridge Basalts, in M. W. McElhinney, ed., *The Earth: Its Origin, Structure, and Evolution*, Academic Press, London, pp. 265-299.

Hamilton, W., 1969, Mesozoic California and the Underflow of Pacific Mantle, *Geol. Soc. America Bull.* **80**:2409-2430.

Johnson, R. W., D. E. Mackenzie, and I. E. M. Smith, 1978, Delayed Partial Melting of Subduction-modified Mantle, *Tectonophysics* **46**:197-216.

Kahle, C. F., ed., 1974, Plate Tectonics—Assessments and Reassessments. *Am. Assoc. Petroleum Geologists Mem.* **23**:1-514.

Miyashiro, A., 1961, Evolution of Metamorphic Belts, *Jour. Petrology* **2**:277-311.

Miyashiro, A., 1972a, Metamorphism and Related Magmatism in Plate Tectonics, *Am. Jour. Sci.* **272**:629-656.

Miyashiro, A., 1972b, Pressure and Temperature Conditions and Tectonic Significance of Regional and Ocean Floor Metamorphism, *Tectonophysics*, **13**:141-159.

Miyashiro, A., 1973, Paired and Unpaired Metamorphic Belts, *Tectonophysics* **17**:241-254.

Nairn, A. E. M., and F. G. Stehli, 1973, *The Ocean Basins and Margins*, vol. 1, *The South Atlantic*, Plenum Press, New York, 583 p.

Nairn, A. E. M., and F. G. Stehli, 1974, *The Ocean Basins and Margins*, vol. 2, *The North Atlantic*, Plenum Press, New York, 598 p.

Nairn, A. E. M., and F. G. Stehli, 1975, The Ocean Basins and Margins, vol.

3, *The Gulf of Mexico and the Caribbean*, Plenum Press, New York, 706 p.

Nairn, A. E. M., M. Churkin, Jr., and F. G. Stehli, 1981, *The Ocean Basins and Margins*, vol. 5, *The Arctic Ocean*, Plenum Press, New York, 672 p.

Narin, A. E. M., W. H. Kanes, and F. G. Stehli, 1977, *The Ocean Basins and Margins*, vol. 4a, *The Eastern Mediterranean*, Plenum Press, New York, 503 p.

Narin, A. E. M., W. H. Kanes, and F. G. Stehli, 1978, *The Ocean Basins and Margins*, vol. 4b, *The Western Mediterranean*, Plenum Press, New York, 447 p.

Presnall, D. C., J. R. Dixon, T. H. O'Donnell, and S. A. Dixon, 1979, Generation of Mid-ocean Tholeiites, *Jour. Petrology* **20**:3–35.

Ross, J. V., J. C. C. Mercier, J. G. Ave Lallement, N. L. Carter, and J. Zimmerman, 1980, The Vourinos Ophiolite Complex, Greece: The Tectonite Suite, *Tectonophysics* **70**:63–83.

Sengor, A. M. C., Y. Yilmaz, and I. Ketin, 1980, Remnants of a Pre-Late Jurassic Ocean in Northern Turkey: Fragments of Permian-Triassic Tethys?, *Geol. Soc. America Bull.* **91**:599–609.

Whitford, D. J., W. Compston, and I. A. Nicholls, 1977, Geochemistry of Late Cenozoic Lavas from Eastern Indonesia: Role of Subducted Sediments in Petrogenesis, *Geology* **5**:571–575.

Wood, D. A., J.-L. Joron, and M. Treuil, 1979, A Re-appraisal of the Use of Trace Elements to Classify and Discriminate Between Magma Series Erupted in Different Tectonic Settings, *Earth and Planetary Sci. Letters* **45**:326–336.

Wyllie, P. J., 1979, Petrogenesis and Physics of the Earth, in H. S. Yoder, Jr., ed., *The Evolution of Igneous Rocks—Fiftieth Anniversary Perspectives*, Princeton University Press, Princeton, N.J., pp. 483–520.

Reprinted from pages 551–552 of Am. Jour. Sci. **272**:551–576 (1972)

EVIDENCE FOR PLATE-TECTONIC REGIMES IN THE ROCK RECORD

WILLIAM R. DICKINSON

ABSTRACT. The plate-tectonic theory requires a fresh evaluation of the rock record of geologic history in terms of the characteristic petrotectonic assemblages associated with different kinds of plate junctures and modified by the tectonic consequences of the evolution of plate margins. As the oceanic lithosphere formed at divergent plate junctures is destroyed systematically by plate consumption at convergent junctures, the fragmentary record that remains from older plate-tectonic regimes is preserved only within the present continental blocks. The key petrotectonic assemblages are ophiolitic sequences of oceanic crust from divergent junctures and volcano-plutonic orogens of magmatic arcs from convergent junctures, but present distributions of these rocks are the result of complex plate motions, juncture migrations, and crustal collisions in the past.

Ophiolitic sequences of peridotite, gabbro, dolerite, basalt, and various sedimentary strata within orogenic belts represent residual upper mantle, partly metamorphosed igneous crust, and overlying sediments of oceanic lithosphere from vanished ocean basins. The rifted margins of continental fragments produced by continental separations that initiated the development of ocean basins are marked by continental terraces of miogeoclinal strata and their offshore facies equivalents deposited on adjacent oceanic crust. Exogeosynclinal wedges spread by dispersal of sediment backward across miogeoclinal assemblages are evidence for orogeny involving juxtaposition of a continental margin and a convergent plate juncture by either activation of or collision with an arc-trench system.

In arc-trench systems, melanges and imbricate slabs are formed by subduction at the trench, both basins and uplifts occur within the arc-trench gap, and volcanic chains are underlain by batholith belts along the magmatic axis; either thrust belts and foreland basins or marginal seas with oceanic structure occur in the backarc area. The belt between the trench and the arc is a sliver plate thermally detached from the main plate behind the arc: transform shear may occur along the arc, and divergent spreading or convergent subduction may occur at the rear of the arc. Although arc orogens include both oceanic and continental varieties, there is a consistent relation between increasing potash content (K) in the magmas with increasing depth (h) to the inclined seismic zone in the mantle beneath. The polarity of volcano-plutonic orogens is indicated by transverse gradients in K for coeval igneous rocks and also by the varying intensity of blueschist metamorphism, which increases toward the arc at the structural levels exposed in uplifted melange belts of old subduction zones. Melange belts may grow by lateral accretion as subduction continues, and migration of arc activity may shift the magmatic axis over prior sites of the subduction zone. Crustal collisions which represent successive steps in continental assembly bring sialic blocks that inhibit plate consumption against arc structures across the melange belts and ophiolitic shreds of subduction zones, which thus become suture belts along which strike-slip is common during and following the collision orogeny.

The nature of the oceanic lithosphere present during different times in the past was dependent on secular trends in the evolution of the crust-mantle system. The evolution of continental lithosphere preserved from consumption may have depended also upon unknown processes of aging with time. Plate-tectonic behavior is the unifying mechanism for crustal evolution by the formation of oceanic crust from mantle-derived magmas melted beneath rises and the formation of continental crust from arc magmas melted primarily off the tops of slabs of oceanic lithosphere descending along the inclined seismic zones.

INTRODUCTION

According to plate-tectonic theory (McKenzie, 1972), major tectonic activity is concentrated within elongate belts of juncture between nearly rigid slabs or plates of a strong outer rind of the Earth. Tectonism is confined mostly to these plate junctures because the plates are in relative motion with respect to one another and also with respect to a weaker and softer layer beneath. The outer rind is the lithosphere, and the undermass is the asthenosphere (Daly, 1940 after Barrell, 1914-1915). The base of the rigid lithosphere is probably coincident with the top of the low-velocity zone, a partly molten and probably mobile region of the mantle (Anderson, Sammis, and Jordan, 1971). The base of the crust at M is everywhere contained within the slabs of lithosphere.

The configurations of present plate margins are defined by belts of seismicity marking active deformation of the crust. The relative motions of slabs of lithosphere in contact along plate junctures can be described in terms of components of divergence, convergence, and shear. These three extremes of behavior have characteristic geologic expression as intra-oceanic rise crests, arc-trench systems, and transform faults, respectively. The fast rates (1-10 cm/yr) of relative plate motions and the unsystematic orientations of plate junctures imply that the geometry and nature of plate margins are not stable with time. The interpretation of geologic history in terms of the plate-tectonic model requires that past plate margins or junctures be identified in the rock record. This can be done by relating certain petrotectonic assemblages to sets of tectonic and associated geologic processes that occur only under specific plate-tectonic regimes (Dickinson, 1971d).

The importance of bringing plate-tectonic logic to bear on the rock record of tectonic history is shown by the fresh insights that the exercise brings to the following major topics of geology (*also see* Dewey and Horsfield, 1970):

1. The nature of the tectonic entities and events called geosynclines (Dickinson, 1971b) and orogenies (Dickinson, 1971c) and their mutual relations in geotectonic cycles or sequences (Coney, 1970) responsible for the development of mountain belts (Dewey and Bird, 1970).

2. The separation of continents by drift (Stewart, 1972), the growth of continents by accretion of materials to active continental margins (Hamilton, 1969), and the assembly of composite continents by crustal collision (Hamilton, 1970).

3. Paleogeographic patterns of landmasses, epeiric seas, and ocean basins with their implications for the history of organic evolution (Valentine and Moores, 1970) and the relations between epeirogeny and orogeny (Armstrong, 1969; Johnson, 1971).

4. The origins of igneous magmas (Dickinson, 1970; Green, 1971) and controls on the distribution of different kinds of magmatic provinces (Lipman, Prostka, and Christiansen, 1971).

5. The geochemical evolution of the crust-mantle system during geologic time (Ringwood, 1969; Dickinson and Luth, 1971).

As well as influencing the most fundamental concepts of geological science, these topics also bear directly on two significant aspects of economic geology: (A) the distribution of metallogenic provinces in space and time, and (B) the occurrence of sedimentary receptacles for fossil fuels.

[*Editor's Note:* In the original, material follows this excerpt. Only the references cited in the preceding excerpt are reproduced here.]

REFERENCES

Anderson, D. L., Sammis, C., and Jordan, T., 1971, Composition and evolution of the mantle and core: *Science,* v. 171, p. 1103–1112.

Armstrong, R. L., 1969, Control of sea level relative to the continents: *Nature,* v. 221, p. 1042–1043.

Barrell, J., 1914–1915, The strength of the earth's crust: *Jour. Geology,* v. 22, p. 655–683, v. 23, p. 425–443.

Coney, P. J., 1971, Cordilleran tectonic transitions and motion of the North American plate: *Nature,* v. 233, p. 462–465.

Daly, R. A., 1940, *Strength and structure of the earth:* Englewood Cliffs, N.J., Prentice-Hall, 434 p.

Dewey, J. F., and Bird, J. M., 1970, Mountain belts and the new global tectonics: *Jour. Geophys. Research,* v. 75, p. 2625–2647.

Dickinson, W. R., 1970, Relations of andesites, granites, and derivative sandstones to arc-trench tectonics: *Rev. Geophysics and Space Physics,* v. 8, p. 813–860.

Dickinson, W. R., 1971b, Plate tectonic models of geosynclines: Earth *Planetary Sci. Letters,* v. 10, p. 165–174.

Dickinson, W. R., 1971c, Plate tectonic models for orogeny at continental margins: *Nature,* v. 232, p. 41–42.

Dickinson, W. R., 1971d, Plate tectonics in geologic history: *Science,* v. 173, p. 107–113.

Dickinson, W. R., and Luth, W. C., 1971, A model for plate tectonic evolution of mantle layers: *Science,* v. 174, p. 400–404.

Green, D. H., 1971, Composition of basaltic magmas as indicators of conditions of origin with application to oceanic volcanisms: *Roy. Soc. London Philos. Trans.,* v. 268A, p. 707–725.

Hamilton, Warren, 1969, Mesozoic California and the underflow of Pacific mantle: *Geol. Soc. America Bull.,* v. 80, p. 2409–2430.

Hamilton, Warren, 1970, The Uralides and the motion of the Russian and Siberian platforms: *Geol. Soc. America Bull.,* v. 81, p. 2553–2576.

Johnson, J. G., 1971, Timing and coordination of orogenic, eperiogenic, and eustatic events: *Geol. Soc. America Bull.,* v. 82, p. 3263–3298.

Lipman, P. W., Prostka, H. J., and Christiansen, R. L., 1971, Evolving subduction zones in the western United States as interpreted from igneous rocks: *Science,* v. 174, p. 821–825.

McKenzie, D. P., 1972, Plate tectonics, in Robertson, E. C., ed., *The Nature of the Solid Earth (Birch Symposium):* New York, McGraw-Hill, N. Y., p.323–360.

Moores, E., 1970, Ultramafics and orogeny, with models of the U.S. Cordillera and Tethys: *Nature,* v. 228, p. 837–842.

Ringwood, A. E., 1969, Composition and evolution of the upper mantle, in Hart, P. J., ed., *The Earth's Crust and Upper Mantle:* Am. Geophys. Union Geophys. Mon. 13, p. 1–17.

Stewart, J. H., 1972, Late Precambrian continental separation in western North America on evidence from sedimentary and volcanic rocks: *Geol. Soc. America Bull.,* v. 83, in press.

36

FRONTAL CHAINS (ANDES) AND GEOSYNCLINAL CHAINS (ALPS)

J. Aubouin

This excerpt was translated expressly for this Benchmark volume by C. A. Schwab and F. L. Schwab, Washington and Lee University, from pages 459–460 of "Châines liminaires (Andines) et châines géosynclinales (Alpines)," in 24th Internat. Geol. Congr. Proc. section 3, 1972, pp. 438–461

CONCLUSIONS

1. The frontal (or marginal) type of mountain chain and the geosynclinal type of mountain chain evidently are idealized "models" susceptible to variations. However, they permit the evolution of chains to be analysed comparatively from the point of view of super-imposed paleogeography and tectonics and the phenomena that link them. Notably these models allow the tectonic history to be retraced permit-ting a comparison of corresponding structures.

2. Aside from the differences and similarities in detail, the most important feature of both examples is the comparatively recent history of Pliocene-Quaternary tectonics and its style of tensional faulting, irrespective of any later horizontal displacement (trans-current faulting). This tensional faulting style in both cases is quite novel in contrast with earlier deformation where compression dominates--field evidence, rather than *a priori* deduction--leads us to conclude that, at least at the surface, the present state of oro-genic belts is extensional. Consequently, observations of present-day deformation apply only to late stage, post-geosynclinal, post-frontal belt deformation and cannot be extrapolated for the past: in short, the principle of uniformitarianism does not appear to be applicable directly to tectonism. Thus, modern earthquakes that reveal so much about modern tectonism tell us nothing about the compressive movements that produced thrusting and overthrusting; in this sense, far from marking chains in the process of formation, they mainly indicate those in the process of being destroyed.

3. Consequently, comparisons of paleogeographic and tectonic models with oceanographic and geophysical models is rendered somewhat tenuous, because the latter are based on modern data and the former on past data. Ideally, it would help to have available all the paleo-geophysical data, but aside from paleomagnetic data that is enormously helpful in reconstructing continental mobility, it is not possible to envisage with any certainty paleogravimetric or paleoseismic data. One can *suppose* that such and such a paleogeographic zone was under-lain by oceanic crust (for example. as we have seen, eugeosynclinal troughs), but there is little chance that a sliver of such oceanic crust will be found preserved since later tectonic movements will

most often have obscured it. One must compare 'the models as they exist
presently, the frontal or marginal belts and the geosynclinal belts,
each with their distinctive types of oceanic and continental crusts
and siting relative to cratonic zones.

4. Based on the above analysis, one can envision frontal belts
and geosynclinal belts in the context of the present theory of sea-
floor spreading and plate tectonics that in any case, furnish a mechanism for the displacement of continents made necessary by the horizontal continuity of structural belts and the existence of a global orogenic chronology.

However, it seems unlikely that both marginal and geosynclinal
belts can be accounted for by an identical mechanism. The difference
is not due only to the collision of two continental masses that characterizes the second case; the two models differ from the very beginning
of their paleogeographical evolution. If, as is contended, marginal
belts (such as the Andes) result from the subduction of oceanic crust
beneath the continents around the periphery of the Pacific, this sliding
down is probably not identical with the process that produces geosynclinal chains. These seem instead to result from the spreading of an
ocean analogous to the Atlantic, with later compression occurring
between continental masses bordering the ocean. In short, one must
conclude that *marginal type belts characterize the Pacific borders
and geosynclinal belts are produced from paleo-oceans of the Atlantic
type*.

Such a reflection, however, calls for some reservations, the most
important being present-day extensional deformation. Extensional tectonism is seldom serious in geosynclinal chains--after having collided,
the continental masses would again simply begin to separate--tensional
rifting is more important in marginal belts by reason of internal logic
or reasoning. In effect, if one succeeds in accounting for the superficial distension in the Cordillera and in the corresponding peri-
Pacific trenches by subduction of oceanic crust beneath the South
American continent, one establishes a logical, albeit delicate, mechanism for Plio-Quaternary time; however, prior to that time the problem
is to establish surface compressions. The same mechanism of deep subduction of oceanic crust cannot produce surface results so diametrical-
ly opposite: tensional in one case, compressive in the other. Therefore, if the "Benioff plane" model is accurate for the present epoch,
it could not have "functioned" prior to the Pliocene. For example,
one cannot contend that the periods of oceanic expansion in the Pacific
through Miocene time had as their consequence the compressive structures of the Cordillera; thus accomplished, the ocean crust then dived
beneath the continent. The mechanism inferred, based on the existence
of a Benioff plane, does not give a satisfactory explanation for the
actual observations. Perhaps it is necessary to come back to the recent
theory of general global expansion, limiting it to Plio-Quaternary
time; prior to this time, in earlier epochs, this expansion was matched by compression in the orogenic belts. After all, it is not unimag-
inable that all the serious crises represented by 200 million years
of orogenic episodes could have been terminated by a period of global
expansion that might also explain many other global peculiarities,
such as climatic anomalies, which generally characterize the ends of
orogenic cycles.

Reprinted from pages 35–38 of *Geol. Soc. America Mem. 132*, 1972, pp. 35–55

Origin of Lithosphere behind
Island Arcs, with Reference
to the Western Pacific

RALPH MOBERLY

Hawaii Institute of Geophysics, Honolulu, Hawaii 96822

ABSTRACT

The sea floor inside island arcs characteristically is less deep and has higher heat flow than the ocean floor outside the arc and trench system. Direct evidence from drilling and indirect evidence based on thin sediment cover, interrupted geologic trends, paleomagnetic studies, and fitting of pre-drift continental margins show that the lithosphere behind island arcs is young and commonly did not form on the mid-oceanic ridge system. The slab of dense lithosphere that flexes and sinks spontaneously through the asthenosphere under arcs is shown to sink at an angle that is steeper than the plane of the earthquakes. As a consequence, the trench and arc migrate seaward against the retreating line of flexure of the suboceanic lithosphere. Part of the warm asthenosphere pushed aside by the plunging slab migrates up by creep and as magma, then cools and forms new lithosphere in the extensional region behind the advancing island arc. Extension is favored where the lithospheric plate behind the arc is moving tangentially or away from the plate outside the arc.

A series of maps shows the tectonic development of the western Pacific from mid-Eocene to the present. The maps are based on concepts developed from sea-floor spreading and the new global tectonics, and incorporate the postulate that new lithosphere can form behind advancing island arcs. The origin and later deformation of arcs and basins are shown as resulting mainly from the great shear between the northward-moving Australian plate and the northwestward-moving Pacific plate.

INTRODUCTION

The radio message reporting the death of Harry Hess was sent to us on board the drilling vessel *Glomar Challenger* by Professor A. G. Fischer, while we were working in the Caroline Basin in August 1969. The shock of the news was intensified by the fact that his name had been on our lips repeatedly. The bathymetry and tectonic elements of that part of the western Pacific were described in a classic paper by Hess (1948). The concept that the ocean basins are formed at, and spread from, the ridge system (Hess, 1962, 1965) had been amply supported by cores obtained during earlier legs of the *Glomar Challenger,* but Leg 6 in the western Pacific (Fischer and others, 1970) and our own Leg 7 (Winterer and others, 1969) had been finding evidence of young oceanic crust that apparently had not formed on part of the mid-ocean ridge system.

During the ensuing months any reminder of Harry's passing brought back to mind that evidence, the topic of this presentation. As a point of departure, it is assumed that the reader is familiar with the main points in the growth of global tectonic theory during the 1960s, as summarized for example by Phinney (1968), or by Isacks and others (1968). I will especially hold to the following two premises. (1) The lithosphere moves as great blocks (McKenzie and Parker, 1967; Morgan, 1968), whose rate and direction of movement can be determined from studies of seismology, geomagnetism, and structural geometry (*see,* for example, Isacks and others, 1968; Vine, 1966; Heirtzler and others, 1968; Menard and Atwater, 1968), so that past positions of the lithospheric blocks or plates can be approximated (Le Pichon, 1968). (2) The lithosphere is denser than the asthenosphere, a suggestion going back at least to Daly (1925) and supported today by interpretations of mechanical properties of the lithosphere (Elsasser, 1968), distribution of satellite-determined gravity anomalies (Moberly and Khan, 1969), stress orientation from earthquake-mechanism solutions (Isacks and Molnar, 1969), and randomly generated earth models (Press, 1969; for a contrary view, *see* Wang, 1970). Bass (1969, unpub. data) and Laubscher (1969) have commented on the general tectonic effects of a dense, spontaneously sinking lithosphere.

YOUTHFUL CRUST BEHIND ISLAND ARCS

Location

In recent years interpretations of the geomagnetic stripe patterns have shown that the greatest part of the deep-sea floor has formed along the oceanic rise or mid-ocean ridge system (Vine and Matthews, 1963; Wilson, 1965; Heirtzler and others, 1968). There are exceptions (Fig. 1).

The computer-fitting of the edges of continental masses now separated by ocean has been a powerful means of reconstructing paleogeography before a particular episode of sea-floor spreading. Often the reconstructions are aided by restoring a smaller block that presumably has rotated (Spain: Bullard and others, 1965) or split away (east Canary Islands: Dietz and Sproll, 1970) from the main continental block. For the Atlantic fit, however, North and South America are close together. Even if the Gulf of Mexico was formed by fragmentation and drifting of small blocks that became Honduras, Cuba, and so on, as proposed by Bass (1968, unpub. data), there is still the embarrassment of accounting for the sea floor of the Caribbean, which must have formed after the initiation of drift (Funnell and Smith, 1968).

Karig (1970) concluded from its thin sediment cover that the Lau-Havre Basin west of the Tonga-Kermadec Ridge was a youthful feature, where new sea floor formed by extensional rifting very late in the Cenozoic. He also suggested that

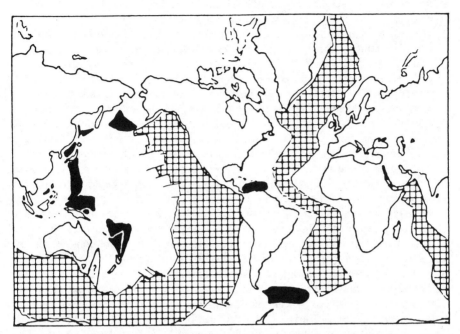

Figure 1. Suboceanic lithosphere formed in Cenozoic times. Cross-ruling: formed on mid-ocean ridge system (after Vine, 1968); black: known or suspected of having formed behind advancing island arcs.

several marginal basins on the concave side of island arcs, characterized by higher heat flow (Vacquier and others, 1966) and thinner sediments than might otherwise be expected, could be other regions of new oceanic crust. What is known of the attitude, amplitude, and nonsymmetry of magnetic anomalies of the sea floor behind the Scotia arc (Kroenke and Woollard, 1968) suggests that it is not forming from a part of the oceanic rise system.

It seems, therefore, that besides the places in the western Pacific where drilling has shown young oceanic crust to be present, there are other places where it is reasonable to conclude that young crust has also formed. These areas are behind island arc systems at the present day, or as will be shown below, almost certainly were behind arcs in mid-Cenozoic time.

Origin

Oceanic crust behind island arcs has been attributed by some to foundering of continental crust (Belousov, 1967), a concept that many others consider difficult to support on petrogenic and isostatic grounds. Several workers have advanced theories that the arcs are shaped around rising convection currents (shallow but powerful: Holmes, 1965; subsidiary circulation: Wright, 1966; the Melanesian Rise: Menard, 1964). In the present-day context of global tectonics, it would follow that the crust (or more properly, the lithosphere, or tectosphere, composed of crust and uppermost mantle) would have formed at the crest of those small rises. The

average bathymetry and heat flow certainly do support the concept that the lithosphere is new, but actual patterns of bathymetric and magnetic trends are difficult
to reconcile with plate theory and orthogonal spreading.

The post-Miocene Andaman Basin is explained by Rodolfo (1969) as a rhombo-
chasm that resulted from a south-southwestward movement of southeast Asia.

Oxburgh and Turcotte (1970) believe that the trench and front part of an arc
migrate oceanward as a consequence of the addition of sedimentary material to
the front part of the island arc system, but that the migration is rather small.

McKenzie and Sclater (1968) discussed two possible sources of the high heat
flow inside the island arcs of the northwestern Pacific. One was the upward movement of magmas and volcanism, but they argued that an enormous volume of
volcanic material would be required to maintain the heat flow anomaly and the
rate of intrusion and extrusion would be very high. They calculated, for instance,
that the crust in the Sea of Japan would have formed in about the last 10 m.y.
Because they believed that the trench and island arc features of the entire north-
western Pacific are unlikely to be so recent, McKenzie and Sclater hesitated to
assign the unexpectedly high heat flow to a source from the upward movement of
magma. Their text is not entirely clear, but apparently they were considering only
andesitic volcanism (McKenzie and Sclater, 1968, p. 3176–3177), and they did
point out the great difficulties of dating deformation in island arcs, where old
deformed rocks may have either been formed by the present arc or have been
carried in from elsewhere by moving lithosphere.

Of special interest is Karig's (1970) paper in which he has demonstrated the
strong probability that only a particular segment ("interarc basin" or the Lau-
Havre Trough) of the region behind the Tonga-Kermadec arc has been newly
formed. The generation and extension of new oceanic lithosphere without mid-
ocean ridges follows behind the "frontal" (volcanic and coralline) island arc and
trench, which are visualized as migrating eastward. Karig postulates the extension
as having been related to the initiation of the present episode of island arc volcanism, which he places in the late Tertiary at no more than 5 m.y. ago. Other
marginal basins behind island arcs of the western Pacific are also held to be active
or previously active extensional areas. Karig does not, however, offer a mechanism
by which the island arc is able to advance against the oceanic plate that is spreading
toward it. Specifically for his area of major work (Karig, 1970), he does not
explain how the Tonga-Kermadec Ridge can migrate eastward from the Australian
plate against the Pacific plate that is spreading west-northwestward from the East
Pacific Rise and into the Tonga-Kermadec Trench.

In order to integrate these observations of Karig, and those of Fischer, and of
others, into the model of global tectonics, two explanations are necessary. One
of these is the development of a generalized account that would explain how new
lithosphere might form behind island arcs. The other is a specific application to
the varied geometry of plate boundaries and spreading vectors and to the geologic
and geophysical patterns that actually exist.

[*Editor's Note:* Material has been omitted at this point. Only the references cited in the preceding excerpt are reproduced here.]

REFERENCES

Beloussov, V. V., 1967, Some problems concerning the oceanic earth's crust and upper mantle evolution: *Geotectonics*, v. 1, p. 1–6.

Bullard, E. C., Everett, J. E., and Smith, A. G., 1965, The fit of the continents around the Atlantic: *Royal Soc. London Philos. Trans.*, Ser. A, v. 258. p. 41–51.

Daly, R. A., 1925, Relation of mountain building to igneous action: *Am. Philos. Soc. Trans.*, v. 64, p. 283–307.

Dietz, R. S., and Sproll, W. P., 1970, East Canary Islands as a microcontinent within the Africa-North America continental drift fit: *Nature*, v. 226, p. 1043–1045.

Elassner, W. M., 1968, Submarine trenches and deformation: *Science*, v. 160, p. 1024.

Fischer, A. G., Heezen, B. C., Boyce, R. E., Bukry, D., Douglas, R. G., Garrison, R. E., Kling, S. A., Krasheninnikov, V., Lisitzin, A. P., and Pimm, A. C., 1970, Geological history of the western North Pacific: *Science*, v. 168, p. 1210–1214.

Funnell, B. M., and Smith, A. G., 1968, Opening of the Atlantic Ocean: *Nature*, v. 219, p. 1328–1333.

Heirtzler, J. R., Dickson, G. O., Herron, E. M., Pitman, W. C., III, and Le Pichon, X., 1968, Marine magnetic anomalies, geomagnetic field reversals, and motions of the ocean floor and continents: Jour. Geophys. Research, v. 73, p. 2119–2136.

Hess, H. H., 1948, Major structural features of the western North Pacific, an interpretation of H. O. 5485, bathymetric chart, Korea to New Guinea: *Geol. Soc. America Bull.*, v. 59, p. 417–466.

Hess, H. H., 1962, History of ocean basins, in Engel, A. E. J, James, H. L., and Leonard, B. F., eds., *Petrologic studies* (Buddington volume): Geol. Soc. America, p. 559–620.

Hess, H. H., 1965, Mid-oceanic ridges and tectonics of the sea-floor, in Whittard, W. F., and Brawshaw, R., eds., *Submarine geology and geophysics*: London, Butterworths, p. 317–333.

Holmes, A., 1965, *Principles of physical geology* (2d ed.): New York, Roland Press Co., 1288 p.

Isacks, B., and Molnar, P., 1969. Mantle earthquake mechanisms and the sinking of the lithosphere: *Nature*, v. 223, p. 1121–1124.

Isacks, B., Oliver, J., and Sykes, L. R., 1968, Seismology and the new global tectonics: *Jour. Geophys. Research*, v. 73, p. 5855–5899.

Karig, D. E., 1970, Ridges and basins of the Tonga-Kermadec island-arc system: *Jour. Geophys. Research*, v. 75, p. 239–259.

Kroenke, L. W., and Woollard, G. P., 1968, *Magnetic investigations in the Labrador and Scotia Seas, USNS Eltanin cruises 1–10, 1962–1963*: Hawaii Inst. Geophysics Rept. 68–4, 59 p.

Laubscher, H., 1969, Mountain building: *Tectonophysics*, 7, p. 551–563.

Le Pichon, X., 1968, Sea-floor spreading and continental drift: *Jour. Geophys. Research*, v. 73, p. 3661–3697.

McKenzie, D. P., and Parker, R. L., 1967, The North Pacific: An Example of tectonics on a sphere: *Nature*, v. 216, p. 1276–1280.

McKenzie, D. P., and Sclater, J. G., 1968. Heat flow inside the island arcs of the north-western Pacific: *Jour. Geophys. Research*, v. 73, p. 3173–3179.

Menard, H. W., 1964, *Marine geology on the Pacific*: New York, McGraw-Hill Book Co., 271 p.

Menard, H. W., and Atwater, T., 1968, Changes in direction of sea-floor spreading: *Nature*, v. 219, p. 463–467.

Moberly, R., Jr., and Khan, M. A., 1969, Interpretation of the sources of the satellite-determined gravity field: *Nature*, v. 223, p. 263–267.

Morgan, W. J., 1968, Rises, trenches, great faults, and crustal blocks: *Jour. Geophys. Research*, v. 73, p. 1959–1982.

Oxburgh, E. R., and Trucotte, D. L., 1970, Thermal structure of island arcs: *Geol. Soc. America Bull*, v. 81, p. 1665–1688.

Phinney, R. A., 1968, Introduction, in Phinney, R. A., ed., *The history of the Earth's crust*, Princeton, Princeton Univ. Press, p. 3–12.

Press. F., 1969. The Sub-oceanic mantle: *Science*, v. 165, p. 174–176.

Vacquier, V., Uyeda, S., Yasui, M., Sclater, J., Corrie, C., and Watanabe, T., 1966, Heat flow measurements in the northern Pacific: *Tokyo Univ. Earthquake Research Inst. Bull.*, v. 44, p. 1519–1635.

Vine, F. J., 1966, Spreading of the ocean floor; new evidence: *Science*, v. 154, p. 1405–1415.

Vine, F. J., 1968, Evidence from submarine geology: *Am. Philos. Soc. Proc.* v. 112, p. 325–334.

Vine, F. J., and Matthews, D. H., 1963, Magnetic anomalies over oceanic ridges: *Nature*, v. 199, p. 947–949.

Wang, C. Y., 1970, Density and constitution of the mantle: *Jour. Geophys. Research*, v. 75, p. 3264–3284.

Wilson, J. T., 1965, A new class of faults and their bearing on continental drift: *Nature*, v. 207, p. 343–347.

Winterer, E. L., Riedel, W. R., Moberly, R., Jr., Resig, J. M., Kroenke, L. W., Gealy, E. L., Heath, G. R., Bronnimann, P., Martini, E., and Worsley, T. R., 1969, Deep Sea Drilling Prospect: Leg 7: *Geotimes*, v. 14, p. 14–15.

Wright, J. B., 1966, Convection and continental drift in the Southwest Pacific: *Tectonophysics*, v. 3, p. 69–81.

38

Reprinted from pages 75, 83–85 of Geol. Soc. America Mem. 132, 1972, pp. 75–85

Uplifted Eugeosynclines and Continental Margins

C. A. Burk

Mobil Oil Corporation, Princeton, New Jersey 08540

ABSTRACT

Mesozoic eugeosynclinal sequences of turbidites with radiolarian cherts, pillowed basalts, and ultramafic rocks, appear to characterize much of the exposed Pacific continental margins and much of the Tethyan tectonic belts. Extremely great stratigraphic thicknesses have been reported for many of these deep-ocean sequences.

These eugeosynclinal rocks have had a complex history of penecontemporaneous deformation and subsequent tectonic displacements, and have been uplifted and added to the margins of the continents. Based on studies of the southern continental margin of Alaska, the apparently great thicknesses and the subsequent uplift of these eugeosynclinal sequences seem best explained by deposition in oceanward-migrating trenches and the repeated landward uplift of the sedimentary fill in these successive trenches.

INTRODUCTION

The uplifted margins of many continents consist of thick sequences of eugeosynclinal sediments, presumed to have been deposited at great oceanic depths. This is particularly true of Mesozoic rocks around the periphery of the Pacific Ocean. Since these deep-water sediments are now widely exposed at continental margins, they must have been uplifted by an amount at least equal to their stratigraphic thickness plus the original oceanic depth of the crust on which they were deposited.

According to present concepts, based on sea-floor spreading (Hess, 1962), much of the Pacific margin is the site of convergent and downward-directed forces, which may have provided both the depressed oceanic trenches in which the eugeosynclinal deposits accumulated and the stresses to account for their severe deformation. Within such a tectonic framework, the nature of any large-scale and repeated uplift becomes particularly important.

325

[*Editor's Note:* Material has been omitted at this point.]

PACIFIC EUGEOSYNCLINAL SEQUENCES AND POSSIBLE
RELATION TO GLOBAL TECTONICS

The regional relations described above seem most reasonably accounted for by successive uplifting of sediments accumulated in deep oceanic trenches, which cumulatively have provided the very thick Mesozoic and earliest Tertiary eugeosynclinal sequences now exposed on the southern Alaska continental margin. Similar

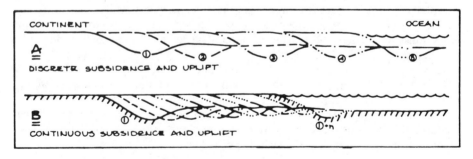

Figure 5. (a) Repeated sequence of tectonic depression of the oceanic trench, filling with sediments, and uplift; and (b) the possibly continuous sequence of oceanward migration of the trench with continuous uplift of the landward side of the trench.

histories may basically be applicable to other Pacific margins (always with necessary local modifications).

It is difficult, however, to reconcile these long steep faults of southern Alaska, spaced at irregular intervals and separating rocks of apparently discrete ages, with the postulated continuous filling of trenches as oceanic crust is continuously thrust underneath the continents. Marine magnetic surveys indicate that this Aleutian trench system has been the site of continued lithospheric plate consumption (for example, Grim and Erickson, 1969) and is not uniquely different from other Pacific margins. The high-pressure blueschist metamorphism typical of much of the peripheral Pacific has not yet been reported in the eugeosynclinal sequence of southwestern Alaska (the regional metamorphic grade does not appear to exceed zeolite and hornfels facies). It is also apparent that the oldest part of the eugeosynclinal sequence is adjacent to contemporaneous, shallow-water early Mesozoic sediments of the Alaska Peninsula.

Holmes and others (1970) have pointed out that recent reflection seismic surveys of the Aleutian Trench south of Amchitka Island have shown deformation of the trench sediments at the northern wall, yielding local folds of 500 m wavelength and 30 m amplitude, comparable to those observed in present exposures (J. C. Moore, 1971); and von Huene and Shor (1969) have shown that the oceanic crust on the seaward edge of the trench has been depressed, with its overlying sequence of pelagic deposits, to unconformably underlie underformed turbidite deposits of the present Aleutian Trench.

All of these factors are compatible with present concepts of sea-floor spreading and global plate tectonics, and with the observed data along the continental margin of the North Pacific Ocean. A major problem would be the apparent effect of seaward migration and uplift of the related trench system, as described above, with the geometry of subduction and Benioff Zones in the deep crust and upper mantle. It is obviously possible that the periodic uplift indicated in the successive trenches off southern Alaska is isostatic and represents the relaxation of tectonic stresses.

In any case, the seaward migration and successive marginal uplift of oceanic trenches and the presumed decrease in the dip of the Benioff zone must be considered as a major factor in any interpretation of global tectonics.

ACKNOWLEDGMENTS

I gratefully acknowledge useful discussions with J. Casey Moore and George Plafker, in addition to discussions with my associates at Princeton University and at Mobil Regional Geology.

REFERENCES CITED

Bailey, E. H., Erwin, W. P., and Jones, D. L., 1964, Franciscan and related rocks, and their significance in the geology of western California: California Div. Mines and Geology Bull. 183, 177 p.

Burk, C. A., 1965, Geology of the Alaska Peninsula, island arc and continental margin: Parts I and II: Geol. Soc. America Mem. 99, 250 p., 3 maps.

—— 1966, The Aleutian Arc and Alaska continental margin, in Continental margins and island arcs: Canada Geol. Survey Pub. 66–15, p. 206–215.

Drake, C. L., Ewing, M., and Sutton, G. H., 1959, Continental margins and eugeosynclines; the east coast of North America north of Cape Hatteras, *in* Ahrens, L. H., and others, eds., Physics and chemistry of the Earth, v. 3: New York, Pergamon Press, p. 110–198.

Grim, P. J., and Erickson, B. H., 1969, Fracture zones and magnetic anomalies south of the Aleutian Trench: Jour. Geophys. Research, v. 74, no. 6, p. 1488–1494.

Hess, H. H., 1955, Serpentines, orogeny, and epeirogeny, *in* Poldervaart, A., ed., Crust of the Earth: Geol. Soc. America Spec. Paper 62, p. 391–407.

—— 1962. History of the ocean basins, *in* Engel, A. E. J., and others, eds., Petrologic studies: A volume in honor of A. F. Buddington: Geol. Soc. America, p. 599–620.

—— 1965, Mid-oceanic ridges and tectonics of the sea-floor *in* Wittard, W. F., and Bardshaw, R., eds., Geophysics: London, Butterworths, p. 335–362.

Holmes, M. L., von Huene, R., and McManus, D. A., 1970, Underthrusting of the Aleutian Arc by the Pacific Ocean floor near Amchitka Island [abs.]: Am. Geophys. Union Trans., v. 51, no. 4, p. 330.

King, P. B., 1969, Tectonic map of North America: Washington, D. C., U. S. Geol. Survey.

Kuenen, Ph. H., 1964, Deep-sea sands and ancient turbidites, *in* Developments in sedimentology, v. 3: Amsterdam, Elsevier, p. 1–33.

Martin, G. C., 1915, Geology and mineral resources of Kenai Peninsula, Alaska: U. S. Geol. Survey Bull. 587, 243 p.

Moffit, F. H., 1954, Geology of the Prince William Sound region, Alaska: U. S. Geol. Survey Bull. 989–E, p. 225–310.

Moody, J. D., 1966, Crustal shear patterns and orogenesis: Tectonophysics, v. 6, no. 3, p. 479–522.

Moore, G. W., 1967, Geologic map of Kodiak Island and vicinity, Alaska: U. S. Geol. Survey open-file map.

—— 1969, New formations on Kodiak and adjacent islands, Alaska: U. S. Geol. Survey Bull. 1274–A, p. 27–35.

Moore, J. Casey, 1971, Geologic studies of the Cretaceous(?) flysch, southwestern Alaska [Ph.D. thesis]: Princeton, Princeton Univ.

—— 1972, Uplifted trench sediments: southwestern Alaska–Bering shelf edge: Science, v. 175, p. 1103–1105.

Murray, C. G., 1969, The petrology of the ultramafic rocks of the Rockhampton district, Queensland: Queensland Geol. Survey Pub. no. 343, 9 p.

Mutch, A. R., 1957, Facies and thickness of the upper Paleozoic and Triassic sediments of Southland: Royal Soc. New Zealand Trans., v. 84, no. 3, p. 499–511.

Plafker, G., 1967, Surface faults on Montague Island associated with the 1964 Alaska earthquake: U. S. Geol. Survey Prof. Paper 543–G, 42 p.

—— 1969, Tectonics of the March 27, 1964 Alaska earthquake: U. S. Geol. Survey Prof. Paper 543–I, 74 p.

Plafker, G., and MacNeil, F. S., 1966, Stratigraphic significance of Tertiary fossils from the Orca Group in the Prince William Sound region, Alaska: U. S. Geol. Survey Prof. Paper 550–B, p. 62–68.

Stoneley, R., 1969, Sedimentary thicknesses in orogenic belts, *in* Kent, P. E., and others, eds., Time and place in orogeny: London Geol. Survey Spec. Pub. 3, p. 215–238.

von Huene, R., and Shor, G. G., Jr., 1969, The structure and tectonic history of the eastern Aleutian Trench: Geol. Soc. America Bull., v. 80, no. 10, p. 1889–1902.

Manuscript Received by the Society March 29, 1971

39

*A paper presented at the 9th International Congress
of Sedimentology, Nice, 1975*

CONTINENTAL MARGIN SEDIMENTATION

R. W. Fairbridge

An attempt is made here to relate the sedimentary suites or associations of continental margins to plate tectonics in terms of traditional "geosynclinal" terminology, using the basic Stille-Kay definitions, i.e. that a geosyncline is any site of thick sedimentary
accumulation. This definition is in line with the original concepts of
Hall and Dana, but in no way excludes the miogeocline and eugeocline
notions of Dietz.

Two distinctive settings are known for continental margins, those
parallel or concordant to the grain of the coastal structures (Cordilleran or Pacific-type), and those transverse or discordant to the
structural grain (Atlantic-type). Nearly a century ago Suess (1885)
recognized that the world's principal continental *coastlines* could be
divided into two categories that he named Atlantic type (or "discordant," faulted and transverse to main structures) and Pacific type
(or "concordant," and parallel to the geotectonic trends). There are
also various petrographic and other tectonic correlations (Stille,
1958; Fairbridge, 1966, 1968).

The same labels can be applied to continental *margins,* provided
it is understood that one may have a "Pacific-type" margin in the
Atlantic (e.g. off the Lesser Antilles), and equally well an "Atlantic-
type" margin in the marginal seas of the Pacific (e.g. off eastern
Australia), as pointed out by Gregory (1932). There is also a "Pacific-
type" margin today off the Hellenic arc in the eastern Mediterranean;
in the Indian Ocean both types are well represented. In the last decade,
since the recognition of continental separation by plate tectonics,
the term *Atlantic-type margin* has become widely used, and contrasted
to *Cordilleran-type margin,* a term proposed long ago by Von Humboldt
for the western borders of the two Americas and possessing some advantages over the more ambiguous term "Pacific-type." (Dewey in 1969 spoke
of the "Andean-type," but Von Humboldt's term has a more general application, inasmuch as active subduction is not required.)

Within these two basic categories of continental margins many finer
distinctions may be recognized--*stages:* abrupt margin, stepped margin,
continental borderland (horst and graben type).

Plate tectonics recognizes continental margins that have been
subject either to rifting (Atlantic-type) or to subduction (Cordilleran or Pacific-type). The first is subject to extensional stresses,
resulting in crustal stretching with warping, block-faulting and even-

tually total rifting and drifting apart (e.g. Red Sea, Western Atlantic, Bahamian types). The second type (e.g. circum-Pacific, Indonesian, Hellenic types) is marked by discontinuous orogenic events (runaway thermal pulses in the mantle and stick-slip mechanisms in subduction), eventually leading to inter-continental collision. Both settings are associated with distinctive sedimentary and igneous suites, developed in recognized sequences. Only the Cordilleran category occurs in pre-orogenic, synorogenic, postorogenic phases: the Atlantic sequence is essentially preorogenic and ceases when subduction is initiated or intercontinental collision takes place. Microplate situations (e.g. in "back-arc basins" as in the Pacific marginal seas, in the Caribbean, the Aegean, etc.) present complex attributes.

SEDIMENTATIONAL CONDITIONS

Geotectonically speaking, by definition a continental margin is a region adjacent to the junction of continental ("sialic") and oceanic ("simatic") crust. It does not refer to shorelines as such or to the boundaries of epicontinental seas. Sediments of the continental margin may include the paralic, neritic and bathyal suites, corresponding therefore to the coastal plain, continental shelf, slope, and rise associations. The hadal or abyssal environment is largely restricted to regions of oceanic crust. By convention the continental margin is taken as 2000m (half the isostatic "freeboard" of the continental crust: Carey, 1958), but attenuated remnants of continental crust may be encountered at any oceanic depth (e.g. in the Seychelles region). During the sedimentary evolution along a continental margin, the shelf break, slope and rise facies eventually prograde across the continental/oceanic crust boundary, except where active subduction leads to down-warping (oceanic trench margins). Nevertheless although most orogenic belt geosynclinal sediments seemed at first sight to be ensialic, their axial zones with traces of ensimatic foundations are now becoming widely recognized (Lockwood, 1972; Helwig, 1974).

Much of the sediment that is now exposed in continental regions was once laid down on continental margins, and systematic study of the geochronology of the continental and oceanic crusts has shown that through geologic time continents have grown by accretion, as well as having been periodically subdivided and sutured by various mechanisms of plate tectonics. One may say therefore that the continental margin sedimentation of today builds, by orogeny, the continental crust of tomorrow (in a time-frame of 10^8 years). Orogeny is understood here as a major compressional diastrophism, as distinct from epeirogeny, that is warping in the vertical sense, and taphrogeny or extensional tectonics, usually represented by block-faulting. Although the question is still controversial, plate tectonic history seems to confirm the concept that periodic orogenic "revolutions" occur, often having several "phases," usually as crescendos or paroxysms that vary somewhat in time or place. Fairbridge and Rice (1974) propose a major rhythm of 85my, engendered by thermal runaway cycles in mantle convection.

STAGES IN MARGINAL GROWTH

In a genetic sense, it seems evident that an Atlantic-type margin develops only an essentially preorogenic phase. A useful symposium on the new approach to the sedimentation of the Atlantic-type margin was presented by Stanley (1969). This type is initiated during an extensional crustal regime, shown in the sediment first by an autogeo-synclinal phase (shallow basin sediments, continental or neritic) followed by a bisymmetric taphrogeosynclinal or aulacogenic sequence (fanglomerates, arkoses, evaporites, etc. cf. Hoffman, 1972). (N.B. The writer employs the terminology for sites of sedimentation recommended by Kay, 1951.) This block-faulting and rifting is followed by continental separation and drifting, sedimentation becoming paraliageosynclinal and an asymmetric polarity develops. As continental separation proceeds, an originally symmetric sedimentary pattern gradually becomes stretched out, so that sooner or later, for each matching continental margin, depending upon the hinterland, relief, and paleoclimates, a large and potentially unstable sedimentary wedge complex ("miogeocline") will develop. Eventually it seems likely that the building of the miogeo-clinal wedge will create a zone of weakness in the lithosphere inviting the initiation of subduction during a subsequent surge of plate tec-tonism. Thus an orthogeosynclinal cycle is initiated, and the region will eventually become an orogenic belt, thus automatically instituting a Cordilleran-type sequence of events (Dietz and Holden, 1967, 1974; Dewey, 1969; Dewey and Bird, 1970). The sedimentary suites or associa-tions would then change progressively from the paralic ("paraliageo-synclinal") and neritic to bathyal, and even abyssal. The Atlantic phase ceases in any case if intercontinental collisions occur.

In contrast, the Cordilleran-type margin is typically *syn-orogenic* or *post-orogenic*, but in reactivated belts (e.g. the Gulf of California) it can be *pre-orogenic*. In the Cordilleran setting we recognize the actualistic analogues of the flysch (Reading, 1972; Stanley, 1974) and molasse associations. Here too, in the orogenic axial zones (subduction sites) come the hadal complexes with their structural melanges, corre-sponding to the synorogenic and early postorogenic stages. The unique characteristics of the circum-Pacific belt with its "andesite line" and distinctive trenches has long been recognized, even in the "fixis-tic" camp; a specific "thalassogeosyncline" was proposed by Bogdanov (1969) to designate its distinctive sedimentologic setting. As noted above, however, these attributes also apply equally to certain limited sectors of the other oceans (e.g. Antillean, Scotian and Hellenic arcs; also Indonesian--Sunda, Banda-arcs). Most of these trenches are eugeo-clines in the sense of Dietz (1963) and eugeosynclines in the sense of Stille and Kay. From the sedimentologic viewpoint the difference is rather semantic, but in the geotectonic sense the difference is significant.

In the "back-arc depressions" which developed subsequent to the initial phases, block-faulting often develops and alternating platform and basin facies appear, as seen today in the Aegean Sea or in the Southern California borderland. In the more complex and evolved cases of the back-arc basins, both inward-facing subduction and renewed sea-floor spreading may occur, and a new cycle is initiated, as in the marginal seas of the Western Pacific (Japan Sea, South China Sea, etc.). In the Tasman Sea, for instance, the orogenic crescendo was late Permian; general taphrogeny began in the Triassic, and sea-floor spread-ing took place during the Cretaceous and Tertiary. In a comparable setting but in warmer latitudes, the Coral Sea underwent the same his-tory, but was marked by the development of reef facies with typical carbonate platforms of Bahamian type.

Examples will be given below of several types of pre-, syn- and postorogenic continental margin situations (See Table 1).

PREOROGENIC, ATLANTIC-TYPE MARGINS

Much attention has been paid to the sedimentary conditions antecedent to and related to the initiation of a sea-floor spreading cycle.
Several types or stages of margin evolution within these cyclic histories may be studied, but the clearest are provided by contemporary
prototypes, and the following examples are presented more or less in
the sense of an evolutionary sequence.

Red Sea (Erythrean) Type: Crustal attenuation in the Red Sea area
began in the mid-Tertiary, perhaps 40my ago. Downwarping led to a
regional transgression with autogeosynclinal facies (continental,
neritic and evaporite), followed by the growth of an aulacogenic wedge
from the Gulf of Aden. Block-faulting especially in the Afar Triangle
was marked by taphrogeosynclinal facies, bioherms, more evaporites
and basic volcanics. Continental separation and drifting have opened
a medial abyssal rift and sea-floor spreading here should, in the
future, produce a Carey-style sphenochasm (1958).

Western Atlantic: Two 85my cycles back, around 170my ago, important aspects of the "Red Sea story" were being enacted between north-
west Africa and the Atlantic margin of North America (Ryan, 1974).
Geophysical work here by Drake et al. (1959) showed a double sedimenta-
tional belt had developed, interpreted in terms of an orthogeosynclinal
complex. Dietz (1963) felt that actualistic interpretation required a
term recognizing its one-sided geometric nature and "miogeocline" was
proposed. The evolutionary record shows, however, that since Jurassic
times there has been sedimentation here controlled by an alteration of
high and low level seas (thalassocratic, epeirocratic: either eustatic
or geoidal--Fairbridge, 1961), so that the twin sedimentary sites alternate between paraliageosynclinal and miogeosynclinal (miogeoclinal)
facies. In the hinterland, the Appalachian belt, peneplanation surfaces
have been discontinuously emergent, the interruptions apparently corre-
lating with minor cycles or pulses in sea-floor spreading (Wilson and
Fairbridge, 1971). This progressive but pulsatory relief increase is
global in effect and is naturally reflected globally by the pelagic/
terrigenous sediment ratios offshore. Regressions of major proportions
(e.g. in mid-Oligocene and late Miocene) have favored sub-aerial canyon
cutting to great depths (Whitaker, 1974) and important phases of turbidite production. It should be stressed, however, that the Western
Atlantic setting does not generate a typical flysch facies as sometimes
claimed.

Bahama Platform Type: In the southern section of the Western
Atlantic margin, two additional factors come into play: a broader,
block-faulted basement (cf. Afar Triangle) and warmer latitudes. A
biohermal, carbonate platform complex has resulted (Dietz et al., 1970).
Initial oil exploration bores and subsequent deep-sea drilling have
disclosed a pattern remarkably suggestive of the northern Mediterranean
Mesozoic situation (Bernoulli and Jenkyns, 1974; Dewey, et al., 1973;
D'Argenio, 1970, 1971).

CORDILLERAN-TYPE MARGINS

The Cordilleran or "Pacific-type" continental margins of the world
at the present time are expressed in three stages: (a) synorogenic,
marked by trenches, with active or recent subduction, (b) postorogenic,
marked by block-faulting of recently developed crust and opening of

"back-arc basins" and (c) preorogenic, a reactivation stage, where sea-floor spreading reappears more or less concordantly to the recently-formed crust.

Aleutian Trench (Synorogenic) Type: In the circum-Pacific belt of today there are many active or recently active subduction sites, represented by trenches, Benioff zones, deep focus seismicity, "andesite-line" (calcalkaline) volcanicity. Axial strike-slip (transcurrent) faulting is very general. There are normally two island arcs or submarine ridges parallel to the trench, as recognized by Brouwer (1920), an inner (volcanic) arc and an outer (non-volcanic) arc. There are three principal sedimentary sites involved: the deep outer trench, an intermediate trough (between the two arcs), and an inner trough (see below, under "back-arc basin"). All appear to be represented in the complex of ancient eugeosynclinal facies.

The Aleutian Trench is selected as a convenient model inasmuch as it contains many of the features of the classical orthogeosyncline (Menard, 1964; Burk, 1965). In its (type) center section its outer ridge is submerged. The same is true of the Indonesian Arcs (Sunda and Banda arcs), where the Java Trench is separated from the volcanic arc by a submerged ridge and the Bali Trough. In both models in the eastern part, a sector of upthrust, non-volcanic crust appears. The sedimentation in such trenches depends much on the source areas; for the Aleutian Trench, it is mainly from the Gulf of Alaska, so the trench is fed longitudinally from one end. Gravitational slides, olistostromes and exotic blocks tend to slide down the short steep-slopes of the trench to mix with spilites, serpentinites, deep-sea cherts and graywackes in melange complexes. In contrast, there are some isolated arcs, for example, the Tonga-Kermadec Trench, where there is virtually no clastic source area and consequently the trench is starved.

California Borderland (Postorogenic) Type: There was widespread orogeny and granitization along the Cordilleran borders of California and Baja California (Mexico) in late Jurassic and early Cretaceous time, in places going on into the Tertiary. After consolidation and planation, general taphrogeny ensued and extensive areas were flooded by the sea. The Great Valley of California is a classic epieugeosyncline, but today is largely sediment-filled and continental. Off Southern California there is a drowned horst and graben complex. Molasse-type conglomerates nearshore pass to turbidites and pelagic deposits in the seaward basins (Emery, 1960). Farther north the borderland basins narrow and are by-passed so that submarine fans (cones) are distributed directly onto the deep-sea floor (Menard, 1964).

Gulf of California (Reactivation) Type: Followed northwards, the spreading axis of the East Pacific Rise is found to enter the Gulf of California like a wedge, almost separating Baja California from the mainland of Mexico. The last orogeny was early Cretaceous but farther north, as in the California Borderland, block-faulting began in the mid-Tertiary, followed by sea-floor spreading. The pattern is in an echelon sequence separated by transform faults, apparently in the same stress system as the San Andreas Fault. The trough is being filled today, largely by terrigenous sediments (turbidites) from the Colorado delta, another example of longitudinal filling. From the concordant setting and sedimentation it could be classified as an epieugeosyncline, and comparable to the late Triassic/early Jurassic basins of the Atlantic margins two 85my cycles ago.

Western Pacific ("Back-Arc") Types: A consideration of the Western Pacific "back-arc basins" (Karig, 1971; Matsuda and Uyeda, 1971; Packham and Falvey, 1971), as well as those of the Indonesian archipelago, the Caribbean and the Aegean (Boccaletti et al., 1974) suggests that these basins are in various stages of crustal reactivation or regeneration (what Stille called "quasicratonic"), but, in contrast to the Gulf of California, for the most part they are not sites of active sea-floor spreading today. In part, the setting is strictly postorogenic and sedimentation has accumulated over downwarped continental crust (the exogeosynclines of Kay, 1951; the idiogeosynclines of Umbgrove), fed mainly by active motion in the rising cordillera (e.g. Sumatra and Java).

In other parts of the Western Pacific type basins, there has been new simatic crust developed (e.g. in the Japan Sea, Flores Sea, Coral Sea) so that some analogous sedimentary associations have both ensialic and ensimatic foundations. In certain examples active subduction is taking place in the reverse sense, that is to say, with the polarity towards the continent, and not the open ocean (e.g. the New Hebrides Trench. The back-arc basins in general seem to be at various stages of development today, but inasmuch as they initiate *new cycles,* in view of the reactivation of spreading and subduction, they may be categorized as primarily preorogenic.

The marginal seas are regions of microplates, platelets, and microcontinents, that characterize the interaction zones between the major plates of the earth (e.g. between the Pacific and Asian plates, or between the European and African plates). Sedimentary processes at minor plate boundaries do not, however, appear to present problems fundamentally different from those of the major plate margins.

REFERENCES

Barberi, F., Bonatti, E., Marinelli, G. and Varet, J., 1973. Transverse tectonics during the split of a continent: data from the Afar rift. *Tectonophysics,* 23, 17-29.

Bernoulli, D. and Jenkyns, H. C., 1974. Mediterranean and Central Atlantic Mesozoic facies in relation to the early evolution of the Tethys. *SEPM* Spec. Pub., 19, 129-160.

Boccaletti, M., Manetti, P. and Peccerillo, A., 1974. The Balkanids as an instance of back-arc thrust belt. *Geol. Soc. Am. Bull.,* 85, 1077-1084.

Bogdanov, N. A., 1969. Thalassogeosynclines of the circum-Pacific ring. *Geotectonics,* 141-147.

Brouwer, H. A., 1920. On the crustal movements in the regions of the curving rows of islands in the eastern part of the East Indian Archipelago. *Proc. Kon. Akad. Wiss. Amsterdam,* 22, 772-782.

Burk, C. A. and Drake, C. L., 1974. *The Geology of Continental Margins,* New York: Springer-Verlag, 1009 p.

Carey, S. W., 1958. The tectonic approach to continental drift, in *Continental Drift--a Symposium,* Hobart, Tasmania: Univ. Hobart, 177-355.

Crowell, J. C., 1974. Origin of late Cenozoic basins in southern California. *SEPM* Spec. Pub., 22, 190-204.

D'Argenio, B., 1970. Evoluzione geotettonica comparata tra alcune piattaforme carbonatiche dei Mediterranei Europeo ed Americano. *Accad. Pontaniana Atti.,* 20, 3-34.

D'Argenio, B., Radoicic, R. and Sgrosso, I., 1971. A paleogeographic section through the Italo-Dinaric external zones during Jurassic and Cretaceous times. *Nafta,* Zagreb, 22, 195-207.

Dewey, J. F., 1969. Continental margins: a model for conversion of Atlantic type to Andean type. *Earth and Planetary Sci. Letters,* 6, 189-197.

Dewey, J. F. and Bird, J. M., 1970. Plate tectonics and geosynclines. *Tectonophysics,* 10, 628-638.

Dewey, J. F., et al., 1973. Plate tectonics and the evolution of the Alpine system. *Geol. Soc. Am. Bull.,* 84, 3137-3180.

Dickinson, W. R., 1971. Clastic sedimentary sequences deposited in shelf, slope, and trough settings between magmatic arcs and associated trenches. *Pacific Geology,* 3, 15-20.

Dickinson, W. R., 1974. Plate tectonics and sedimentation. *SEPM* Spec. Pub. 22, 1-27.

Dietz, R. S. and Holden, J. C., 1967. Deep sea sediments *in*-- but not on continents. *Am. Assoc. Pet. Geol. Bull.,* 50, 351-362.

Dietz, R. S., 1963. Collapsing continental rises: an actualistic concept of geosynclines and mountain-building. *Jour. Geology,* 71, 314-333.

Dietz, R. S. and Holden, J. C., 1974. Collapsing continental rises: actualistic concept of geosynclines--a review. *SEPM* Spec. Pub., 19, 14-25.

Drake, C. L., Ewing, M., and Sutton, G. H., 1959. Continental margins and geosynclines; the east coast of North America north of Cape Hatteras, in *Physics and Chemistry of Earth,* London: Pergamon Press, 110-198.

Emery, K. D., 1960. *The Sea Off Southern California,* New York: Wiley, 366 p.

Evans, I., Kendall, C. G. S. C., and Warme, J. E., 1974. Jurassic sedimentation in the High Atlas Mountains of Morocco during early rifting of Africa and North America. *Geology,* 2(7), 295-296.

Fairbridge, R. W., 1961. Eustatic changes in sea level, in *Physics and Chemistry of Earth,* London: Pergamon Press, 99-185.

Fairbridge, R. W., 1968. Atlantic and Pacific type coasts, in *Encyclopedia of Geomorphology,* New York: Reinhold, 34-35.

Fairbridge, R. W., 1966. Trenches and related deep sea troughs, in *Encyclopedia of Geomorphology,* New York: Reinhold, 929-939.

Galloway, R. W., 1970. Coastal and shelf geomorphology and late Cenozoic sea levels. *Jour. Geology,* 78, 603-610.

Gregory, J. W., 1912. The structural and petrographical classification of coast-types. *Scientia,* 2, 36-63.

Hart, M. B. and Tarling, D. H., 1974. Cenomanian paleogeography of the North Atlantic and possible mid-Cenomanian eustatic movements and their implications. *Palaeogeogr., Palaeoclimat., Palaeoecol.,* 15(2), 95-108.

Helwig, J., 1974. Eugeosynclinal basement and a collage concept of orogenic belts. *SEPM* Spec. Pub., 19, 359-376.

Hoffman, P., Dewey, J. F., and Burke, K., 1974. Aulacogens and their genetic relation to geosynclines, with a proterozoic example from Great Slave Lake, Canada. *SEPM* Spec. Pub., 19, 38-55.

Karig, D. E., 1971. Origin and development of marginal basins in the western Pacific. *Jour. Geophys. Research,* 76, 2542-2561.

Kay, M., 1971. North American geosynclines. *Geol. Soc. Am. Mem.,* 48, 143 p.

Kuenen, P. H., 1970. The turbidite problem: some comments. *New Zealand Jour. Geol. Geophys.,* 13, 852-857.

Lockwood, J. P., 1972. Possible mechanisms for the emplacement of alpine-type serpentinite. *Geol. Soc. Am. Mem.,* 132, 273.

Matsuda, T. and Uyeda, S., 1971. On the Pacific-type orogeny and its model extension of the paired belts concept and possible origin of marginal seas. *Tectonophysics,* 11, 5-27.

Menard, H. W., 1964. *Marine Geology of the Pacific,* New York: McGraw-Hill, 271 p.

Mitchell, A. H. and Reading, H. G., 1969. Continental margins, geosynclines, and ocean floor spreading. *Jour. Geology,* 77, 629-646.

Moore, J. C., 1973. Cretaceous continental margin sedimentation, southwestern Alaska. *Geol. Soc. Am. Bull.,* 84, 595-614.

Packham, G. H. and Falvey, D. A., 1971. An hypothesis for the formation of marginal seas in the western Pacific. *Tectonophysics,* 11, 79-109.

Paulus, F. J., 1972. The geology of site 98 and the Bahama platform, in Holister, C. D., Ewing, J. I. and others, eds., "Initial reports of the Deep Sea Drilling Project," Washington, D. C.: U. S. Govt. Printing Office, 11, 877-897.

Reading, H. G., 1972. Global tectonics and the genesis of flysch successions. *24th Int. Geol. Cong.,* 6, 59-65.

Renz, O., Imlay, R., Lancelot, Y. and Ryan, W. B. F., 1975. Ammonite-rich Oxfordian limestones from the base of the continental slope of northwest Africa. *Eclog. Geol. Helv.,* 68, in press.

Rona, P. A., 1973. Worldwide unconformities in marine sediments related to eustatic changes of sea level. *Nature Phys. Sci.,* 244, 25-26.

Rona, P. A., 1974. Subsidence of Atlantic continental margins. *Tectonophysics,* 22, 283-299.

Ryan, W. B. F., 1974. Personal communication. (see also, Renz, et al., 1975).

Schneider, E. D., 1972. Sedimentary evolution of rifted continental margins. *Geol. Soc. Am. Mem.,* 132, 109-118.

Schwab, F. L., 1974. Ancient geosynclinal sedimentation, paleogeography, and provinciality: a plate tectonics perspective for British Caledonides and Newfoundland Appalachians. *SEPM* Spec. Pub., 21, 54-74.

Stanley, D. J. (ed.), 1969. *The New Concept of Continental Margin Sedimentation,* Washington, D. C.: AGI.

Stanley, D. J., 1974. Modern flysch sedimentation in a Mediterranean island arc setting. *SEPM* Spec. Pub., 19, 240-259.

Stille, H., 1958. Einiges über die Weltozeane und ihre Umrahmungsräume. *Geologie,* 7, 237-306.

Suess, E., 1885-1909. *Das Antlitz der Erde,* Vienna: F. Temsky; English trans. *The Face of the Earth,* Oxford: Clarendon, 1902-1924.

Uchupi, E., Milliman, J. D., Luyendyk, B. P., Bowin, C. O., and Emery, K. O., 1971. Structure and origin of southeastern Bahamas. *Am. Assoc. Pet. Geol. Bull.,* 55, 687-704.

Wang, C. S., 1972. Geosynclines in the New Global Tectonics. *Geol. Soc. Am. Bull.,* 83, 2105-2110.

Whitaker, J. H. McD., 1974. Ancient submarine canyons and fan valleys, in Dott, R. H. and Shaver, R. H., eds., Modern and ancient geosynclinal sedimentation. *SEPM* Spec. Pub., 19, 106-135.

Wilson, J. T. and Fairbridge, R. W., 1971. Appalachian peneplains, paleosols and plate tectonics. *EOS (Trans. Am. Geophys. Union),* 52, 350 (abs.).

Addendum: Fairbridge, R. W. and Rice, A., 1975. In preparation.

Table 1. Classification of Continental Margin Sedimentation.
This table is designed in a sequential plan, embracing two 85my crustal
cycles. There is often a transition from one sedimentary association to
the next, so that boundaries are rarely sharp. Basins are frequently
multiple at any one time, and more than one basin type can coexist at
any point in time.

SETTING	BASIN TYPE	SEDIMENT TYPE
Atlantic-type *Preorogenic*	Autogeosyncline (continental interior: *NOT* marginal)	Continental facies evaporites limestones
	Taphrogeosyncline *OR* Aulacogen (e) e.g. Red Sea (Eryth- rean type)	Fanglomerates arkosic sands evaporites (anhydrite halite ± sed. ore deposits)
	Paraliageosyncline e.g. U.S. Atlantic Coastal type	Paralic lagoon, swamp, fluvial, barrier, and neritic associations
	OR: in warmer lati- tudes-- Bahamian Platform type	Barrier reef and carbonate platform associations
	Orthogeosynclinal (a) miogeosyncline (miogeocline), e.g. western Atlantic continental rise	Bathyal slope submarine fan and turbidite associations
Cordilleran or *Pacific-type*		
Synorogenic	Orthogeosynclinal (b) eugeosyncline (eugeocline) e.g. Aleutian Trench (multiple troughs)	Rhythmites: flysch associations, includ- ing wild-flysch (olistostromes), tran- sitional to melanges, ophiolites in axial or hadal zone
Postorogenic	Exogeosyncline or zeugeosyncline e.g. Sumatra-Jave (Sunda Sea) "idiogeosynclinal" type (also "back-arc basins" of western Pacific and Aegean Sea).	External "foredeep" molasse association
Preorogenic *(reactivation)*	Epieugeosyncline e.g. Gulf of California	Internal molasse association (initia- tion of *New Cycle*)

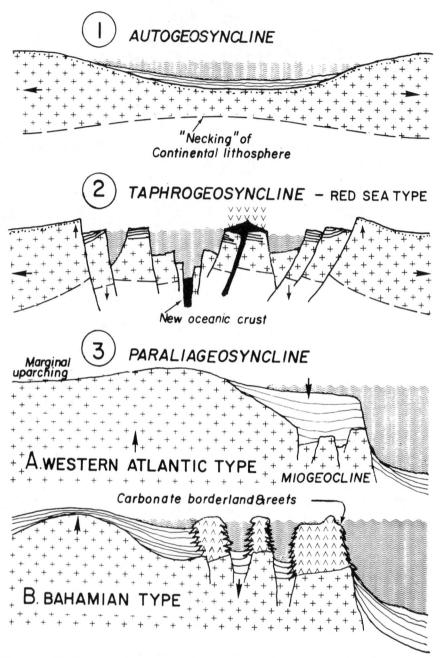

① AUTOGEOSYNCLINE

"Necking" of
Continental lithosphere

② TAPHROGEOSYNCLINE – RED SEA TYPE

New oceanic crust

③ PARALIAGEOSYNCLINE

Marginal
uparching

A. WESTERN ATLANTIC TYPE

MIOGEOCLINE

Carbonate borderland & reefs

B. BAHAMIAN TYPE

Fig. 1. Evolution of an extensional, Atlantic-type continental margin.
No. 1 represents the stretching phase that produces lithosphere "necking"
and creates an autogeosyncline (usually with evaporites). No. 2 is the
"Red Sea" phase when taphrogeny takes over. No. 3A is the now-isolated
asymmetric margin with twin geosynclines--a shallow paralic one and a
deep-water miogeoclinal one offshore. No. 3B is the specialized Bahamian
type reflecting carbonate bank and reef accumulation.

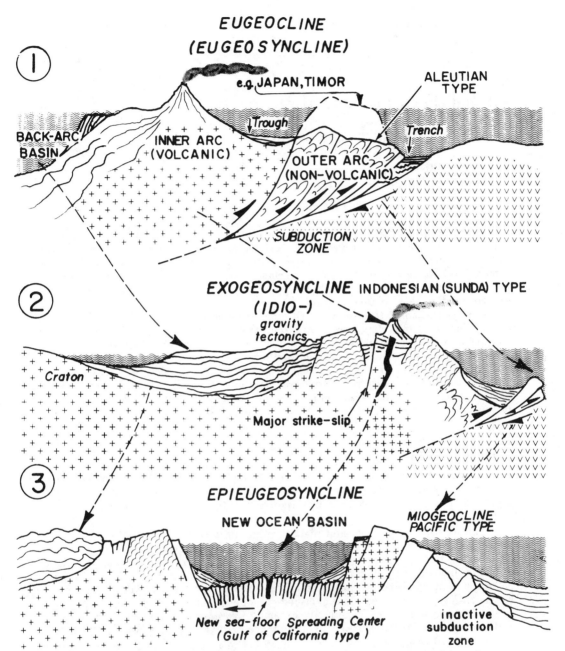

Fig. 2. Evolution of the Cordilleran or Pacific type margin. Profile 1 shows a double island arc, volcanics to the interior, non-volcanics (ridge or island) in front of it, bounded by a trench and subduction zone. No. 2 shows the back-arc basin evolving into an exogeosyncline. No. 3 suggests a new cycle of sea-floor spreading, with analogies to Fig. 1 (no. 2) but in a Cordilleran setting.

40

Reprinted from *24th Internat. Geol. Congr. Proc.* section 6, 1972,
pp. 59–66

Global Tectonics and the
Genesis of Flysch Successions

H. G. READING,
United Kingdom

ABSTRACT

Global tectonic features occur on three scales. First-order plate and continental margins, geosynclines and orogenic belts extend for several thousand kilometres and last for at least 100 m.y. Second-order plate and continental margins and basins extend hundreds of kilometres and normally last 10-20 m.y. Third-order tectonic features include small basins usually less than 100 kms in extent and may be developed on plates rather than at their margins.

A consideration of first- and second-order tectonic features enables plate margins to be divided into overriding, destructive, conservative and constructive, and continental margins into passive and active. Associated flysch successions are: Atlantic, Japan Sea, Island Arc, Andean, Mediterranean and Californian.

Flysch formed in each of these subdivisions has recognizable characteristics. It is difficult to relate successions to first-order tectonic features. They can more easily be interpreted in terms of second-order tectonic features. However, third-order tectonic features are required to explain satisfactorily the detailed development of most flysch basins.

INTRODUCTION

THE REVOLUTION IN GEOLOGICAL THOUGHT brought about by the recent advance in our understanding of present crustal phenomena has probably affected sedimentologists and paleontologists less than structural geologists and igneous and metamorphic petrologists.

This is because igneous and metamorphic activity results directly from geophysical processes, and it is generally easier to relate quantitative petrological measurements to geophysical data. One can, for example, identify ancient island arcs and determine the polarity of their related Benioff zones by the location of paired metamorphic belts, the variation of basalt types or the determination of K/Na ratios in andesites. Sedimentological phenomena, on the other hand, are secondary and are affected more by external processes such as climate, salinity and temperature of waters, and the abundance or diversity of fauna and flora.

Of all sedimentary facies concerned with the relationship of stratigraphy to tectonics, the term "flysch" is the best known and most important. In spite of repeated attempts to kill it, the term survives. That it not only survives but is expanding is shown by the recent publication of the Geological Association of Canada Special Paper on "Flysch Sedimentology in North America". In this volume, the historical usage of "flysch" is summarized by Hsu (1970) and both he and Stanley (1970) suggest current definitions and usages.

340

I do not intend to discuss the use of the term except to say that although I agree with de Raaf (1968) that turbidites formed in fluvio-lacustrine or paralic-deltaic environments should be excluded from flysch, his distinction between true flysch of orthogeosynclines and flysch-like or flyschoid successions formed at later geosynclinal stages is now inappropriate. In order to fit flysch into its tectonic setting it is preferable, initially, to embrace all possible flysch and flysch-like successions and then to qualify them according to their genesis.

I therefore suggest that the term flysch be used for "any thick succession of alternations of sandstone, calcarenite or conglomerate with shale or mudstone, interpreted as having been largely deposited by turbidity currents or mass flow in a deep water environment, within a geosynclinal belt". This definition is thus firstly descriptive (although the question of how thick is left unanswered); secondly, it depends on a sedimentological interpretation which is not too difficult for most, if not quite all, sedimentologists; thirdly it brings in the tectonic background, but of course begs the question of what is a geosyncline.

Starting with as broad a definition of flysch as possible, I will now suggest potential flysch models and the features by which they may be recognized so that others may interpret particular flysch successions, with which they are familiar, and so establish whether, on the one hand, global tectonics can help explain the genesis of flysch successions and, on the other hand, flysch successions can advance our knowledge of ancient global tectonics.

GLOBAL TECTONICS

Plate junctions, plate and continental margins

During the last three years, attempts have been made to use the concept of plate tectonics to unify and explain geological phenomena. Mitchell and Reading (1969) and Dewey and Bird (1970) have shown how many features of geosynclines and orogenic belts can be explained in terms of modern plate and continental margins; Dickinson (1970) and Mitchell and Reading (1971) have done the same for island arcs and Inman and Nordstrom (1971) for coasts. Most of these writers have also interpreted ancient rock successions in terms of the new global tectonics; specific orogenic belts and formations have been considered by Dewey and Bird (1970), Hughes (1970), Hamilton (1969), Ernst (1970), Page (1970), Fitton and Hughes (1970) and Hamilton (1970).

Tectonic activity is today mainly limited to regions adjacent to plate junctions, of which there are three types: (1) *divergent*, where two lithospheric plates are moving apart, (2) *convergent*, where one lithospheric plate descends beneath another and (3) *strikeslip*, where two plates are moving laterally without substantial divergence or convergence.

At divergent junctions, both plate margins are constructive as addition of lithosphere is taking place. At convergent junctions, the margin of the underriding plate is destructive as it is being consumed. This has led to the commonly used terms destructive or consuming plate margins for these junctions. However, it is important to distinguish the *junctions* between two plates from the *margins* of individual plates. The term 'consuming plate margin' for a 'converging junction' is misleading, as only one plate is being lost; the other, overriding plate, is not and is probably accreting by the igneous activity associated with the underlying Benioff zone and the tectonic piling up of oceanic material from the underriding plate. At strikeslip junctions, plate margins are conserved as lithosphere is neither created nor destroyed.

TABLE 1 — Relationship of plate junctions, plate and continental margins to flysch successions

PLATE JUNCTION	PLATE MARGIN	TYPE OF CONTINENTAL MARGIN, IF PRESENT		ASSOCIATED FLYSCH SUCCESSION
None	None	Passive	Atlantic	Atlantic
			Japan Sea	Japan Sea
Convergent	Overriding and Destructive	Active		Island Arc
			Andean	Andean
			Himalayan	Mediterranean
Strikeslip	Conservative	Active		Californian
Divergent	Constructive	Active		None

Plate margins are thus either *constructive, destructive, conservative* or *overriding*. Continental margins, are either *active,* when they are situated at or near plate junctions, or *passive,* when situated within lithospheric plates (Table 1).

Scale of global tectonic features

As is now widely accepted, the lithosphere can be divided into a number of plates: Le Pichon (1968) distinguished 6 major blocks; Morgan (1968) discerned 20 smaller plates.

First-order plate and continental margins. The original six plates of Le Pichon (1968) are bounded by first-order plate junctions which are mainly either divergent or convergent. The only major present strikeslip junction is that of western North America. The margins of these plates may be constructive (mid-Atlantic ridge) or destructive/overriding, as with the Andean-Central American or western Pacific margins or the Alpine-Himalayan belt of continental collision. The Atlantic is bounded by two passive continental margins where no plate junctions occur. These plate and continental margins not only extend for distances up to 10,000 kms, but they have existed since at least Cretaceous time (> 100 m.y.).

Second-order plate and continental margins. These are the boundaries of the smaller plates of Morgan (1968), such as the Caribbean, Cocos, Chilean and Philippine plates, and include more recently discovered ones such as the Scotian plate immediately behind and to the west of the Scotia arc. Strikeslip junctions formed by transform faults are of major importance. Mid-oceanic ridges consist, on this scale, of alternating constructive margins, the rifts, and conservative margins, the transform faults. In the northeast Pacific, transform faults, such as the Mendocino, separate plates with constructive eastern margins from those which are conservative. The southern Andean margins show comparable diversity and the southwest Pacific is an even more complex pattern of destructive, overriding and conservative plate margins. In the Mediterranean, a probable destructive margin in the Eastern Mediterranean lies adjacent to a possible constructive or conservative margin in the Western Mediterranean.

These margins extend for hundreds (occasionally over a thousand) kilometres. Individual plate margins change their position after a comparatively short time, either every few million years, as in the northeast Pacific, or after about 10-20 m.y., as in island arcs.

Third-order tectonic features. Innumerable topographic irregularities occur along or close to plate margins. For example, there are the individual islands of island arcs, the chain of disconnected islands such as Puerto Rico and Jamaica that occur along the strikeslip fault margin of the Greater Antilles, the islands of Corsica, Crete and Cyprus in the Mediterranean and the southern Californian "continental borderlands". The various islands and ridges are separated by basins generally 10-100 kms across.

There is no consensus of opinion as to the origin of these tectonic highs and basins and it is not known to what extent the proximity of plate margins is connected with this tectonic activity. They may be due indirectly to movements at plate junctions or they may be caused by other processes. Nevertheless, these structural highs and lows 10-100 kms across are of extreme importance in the development of depositional basins and in the provision of local sediment sources.

Discussion

As with any arbitrary division of natural phenomena, there are gradations between these scales and there is scope for discussion about the order to which some features belong. For example, should the Chile — East Pacific plate be classed as a first- or second-order feature?

First-order tectonic features help to explain the history of major fold belts such as the Appalachians or Caledonides in terms of the opening and closing of major oceans such as the present Atlantic or Pacific.

Second-order tectonic features assist the appraisal of individual regions, individual thrust blocks or changes over periods of about 10-20 m.y. For example, within the Caledonides, Ireland, Great Britain, Central Norway and Northern Norway display similar events which occurred in different ways at slightly different times. Within any one of these regions there are smaller areas such as the Welsh Basin, the Lake District and the Southern Uplands which each have different features and which can be interpreted respectively as a Japan Sea type basin, an island arc and an ocean in middle Ordovician times.

First- and second-order tectonic features can be used to divide plate and continental margins into the following genetic types (Table 1):

1. *Atlantic* — formed at passive continental margins where there is no relative movement between plates.

2. *Japan Sea* — formed in small ocean basins between a continental margin and an island arc. These are the marginal seas of Packam and Falvey (1971) and are formed by oceanization of continental crust (Beloussov, 1968), by the "trapping" of normal oceanic crust from the main ocean by the creation of an island arc or by some extensional process, probably involving the formation of new crust between the continent and island arc (Karig, 1971; Packam and Falvey, 1971).

3. *Island Arc* — formed away from the continental crust above a Benioff zone, where oceanic crust descends beneath intermediate or oceanic crust.

4. *Andean* — formed close to an active continental margin beneath which oceanic crust descends along a Benioff zone.

5. *Mediterranean* — formed at convergent plate junctions between two colliding continental plates. Similar margins have been called "Himalayan" by Mitchell and

Reading (1969) to emphasize the type of orogenic belt resulting from continental collision.

6. *Californian* — formed at conservative plate margins where displacement is essentially lateral as strikeslip or transform fault movement.

Atlantic-type margins show constant features over large distances, although even they (Inman and Nordstrom, 1971) can be further subdivided. Other types are much more variable and may not have similar features over their whole extent. For example, Japan Sea basins show features transitional from those of an Atlantic-type margin on their continental side to those of an island arc on their oceanic side. The features of a Californian margin differ between California itself, which is continental, and the Greater Antilles, which is not. Mediterranean-type basins may show even greater differences as, during continental collision, local junctions may, on a second-order scale, be continent to ocean, with basins formed of trapped ocean floor or even newly created oceanic crust.

FLYSCH SUCCESSIONS

The principal features of first- and second-order flysch successions are shown in Table 2.

Atlantic flysch successions contain compositionally mature turbidites, with a high proportion of quartz and some feldspar. Metamorphic rock fragments are abundant, showing derivation from an extensive continental source area. Silty mudstones deposited by oceanic bottom currents may be important. Igneous rocks are very rare or absent. The flysch successions are probably associated laterally with extensive, mature, orthoquartzite-carbonate shallow marine sediments or with deltaic piles. Unconformities show little angularity. No orogenic event is associated with the flysch which will appear to be pre-orogenic. It may be overlain by paralic sediments in the form of a deltaic alluvial succession, but this will be relatively fine grained, with little of the coarser, conglomeratic facies associated with molasse resulting from orogenic activity. However, if the nature of the margin changes, by the development of a Benioff zone, to one of Andean type, there will be a subsequent orogeny and it may be followed by syn-orogenic flysch or true molasse. This flysch overlies pre-flysch, comprised of ocean-floor pelagic sediments and ophiolites consisting of greenschists, serpentinites, peridotites, gabbros and spilites.

Japan Sea flysch successions compare closely with those of Atlantic flysch successions on their continental side in consisting of mature turbidites which pass landward into shelf and deltaic sediments. Oceanic bottom currents are, however, unlikely to be present. Tuff horizons are more important and become abundant as the island arc, from which volcaniclastic sediments are also derived, is approached.

As the origin of Japan Sea basins is still debated, the nature of the underlying facies cannot be postulated with certainty, but it seems probable that the flysch is underlain by pelagic sediments and either by basalts formed by some process similar to that at oceanic ridges or by alkali-olivine basalts associated with the underlying Benioff zone.

Island Arc flysch successions are dominated by compositionally immature turbidites, lacking quartz and full of acid-intermediate volcanic debris and granodioritic clasts. They do not pass laterally into a shelf facies or upward into molasse, although they may be overlain by reefs formed when the flysch is later uplifted. They lie upon either ocean floor sediments and ophiolites or earlier island arc volcanics.

TABLE 2 — Principal features of flysch successions

	AT-LANTIC	JAPAN SEA	ISLAND ARC	ANDEAN	MEDITER-RANEAN	CALI-FORNIAN
PLATE JUNCTION	None	None; possibly divergent	Convergent	Convergent	Convergent	Strikeslip
CRUSTAL JUNCTION	Continental to oceanic	None	Oceanic or intermediate to oceanic	Continental to oceanic	Probably continent to continent	Any type
NATURE OF UNDERLYING CRUST	Continental and oceanic	Intermediate, oceanic or possibly continental	Oceanic or intermediate	Continental and oceanic	Continental or possibly oceanic	Continental, intermediate or oceanic
UNDERLYING FACIES (PRE-FLYSCH)	Ophiolites and oceanic sediments or earlier shelf and continental rise sediments	Possibly ophiolites and oceanic sediments, or an earlier island arc	Ophiolites or earlier island arc	Ophiolites or earlier shelf and continental rise sediments	Varied	Varied
COMPOSITIONALLY MATURE TURBIDITES	Abundant	Abundant to rare	Rare	Rare to Common	Common to abundant	Common to abundant
ACID-INTERMEDIATE VOLCANIC TURBIDITES	Absent	Rare to Abundant	Abundant	Common to Abundant	Absent to Common	Absent
ASSOCIATED IGNEOUS ROCKS	Rare, except as ophiolites underlying flysch	Tuffs common	Abundant; extrusive and intrusive	Abundant; extrusive and intrusive, especially granites	Rare to common	Rare
MOLASSE	Absent	Absent	Absent	Present	Present	Present

Andean flysch successions are characterised by turbidites which are composition-ally relatively immature, but both quartzose and volcanic-rich turbidites may occur and slide deposits in the form of wild-flysch are particularly common. The flysch forms contemporaneously with molasse into which it passes laterally. Shelf sediments are not extensive. The flysch may either overlie earlier continental rocks, if it formed on the overriding block, or be associated with ophiolites and ocean floor sediments if it formed in the trench and was subsequently tectonically deformed along the Benioff zone.

Mediterranean flysch successions are probably very varied and there is a greater degree of uncertainty about their features, as too little is yet known about how Mediterranean basins developed. They contain relatively mature turbidites, as the depositional basins are surrounded by continents and young mountain ranges. Wild-flysch is common. Volcanic activity is not very extensive, but locally results in tuffs and volcaniclastic turbidites. The underlying rocks may be continental or oceanic, depending on whether the basin originated by rifting or subsidence. The flysch is syn-orogenic and is overlain by molasse into which it may pass laterally.

Californian flysch successions are also varied, as they may occur at oceanic or intermediate crustal junctions such as the Greater Antilles or at continental to oceanic crustal junctions as in California. However, as volcanic activity is absent, turbidites are relatively mature and will lack acid-intermediate volcanic debris, unless older volcanic arcs, as in the Greater Antilles, are uplifted. The underlying rocks may be of any type, depending on the nature of the underlying crust. The flysch will appear to be synorogenic and successions are short in duration and not of great extent. They pass upward and laterally into molasse.

CONCLUSIONS

Several authors have used global tectonics to explain flysch successions (e.g. Mitchell and Reading, 1969; Dewey and Bird, 1970; Stephens, 1970). This paper attempts to set out a provisional framework within which flysch successions may be fitted. Atlantic, Japan Sea, Island arc and Andean flysch successions are reliably based on plate tectonic theory, but Mediterranean and Californian types may have to be modified as our knowledge of continental collision and strikeslip junctions increases. Like any geological framework, it will have to be flexible and should be considered an aid to conceptual thinking rather than as a set of boxes into which each flysch can be firmly slotted. The framework is mainly applicable to second-order tectonic features. First-order tectonic features are generally too varied. Some flysch successions may not be fitted into the framework. This is because either individual flysch successions are considerably modified by third-order tectonic features or flysch forms upon plates in situations unrelated to plate junctions. A full comprehension of the relationship of flysch to tectonics must await both a convincing explanation for small-scale tectonic patterns and an understanding of movements within plates.

ACKNOWLEDGMENTS

I wish to thank Andrew Mitchell for many stimulating discussions during the preparation of this paper and for his constructive criticism of the manuscript.

REFERENCES

Beloussov, V. V., 1968. Some problems of development of the earth's crust and upper mantle of the oceans. *In* Knopoff, L., Drake, C. L., and Hart, P. J. *(Editors)*, The Crust and Upper Mantle of the Pacific Area. Geophys. Monogr., 12, p. 449-459.

Dewey, J. F., and Bird, J. M., 1970. Mountain belts and the new global tectonics. J. Geophys. Res., 75, p. 2625-2647.

Dickinson, W. R., 1970. Relations of andesites, granites and derivative sandstones to arc-trench tectonics. Revs. Geophys. and Space Phys., 8, p. 813-860.

Ernst, W. G., 1970. Tectonic contact between the Franciscan mélange and the Great Valley sequence — crustal expression of a late Mesozoic Benioff zone. J. Geophys. Res., 75, p. 886-901.

Fitton, J. G., and Hughes, D. J., 1970. Volcanism and plate tectonics in the British Ordovician. Earth Planet. Sci. Lett., 8, p. 223-228.

Hamilton, W., 1969. Mesozoic California and the underflow of Pacific mantle. Geol. Soc. Am. Bull., 80, p. 2409-2430.

———, 1970. The Uralides and the motion of the Russian and Siberian platforms. Geol. Soc. Am. Bull., 81, p. 2553-2576.

Hsü, K. J., 1970. The meaning of the word flysch — a short historical search. *In* Lajoie, J. *(Editor)*, Flysch Sedimentary in North America. Geol. Assoc. Can. Spec. Pap. 7, p. 1-11.

Hughes, C. J., 1970. The late Precambrian Avalonian orogeny in Avalon, southeast Newfoundland. Am. J. Sci., 269, p. 183-190.

Inman, D. L., and Nordstrom, C. E., 1971. On the tectonic and morphologic classification of coasts. J. Geol., 79, p. 1-21.

Karig, D. E., 1971. Origin and development of marginal basins in the western Pacific. J. Geophys. Res., 76, p. 2542-2561.

Le Pichon, X., 1968. Sea floor spreading and continental drift. J. Geophys., Res., 73, p. 3361-3697.

Mitchell, A. H., and Reading, H. G., 1969. Continental margins, geosynclines and ocean floor spreading. J. Geol., 77, p. 629-646.

————, 1971. Evolution of island arcs. J. Geol., 79, p. 253-284.

Morgan, W. J., 1968. Rises, trenches, great faults and crustal blocks. J. Geophys. Res., 73, p. 1959-1982.

Packham, G. H., and Falvey, D. A., 1971. An hypothesis for the formation of marginal seas in the western Pacific. Tectonophysics, 11, p. 79-109.

Page, B. M., 1970. Sur-Nacimiento fault zone of California: continental margin tectonics. Geol. Soc. Am. Bull., 81, p. 667-690.

de Raaf, J. F. M., 1968. Turbidites et associations sédimentaires apparentées. Koninkl. Nederlandse Akad. Wetensch. Proc., 71, p. 1-23.

Stanley, D. J., 1970. Flyschoid sedimentation on the outer Atlantic margin of northeast North America. In Lajoie, J. (Editor), Flysch Sedimentology in North America. Geol. Assoc. Canada Spec. Pap. 7, p. 179-210.

Stephens, R. K., 1970. Cambro-Ordovician flysch sedimentation and tectonics in west Newfoundland and their possible bearing on a proto-Atlantic ocean. In Lajoie, J. (Editor), Flysch Sedimentology in North America. Geol. Assoc. Can. Spec. Pap. 7, p. 165-177.

41

Framework mineralogy and chemical composition of continental margin-type sandstone

Frederic L. Schwab
Department of Geology
Washington and Lee University
Lexington, Virginia 24450

ABSTRACT

Compositional data for modern deep-sea sand and ancient graywacke were used by Crook to postulate that Atlantic, Andean, and western Pacific continental margin types could be distinguished on the basis of unique composition—a progressive decrease in framework quartz, total SiO_2, and K_2O/Na_2O ratio. A consideration of the major source areas generated by plate tectonics mechanisms suggests that the premise is valid and can be expanded to include sandstone varieties other than graywacke, but more data on modern sand are needed to refine the concept. The potential usefulness of such contrasts in sandstone mineralogy and chemistry is demonstrated by comparing the composition of several ancient sandstone units (mainly from the Appalachian-Caledonian belt) for which plate tectonics basin models have been proposed.

INTRODUCTION

Sedimentary petrologists have repeatedly demonstrated a close relationship between sandstone composition and tectonic setting. For example, Middleton (1960) and Blatt and others (1972) have shown that taphrogeosynclinal, eugeosynclinal, exogeosynclinal, and miogeosynclinal (quartzite) sandstone clans differ from one another in chemical composition, mainly as a result of

different kinds of sources. Because of the sudden replacement of conventional geosynclinal theory by plate tectonics theory, I (Schwab, 1971a) compared the bulk lithologic and chemical composition of "geosynclinal" sedimentary sequences with the composition of their supposed modern analogues—sediments of the present ocean accumulating in trenches, along the continental shelves and rises, and on the abyssal plains.

Crook (1974) suggested elegantly a direct correlation between the composition (framework components and volatile-free chemistry) of flysch arenite (graywacke) and its geotectonic setting. Crook proposed that the three distinct types of continental margins described by Mitchell and Reading (1969) have characteristic compositional varieties of graywacke associated with them, as follows:

1. Quartz-rich modern deep-sea sand and ancient graywacke (>65 percent quartz, average $SiO_2 = 70$ percent, $K_2O/Na_2O \geq 1$) only accumulate as continental rise deposits located along the trailing edges of continental blocks, which is typical of Atlantic-type continental margins.

2. Quartz-intermediate modern sand and ancient graywacke (15 to 65 percent quartz, average 68 to 74 percent SiO_2, $K_2O/Na_2O < 1$) characterize trench deposits along Andean-type margins, where subduction of an oceanic lithosphere plate occurs adjacent to and beneath a continent.

3. Quartz-poor modern sand and ancient graywacke (<15 percent quartz, average 58 percent SiO_2, $K_2O/Na_2O \ll 1$) occur only along western Pacific margins in trenches located externally to volcanic island-arc systems.

Crook based his conclusions on relatively few observations of modern marine sand. Modern "Atlantic-type" quartz-rich graywacke sand has been analyzed only in the western North Atlantic province (Hubert and Neal, 1967) and certain areas off the coast of Australia (Conolly, 1969; Eade and van der Linden, 1970). Modern "Andean-type" quartz-intermediate sand has been studied in the Astoria Fan off the Oregon coast (Kulm and Fowler, 1974) and in the Coral Sea Basin off Papua, New Guinea (Crook, 1974), but no truly "Andean" sands off western South America have been analyzed. Modern quartz-poor "western Pacific–type" sand has only been qualitatively studied in a few localities, mainly adjacent to the Solomon, Marianas, and Papua volcanic arcs.

The need for more data on modern sand is obvious from the above discussion. In fact, because almost no chemical analyses of modern marine sand existed, Crook (1974) could only define the chemical parameters (percentage of SiO_2, K_2O, and Na_2O) on the basis of existing chemical analyses of ancient sandstone units that are petrologically similar to the major varieties of modern sand. Nevertheless, the systematic

F. L. Schwab

changes in sandstone composition that he suggested are logical in terms of the variation in source rocks along the three types of continental margins. The plate tectonics paradigm provides essentially two contrasting source materials, quartz-rich cratons and quartz-poor volcanic arcs. Deposition of sandstone in evolving mobile belts may occur initially in rift valleys and later along Atlantic-type margins, with the craton as the principal source in both cases. Otherwise, deposition occurs along Andean and western Pacific margins, with volcanic-arc source areas either as important as the craton (Andean) or dominant (western Pacific). Therefore, for purposes of comparison, this paper accepts Crook's hypothesis. The composition of several additional sandstone units is listed to show the geotectonic setting that Crook's hypothesis indicates for them. Several of these units were deliberately chosen because plate tectonics mechanisms have been proposed to explain the origin of their basins. Such an approach necessarily involves some autocorrelation. I believe this is justified in the interest of demonstrating the potential usefulness of such an approach.

MINERAL COMPOSITION

Table 1 lists the abundance of quartz (as a percentage of the total framework) for a number of modern and ancient sandstone sequences. The boundaries used by Crook to infer continental margin types are used to separate the sandstone clans. (Several of the number designations of Table 1 are referred to below.)

Modern deep-sea flysch arenites (2) from the western North Atlantic help define the Atlantic-type margin field. Other sandstones of this class include several late Precambrian stratified sequences exposed along the axis, northwestern flank, and southeastern flank of the Blue Ridge anticlinorium in Virginia: respectively, the Mechum River Formation (6), the Mount Rogers Formation (5), and the Lynchburg Formation (4). Recent plate tectonics models for the central and southern Appalachians infer that these units accumulated in rift valleys and along a stable, Atlantic-type continental margin developed immediately after the separation of North America, Europe, and Africa (Brown, 1970; Schwab, 1973; Rankin, 1974). Younger (Eocambrian?) sandstones of the Chilhowee Group, the basal Unicoi Formation (1), and the overlying Harpers Formation (3) have

been interpreted as Atlantic-type continental rise deposits.

The only modern example of quartz-intermediate sandstone quantitatively analyzed is the sand of the Astoria Fan (12), which is accumulating off the Oregon-Washington coast (Nelson and Nilsen, 1974). More work must be done on modern sand along true Andean-type margins to better delineate their composition. Assuming that Crook's (1974) 15 to 65 percent quartz boundaries are valid, ancient compositional analogues of this variety include the Jurassic-Cretaceous Franciscan Formation of California (8 and 9), the Ordovician Martinsburg Formation (10), and classical flysch assemblages of New Zealand (13), continental Europe (11 and 15), and Great Britain (7 and 14). Several convincing models, based largely on other evidence, interpret the Franciscan Formation as a trench deposit accumulating along a Mesozoic Andean-type margin (Hamilton, 1969; Dickinson, 1970). Similar models have been proposed for the Ordovician and Silurian grits of the Caledonian belt (Schwab, 1974b). No generally accepted plate tectonics models exist that explain the genesis of the basins in which the other units accumulated.

The compositional field for sandstones indicative of western Pacific margins also

needs better definition. A logical consideration of the source rocks available in volcanic arcs, together with Crook's observations on Holocene sediment in the western Pacific and quantitative analyses of two Tertiary units (19 and 20), suggests that such sandstone should be quartz poor.

CHEMICAL COMPOSITION

Table 2 compares the volatile-free K_2O, Na_2O, and SiO_2 content of a number of sandstone samples (as well as different components of the Earth's crust). Several of the ancient units that Crook used in establishing the chemical characteristics of quartz-rich, quartz-intermediate, and quartz-poor sandstones appear for reference. Crook advocated making distinctions between types of continental margins mainly on the basis of K_2O/Na_2O ratio and overall SiO_2 content. I accept his compositional criteria for the sake of comparison, realizing that without chemical data on *modern* sand the results are not totally conclusive. Nevertheless, the results are intriguing.

Quartz-rich sandstone (rift valleys and Atlantic-type margins?) has a high SiO_2 content and a K_2O/Na_2O ratio that equals or exceeds 1.0. Sandstones with this composition include Middleton's (1960) "taphro-

TABLE 1. PERCENTAGE OF QUARTZ IN FRAMEWORK OF REPRESENTATIVE SANDSTONE UNITS

Sandstone class	No.	Unit	Quartz (%)	No. of samples	Reference
Quartz ≥ 65%	1	Unicoi Formation, Eocambrian, central Virginia	88	30	Schwab (1972)
	2	Modern deep-sea sands, western province, North Atlantic	86	92	Hubert and Neal (1967)
	3	Harpers Formation, Eocambrian, central Virginia	86	8	Schwab (1971b)
	4	Lynchburg Formation, late Precambrian, central Virginia	73	6	Brown (1958)
	5	Mount Rogers Formation, late Precambrian, southwestern Virginia	66	5	Schwab (unpub. data)
	6	Mechum River Formation, late Precambrian, central Virginia	66	30	Schwab (1974a)
Quartz 15-65%	7	Aberystwyth Grit, Silurian, Wales	57	1	Okada (1967)
	8	Franciscan Formation, Jurassic-Cretaceous, California	56	17	Taliaferro (1943)
	9	Franciscan Formation, Jurassic-Cretaceous, California	45	80	Bailey and others (1964)
	10	Martinsburg Formation, Ordovician, central Appalachians	45	5	McBride (1962)
	11	Carboniferous, Harz Mountains, Germany	39	88	Huckenholz (1963)
	12	Modern deep-sea sands, Astoria sea fan	30	14	Nelson and Nilsen (1974)
	13	Axial facies, upper Paleozoic and Mesozoic, New Zealand geosyncline	30	7	Dickinson (1971)
	14	Southern uplands, Ordovician, Scotland	27	several	Dzulynski and Walton (1965)
	15	Tanner Graywacke, Devonian, Harz Mountains, Germany	23	1	Huckenholz (1959)
Quartz < 15%	16	Cretaceous of Papua	14	4	Pettijohn and others (1973)
	17	Plagioclase graywacke, Carboniferous, Australia	11	1	Crook (1960)
	18	Devonian graywacke, Australia	0.6	1	Crook (1974)
	19	Tertiary sand, North Coast Basin, Colombia	0	1	Crook (1974)
	20	Miocene graywacke, Papua	0	1	Pettijohn (1957)

TABLE 2. CHEMICAL CHARACTERISTICS OF REPRESENTATIVE SANDSTONE UNITS

Compositional class	No.	Unit	K_2O/Na_2O	K_2O (%)	Na_2O (%)	SiO_2 (%)	No. of samples	Reference
$K_2O/Na_2O \geq 1$; SiO_2 usually > 70%	1	Torridonian arkose, late Precambrian, Scotland	10.5	5.27	0.50	75.6	3	Pettijohn (1957)
	2	Littleton Formation, Devonian, New England	9.4	1.32	0.14	86.9	1	Billings and Wilson (1964)
	3	"Taphrogeosynclinal" sandstone	4.0	2.88	0.72	82.8	28	Middleton (1960)
	4	Average sedimentary cover, Russian and North American platforms (excluding carbonate)	3.2	3.2	1.0	73.0	624	Ronov and Migdisov (1971)
	5	Grandfather Mountain Formation, late Precambrian, North Carolina	3.12	4.53	1.45	75.9	7	Bryant and Reed (1970)
	6	"Cratonic" (quartzite) sandstone	2.65	0.82	0.31	93.3	54	Middleton (1960)
	7	Clastic component, average (preorogenic) miogeosyncline	2.10	2.30	1.10	78.8	several	Schwab (1971a)
	8	Orfordville Formation, Ordovician, New England	2.10	3.22	1.13	66.3	1	Billings and Wilson (1964)
	9	Cambrian-Quaternary sand, North American platform	1.93	1.99	1.03	86.3	152	Ronov and Migdisov (1971)
	10	"Exogeosynclinal" sandstone	1.81	1.51	0.83	78.7	45	Middleton (1960)
	11	Typical quartz-rich sandstone (Tasman geosyncline, Australia)	1.7	1.18	0.69	88.8	24	Crook (1974)
	12	Triassic arkose, Connecticut	1.65	5.43	3.30	69.9	1	Pettijohn (1957)
	13	Thunderhead Sandstone (Ocoee Series), late Precambrian, Tennessee	1.42	3.35	2.36	75.8	1	King (1970)
	14	Great Smoky Group (Ocoee Series), late Precambrian, Tennessee	1.00	2.62	2.62	73.7	7	King (1970)
	15	Metagraywacke, Middle Cambrian(?), Piedmont of North Carolina	1.00	2.80	2.80	65.2	5	Sundelius (1970)
$K_2O/Na_2O < 1$; SiO_2 68-74%	16	Evington Group, Piedmont of Virginia	0.90	1.88	2.08	71.3	1	Smith and others (1964)
	17	Rennselaer Grit, Eocambrian, New York	0.80	1.96	2.69	71.1	3	Balk (1953)
	18	Wissahickon Formation, Maryland Piedmont	0.67	1.80	2.70	71.8	5	Hopson (1964)
	19	Typical quartz-intermediate sandstone (Harz Mountains, Germany)	0.60	1.40	2.50	73.5	20	Huckenholz (1963)
	20	"Eugeosynclinal" sandstone	0.58	1.79	3.11	71.4	41	Middleton (1960)
	21	Typical quartz-intermediate sandstone (Franciscan Formation)	0.48	1.80	3.70	70.1	21	Bailey and others (1964)
$K_2O/Na_2O \ll 1$; average SiO_2 57-59%	22	Schists of the Inner Piedmont, North Carolina	0.17	0.60	3.60	58.9	2	Bryant and Reed (1970)
	23	Typical quartz-poor sandstone (volcanic graywacke, Devonian, Australia)	0.14	0.70	5.05	57.6	10	Crook (1974)
	24	Typical quartz-poor sandstone (plagioclase graywacke, Carboniferous, Australia)	0.07	0.42	6.89	59.1	1	Crook (1974)
Various components of the Earth	25	Average composition of shields	0.96	3.09	3.21	78.8	several	Ronov and Migdisov (1971)
	26	Tholeiitic andesite, Fiji	0.96	0.44	4.29	58.6	1	Crook (1974)
	27	Composition of "developed island arc"	0.23	0.82	3.39	58.8	estimate	Jakeš and White (1971)
	28	Catoctin Greenstone, late Precambrian-Eocambrian, Virginia	0.10	0.40	3.90	48.1	7	Reed and Morgan (1971)
	29	Average upper mantle composition	0.06	0.015	0.25	44.5	estimate	Wyllie (1971)

geosynclinal" (3), "miogeosynclinal" (quartzite; 6), and "exogeosynclinal" (10) clans, plus a number of units from the Appalachian-Caledonian belt. The Triassic basin deposits of Connecticut (12), the Grandfather Mountain Formation of North Carolina (5), and the Torridonian arkoses of Scotland (1) all belong to this compositional clan. Plate tectonics models infer that each of these deposits accumulated in tensional rift valleys produced along or adjacent to Atlantic-type continental margins. Other "Atlantic-type" sandstones include the Cambrian through Quaternary sandstones of the Russian and North American platforms (4 and 9), the average clastic residue of preorogenic miogeosynclines (7), two units of the late Precambrian Ocoee Series of Tennessee (13 and 14), metagraywacke (Middle Cambrian?) from the North Carolina Piedmont (15), and metagraywacke from the Appalachian "eugeosynclinal" terrane of New England—the Ordovician Orfordville Formation (8) and the Devonian Littleton Formation (2). Sediment of the North American and Russian platforms (4 and 9) and typical sandstone in miogeosynclinal belts (7; prior to deformation) are derived from the craton, and their composition mirrors that source. Hatcher (1972) proposed a model that shows the Ocoee

Series accumulating along the trailing edge of the late Precambrian North American continent within an inland-sea basin. The same model infers that during Cambrian time, parts of the Piedmont were also an inland-sea area juxtaposed between the craton and a volcanic island arc. Bird and Dewey (1970) inferred that the Orfordville Formation is an Atlantic-type continental rise deposit.

Ancient sandstones that correspond compositionally to the variety thought to be characteristic of Andean-type margins ($K_2O/Na_2O < 1$; average SiO_2 68 to 74 percent) include Middleton's "eugeosynclinal" sandstone clan (20), the late Precambrian or Cambrian Rennselaer Grit (17), and two

metagraywacke units of uncertain age from the Appalachian Piedmont—the Wissahickon Formation (18) and the Evington Group (16). Brown (1970) interpreted the Evington Group as trench deposits along an Andean-type margin. No generally accepted plate tectonics model exists for the Wissahickon Formation specifically or the Glenarm Series generally. Bird and Dewey (1970) interpreted the Rennselaer Grit as a rift-valley deposit rather than trench fill, and the field evidence is convincing. The compositional boundaries for this class obviously need better definition.

The only additional inferred western Pacific-type sandstone (low SiO_2, very low K_2O/Na_2O) other than those originally pro-

350

vided by Crook (1974) is from the metamorphic terrane of the Inner Piedmont Belt of North Carolina (22).

For purposes of comparison, Table 1 also lists the calculated or estimated composition of two major crustal units, shield areas (25), and a developed island arc (27), plus a typical modern andesite (26), the Catoctin Greenstone of Virginia (28), and the upper mantle (29). Shield areas (25) and volcanic arcs (27) presumably represent the two principal source areas for continental margin sediment.

CONCLUSIONS

The premise that the mineralogy (percentage of quartz) and chemical composition (K_2O/Na_2O ratio and total percentage of SiO_2) of ancient sandstone can be used to discriminate among rift-valley, Atlantic, Andean, and western Pacific continental margin deposits requires additional data obtained principally from modern sand. If verified, the model provides a useful tool for recognizing ancient continental margin types by means of simple petrographic or chemical analyses of sandstone.

REFERENCES CITED

Bailey, E. H., Irwin, W. P., and Jones, D. L., 1964, Franciscan and related rocks and their significance in the geology of western California: California Div. Mines and Geology Bull. 183, 178 p.

Balk, R., 1953, Structure of graywacke areas and Taconic Range, east of Troy, New York: Geol. Soc. America Bull., v. 64, p. 811–864.

Billings, M. P., and Wilson, J. R., 1964, Chemical analyses of rocks and rock-minerals from New Hampshire: Pt. XIX, Mineral resources survey, Concord, New Hampshire: New Hampshire Dept. Resources and Econ. Devel., 104 p.

Bird, J. M., and Dewey, J. F., 1970, Lithosphere plate-continental margin tectonics and the evolution of the Appalachian orogen: Geol. Soc. America Bull., v. 81, p. 1031–1060.

Blatt, H., Middleton, G., and Murray, R., 1972, Origin of sedimentary rocks: Englewood Cliffs, N.J., Prentice-Hall, Inc., 634 p.

Brown, W. R., 1958, Geology and mineral resources of the Lynchburg quadrangle, Virginia: Virginia Div. Mineral Resources Bull. 74, 99 p.

— 1970, Investigations of the sedimentary record in the Piedmont and Blue Ridge of Virginia, in Fisher, G. W., Pettijohn, F. J., Reed, J. C., Jr., and Weaver, K. N., eds., Studies of Appalachian geology: Central and southern: New York, Interscience Pubs., Inc., p. 335–349.

Bryant, B., and Reed, J. C., Jr., 1970, Geology of the Grandfather Mountain Window and vicinity, North Carolina and Tennessee: U.S. Geol. Survey Prof. Paper 615, 190 p.

Conolly, J. R., 1969, Western Tasman Sea floor: New Zealand Jour. Geology and Geophysics, v. 12, p. 310–343.

Crook, K. A., 1960, Petrology of Parry Group, Upper Devonian–Lower Carboniferous, Tamworth-Nundle district, New South Wales: Jour. Sed. Petrology, v. 30, p. 538–552.

— 1974, Lithogenesis and geotectonics: The significance of compositional variation in flysch arenites (graywackes), in Dott, R. H., Jr., and Shaver, R. H., eds., Modern and ancient geosynclinal sedimentation: Soc. Econ. Paleontologists and Mineralogists Spec. Pub., no. 19, p. 304–310.

Dickinson, W. R., 1970, Global tectonics: Science, v. 168, p. 1250–1259.

— 1971, Detrital modes of New Zealand graywackes: Sed. Geology, v. 5, p. 37–56.

Dzulyński, S., and Walton, E. K., 1965, Sedimentary features of flysch and greywackes: Amsterdam, Elsevier Pub. Co., 274 p.

Eade, J. V., and van der Linden, J. M., 1970, Sediments and stratigraphy of deep-sea cores from the Tasman Basin: New Zealand Jour. Geology and Geophysics, v. 13, p. 228–268.

Hamilton, W. B., 1969, Mesozoic California and the underflow of Pacific mantle: Geol. Soc. America Bull., v. 80, p. 2409–2430.

Hatcher, R. D., 1972, Developmental model for the southern Appalachians: Geol. Soc. America Bull., v. 83, p. 2735–2760.

Hopson, C. A., 1964, The crystalline rocks of Howard and Montgomery Counties, in The geology of Howard and Montgomery Counties: Baltimore, Maryland Geol. Survey, p. 27–215.

Hubert, J., and Neal, P. F., 1967, Mineral composition and dispersal patterns of deep-sea sands in the western North Atlantic petrologic province: Geol. Soc. America Bull., v. 78, p. 749–772.

Huckenholz, H. G., 1959, Sediment-petrographische Untersuchungen an Gesteinen der Tanner Grauwacke: Beitr. Mineralogie u. Petrologie, v. 6, p. 261–298.

— 1963, Mineral composition and texture in graywackes from the Harz Mountains (Germany) and in arkoses from the Auvergne (France): Jour. Sed. Petrology, v. 33, p. 914–918.

Jakeš, P., and White, A.J.R., 1971, Composition of island arcs and continental growth: Earth and Planetary Sci. Letters, v. 12, p. 224–230.

King, P. B., 1970, Precambrian of the United States of America: Southeastern United States, in Rankama, K., ed., The Precambrian, Vol. 4: New York, Interscience Pubs., Inc., p. 1–71.

Kulm, L. D., and Fowler, G. A., 1974, Cenozoic sedimentary framework of the Gorda-Juan de Fuca plate and adjacent continental margin—A review, in Dott, R. H., Jr., and Shaver, R. H., eds., Modern and ancient geosynclinal sedimentation: Soc. Econ. Paleontologists and Mineralogists Spec. Pub., no. 19, p. 304–310.

McBride, E. F., 1962, Flysch and associated beds of the Martinsburg Formation (Ordovician), central Appalachians: Jour. Sed. Petrology, v. 32, p. 39–91.

Middleton, G. V., 1960, Chemical composition of sandstones: Geol. Soc. America Bull., v. 71, p. 1011–1026.

Mitchell, A. H., and Reading, H. G., 1969, Continental margins and ocean floor spreading: Jour. Geology, v. 77, p. 629–646.

Nelson, C. H., and Nilsen, T. H., 1974, Depositional trends of modern and ancient deep-sea fans, in Dott, R. H., Jr., and Shaver, R. H., eds., Modern and ancient geosynclinal sedimentation: Soc. Econ. Paleontologists and Mineralogists Spec. Pub., no. 19, p. 69–91.

Okada, H., 1967, Composition and cementation of some Lower Paleozoic grits in Wales: Kyushu Univ. Fac. Sci., ser. D., Geology, v. 18, p. 261–276.

Pettijohn, F. J., 1957, Sedimentary rocks: New York, Harper & Row Pubs., 718 p.

Pettijohn, F. J., Potter, P. E., and Siever, R., 1973, Sand and sandstone: New York, Springer-Verlag New York, Inc., 618 p.

Rankin, D. W., 1974, Repeated opening and closing of the Atlantic Ocean basin: Evidence from the southern Appalachian Mountains: Geol. Soc. America Abs. with Programs, v. 6, p. 920.

Reed, J. C., Jr., and Morgan, B. A. 1971, Chemical alteration and spilitization of the Catoctin Greenstone, Shenandoah National Park, Virginia: Jour. Geology, v. 79, p. 526–548.

Ronov, A. B., and Migdisov, A. A., 1971, Geochemical history of the crystalline basement and the sedimentary cover of the Russian and North American platforms: Sedimentology, v. 16, p. 137–187.

Schwab, F. L., 1971a, Geosynclinal compositions and the new global tectonics: Jour. Sed. Petrology, v. 41, p. 928–938.

— 1971b, Harpers Formation, central Virginia: A sedimentary model: Jour. Sed. Petrology, v. 41, p. 139–149.

— 1972, The Chilhowee Group and the late Precambrian–Early Paleozoic sedimentary framework in the central and southern Appalachians, in Lessing, P., Hayhurst, R. I., Barlow, J. A., and Woodfork, L. D., eds., Appalachian structures: Origin, evolution, and possible potential for new exploration frontiers: Morgantown, West Virginia Geol. and Econ. Survey, p. 59–94.

— 1973, Plate tectonics models: British Caledonides, Newfoundland Appalachians, and central Appalachians: Geol. Soc. America Abs. with Programs, v. 5, p. 432.

— 1974a, Mechum River Formation: Late Precambrian(?) alluvium in the Blue Ridge province of Virginia: Jour. Sed. Petrology, v. 44, p. 862–871.

— 1974b, Ancient geosynclinal sedimentation, paleogeography, and provinciality: A plate tectonics perspective for British Caledonides and Newfoundland Appalachians, in Ross, C. A., ed., Paleogeographic provinces and provinciality: Soc. Econ. Paleontologists and Mineralogists Spec. Pub., no. 21, p. 54–74.

Smith, J. W., Milici, R. C., and Greenburg, S. S., 1964, Geology and mineral resources of Fluvana County: Virginia Div. Mineral Resources Bull. 79, 62 p.

Sundelius, H. W., 1970, The Carolina Slate Belt, in Fisher, G. W., Pettijohn, F. J., Reed, J. C., Jr., and Weaver, K. N., eds., Studies of Appalachian geology: Central and southern: New York, Interscience Pubs., Inc., p. 351–367.

Taliaferro, N. L., 1943, Franciscan-Knoxville problem: Am. Assoc. Petroleum Geologists Bull., v. 27, p. 109–219.

Wyllie, P. J., 1971, The dynamic Earth: New York, John Wiley & Sons, Inc., 416 p.

ACKNOWLEDGMENTS

Partly supported by two Penrose Research Grants from the Geological Society of America and a Glenn Grant from Washington and Lee University.

M. T. Roberts constructively criticized the manuscript.

MANUSCRIPT RECEIVED DEC. 16, 1974

MANUSCRIPT ACCEPTED JUNE 23, 1975

42

Reprinted from Geology 4:723–727 (1976)

Modern and ancient sedimentary basins: Comparative accumulation rates

Frederic L. Schwab
Department of Geology
Washington and Lee University
Lexington, Virginia 24450

ABSTRACT

The rate at which sedimentary fill accumulates in both modern and ancient depositional basins is related to plate tectonic setting, among other factors. The tectonic setting of any basin can be specified in terms of type of substratum (oceanic, continental, or transitional), proximity to plate margins (midplate versus interplate), and the nature of the nearby plate margins (constructive, destructive, and conservative).

Midplate basins generally accumulate sedimentary fill at much slower rates than basins located along or proximal to plate margins. Midplate basins situated entirely above a single substratum type (that is, cratonic basins) ordinarily accumulate sediment cover very slowly (usually less than 0.006 m/1,000 yr), although more rapidly subsiding negative areas within continental blocks and near their margins adjacent to orogenic areas

exhibit higher rates. Midplate basins that straddle both continental and oceanic crust (modern continental terraces and rises, ancient miogeoclines and the nonorogenic portion of ancient eugeosynclines) accumulate sediment several times as rapidly as cratonic basins (generally 0.01 to 0.04 m/1,000 yr).

Accumulation rates for sediment deposited in basins along or adjacent to plate margins almost invariably exceed 0.04 m/1,000 yr and are commonly much higher than the accumulation rates for midplate basins. However, the accumulation rates for basins along convergent, divergent, and conservative plate margins are not sufficiently different from one another to be distinctive. Sediment filling the successor basins that commonly develop over the site of sutured plate margins has accumulation rates comparable to those of interplate basins.

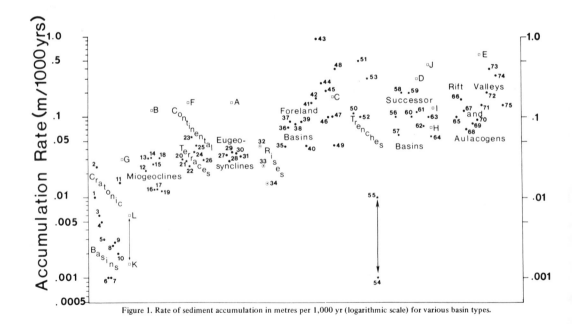

Figure 1. Rate of sediment accumulation in metres per 1,000 yr (logarithmic scale) for various basin types.

INTRODUCTION

Many geologists now agree that lithospheric plate construction, transformation, and destruction are major mechanisms controlling the origin and evolution of many modern sedimentary basins (that is, areas of subsidence containing more than 1 km of sediment). Consequently, several attempts have been made to classify and name both modern and ancient basins in a manner consistent with the plate-tectonics paradigm, characterizing as far as possible the unique features of the sedimentary fill in each type (see, for example, Dott, 1974; Dickinson, 1971, 1974a; Bally, 1975; Mitchell and Reading, 1969). These new classification schemes often differ markedly from those of Kay (1951) that mainly covered ancient sedimentary basins, considering each as a variation of geosyncline.

Disagreements persist as to the specific varieties of basins, their origin, number, and most appropriate names, but I believe that a consensus now exists that most modern sedimentary basins belong to one of the following basic varieties, each of which occupies a distinctive plate tectonic setting:

1. Midplate basins are situated entirely within the boundaries of a single lithospheric plate and can be subdivided into two categories: (1) cratonic basins and (2) Atlantic-type continental margin basins. These two varieties differ from each other in one important aspect. Individual cratonic basins are "floored" totally by either continental or oceanic crust, but single (individual) Atlantic-type continental margin basins invariably straddle (are floored by) both types of crust. Cratonic basins include two markedly different sedimentary assemblages: (a) marine and continental platform deposits of the continental blocks and (b) the sedimentary cover of the abyssal ocean floor. Atlantic-type continental margin basins also consist of two distinct sedimentary prisms: (a) a marginal continental shelf (terrace) area juxtaposed on the oceanward flank by (b) a continental rise.

2. Interplate basins are located at or near plate boundaries and include three major types separated on the basis of whether the particular proximal boundary is constructive (divergent), destructive (convergent), or conservative. (1) Divergent margin basins (rift valleys) straddle extensional (constructive) plate margins and their genesis is related to the initial break-up of continental blocks due to sea-floor spreading. (2) Convergent margin basins constitute a very complex and heterogeneous group of basins that straddle or are proximal to destructive plate margins. They are genetically related to the descent of one lithospheric plate beneath and adjacent plate along a subduction zone, although individual basins may be produced by the growth of smaller ocean basins adjacent to island arcs. Examples of the many individual varieties can be found along the modern western Pacific and Andean continental margins. Individual varieties include modern deep-sea trenches (such as the Japanese and Aleutian trenches) and the complex of basins within and peripheral to island arcs (forearc basins such as the Anchorage Basin, intra-arc basins, and retroarc basins). The retroarc basins may include both foredeep flysch and molasse "clastic wedges," as well as backdeep basins (Sumatra and Sunda Basins) and marginal basins (Sea of Japan). (3) Conservative margin basins are exemplified by the Cenozoic basins of California that are located adjacent to the San Andreas fault and its subsidiaries. These basins are generated by horizontal and vertical movements associated with the transform systems bounding plates.

3. Successor basins occur within deformed mobile belts. They straddle recently active but now passive plate boundaries. Many, but not all, modern successor basins are developed on sutured margins following the closure of ocean basins and the development of a mobile belt by plate convergence.

Refinement of the above (and similar) classifications and an improved understanding of the genetic contrasts among basin types is obviously contingent on clarifying the differences in sediment composition, texture, depositional environment, and provenance among modern basins. It is particularly important to evaluate how effectively plate tectonic mechanisms and models can be applied to ancient mobile belts by comparing modern basins with their presumed ancient analogues. Recent publications by Dott and Shaver (1974) and Dickinson (1974b) probably represent the most comprehensive compilations of data on a variety of modern and ancient basins. Despite such efforts, and despite the availability of the raw data necessary for the calculations, little comparative data have been compiled to document the contrasts in (inferred) tectonic mobility and the accumulation rate of the sedimentary fill for either modern sedimentary basins or ancient, at least partially analogous, geosynclinal belts. This paper provides such a compilation from the literature and discusses aspects of the data interpretatively.

PROCEDURE

Table 1 and Figure 1 summarize the accumulation rates (in m/1,000 yr) for sedimentary successions in a variety of modern and ancient sedimentary basins. The data are intentionally grouped by major basin type to emphasize similarities and differences. Each accumulation rate was calculated using two kinds of raw data: (1) the published overall thickness (in metres) of the "average" or "typical" sedimentary section as determined directly by field measurement (or, in a few cases, estimated) and (2) the estimated interval of time (in millions of years) during which an individual basin persisted both as a geographical and tectonic entity as defined in the preceding classification scheme.

The calculated accumulation rates, unless otherwise noted, were deliberately chosen when represented in the literature as average or typical rather than maximum or minimum. The accumulation time is net accumulation time, that is, the interval of time during which individual sections accumulated includes not only intervals of actual deposition but also episodes of uplift, nondeposition, and even erosion. No attempt was made to quantitatively evaluate that portion of a section that is missing now. I emphasize that these rates are sediment accumulation rates, not rates of subsidence or, strictly speaking, rates of sedimentation. I have, therefore, deliberately not listed the rates in Bubnoff units (metres/million years), but instead rates are shown in metres per 1,000 yr. The rate at which a sediment accumulates may closely reflect rate of subsidence and (or) rate of sedimentation of course. However, it might equally reflect rate of sediment supply, eustatic changes in sea level, and so on. The data, as a result of these considerations, are only a first approximation of the tectonic history of these various basins.

RESULTS

Cratonic basins, as might be expected, show the lowest accumulation rates of any of the major basin types. The more rapidly subsiding negative areas of the continental craton (such as the Michigan and Williston Basins; 1, 2, 3) and the cratonic margins adjacent to mobile ocean basins (4, 5) commonly accumulate sediment several times as fast as neutral areas of the continental craton (6) or the abyssal plains of the ocean basins (7 through 11).

The neutral (nonbasinal) areas of the Paleozoic continental platform accumulated a sedimentary rock cover at a rate similar to the slow rates estimated for modern abyssal plains.

The quartz-arenite and carbonate successions of ancient miogeosynclines (12 through 19) accumulated at an average rate of 0.022 m/1,000 yr. This mean and the range in rates resembles the mean (0.036 m/1,000 yr) and range of rates of sediment being deposited along modern continental terraces of Atlantic-type continental margins (20 through 26). Drake and others (1959) and Dietz (1963) originally equated Kay's (1951) concept of a miogeosyncline with modern continental shelf deposits. However, the miogeosyncline as originally perceived by Marshall Kay and Hans Stille encompassed clastic-wedge (in part, exogeosynclinal) deposits as well as the classical miogeosynclinal carbonates and quartz arenites. Dietz and Holden (1966, 1974) later modified this concept by substituting the term "miogeocline" for miogeosyncline in order to refer only to the seaward-thickening, nonorogenic sequence of shallow-water sedimentary rocks in geosynclinal belts that are most directly analogous to modern continental shelf deposits. Consequently, the data listed for ancient miogeosynclines in Table 1 and Figure 1 actually represent only the "miogeoclinal" portion of the specific miogeosynclinal successions that were analyzed.

The sedimentary successions in a number of ancient eugeosynclinal belts (27 through 31) exhibit a mean accumulation rate of 0.034 m/1,000 yr. The sections analyzed are classical eugeosynclinal sequences according to Kay's (1951) conception, that is, the sections include both older, nonorogenic deep-water marine sediment (predominantly mudstone, turbidites, and submarine slide deposits) as well as younger volcanic and volcaniclastic sediments deposited within and adjacent to island arc–trench complexes. Dietz (1963) and Dietz and Holden (1966, 1974) also modified their original analogy that directly compared ancient eugeosynclines and modern continental rises by proposing the concept of the eugeocline. The eugeocline includes only the nonorogenic portion of eugeosynclinal fill and is directly equated with modern continental rise prisms. The mean accumulation rate along several modern continental rise areas (32, 33, 34) is 0.028 m/1,000 yr.

Sedimentologists have not sufficiently differentiated the many varieties of convergent margin basins in the literature so that I could calculate accumulation rates for every variety. Many of these basin types were virtually unrecognized prior to the development of plate tectonics; many species are still undecipherable within ancient mobile belts. However, I was able to obtain data on two species of convergent margin basins: foreland basins and trenches. Both have been accurately recognized and described in considerable detail, even though our understanding of their origin and evolution has changed greatly with the development of plate tectonic theory.

Foreland basin deposits are comparable to Kay's (1951) "exogeosyncline" and the classical flysch and molasse clastic wedges. The collective mean accumulation rate for foreland basin deposits is 0.186 m/1,000 yr. Foreland basin molasse sequences in the Appalachian and Cordilleran geosynclines (35 through 40) show a mean accumulation rate of 0.070 m/1,000 yr. Comparable flysch deposits (41 through 45), typical of those that underlie molasse, have a higher mean accumulation rate (0.345 m/1,000 yr). Younger, Cenozoic molasse in basins fronting the Pyrenees (46, 47) and the Alps (48) have rates intermediate between the above two groupings. The Lau Basin (49), a modern marine basin behind the Tonga arc, is included for comparison.

The data for trenches includes overall accumulation rates for specific trench systems (52, 53), as well as typical rates for the faster growing portion of trench fill, modern (51) and ancient (50) submarine fan deposits. I have also included two very general estimates from modern Pacific trenches (54, 55). Without additional data, it is unrealistic to develop a meaningful average accumulation rate for trench fill, because so many trenches seem to be nearly empty and considerable doubt exists as to how directly comparable modern trench systems are to ancient eugeosynclinal belts (Scholl and Marlow, 1974).

Successor basins (Kay's [1951] epieugeosynclines?; 56 through 64) show high accumulation rates compared to cratonic basins, modern continental rise and shelf areas, ancient miogeoclines, and ancient eugeosynclines. Surprisingly, however, the mean accumulation rate for successor basins (0.113 m/1,000 yr) is lower than the mean rate estimated for foreland basins (0.186 m/1,000 yr).

I have deliberately grouped two types of divergent margin basins together: ancient aulacogens and rift valleys. Aulacogens are genetically related to "failed" arms of rift valleys, and the initial deposits in them accumulate in a setting identical to that of rift valleys (taphrogeosynclines). Aulacogens (68, 69, 70), modern rift valley fill (71, 72, 75), and ancient rift valley fill (65 through 67, 73, 74) show considerable ranges in accumulation rates. The overall accumulation rate for divergent margin basins is high (0.169 m/1,000 yr) and approaches that of foreland basins.

For comparative purposes I have also listed the maximum or (if not noted) mean accumulation rates published by Kay (1955) for a number of geosynclinal basin assemblages (A-L). Kay's data include the Cenozoic sediments of the Ventura basin (J), which can be interpreted as a conservative margin basin.

All sediment accumulation rates should also be compared with estimated rates of lateral sea-floor spreading, which range from 20 to 180 m/1,000 yr.

CONCLUSIONS

Sedimentologists concur that the major control of all sedimentation is tectonics. Tectonic mobility controls not only the rate of source uplift and erosion (and consequently the rate of sediment supply) but also the rate of basin subsidence. Sediment accumulation rates are in turn essentially a function of the interplay between rate of sediment supply and rate of basin subsidence (plus or minus uplift [or eustatic fall], erosion, and [or] nondeposition).

Dickinson (1974a) suggested that if plate tectonic mechanisms were applicable, then the settings of sedimentary basins were controlled by only three factors: (1) type of crustal substratum (oceanic, continental, or several intermediate varieties), (2) proximity of the basin to a plate margin (basically midplate versus near-margin), and (3) the nature of the plate junction(s) nearest the basin (divergent, convergent, transform).

Figure 1 and Table 1 suggest a crude correlation between sediment accumulation rates and the plate tectonic position of individual basins. Basins situated far from plate margins of any type (cratonic basins, continental terraces and rises and their ancient analogues) show slow sediment accumulation rates (≤0.060 m/1,000 yr and generally less than 0.030 m/1,000 yr). Basins proximal to plate margins almost invariably show faster rates of sediment accumulation (almost always at least 0.070 m/1,000 yr, usually several times faster). However, no statistically significant differences exist between sediment accumulation rates along divergent (rift valleys and aulacogens) and convergent (foreland basins and trenches) margin basins despite the very great known differences in sediment composition, texture, and depositional environment.

TABLE 1. ACCUMULATION RATE DATA FOR SELECTED SEDIMENTARY SUCCESSIONS

Section	Age and geographical setting	Thickness (m)	Time interval (m.y.)	Accum. rate (m/1,000 yr)	Reference
Cratonic basins					
1	Paleozoic cratonic basin (Michigan Basin)			0.010	General estimate
2	Paleozoic cratonic basin (Michigan Basin)			0.024	Fischer (1975)
3	Paleozoic cratonic basin (Williston Basin)			0.006	Fischer (1975)
4	Paleozoic Wyoming cratonic shelf	2,156	400	0.005	Schwab (1969a)
5	Paleozoic Arizona cratonic shelf			0.003	Fischer (1975)
6	Paleozoic cratonic "neutral" area			0.001	General estimate
7	Typical modern abyssal plain sedimentation (1 mm/1,000 yr)			0.001	General estimate
8	Modern western North Atlantic abyssal plain			0.0025	Lisitzin (1972)
9	Modern abyssal plain off Peru and Chile			0.0027	Lisitzin (1972)
10	Modern abyssal plain off Baja			0.002	von Huene (1974)
11	Modern Pacific border off California			0.015	Lisitzin (1972)
Ancient miogeosynclines					
12	Cambrian-Ordovician carbonates, central and southern Appalachians	2,464	115	0.0214	Colton (1970)
13	Silurian-Devonian carbonates, central and southern Appalachians	770	25	0.0308	Colton (1970)
14	Paleozoic miogeosyncline, Vermont	4,900	160	0.031	Schwab (1969b)
15	Paleozoic miogeosyncline, Tennessee	3,400	150	0.026	Schwab (1969b)
16	Paleozoic and Mesozoic Wyoming miogeosyncline	5,000	400	0.0125	Schwab (1969b)
17	Paleozoic miogeosyncline, southeastern Idaho	4,800	375	0.0128	Schwab (1969b)
18	Paleozoic miogeosyncline, central Nevada	6,400	160	0.031	Schwab (1969b)
19	Precambrian Coronation miogeosyncline, Canada	3,500	300	0.012	Hoffman and others (1974)
Modern continental terraces (shelves)					
20	Modern Brazilian marginal basins (miogeosyncline)	3,000	100	0.030	Asmus and Ponte (1973)
21	Cenozoic Blake-Bahama Platform	2,000	70	0.029	Asmus and Ponte (1973)
22	Upper Jurassic–Cenozoic Scotia Shelf	4,000	165	0.024	Keen (1974)
23	Cenozoic east African continental shelf	4,000	70	0.057	Kent (1974)
24	Jurassic-Tertiary Newfoundland shelf	6,000	160	0.0375	Sheridan (1974)
25	Triassic-Cenozoic Gulf of Maine–Georges Bank	10,000	225	0.044	Sheridan (1974)
26	Jurassic-Tertiary terrace, off Cape Hatteras	4,000	135	0.0296	Stewart and Poole (1974)
Ancient eugeosynclines					
27	Cambrian-Devonian eugeosyncline, New England	8,400	250	0.034	Schwab (1969b)
28	Cambrian-Jurassic eugeosyncline, central Nevada	12,000	420	0.029	Schwab (1969b)
29	Paleozoic eugeosyncline, Klamath Mountains, Oregon	13,500	365	0.037	Schwab (1969b)
30	Precambrian-Devonian eugeosyncline, Great Basin, Nevada	11,000	300	0.036	Stewart and Poole (1974)
31	Paleozoic eugeosyncline, central Nevada	10,000	300	0.033	Eardley (1962)
Modern continental rises					
32	Cretaceous-Cenozoic rise off Cape Hatteras	6,000	135	0.0444	Stewart and Poole (1974)
33	Mesozoic-Cenozoic rise off Newfoundland	4,000	160	0.025	Sheridan (1974)
34	Cretaceous-Cenozoic rise off the Gulf of Maine	2,000	135	0.015	Sheridan (1974)
Foreland basins—Flysch and molasse clastic wedges					
35	Mesozoic-Cenozoic exogeosyncline, Jackson, Wyoming	4,700	110	0.0427	Schwab (1969a)
36	Taconic clastic wedge, New England	740	10	0.0740	Colton (1970)
37	Acadian clastic wedge, New England	3,080	35	0.088	Colton (1970)
38	Antler clastic wedge, Nevada	4,500	55	0.081	Poole (1974)
39	Antler-Sonoma clastic wedge, Nevada	12,250	140	0.088	Bissel (1974)
40	Cretaceous-Tertiary molasse, Canadian Cordillera	6,000	135	0.044	Eisbacher and others (1974)
41	Upper Ordovician Martinsburg flysch	3,000	20	0.150	Colton (1970)
42	Carboniferous wildflysch, Oklahoma			0.169	Morris (1974)
43	Carboniferous flysch, Oklahoma			0.927	Morris (1974)
44	Jurassic-Cretaceous Franciscan Formation, California	20,000	75	0.266	Schwab (1969b)
45	Jurassic-Cretaceous Great Valley Sequence, California	16,000	75	0.213	Blake and Jones (1974)
46	Ebro molasse basin, Pyrenees, late Eocene–Miocene	3,000	30	0.100	Van Houten (1974)
47	Aquitane molasse basin, Pyrenees, Cenozoic	3,000	30	0.100	Van Houten (1974)
48	Cenozoic Alpine molasse basins	6,000	15	0.400	Van Houten (1974)
49	Cenozoic Lau Basin—marginal basin behind Tonga arc	500	11	0.045	Hawkins (1974)
Trench fill					
50	Eocene Butano submarine fan	2,200	20	0.110	Nelson and Nilsen (1974)
51	Modern Astoria sea fan, Oregon	1,000	2	0.500	Nelson and Nilsen (1974)
52	Japanese arc-trench system, Jurassic–early Tertiary	7,000	70	0.100	Okada (1974)
53	Modern Kurile-Kamchatka trench			0.300	Lisitzin (1972)
54	Modern Pacific trenches—essentially empty	200	200	0.001	Scholl and Marlow (1974)
55	Modern Pacific trenches—full	2,000	200	0.010	Scholl and Marlow (1974)
Successor basins					
56	Cenozoic Canadian Cordillera (Laberge Formation)	2,000	20	0.100	Eisbacher (1974)
57	Cenozoic Canadian Cordillera (Bowser Formation)	2,000	30	0.066	Eisbacher (1974)
58	Cenozoic Uinta Basin, Utah	800	4	0.200	Anderson and Picard (1974)
59	Cenozoic Hoback Basin, Wyoming	4,000	20	0.200	Schwab (1969a)
60	Cenozoic Powder River Basin, Wyoming	3,000	30	0.100	Eardley (1962)
61	Cenozoic Huerfano Park Basin, Colorado	2,300	20	0.115	Eardley (1962)
62	Cenozoic Wind River Basin, Wyoming	3,500	45	0.077	Eardley (1962)
63	Cenozoic Big Horn Basin, Wyoming	3,000	30	0.100	Eardley (1962)
64	Cretaceous-Cenozoic Los Angeles Basin, California			0.058	Fischer (1975)
Rift valleys and aulacogens					
65	Upper Precambrian Grandfather Mountain Formation, North Carolina	6,000	60	0.100	Schwab (1976a)
66	Upper Precambrian Mount Rogers Formation, Virginia	10,000	60	0.166	Schwab (1976b)
67	Upper Precambrian Ocoee Series, Tennessee	12,000	100	0.120	General estimate
68	Ouachita aulacogen, Paleozoic, Oklahoma	5,000	70	0.071	Hoffman and others (1974)
69	Paleozoic Anadarko-Ardmore aulacogen, Oklahoma	7,000	85	0.082	Hoffman and others (1974)
70	Cenozoic Benue graben, west Africa	6,000	65	0.092	Machens (1973)
71	Cenozoic Brazilian rift basins	6,000	40	0.150	Asmus and Ponte (1973)
72	Triassic rift valley fill, Cape Hatteras	2,000	10	0.200	Dietz and Holden (1974)
73	Triassic Newark Series, New Jersey	6,000	10–20	0.400	Eardley (1962)
74	Triassic Newark Basin, New Jersey			0.330	Fischer (1975)
75	Modern Red Sea	6,000	40	0.150	Coleman (1974)
Classical geosynclinal basins					Kay (1955)
A	Average maximum eugeosyncline			0.150	
B	Average maximum miogeosyncline			0.120	
C	Average maximum exogeosyncline			0.180	
D	Average maximum epieugeosyncline (successor basin)			0.300	
E	Average maximum taphrogeosyncline (rift basin)			0.600	
F	Mesozoic-Cenozoic Gulf Coast (parageosyncline)			0.150	
G	Cambrian-Pennsylvania Michigan Basin (autogeosyncline)			0.030	
H	Pennsylvania zeugogeosynclines (yoked basins), Colorado			0.075	
I	Paleozoic-Eocene Hannah intermontane basin, Wyoming			0.129	
J	Cenozoic Ventura Basin, California			0.455	
K					
L	Average rates of continental subsidence			0.0015–0.006	

REFERENCES CITED

Anderson, D. W., and Picard, M. D., 1974, Evolution of synorogenic clastic deposits in the intermontane Uinta Basin of Utah, *in* Dickinson, W. R., ed., Tectonics and sedimentation: Soc. Econ. Paleontologists and Mineralogists Spec. Pub. 22, p. 167-189.

Asmus, H. E., and Ponte, F. C., 1973, The Brazilian marginal basins, *in* Nairn, A.E.M., and Stehli, F. G., eds., The ocean basins and margins, the South Atlantic, Vol. 1: New York, Plenum Press, p. 87-133.

Bally, A. W., 1975, A geodynamic scenario for hydrocarbon occurrences, *in* Proceedings 9th World Petroleum Congress, Tokyo, Vol. 2 — Geology: Barking, England, Applied Science Pubs. p. 33-44.

Bissel, H. J., 1974, Tectonic control of late Paleozoic and early Mesozoic sedimentation near the hinge line of the Cordilleran miogeosynclinal belt, *in* Dickinson, W. R., ed., Tectonics and sedimentation: Soc. Econ. Paleontologists and Mineralogists Spec. Pub. 22, p. 83-97.

Blake, M. C., Jr., and Jones, D. L., 1974, Origin of Franciscan melanges in northern California, *in* Dott, R. H., Jr., and Shaver, R. H., eds., Modern and ancient geosynclinal sedimentation: Soc. Econ. Paleontologists and Mineralogists Spec. Pub. 19, p. 345-357.

Coleman, R. G., 1974, Geological background of the Red Sea, *in* Burk, C. A., and Drake, C. L., eds., The geology of continental margins: New York, Springer-Verlag, p. 743-752.

Colton, G. W., 1970, The Appalachian Basin — Its depositional sequences and their geological relationships, *in* Fisher, G. W., Pettijohn, F. J., Reed, J. C., Jr., and Weaver, K. N., eds., Studies of Appalachian geology: Central and southern: New York, John Wiley Interscience, p. 5-47.

Dickinson, W. R., 1971, Plate tectonic models of geosynclines: Earth and Planetary Sci. Letters, v. 10, p. 165-174.

———1974a, Plate tectonics and sedimentation, *in* Dickinson, W. R., ed., Tectonics and sedimentation: Soc. Econ. Paleontologists and Mineralogists Spec. Pub. 22, p. 1-27.

———ed., 1974b, Tectonics and sedimentation: Soc. Econ. Paleontologists and Mineralogists Spec. Pub. 22, 204 p.

Dietz, R. S., 1963, Collapsing continental rises: An actualistic concept of geosynclines and mountain building: Jour. Geology, v. 71, p. 314-333.

Dietz, R. S., and Holden, J. C., 1966, Miogeoclines (miogeosynclines) in space and time: Jour. Geology, v. 74, p. 566-583.

———1974, Collapsing continental rises: Actualistic concept of geosynclines — A review, *in* Dott, R. H., Jr., and Shaver, R. H., eds., Modern and ancient geosynclinal sedimentation: Soc. Econ. Paleontologists and Mineralogists Spec. Pub. 19, p. 14-25.

Dott, R. H., Jr., 1974, The geosynclinal concept, *in* Dott, R. H., Jr., and Shaver, R. H., eds., Modern and ancient geosynclinal sedimentation: Soc. Econ. Paleontologists and Mineralogists Spec. Pub. 19, p. 1-13.

Dott, R. H., Jr., and Shaver, R. H., eds., 1974, Modern and ancient geosynclinal sedimentation: Soc. Econ. Paleontologists and Mineralogists Spec. Pub. 19, 380 p.

Drake, C. L., Ewing, M., and Sutton, G., 1959, Continental margins and geosynclines: The east coast of North America north of Cape Hatteras, *in* Physics and chemistry of the Earth: London, Pergamon Press, v. 3, p. 110-198.

Eardley, A. J., 1962, Structural geology of North America (2nd ed.): New York, Harper and Row, 743 p.

Eisbacher, G. H., 1974, Evolution of successor basins in the Canadian Cordillera, *in* Dott, R. H., Jr., and Shaver, R. H., eds., Modern and ancient geosynclinal sedimentation: Soc. Econ. Paleontologists and Mineralogists Spec. Pub. 19, p. 274-291.

Eisbacher, G. H., Carrigy, M. A., and Campbell, R. H., 1974, Paleodrainage pattern and late-orogenic basins of the Canadian Cordillera, *in* Dickinson, W. R., ed., Tectonics and sedimentation: Soc. Econ. Paleontologists and Mineralogists Spec. Pub. 22, p. 143-166.

Fischer, A. G., 1975, Origin and growth of basins, *in* Fischer, A. G., and Judson, S. J., eds., Petroleum and global tectonics: Princeton, N. J., Princeton Univ. Press, p. 47-79.

Hawkins, J. W., Jr., 1974, Geology of the Lau Basin, a marginal sea behind the Tonga Arc, *in* Burk, C. A., and Drake, C. L., eds., The geology of continental margins: New York, Springer-Verlag, p. 505-520.

Hoffman, P., Dewey, J. F., and Burke, K., 1974, Aulacogens and their genetic relation to geosynclines with a Proterozoic example from Great Slave Lake, Canada, *in* Dott, R. H., Jr., and Shaver, R. H., eds., Modern and ancient geosynclinal sedimentation: Soc. Econ. Paleontologists and Mineralogists Spec. Pub. 19, p. 38-55.

Kay, M., 1951, North American geosynclines: Geol. Soc. America Mem. 48, 143 p.

———1955, Sediments and subsidence through time, *in* Poldervaart, A., ed., Crust of the Earth: Geol. Soc. America Spec. Paper 62, p. 665-684.

Keen, M. J., 1974, The continental margin of eastern North America, Florida to Newfoundland, *in* Nairn, A.E.M., and Stehli, F. G., eds., The ocean basins and margins, Vol. 2, The North Atlantic: New York, Plenum Press, p. 41-78.

Kent, P. E., 1974, Continental margin of East Africa — A region of vertical movement, *in* Burk, C. A., and Drake, C. L., eds., The geology of continental margins: New York, Springer-Verlag, p. 313-320.

Lisitzin, A. P., 1972, Sedimentation in the world ocean: Soc. Econ. Paleontologists and Mineralogists Spec. Pub. 17, 218 p.

Machens, E., 1973, The geological history of the marginal basins along the north shore of the Gulf of Guinea, *in* Nairn, A.E.M., and Stehli, F. G., eds., The ocean basins and margins, Vol. 1, The South Atlantic: New York, Plenum Press, p. 351-390.

Mitchell, A. H., and Reading, H. G., 1969, Continental margins, geosynclines, and ocean-floor spreading: Jour. Geology, v. 77, p. 629-646.

Morris, R. C., 1974, Sedimentary and tectonic history of the Ouachita Mountains, *in* Dickinson, W. R., ed., Tectonics and sedimentation: Soc. Econ. Paleontologists and Mineralogists Spec. Pub. 22, p. 120-142.

Nelson, C. H., and Nilsen, T. H., 1974, Depositional trends of modern and ancient deep-sea fans, *in* Dott, R. H., Jr., and Shaver, R. H., eds., Modern and ancient geosynclinal sedimentation: Soc. Econ. Paleontologists and Mineralogists Spec. Pub. 22, p. 69-91.

Okada, Hakuyu, 1974, Migration of ancient arc-trench systems, *in* Dott, R. H., Jr., and Shaver, R. H., eds., Modern and ancient geosynclinal sedimentation: Soc. Econ. Paleontologists and Mineralogists Spec. Pub. 19, p. 311-320.

Poole, F. G., 1974, Flysch deposits of Antler foreland basin, western United States, *in* Dickinson, W. R., ed., Tectonics and sedimentation: Soc. Econ. Paleontologists and Mineralogists Spec. Pub. 22, p. 58-82.

Scholl, D. W., and Marlow, M. S., 1974, Sedimentary sequence in modern Pacific trenches and the deformed circum-Pacific eugeosyncline, *in* Dott, R. H., Jr., and Shaver, R. H., eds., Modern and ancient geosynclinal sedimentation: Soc. Econ. Paleontologists and Mineralogists Spec. Pub. 19, p. 193-211.

Schwab, F. L., 1969a, Cyclic geosynclinal sedimentation: A petrographic evaluation: Jour. Sed. Petrology, v. 39, p. 1325-1343.

———1969b, Lexington, Va., Washington and Lee Univ. Open-File Rept. FLS-1.

Sheridan, R. E., 1974, Atlantic continental margin of North America, *in* Burk, C. A., and Drake, C. L., eds., The geology of continental margins: New York, Springer-Verlag, p. 391-408.

Stewart, J. H., and Poole, F. G., 1974, Lower Paleozoic and uppermost Precambrian Cordilleran miogeocline, Great Basin, western United States, *in* Dickinson, W. R., ed., Tectonics and sedimentation: Soc. Econ. Paleontologists and Mineralogists Spec. Pub. 22, p. 28-57.

Van Houten, F. B., 1974, Northern Alpine molasse and similar Cenozoic sequences of southern Europe, *in* Dott, R. H., Jr., and Shaver, R. H., eds., Modern and ancient geosynclinal sedimentation: Soc. Econ. Paleontologists and Mineralogists Spec. Pub. 19, p. 260-273.

von Huene, R. E., 1974, Modern trench sediments, *in* Burk, C. A., and Drake, C. L., eds., The geology of continental margins: New York, Springer-Verlag, p. 207-211.

MANUSCRIPT RECEIVED MAY 24, 1976
MANUSCRIPT ACCEPTED SEPTEMBER 27, 1976

Part X

GEOSYNCLINES AND PLATE TECTONICS: THE SCIENTIFIC REVOLUTION

Editor's Comments
on Papers 43 Through 46

43 DEWEY and BIRD
Plate Tectonics and Geosynclines

44 GREEN
Excerpts from *The Evolution of the Earth's Crust and Sedimentary Basin Development*

45 WANG
Geosynclines in the New Global Tectonics

46 DOTT
The Geosynclinal Concept

Papers 43, 44, 45, and 46 represent various attempts to understand and/or redefine classical geosynclinal theory within the context of modern plate tectonics mechanisms. Dewey and Bird pioneered efforts to reinterpret ancient mobile belts in terms of sea-floor spreading, continental drift, and lithosphere plate convergence and divergence. It is therefore not surprising that they conclude that the bulk of the sedimentological, volcanic, and structural properties attributed to characterize conventional geosynclines (as they were originally conceived by Hall and Dana, and later expanded upon, classified, and refined largely by Stille and Kay) can be readily accommodated within a developmental framework of ocean basins that open and close as a result of lithosphere plate accretion and destruction (Paper 43).

Dewey and Bird systematically demonstrate that the commonly recognized varieties of orthogeosynclinal* basins (that is those geosynclines which eventually become the location of orogenies) can each be equated with modern ocean basin-continental margin complexes seen today at various stages of development around the globe. These include ocean basins in the initial rifting stage like the Red Sea (taphrogeosynclines); ocean basins like the Atlantic where broad abyssal plain areas (leptogeosynclines) are flanked by aseismic, passively subsiding continental terrace and continental rise areas (miogeocline-eugeocline couples); and complicated ocean basins and marginal sea-island arc complexes like those around the periphery of the Pacific (trenches, inland sea areas, idiogeosynclines as defined by Umgrove, 1933, and kinegeoclines as defined by Dewey and Bird in Paper 43, p. 634). Dewey and Bird relate other, nonorthogeosynclinal basins (for example epieugeosynclines and zeugogeosynclines) to positions physically within stabilized orthotectonic orogens. Conversely, exogeosynclines and autogeosynclines are typical intracratonic basins that are essentially identical to those prescribed by Kay within the context of conventional, stabilist tectonism. Dewey and Bird discard as invalid the notion that geosynclines (miogeoclinal-eugeoclinal couples) ever develop within totally ensialic crustal regions. While their scheme largely retains the basic terminological scheme that existed prior to the formulation of plate tectonics, they find it necessary to refine and expand this terminology to some extent.

Paper 44 presents several beautifully drafted diagrams extracted from a paper by A. R. Green, schematically differentiating the various types of geosynclinal basins by using their modern, actualistic equivalents—the various kinds of existing ocean basins—as examples. Like Dewey and Bird, Green contends that these various types of ocean basins with their bordering continental margins can be juxtaposed temporally and spatially to generate the various examples or orogenic belts *and* the bulk of the earth's continental crust. Green argues that basin type (geosynclinal variety), as well as the physical setting, tectonic stability, structural style, and sedimentation in any region are fundamentally controlled by the nature of the crust underlying that area. He considers crustal evolution an orderly process. Various transitional, metastable crustal types occur in a continuum between thin, basic oceanic crust and thick, acidic continental crust. These various types of transitional crusts "host" (that is, control) the various kinds of sedimentary basins occurring around the globe. "The Earth's crust is interpreted as dynamic, changing in both thickness and composition through time. . . . It is postulated that the changes that alter the crust are primarily the result of subcrustal processes and lateral interactions of crustal plates" (Paper 44, p. 1). To Green, geosynclines are transitional in every sense—their crustal position as well as their position in the stage of the earth's development.

Wang (Paper 45) also urges accomodation of geosynclinal principles within the framework of plate tectonics theory, contending that the new global tectonics provides a viable mechanism for generating and deforming the various types of geosynclinal basins. Wang maintains enthusiasm for the reality of the geotectonic cycle and argues that the systematic spatial patterns of sedimentation described by Bertrand and Krynine occur even more commonly than was originally perceived. Their occasional absence is attributed to be the logical result of the imperfections in the geotectonic cycle that can now be understood as due in turn to the somewhat random nature in the timing and location of active lithospheric plate boundaries. Wang is quite obviously at odds with Coney (Paper 34).

Finally Dott (Paqer 46) reviews the geosynclinal concept, tracing its developmnt from Hall's original statements up through the 1960s. Dott's 1974 study (Paper 46) closely followed an earlier analysis by Hsu (1973). Dott concludes that the geosynclinal concept has suffered largely because of an "analogue syndrome"— inaccurate and misleading generalizations about geosynclines because these generalizations have been based on an incomplete list of poorly understood modern analogues. Dott optimistically predicts that geological science will ultimately enjoy a great advance thanks to the melding of geosynclinal theory with modern plate tectonics concepts. Armed with the more accurate understanding of modern deformation provided by plate tectonics, we can greatly refine and improve our undertanding of the classical geosyncline. Rather than discard the geosynclinal concept and all its trappings and inherent terminology, Dott contends that more progress will be made by sharpening the concept using the new knowledge and insights of an ancient earth made more truly actualistic thanks to plate tectonics theory. Two later studies written in a similar vein expand on this thesis (Dott, 1978, 1979).

REFERENCES

Dott, R. H., Jr., 1978, Tectonics and Sedimentation A Century Later, *Earth Sci. Rev.* **14**:1034.

Dott, R. H., Jr., 1979, The Geosyncline—A First Major Geological Concept "Made in America," *Two Hundred Years of Geology in America*, University Press of New England, Hanover, N.H., pp. 239–264.

Hsu, K. J., 1973, *The Odyssey of Geosyncline: Evolving Concepts in Sedimentology*, Johns Hopkins University Press, Baltimore, pp. 66–92.

Umbgrove, J. H. F., 1933, Verschillende Typen van Tertiare Geosynclinalen in den Indischen Archipel, *Leidse Geol. Meded.* **5**:33–43.

43

Reprinted from Tectonophysics **10**:625–638 (1970)

PLATE TECTONICS AND GEOSYNCLINES

JOHN F. DEWEY and JOHN M. BIRD

Department of Geology, Cambridge University, Cambridge (Great Britain)
Department of Geological Sciences, State University of New York,
Albany, N.Y. (U.S.A.)

SUMMARY

Kay's (1951) classification of geosynclines, involving bulk sedimentary, volcanic and tectonic assemblages, is accommodated within the mega-framework of oceanic expansion and contraction by lithospheric accretion and consumption. Apparently, entirely continental eugeosynclines do not exist; geosynclines occur in oceans with marginal continental shelves, continental rise, deep ocean basins, small ocean basins and island arcs. An orogen, resulting from crustal loss in trenches at Benioff zones, grows progressively away from the trench, either on the continental margin or as an island arc. The term, kinegeosyncline, is proposed for the contracting trough, trapped between continental margins and growing orogens. The arrival of a continental mass, with its continental margin sediments, at a trench results in collision and an orogen, which may suture continents together.

INTRODUCTION

The driving mechanisms of geosynclines and orogenic belts has long been an important and controversial topic. The various models which have been proposed for geosyncline and orogen evolution all suffer the disadvantage of being ad hoc without a unifying connection to present-day tectonic processes. A major result of the world-wide exploration of the oceans over the past decades is a more realistic model of world-wide tectonics.

Geosynclines, in general terms, are extensive linear features of unusually thick sediments, that may undergo complex deformation, metamorphism and volcanism. Cratonic plate tectonics involving large tangential movements and linear sites of plate accretion and consumption, appealingly accounts for the linearity of geosynclinal belts and the lateral compression characteristics of their deformation. In this paper, we briefly examine the general characteristics of geosynclines in terms of accreting and consuming crustal plate margins. We propose schematic models based on present plate configurations to account for various time, space and lithologic relations of geosynclines.

Hall (1859) recognized great accumulations of Devonian clastic sediments in eastern New York; Dana (1873) proposed the term "geosynclinal" for the depression he realized would be necessary to receive these sediments. Since that time the use of the term geosyncline has been expanded and modified. Kay (1951), brought together the various terms and concepts of geosynclines, and attempted to relate these to present sedimentary and tectonic processes. He pointed out the necessity of considering interrelated facies distributions and volcanic and tectonic events in classifying geosynclines, and recognized the existence of geosynclinal couples and their probable relationships with continental margins, island arcs and orogenic belts. Kay (1944, 1945, 1947) made the important distinction between non-volcanic, shelf-carbonate (miogeosyncline) assemblages, and the volcanic-clastic (eugeosynclinal) assemblages, particularly those of Lower Paleozoic age bordering the North American continent. Drake et al. (1959) pointed out the similarities of Kay's Lower Paleozoic Appalachian example with the eastern North American continental shelf/slope/rise/abyss.

For the purposes of this paper we rely on the following basic aspects of crustal plates:

(1) The lithosphere is segmented into more or less rigid slabs ranging in thickness from about 10 km in the vicinity of the Mid-Oceanic Ridge, to as much as 150 km in shield areas.

(2) Crustal plates are created by mantle-derived magmatic accretion along the Mid-Oceanic Ridge, and are consumed in oceanic trenches where they descend along Benioff zones into the mantle as cold, seismically active slabs.

(3) Continents ride as superficial passive passengers, however, continents impose important restraints on plate motion and, because of buoyancy considerations, are not appreciably consumed (Dewey, 1969b).

(4) All motions of plates are relative and can be described in terms of great circle vectors.

We believe that plates have been in existence for at least the past 10^9 years, that orogenic belts result from their movement, and that pre-orogenic "geosynclinal" accumulations of sediment are analogous to those now found in oceans and continental margins of the world. Dewey (1969a) and Bird and Dewey (1970) have developed, respectively, a generalized and a specific plate model for the cyclic development of an ocean, continental margin and orogen, involving the following: (1) opening of an ocean basin (expanding phase) by lithosphere plate accretion with the development of graben and continental terrace − continental rise sedimentary accumulations, followed by; (2) initiation of plate consumption by the development of a trench-Benioff zone complex (beginning of spreading-contracting phase). Heating above the descending lithosphere slab leads to collapse of the upper parts of the leading edge of the plate on the continent side of the trench, developing an orthotectonic orogen; (3) the eventual driving of the accreting

---

Fig.1. Epieugeosynclines: A−D. Suggested stages in the distension and rupture of a continent and the development of an accreting plate margin. E. Section across the Gulf of Suez (Picard, 1966). F. Section across the northern end of the Red Sea (Girdler, 1966). G and H. Sections across the continental margin of the eastern United States (Drake, 1966).

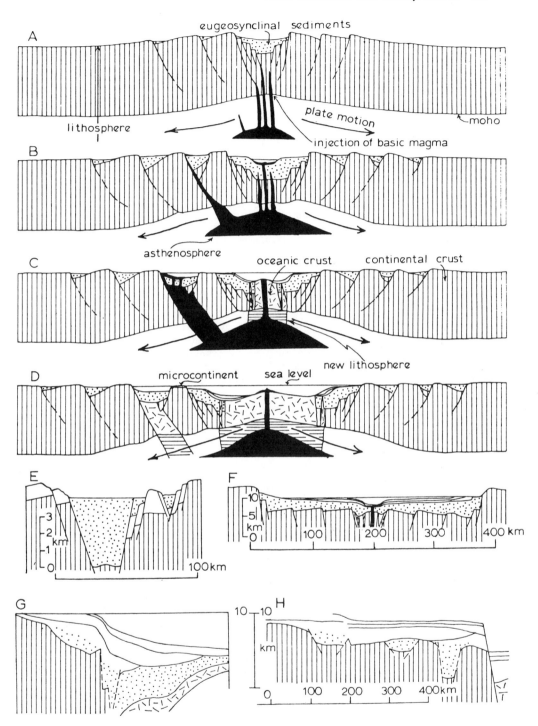

plate margin into the trench, and the development of a contracting ocean phase, during the final stages of which a continent/island arc or continent/ continent collision occurs to form a paratectonic orogen. Consideration of this generalized model leads us to propose general relations between geo- synclines and epicontinental seas, and to fit these to the Kay terminology for geosynclines.

GEOSYNCLINES AND CRUSTAL ACCRETION

There are several present-day oceans with essentially aseismic margins that are expanding by crustal accretion. The Atlantic Ocean has undergone a complex opening history since Late Triassic times and is in an advanced state of opening. The Red Sea/Gulf of Aden system is in a youth- ful state of plate accretion, having started opening in Miocene times (Girdler, 1966; Laughton, 1966). The Baikal Rift (Florensov, 1966), with depths of over 4,000 ft. in Lake Baikal, may represent the initiation of tensional rupture of the Asian continent. We suggest that, during the earliest phases of continental rupture, doming and distension of the conti- nental crust occurs by necking and gravity faulting (Fig.1A) as the base of the lithosphere rises. This gives rise to a complex of graben into which coarse clastics are supplied from intervening horsts. Initially, the graben are floored by continental basement but, as distension continues, the con- tinental crust probably suffers mega-boudinage and simatic prisms are injected. This process may, with continued distension, provide a mechanism for the production of microcontinents on the margins of expanding oceans. Eventually, a systematically accreting plate margin is formed and an expanding ocean develops (Fig.1B). In Fig.1 (A–D), a probable evolutional sequence of geometrical relationships between distending continental crust, graben, horsts, microcontinents, and new ocean floor are shown. The Carboniferous/Permian fault-bounded troughs of the Canadian Maritime Provinces (Belt, 1968) were probably formed during the earliest disten- sional phases preceding the Triassic to Recent crustal accretion of the Atlantic. The Canadian Maritime graben are bounded by faults that commonly show a strong strike-slip component (Belt, 1968) suggesting that, as in the Red Sea/Dead Sea rift system, the expansion vector was narrowly oblique

---

Fig.2. Terminology, environmental interpretation, and examples, of geosynclines. Column A indicates polarity and provenance of sedimentary sequences: Horizontal lines indicate sediment derivation from continental crust; dots from orthotectonic orogens resulting from the collapse of continental rises; southwest to northeast oblique lines from island arcs; black by scraping up of oceanic sediments from descending lithosphere; northwest to southeast oblique lines from paratectonic orogen. Column B indicates nature of volcanics; black squares = mixed acidic and basic volcanism; black triangles = mainly basaltic volcanics of oceanic ridges; open circle = mainly basic volcanics of collapsing continental rise; encircled dots = calc-alkaline volcanics; black circle = ophiolite suite. Column C indicates nature of basement: vertical lines = continental crust; black = oceanic crust.

TERMINOLOGY	INTERPRETATION	SEDIMENTS	A	B	C	MACROTECTONICS	D	EXAMPLES
AUTOGEOSYNCLINE	non-linear intracratonic basin	carbonates-shales		■		CONTINENT		North Sea Basin
EPIEUGEOSYNCLINE	fault-bounded trough formed during continental distension	mainly clastics-red beds						Newark Basin (Triassic)
MIOGEOSYNCLINE	trough adjacent to continental margin	orthoquartzites-carbonates-shales				CONTINENTAL SHELF		Dauphinois-Valais trough of Alps
MIOGEOCLINE	oceanward thickening wedge adjacent to continental margin	as above						Blake Plateau
EPIEUGEOSYNCLINE	as above (=taphrogeosyncline)	as above		■				Gulf of Suez
PARALIOGEOSYNCLINE	thick wedge building from continental margin (supplied rise)	deltaic sediments-slumps-turbidites				CONTINENTAL RISE		Mississippi Delta
LEPTOGEOSYNCLINE	starved rise	thin pelagic sediments-carbonate slides		▲				Atlantic continental rise
	oceanic abyssal plain and mid oceanic ridge and seamounts	pelagic sediments- chert distal turbidites-carbonates				EXPANDING OCEAN		Atlantic Ocean
KINEGEOCLINE	contracting trough or trough complex trapped between continental margin and growing orthotectonic orogen	mainly flysch from growing orthotectonic orogen		○		ORTHOTECTONIC OROGEN		Albee-Pawlet flysch trough of NW Appalachian region (Ord.)
EXOGEOSYNCLINE	trough or trough complex developed on site of earlier miogeosyncline or miogeocline	thin-bedded limestones shales-flysch-wildflysch gravity slides						Normanskill trough of North-western Appalachian region (Ord.)
ZEUGOGEOSYNCLINE	fault-bounded trough within orthotectonic orogen	turbidites-deltaic sediments		⊙				Basin and range province of western USA
IDIOGEOSYNCLINE	static small ocean basin trapped between continental margin and island arc	pelagic sediments-distal turbidites from continent-flysch & tephra from arc		⊙		STATIC SMALL OCEAN BASIN		Sea of Japan
	contracting small ocean basin trapped between island arcs	pelagic sediments-flysch and tephra from arcs		⊙		CONTRACTING SMALL OCEAN BASIN		Philippine Sea
ISLAND ARC	ensimatic orthotectonic orogen developed by scraping up of oceanic sediments,vulcanism, and deformation and metamorphism	pelagic sediments-argille scagliose-flysch		⊙		ORTHOTECTONIC OROGEN		Japan-Kuril-Aleutian arcs
OCEANIC TRENCH	site of plate consumption			●				Japan-Kuril-Peru Chile trenches
LEPTOGEOSYNCLINE	oceanic abyssal plain and mid-oceanic ridge and seamounts	pelagic sediments-carbonates		▲				Pacific Ocean
SMALL OCEAN BASIN	ocean in final stages of contraction by plate consumption on opposing margins					DEVELOPING PARATECTONIC OROGEN		Appalachian Atlantic in Silurian times
	ocean in final stages of contraction by plate consumption on one margin only							Levantine Sea
EXOGEOSYNCLINE	exterior trough	flysch-molasse						Miocene molasse trough of Swiss Plain

PLATE ACCRETION — PLATE CONSUMPTION (column D)

CONTRACTING OCEAN

EUGEOSYNCLINE — ORTHOGEOSYNCLINE — EUGEOSYNCLINE — ORTHOGEOSYNCLINE to the accreting plate margin. Thus, faults such as the Guysborough and Hollow faults (Belt, 1968) may well be fracture zones (transform faults). The Triassic Newark and Connecticut basins are generally asymmetric, being bounded by large faults only on their western margins. This asymmetry is reflected in sedimentary facies distributions: generally, conglomerates along the western fault scarps pass eastwards into finer clastics. This asymmetry is probably a reflection of antithetic faulting on the flanks of an arch associated with the earliest phases of continental distension (Fig.1B). Kay (1951) used the Carboniferous/Permian graben of the Canadian Maritimes as examples of epieugeosynclines and we propose that this term can be used for the coarse-clastic filled graben produced during the continental distension preceding crustal-accretion (Fig.2). Dewey (1969b), has argued that successive phases of oceanic opening and closing leads to successive orogenic belts lying adjacent to, and overlapping with, one another. Thus, etymologically, the term epieugeosyncline (lying upon a eugeosyncline) is likely to be correct where this situation holds good; for example the Upper Paleozoic and Lower Mesozoic troughs of eastern North America lie on the deformed sediments and volcanics of the Appalachian orogen.

Volcanicity is commonly an integral feature of the evolution of epieugeosynclines. Initially, mixed acidic and basic volcanics are erupted (e.g. volcanic suites of the Upper Paleozoic troughs of the Alpine Briançonnais). As an accreting plate margin is formed, volcanicity becomes typically oceanic. It is suggested that the gradual shift from mixed to basic volcanicity accompanies the progressive thinning of the continental crust, so that the earliest graben, and those on the flanks of the arch, form on continental crust and, subsequently, graben floors, either between continent and microcontinent or over the accreting crustal margin, become simatic (Fig.1A–D).

It is clear that the whole process of continental rupture varies considerably. The opening of the Red Sea (Fig.1F) was relatively quiet, accompanied by faulting and volcanicity. The Paleocene distension and opening of the North Atlantic between Greenland and Scotland was accompanied, in addition, by granite intrusion. The Late Carboniferous/Permian distension and opening of the northern margin of Tethys was accompanied by Hercynian granitic intrusion, regional high temperature/low pressure metamorphism and, locally, intense deformation.

As oceanic crustal accretion proceeds and the ocean expands, the epieugeosynclinal stage changes to a new régime related to a now distinct continental margin (Fig.1D, G, H; Fig.3). During times of low sea level the continental rise would be a site of relatively rapid sedimentation, for example, the inner edge of the Gulf of Mexico (paraliogeosyncline, Fig.2, 3C). Sections of the continental terrace and continental rise of the eastern United States (Drake et al. 1959) show that the continental terrace may be a simple sedimentary wedge (miogeocline) or a site of geosynclinal subsidence (miogeosyncline). The paraliogeosynclinal sequences grade out into pelagic oceanic sediments (leptogeosynclinal) with occasional distal turbidites

---

Fig.3. Geosynclines associated with expanding oceans and Atlantic-type continental margins.

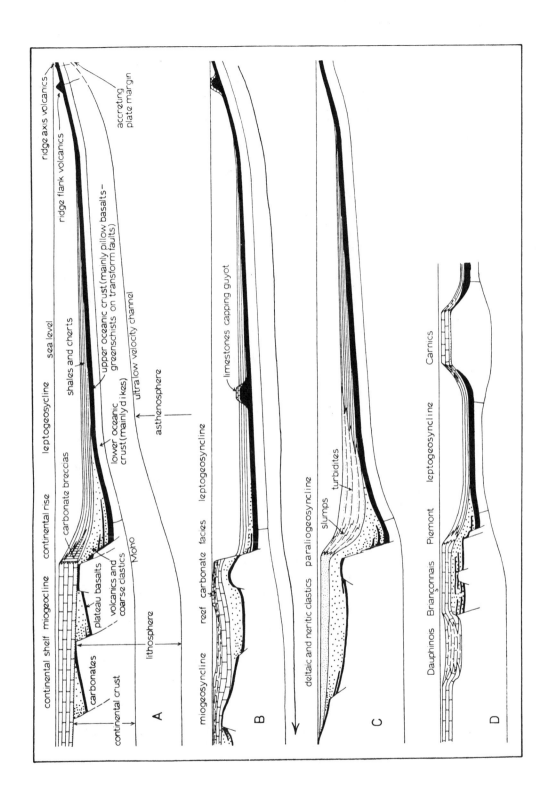

(Fig.2, 3C) supplied by the paraliogeosynclinal continental rise. The lepto-geosynclinal oceanic sediments overlap onto new ocean floor as plate accretion continues. During times of continental submergence an entirely dissimilar facies assemblage will develop (Fig.3A, D). Good examples of this situation are the Cambrian-Ordovician sedimentary assemblages of the western Appalachians (Bird and Dewey, 1970) and the Mesozoic assemblages of the Alps (Ramsay, 1963). Carbonates and calcareous shales accumulate on the continental terrace either as oceanward thickening wedges (miogeo-clines, Fig.3A) (Dietz and Holden, 1967) or as miogeosynclines (Fig.3D) such as the Dauphinois Trough (Ramsay, 1963). The carbonate facies effectively replaces significant sediment supply for the continental rise (Bird and Dewey, 1970) and leptogeosynclinal starved pelagic sequences accumulate (Fig.2; 3A, D).

The transition from epieugeosynclinal to continental margin conditions is accompanied by a rapid decline in volcanic activity as the continental margin is carried further from the accreting plate margin. The movement of the newly created continental margin down the flanks of the arch associated with the accreting margin is possibly a major contributing factor in the sub-sidence which characteristically affects continental margins during the early phases of oceanic expansion (Sheridan, 1969).

Sedimentary polarity during the epieugeosynclinal stage is from horst to graben and, as the ocean expands, from continent to ocean, apart from the distribution of tephra from the Mid-Oceanic Ridge and the accumulation of carbonate screes around seamounts.

GEOSYNCLINES AND PLATE CONSUMPTION

At some advanced stage in oceanic expansion by crustal accretion (Dewey, 1969b), marginal oceanic trenches develop by the consumption of lithosphere (Isacks et al., 1968) leading to a complex series of diachronous processes and the eventual establishment of an orthotectonic orogen (Bird and Dewey, 1970). Dewey, 1969b, has suggested, following Drake et al. (1959), the following outline sequence of events:

(1) The production of the argille scagliose facies in the inner side of the trench by the scraping up of oceanic layers 1 and 2 from the descending lithosphere, and by slumping down the inner trench wall.

(2) The low temperature/high pressure deformation of the trench sediments to form a belt of blueschists.

(3) The progressive inward inversion and collapse of the continental rise accompanied by a diachronous continentward, high temperature/low pressure metamorphism. This sequence on the leading edge of the crustal plate above a Benioff zone is schematically shown in Fig.4A, B.

We are not concerned in this paper with the causal factors for the high surface heat flow behind trenches (McKenzie and Sclater, 1968; Oxburgh and Turcotte, 1968) nor with the problem of trench stability, that is whether the leading plate edge pushes the trench back into the ocean or whether the trench position is fixed by the descending slab of lithosphere.

Fig.4. Geosynclines associated with orogenic belts and contracting oceans.

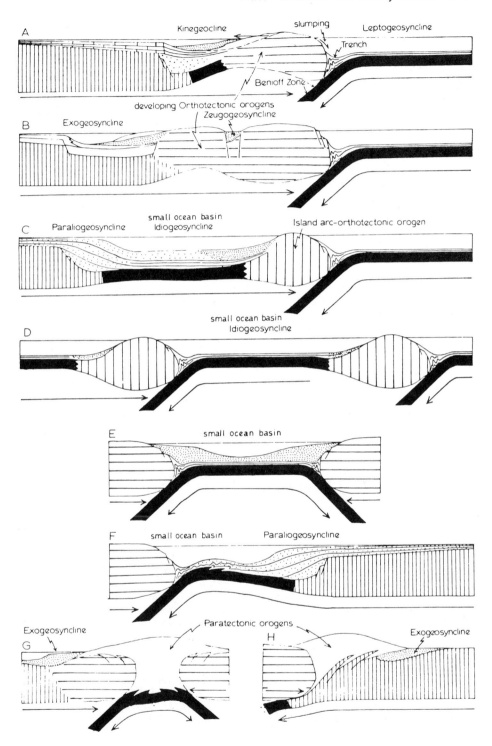

We take all plate motions as relative, but take the view that no significant consumption of plate leading edges occurs.

The progressive continentward collapse of the continental rise, with the concurrent continentward growth of an orthotectonic orogen, leads to a situation excellently displayed during the Ordovician orogeny of the western Appalachians (Bird and Dewey, 1970), where a contracting trough, or trough complex, was trapped between the continental margin and the continentward migrating orthotectonic front. This trough was a synorogenic flysch environment into which clastics poured from the migrating deformation front and became progressively younger towards the continent as the trough complex contracted. We propose the term "kinegeocline" for the continentward-migrating syn-orogenic flysch environment, to indicate the dynamic moving front of an orthotectonic orogen. As the deformation wave arrives at the continental margin the kinegeoclinal complex is destroyed and subsidence now affects the continental terrace to form an exogeosyncline (Fig.2, 4B), in which flysch, wildflysch, gravity slide, and molasse, sequences are deposited (for example, the Ordovician Normanskill exogeosyncline of the western Appalachians, Bird and Dewey, 1970). Practically no sediment is supplied to the kinegeocline and exogeosyncline from the continent; sediments of these environments are dominantly derived from the growing orthotectonic orogen. Thus, for the first time in the history of the evolving ocean, sedimentary polarity is towards the continent.

Zeugogeosynclines (Fig.2, 4B), or fault-bounded troughs yoked to internal cordilleras or uplifts, form after the stabilization of an orthotectonic orogen. Good examples of such troughs are the post-Dalradian South Mayo Trough in western Ireland (Dewey, 1963) and the Ordovician Baie Verte Trough in Newfoundland (Bird and Dewey, 1970).

It is clear from the distribution of consuming crustal plate margins in the Western Pacific that trenches do not always form on distal continental rises but may develop within the ocean. We believe that, in this situation, island arcs are built on a simatic floor by the scraping up of oceanic sediments, by the eruption of basaltic and calc-alkaline volcanics, by the generation of blueschists in trenches, and by the high temperature collapse of scraped-up sediments, perhaps nucleated on microcontinents, above the descending lithosphere slab. The development of an island arc may trap a small ocean basin between continent and arc. This type of small ocean basin (idiogeosyncline, Fig.2, 4C) has an effectively constant size (e.g., Sea of Japan) but becomes progressively filled with sediment (Menard, 1967) derived partly from the island arc. The thick sediment accumulations, combined with the high geothermal gradients metamorphosing the lower parts of the sedimentary sequences of basins trapped behind island arcs, seem to be an effective method of developing a thin continental crust from ocean crust. In theory it would appear that all gradations might exist between a kinegeocline and an idiogeosynclinal small ocean basin, depending on the position of trench development. If the trench forms on the seaward edge of the continental rise, a kinegeocline contracts in advance of an orthotectonic front. If the trench forms well out in the ocean, an idiogeosyncline is trapped between continental margin and the island arc.

Another type of idiogeosynclinal small ocean basin occurs where a segment of oceanic lithosphere is trapped between two island arcs (e.g., Philippine Sea). This type (Fig.2, 4D) must progressively contract if both trenches concurrently consume lithosphere. Thin sedimentary sequences are supplied by the island arcs. A good ancient example of such an idiogeosyncline is the Matapedia Basin (Bird and Dewey, 1970) trapped in Ordovician and Silurian times between the Oliverian and Tetagouche volcanic arcs in the northern Appalachians. The progressive approach, and ultimate suturing, of two island arcs by the contraction of a small ocean basin may be the mechanism by which the double paired metamorphic belts of Japan (Miyashiro, 1967) developed.

During the early stages in the contraction of an ocean, such as the Pacific, by plate consumption in marginal trenches, the trenches block the passage of coarse continent-derived clastics. As the ocean reaches the final stages of contraction, one of two situations may pertain. Dewey (1969a) has suggested that, during Late Silurian times, an Appalachian/Caledonian Atlantic Ocean reached the last stages of contraction by plate loss on both margins. Flysch wedges built out across the trenches onto the ocean floor (Fig.4E) and were deformed in a Late Silurian/Middle Devonian continent/ continent collision to form a paratectonic orogen (Fig.4G). An alternative situation is exemplified by the present day eastern Mediterranean where the oceanic lithosphere of the African Plate is being consumed in the Pliny/ Strabo trench complex south of Crete. The paraliogeosynclinal Nile Delta lies on the aseismic North African continental margin and the distal portions of the delta are being scraped up and deformed to form the Mediterranean Ridge (W.B.F. Ryan, personal communication, 1969). This is schematically shown in Fig.4F. Inevitably, the whole of the Nile Cone will be driven against the Cretan Arc and the North African Shield may underthrust the arc to develop a paratectonic orogen such as the Himalayas (Fig.4H). Exogeosynclines may be associated with both symmetric and asymmetric paratectonic contraction. The Catskill Basin accumulated a thick flysch/ molasse sequence derived from the Middle Devonian paratectonic Acadian contraction of the Appalachian Atlantic (Bird and Dewey, 1970), (Fig.4G), and the Siwalik Trough accumulated flysch/molasse sequence currently being overridden by the Himalayan thrust sheets (Fig.4H).

CONCLUSIONS

We contend that Kay's (1951) classification and analysis of geosynclines, in terms of their facies assemblages and tectonic patterns, may be accommodated within a framework of oceans expanding and contracting by lithospheric plate accretion and consumption. Kay's orthogeosynclines, that is, those geosynclines destined to become the sites of orogeny, constitute the ocean basins and continental margins evolving through complex oceanic expansion and contraction cycles. Kay's eugeosynclines lie entirely on the oceanic side of the continental terrace and are floored mainly by oceanic crust, although the marginal early epieugeosynclines, paraliogeosynclines, and leptogeosynclines may lie partly on thinned and distended continental crust. Eugeosynclines consist of the whole oceanic complex up to the outer edge of the continental terrace and include the continental rise, the main ocean basin, island arcs and small ocean basins. Eugeosynclines are characterized by a specific spatial and temporal distribution of volcanics.

Acidic and basic volcanicity occurs in the epieugeosynclinal stage of continental distension. Throughout the miogeoclinal/miogeosynclinal/paraliogeosynclinal/leptogeosynclinal stages on the continental margin, volcanicity is limited to the accreting plate margin, the site of which becomes progressively further from the continental margin during oceanic expansion. At the time of decoupling at continent and ocean, oceanic crust is segmented and thrust oceanward, behind the inner walls of trenches. During, and after, the growth of marginal orthotectonic orogens calc-alkali volcanic suites are erupted. Thus, for much of the period of accumulation of sediments in the eugeosynclinal environment of the continental rise, volcanicity is completely lacking (for example, the east coast of the United States; Drake et al., 1959).

The epieugeosynclinal, miogeoclinal and miogeosynclinal sequences of the continental terrace are floored by continental basement and are only characterized by volcanicity during the epieugeosynclinal stage, and become the site for both orthotectonic, and paratectonic non-volcanic exogeosynclines.

Fault-bounded zeugogeosynclines form within stabilized orthotectonic orogens and epieugeosynclines may form in a similar position. Furthermore, the facies assemblages of both may be very similar, consisting of coarse clastics derived from adjacent cordilleras and horsts, both are characterized by volcanicity. They differ, however, in their temporal position in an oceanic expansion/contraction cycle, and in the style of their deformation. Epieugeosynclines form at the initiation of oceanic expansion under a tension régime and are blanketed by the sedimentary accumulations of continental shelf and rise, whereas zeugogeosynclines form immediately following orthotectonic orogeny shortly after the beginning of oceanic contraction, the marginal faults commonly have associated serpentinite screens (for example, the South Mayo Trough). The first major deformation of epieugeosynclines occurs during orthotectonic orogeny whereas that of zeugogeosynclines occurs during paratectonic orogeny.

We entirely discard the commonly accepted notion of wholly ensialic miogeosynclinal/eugeosynclinal couples lying between stable forelands, and consider such orogens as the Appalachians, Caledonides, and Urals, to have evolved by the expansion and contraction of ocean basins. To our knowledge, continuity of continental crust under an orogen throughout its history from foreland to foreland, can nowhere be demonstrated. We believe that crustal consumption is the only viable mechanism for providing the large contractions of orogenic belts and it seems to us untenable that a contractional margin could develop within a continent to eventually consume large amounts of continental crust. Clearly continental crust is now being consumed along the Zagros crush zone (Wells, 1970), and under the Himalayas, but this followed the driving out of the oceanic lithosphere of Tethys and may involve the splintering and interfingering of slices of continental crust rather than the steady consumption of crust carried by a descending lithosphere slab. By the very nature of lithosphere plate consumption, leading eventually to continental collision and underthrusting, it will be rare for ocean floor to be extensively preserved beneath the sedimentary sequences of orogenic belts, except as occasional splinters and wedges constituting the ophiolite suite in suture zones such as the Indus suture of the Himalayas and the median suture of the Urals.

REFERENCES

Belt, E.S., 1968. Post-Acadian rifts and related facies, eastern Canada. In: E-an Zen, W.S. White, J.B. Hadley and J.B. Thompson (Editors), Studies of Appalachian Geology — Northern and Maritime. Interscience, New York, N.Y.

Bird, J.M. and Dewey, J.F., 1970. Lithosphere plate-continental margin tectonics and the evolution of the Appalachian Orogen. Geol. Soc. Am., Bull. (in press).

Dana, J.D., 1873. On some results of the earth's contraction from cooling including a discussion of the origin of mountains and the nature of the earth's interior. Am. J. Sci., 5: 423—443.

Dewey, J.F., 1963. The Lower Palaeozoic stratigraphy of central Murrisk, Country Mayo, Ireland, and the evolution of the South Mayo Trough. Quart. J. Geol. Soc. London 119: 313—344.

Dewey, J.F., 1969a. Evolution of the Appalachian/Caledonian orogen. Nature, 22: 124—129.

Dewey, J.F., 1969b. Continental margins: a model for the transition from Atlantic type to Andean type. Earth Planetary Sci. Letters, 6 (3): 189—197.

Dewey, J.F. and Bird, J.M., 1970. Mountain belts and the new global tectonics. J. Geophys. Res. (in press).

Dietz, R.S. and Holden, J.C., 1967. Miogeoclines in space and time. J. Geol., 65: 566—583.

Drake, C.L., 1966. Recent investigations on the continental margin of eastern United States. In: Continental margins and island arcs — Can. Geol. Surv., Paper, 66—15: 33—47.

Drake, C.L., Ewing, M. and Sutton, J., 1959. Continental margins and geosynclines: the east coast of North America north of Cape Hatteras. In: L.H. Ahrens, K. Rankama and S.K. Runcorn (Editors), Physics and Chemistry of the Earth. Pergamon, London, 5: 110—198.

Florensov, N.A., 1966. The Baikal Rift Zone, In: The World Rift System — Can. Geol. Surv., Paper, 66—14: 173—180.

Girdler, R.W., 1966. The role of translational and rotational movements in the formation of the Red Sea and Gulf of Aden. In: The World Rift System — Can. Geol. Surv., Paper, 66—14: 65—77.

Hall, J., 1859. Description and figures of the organic remains of the lower Helderberg group and the Oriskany sandstone. Natural History of New York; Palaeontology. Geol. Surv., N.Y., Albany, N.Y., 3: 532 pp.

Isacks, B., Oliver, J. and Sykes, L.R., 1968. Seismology and the new global tectonics. J. Geophys. Res., 73: 5855—5900.

Kay, M., 1944. Geosynclines in continental development. Science, 99: 461—462.

Kay, M., 1945. Palaeogeographic and palinspastic maps. Bull. Am. Assoc. Petrol. Geologists, 29: 426—450.

Kay, M., 1947. Geosynclinal nomenclature and the craton. Bull. Am. Assoc. Petrol. Geologists, 31: 1289—1291.

Kay, M., 1951. North American geosynclines. Geol. Soc. Am., Mem., 48: 143 pp.

Laughton, A.S., 1966. The Gulf of Aden, in relation to the Red Sea and the Afar Depression of Ethiopia. In: The World Rift System — Can. Geol. Surv., Paper, 66—14: 78—97.

McKenzie, D.P. and Sclater, J.G., 1968. Heat flow inside the island arcs of the northwestern Pacific. J. Geophys. Res., 73: 3173—3179.

Menard, H.W., 1967. Transitional types of crust under small ocean basins. J. Geophys. Res., 72: 3061—3073.

Miyashiro, A., 1967. Orogeny, regional metamorphism, and magmatism in the Japanese islands. Medd. Dansk Geol. Foren., 17: 390—446.

Oxburgh, E.R. and Turcotte, D.L., 1968. Problem of high heat flow and volcanism associated with zones of descending mantle convective flow. Nature, 218 (5146): 1041—1043.

Picard, L., 1966. Thoughts on the graben system in the Levant. In: The World Rift System — Can. Geol. Surv., Paper, 66—14: 22—32.

Ramsay, J.G., 1963. Stratigraphy, structure and metamorphism in the Western Alps. Proc. Geol. Assoc., London, 74: 357–392.

Sheridan, R.E., 1969. Subsidence of continental margins. Tectonophysics, 7: 219–229.

Wells, A.J., 1969. The Zagros Crush Zone and its tectonic implication. Geol. Mag. (in press).

44

Reprinted from pages 5, 6, 7, 8, 9, and 12 of *Geophys. Monogr.* **20**:1–17 (1977)

THE EVOLUTION OF THE EARTH'S CRUST AND SEDIMENTARY BASIN DEVELOPMENT

A. R. Green
Exxon Production Research Company

[*Editor's Note:* Only figures 4 through 8 and 10 from this article are reprinted here.]

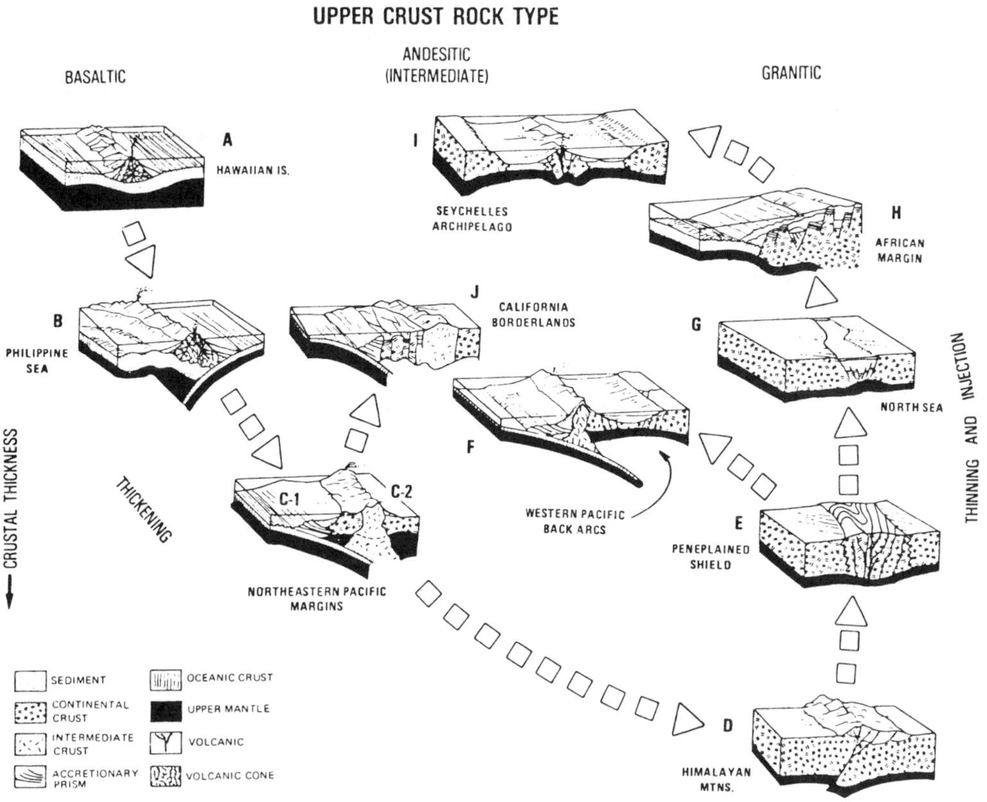

Fig. 4. Schematic cycle of crustal evolution.

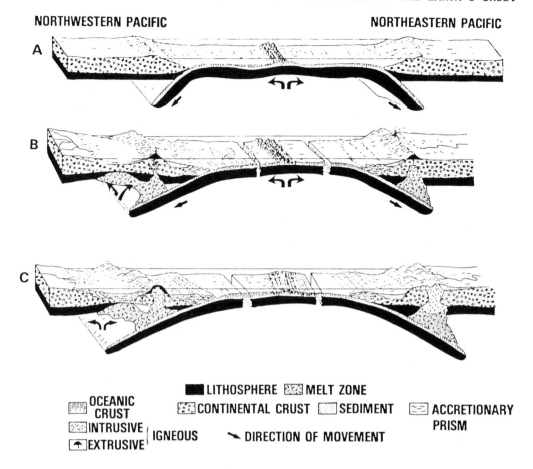

Fig. 5.   Schematic evolution of trench margins of the northwestern and northeastern Pacific.

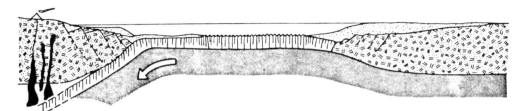

STAGE 1 - OCEANIC CRUST SUBDUCTS UNDER CONTINENT

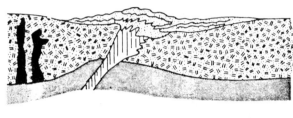

STAGE 2 -
CONTINENTS COLLIDE -
YOUNG MOUNTAIN RANGE

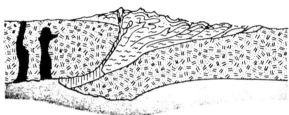

STAGE 3 -
METAMORPHISM AND
THRUSTING FORMS A
MATURE MOUNTAIN RANGE

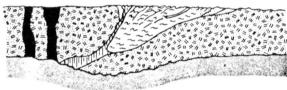

STAGE 4 -
EROSION TO A SHIELD

OCEANIC CRUST   INTRUSIVES   SEDIMENT
CONTINENTAL CRUST   METAMORPHICS   MANTLE

Fig. 6.   Evolution of Himalayan type orogenic belt [after
Dewey and Bird, 1970].

Dewey, J. F., and J. M. Bird, Mountain belts and the new global tec-
    tonics, J. Geophys. Res., 75, 2625-2647, 1970.

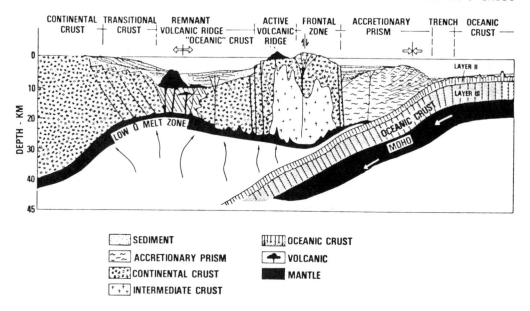

Fig. 7.   Diagrammatic cross section of the western Pacific island arc system [after Karig, 1971].

Karig, D. E., Origin and development of marginal basins in the western
    Pacific, J. Geophys. Res., 76, 2542, 1971.

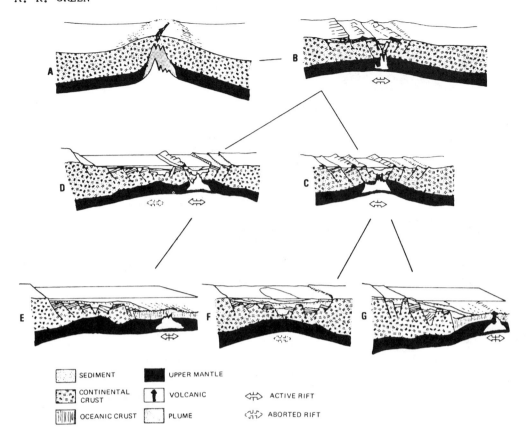

Fig. 8.   Schematic models of rift basin development.

379

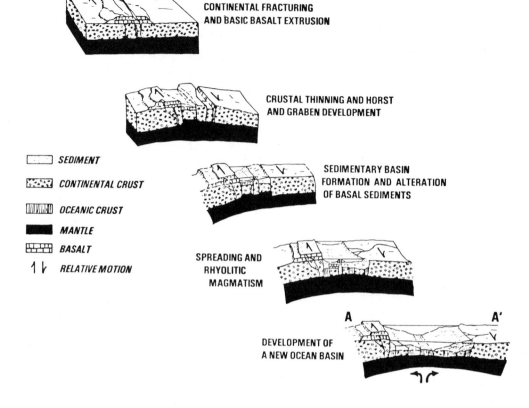

Fig. 10. Evolution of transverse shear basins. (See Figure 9 for location of cross section.)

45

Geosynclines in the New Global Tectonics

CHAUCER S. WANG   *Department of Geology, National Taiwan University, Taipei, Taiwan, China*

ABSTRACT

The concept of the geosyncline has been a controversial subject ever since it was proposed. It appears, however, that in the "new global tectonics," geosynclinal theory could be rationalized into a sound unified concept of geosynclinal genesis and evolution. Starting with continental rifting, or at least stretching, or on continental margins not clearly affected by lithosphere plate movements, an open series of tensile geosynclines will form as extensional trough-ridge complexes or continental terraces, or both, skirted by continental aprons (rises). Then with subsequent ocean closing, or with crustal flexuring and decoupling, a closed series of compressional mio-eugeosynclinal ridge-trough complexes forms, followed by exogeosynclines and molasse basins to complete a geotectonic cycle. In addition to these mature mother geosynclines (orthogeosynclines), which develop through the complete cycle to give rise to Alpinotype mountains, there are also immature geosynclines (parageosynclines), which form in interrupted geotectonic cycles to produce only Germanotype mountains. Endless variations in modes of geosynclinal evolution and mountain building, due to retarded or prolonged geosynclinal opening or to similarly affected flexuring and decoupling in the closing stage, are expected within a geotectonic cycle. These variations are influenced by the vagaries of lithosphere plate movements, as well as those of polycyclic intercontinental and intracontinental geosynclines, some of which are unaffected by lithosphere plate movements.

INTRODUCTION

Throughout the last decade, we have seen with astonishment the recognition of features of ocean-floor spreading and continental drift, from the rift valleys on land down to the ocean floor along the mid-ocean ridges (Hess, 1962; Dietz, 1962). And only three years ago, the theory of the "new global tectonics" was introduced (Isacks and others, 1968; Dickinson, 1970). This theory, which explains existing tectonic conditions under the sea and close to the continent, is considered by many earth scientists, young and veteran, to be the most powerful and workable theory of geotectonics. Certain extreme views, however, even consider that this theory totally revolutionizes the basic concepts of our science—not only concepts like geosynclinal theory, but its corollary, the geotectonic cycle theory, should also be abandoned (Coney, 1970; Ahmad, 1968). I would rather maintain a moderate view of remodeling the old concepts in face of the newly discovered facts. The geosynclinal theory, in particular, appears as valid as ever and now appears as a new unified concept in the light of the theories of ocean-floor spreading and continental drift.

GEOSYNCLINAL GENESIS

There are indeed many kinds of sedimentary depressions on the earth's surface. James Hall first discovered that sedimentary troughs with thick sediment-fill were eventually folded into mountains like the Appalachians. Dana named these sedimentary troughs geosynclinal, yet he regarded the Triassic continental Newark trough, which remained unfolded, also as a geosyncline (Glaessner and Teichert, 1947). Later workers, such as Stille (1940) followed by Kay (1944) classified geosynclines on the basis of sedimentary attributes including thickness, facies, provenance, and composition with further consideration of their folding, spatial arrangements, temporal sequences, and magmatism. Until 1951, at least thirty names had been coined for the different categories of geosynclines (Glaessner and Teichert, 1947; Knopf, 1960).

When Dana referred to the sedimentary trough that gave rise to the Appalachians as geosynclinal, he called the complementary uplift that supplied the sediments a "geanticline," believing both to be formed by compression in a contracting earth. This compressive origin of the geosyncline later evolved into the popular tectogene or mountain-root theory, which postulated that the compressive forces were due to convection currents (Vening Meinesz, 1955). The opposed notion, that geosynclines were formed by crustal stretching, was probably first advocated by Bucher (1933) and has been elaborated by Hsü (1957). Trümpy (1960), in his discussion of the Alps, also expressed the opinion that during an early stage the geosyncline was separated into troughs and horsts by fracturing, contrary to the traditional concept of embryonic tectonics of Argand. Only during the later stage in Cretaceous and Eocene times, when the Alpine geosyncline began to be compressed, were flysch sediments continuously shed from the rising ridges and prograded into the deepening troughs. Thus amidst the different, seemingly unrelated names, we find in addition intracratonic versus intercratonic, ensialic versus ensimatic, tensile versus compressive, and mature "mother" geosynclines (orthogeosynclines) versus immature "sterile" geosynclines (parageosynclines). All in all, they appear to be a group of orderless unrelated features.

However, if we accept the theories of ocean-floor spreading, continental drift, and plate tectonics, the Atlantic-type ocean has been formed from continental rifting starting from rift valleys, evolving through troughs like the Gulf of California or the Red Sea, and finally forming the "Atlantic" ocean. Some geologists and geophysicists for many years have thought that the North American Atlantic continental margin was similar to the Paleozoic Appalachian mio-eugeosyncline couple, although the deep-sea fan apron skirting the Atlantic continental terrace is, if at all, much less intensely volcanic than the Paleozoic eugeosynclinal sediments (Drake and others, 1959). According to this Paleozoic Appalachian model of geosynclinal development (Dietz, 1963; Dietz and Holden, 1966), the North American continental margin might be expected to evolve in the direction of the western Pacific island-arc type of continental margin, and finally again into an Atlantic-type margin (Wang, 1968; Mitchell and Reading, 1969; Dewey and Bird, 1970). The details of such a sequential development will be further discussed by the author in a forthcoming paper concerning continental margins and geosynclines. But it may be said that starting from continental rifting and ending in an eastern Pacific mountain arc or perhaps an Atlantic-type margin, and given the necessary space-time distribution of global mechanisms,

the full sequence of events of a complete orogenic or geotectonic cycle could take place.

Various kinds of geosynclines, recognized and documented as the seemingly unrelated features mentioned above, may well fit into the succession of sedimentary troughs formed within geotectonic cycles that began with extension, which may or may not have been followed by continental rifting. Thus during either rifting and ocean-floor spreading, or during stretching of continental crust, tensile and finally ridge-separated parallel sedimentary troughs formed which may be termed open-series geosynclines. In the Alps, for example, early fracturing, but little separation, took place with development of the Piedmont trough, Brianconnaise platform, and the Valais trough, accompanied by deposition of schistes lustres in the Alpine geosyncline (Trümpy, 1960). Another example, in which continental rifting is clear, is along the Atlantic margin of North America, where the Triassic continental Newark trough and also the Mesozoic-Holocene continental-terrace and continental-rise sediments occur (Drake and others, 1959; Emery and others, 1970). Some geosynclines, such as those of the Paleozoic of western North America, were formed on perhaps accretionary continental margins that were not clearly affected by lithosphere plate movements. When and if an ocean closes, or when tectogenesis begins as a result of various degrees of crustal flexuring or decoupling, a series of events ensues from creation of compressional ridge-trough geosynclinal complexes through development of exogeosynclines and eventually molasse basins (Dietz and Holden, 1966; Trümpy, 1960). The latter events involve sedimentary troughs here termed compressional series geosynclines.

Tensional and compressional events occurred successively in the Paleozoic Appalachian geosyncline. Fracturing took place and extensional trough-ridge complexes developed along the Appalachian trend from the late Precambrian to Early Ordovician. Compressional ridge-trough complexes formed within the crystalline Appalachians after the Early Ordovician during the Taconian-Acadian deformation. Exogeosynclines and molasse basins, which developed in front of the Taconian mountains, were widely superimposed on the originally miogeosynclinal area or encroached on the platform (Zen and others, 1968; King, 1959). Thus with progress both of time and away from the continental nucleus or craton, the open series includes taphrogeosyncline, miogeosyncline, and eugeosyncline, while the closed series, also with passage of time but toward the craton, consists of eugeosynclinal ridge-trough complex, miogeosynclinal ridge-trough complex, epieugeosynclines or exogeosynclines (or both), and molasse basins.

Considered in this way, geosynclines are both tensional and compressional in origin and may be spatially intracratonic, pericratonic, or intercratonic. Thus, apparent discrepancies·disappear, controversies are smoothed over, and the whole concept is unified (Table 1).

GEOSYNCLINAL SEDIMENTATION

The idea that geosynclines formed as open and closed series in a complete geotectonic cycle is consistent with the documented spatial patterns of sedimentation and temporal sequences within the cycle (Krynine, 1951; Pettijohn, 1957; Aubouin, 1965)—in addition to explaining apparent discrepancies in geosynclinal genesis.

In the open-series geosynclines, the taphrogeosynclinal trough is characterized by thick continental coarse clastics of molasse nature, such as those of the Triassic Newark trough in the Appalachian Piedmont area or those of the late Precambrian Torridonian trough in the

TABLE 1. GENETIC SERIES OF GEOSYNCLINES IN A COMPLETE GEOTECTONIC CYCLE

Tensile Open Series	
Geotectonic units	Physiographic equivalents
hedreocraton and parageosynclines	continent
platform	continental
extensional miogeosynclinal ridge and trough	terrace
extensional eugeosynclinal ridge and trough	continental rise
tiefcraton	ocean basin (abyssal plain and seamount complex)
Compressional Closed Series	
(distal to craton)	
tiefcraton	ocean basin and trench
compressional eugeosynclinal ridge	sedimentary outer arc
compressional eugeosynclinal furrows	inter-deeps
	volcanic inner arc
compressional miogeosynclinal ridges	
(proximal to craton)	
miogeosynclinal furrows	idiogeosyncline
exogeosyncline	
molasse basin	shelf or epicontinental seas
platform	continent
hedreocraton	

Northwest Highlands of Scotland. Both the Newark and the Torridonian taphrogeosynclines may be considered to be either the end-stage products of a previous geotectonic cycle or an embryonic feature of a new cycle. Generally speaking, well-developed open-series geosynclines also include existing wide continental terrace deposits skirted by the continental rise (or apron). They also include tectonic elements more distal to the strictly continental setting, notably miogeosynclines and eugeosynclines.

The sediments of the open-series geosynclines are derived chiefly from the continent and those of the continental terrace and rise commonly prograde outward into the open ocean. The continental terrace, cut on basement rocks by both wave and subaerial erosion during a relatively long period of time, is covered by platform sediments that cover its beveled surface at a cleancut, smooth unconformity like the one below the Cambrian quartzite on the Torridonian and Lewisian rocks in northwestern Scotland. The platform sequence is interrupted by numerous diastems, but the sediments are texturally clean and compositionally mature, and are commonly represented by orthoquartzite and limestone. Farther away from the continent, on the outer part of the continental terrace, the sediments gradually thicken. They are in some places cyclic and more feldspathic, with occasional interrelations of limestone and coal. Farthest out, on the continental rise they merge into a thick graywacke section with or without volcanics. Thus in the open series of sedimentary troughs, the sediments comprise platform orthoquartzite and limestone which prograde oceanward over a sequence of turbidites, pelagic lutites, and radiolarian chert.

·The open-series, geosynclinal, graywacke section of the continental rise includes chiefly turbidites conducted down from the neritic-littoral area through submarine canyons, verified by Shepard and others (1969). Because they are displaced neritic-littoral sediments, these turbidites are compositionally and texturally fairly mature, and are commonly more a quartz wacke than a true graywacke. Also, such quartz-wacke turbidites in sections with pelagic sediments have often been considered by some authors to be flysch (Mitchell and Reading, 1969). But inasmuch as the flysch proper in the Alps is cannibalized synorogenic sediment emplaced only after the first geanticlinal ridge or cordillera was raised above water and subaerially eroded (Trümpy, 1960), it is genetically and significantly different from the continent-derived continental-rise (apron) section. Therefore, it seems to me that if a name is needed especially for the continental-rise section, we might coin the term "apron" for it, reserving flysch strictly for the cannibalized detritus deposited near the close of geosynclinal sedimentations.

The sediments of the closed series of geosynclines are commonly cannibalized from ridges or cordillera raised within the geosyncline. They prograde continentward, overlapping the open-series geosynclinal sediments until the geotectonic cycle is ended. Thus as the cycle comes to a close, flysch or molasse is superimposed on the orthoquartzite-limestone section in platform miogeosynclinal areas, followed by flysch superimposed on the "apron" section and commonly accompanied by extrusion of volcanic rocks and emplacement of ultramafic plutons in eugeosynclinal areas distal to the continent (Table 2).

TABLE 2. SEDIMENTATION AND IGNEOUS ACTIVITIES
IN A COMPLETE GEOTECTONIC CYCLE

| Tectonic units | *Tensile Open Series* | |
	Sedimentation (oceanward prograding sedimentation)	Igneous activities
hedreocraton platform	thin quartzite-limestone suite	none
miogeosyncline	quartzite-limestone and feldspathic sandstone-subgraywacke	generally none
eugeosyncline	quartzwacke turbidite and pelagic sediments (apron)	with or without oceanic tholeiitic lava
tiefcraton	pelagic sediments	ophiolite

| Tectonic units | *Compressional Closed Series* | |
	Sedimentation (continentward prograding flysch sedimentation)	Igneous activities
tiefcraton and trench	pelagic and cannibalized turbidites	ophiolite
eugeosynclinal ridge	erosion or neritic carbonate sedimentation (phosphate etc.)	ophiolite
eugeosynclinal furrow	cannibalized flysch sedimentation upon apron	ophiolite and other volcanics
miogeosynclinal ridge	erosion or neritic carbonate sedimentation	andesitic volcanics
miogeosynclinal furrow	flysch emplacement upon quartzite-limestone and subgraywacke sequence	generally none
exogeosyncline	"clastic wedge" to molasse type sediments on quartzite-limestone suite	plutonism in orthogeosyncline area
molasse basin	molasse sedimentation	plateau basalt flow
platform and craton		

IMPERFECTIONS OF THE GEOTECTONIC CYCLE

In the above discussions, both of the genesis of geosynclines and of geosynclinal sedimentation, I have been outlining ideal circumstances during a complete cycle of geosynclinal development, starting from continental fragmentation or at least extension and ending in "orogenesis" and molasse sedimentation. In between the two ends of the complete cycle, there are innumerable forms of incomplete, or apparently incomplete, cycles due to interruptions, delays, or masking of geosynclinal development. Many people cannot appreciate this incomplete nature of geosynclinal development and argue that the concept of the geotectonic cycle should be abandoned (Coney, 1970; Ahmad, 1968).

As we shall see, Stille's geotectonic-cycle concept as interpreted by Cady (1950) and as related to both the intercontinental and intracontinental setting, is easily correlated with the geotectonic cycle based upon ocean-floor spreading and closing. Thus Stille's primary geosynclinal stage (1), marked by the subsidence of longitudinal troughs to leave parallel intervening ridges, refers to the extensional, open series of geosynclinal development. His primary mountain building stage (2), marked by folding and thrusting of the eugeosynclinal-miogeosynclinal ridge-and-trough complex to form a mountain chain with marginal exogeosynclines refers to the compressional, closed series of geosynclinal development. His secondary mountain building stage (3), marked by folding and thrusting of exogeosynclines as well as of primary geosynclinal rocks, includes the deformation of miogeosynclinal flysch and the development of molasse basins in the late closed series. His stage of final differential uplift and local subsidence (4), marked by block faulting, includes the deformation of molasse (Table 3), and may presage development of open-series geosynclines at the beginning of a new geotectonic cycle. When compared to the present-day geomorphic features near continental margins, stage 1 is evidently represented by the extensive continental terrace and high continental rise at the North American Atlantic margin; stage 2 is represented generally by trenches, island arcs and their interspersed furrows, and the extensive epicontinental seas around the margin of the western Pacific Ocean; and stage 3 and probably stage 4 are represented by the mountain arc and trenches along the margin of the eastern Pacific.

In many ways, the concept of the geotectonic cycle is very much like the Davis concept of the geomorphic cycle. We commonly see no complete cycle of geomorphic changes within one region, but the cycle can be appreciated through study of the geomorphic forms preserved at various stages and in different regions. In biological evolution, the sequence of development also becomes obvious only after every link between species has been established. A sea islander, for example, cannot deny evolution merely because he has seen only fishes and birds. To abandon the concept of the geotectonic cycle, because some geosynclines lack primary geosynclinal features and are hence immature inasmuch as they transect a terrane tectonically consolidated in a previous cycle, or because some mature geosynclines have not evolved into orogens inasmuch as compression did or has not yet followed extension, is ridiculous.

There are interruptions within a geotectonic cycle that result either in prolongations of some stages or shortening or inhibition of other stages, just as within a geomorphic cycle there

TABLE 3. COMPARISON OF GEOTECTONIC CYCLES

Classification*	Ocean-floor spreading cycle
1. *primary geosynclinal stage*	*open-series geosynclines*
subsidence between stable elements, ridge occurs close to the end; orthoquartzite-dolomite sedimentation merges to chert, argillite graywacke sequence; spilite and ultramafite	hedreocraton platform ridge miogeosyncline eugeosyncline
2. *primary mountain building stage*	*closed-series geosynclines*
ridge developing to extensive geanticline with secondary geosynclines in front and forward emplacement of nappes. Cannibalized graywacke (arkose) on primary geosynclinal sediments; synorogenic chiefly calc-alkalic[†] plutonism and volcanism	eugeosyncline ridge eugeosyncline furrow miogeosyncline ridge miogeosyncline furrow exogeosyncline
3. *secondary mountain building stage*	
folding of secondary geosynclinal sediments; calc-alkalic[†] intrusion	molasse basin
4. *final differential uplift and local subsidence*	
block faulting within orogens; continental clastic sedimentation and plateau basalt flows, alkalic intrusion[†]	taphrogeosyncline

* After Cady (1950).

[†] W. M. Cady (1971, written commun.).

may be interruptions due to rejuvenation or downwarping. In 1922 Argand had already distinguished the different types of mountains formed by geosynclinal folding; later Stille (1924) pointed out that Alpinotype mountains develop from orthogeosynclines and Germanotype mountains from parageosynclines. Therefore, these differences originate from the unequal development of geosynclinal troughs and have nothing to do with the validity of the geotectonic cycle. Moreover, we can now conclude that the geotectonic cycle may die early due to failure of compression following primary geosynclinal extension, as along the Atlantic Coast of North America where the continental and oceanic lithosphere remain coupled in the American plate. Or the geotectonic cycle in the eastern Pacific plate may be hidden and lost and the orogenic phase in the encroaching North American plate much prolonged because of westward drift of the North American plate which began in the Mesozoic period.

Another aspect of the geotectonic cycle, hotly discussed between 1930 and 1950, was whether orogenesis is cyclic and global. Both the periodicist (Umbgrove, 1947) and the non-periodicist (Gilluly, 1949) are partly right: a full-grown mountain chain developed from a mature geosyncline probably shows a full set of cyclic features reflected in sedimentary and igneous activities, but the life span of an orogen, especially if polycyclic, is obviously not necessarily the same in different places nor does its geographic distribution remain the same. The Appalachian-Caledonian orogen started to develop from a mature geosyncline in the late Precambrian and ended in the Middle Devonian, whereas the Appalachian-Hercynian orogen developed from an immature geosyncline in North America and a relatively mature geosyncline in Europe from the Devonian to the Permian.

ACKNOWLEDGMENTS

I wish to express my sincere thanks for beneficial discussion with K. J. Hsü at Zurich and E. K. Walton at St. Andrews. D. H. Matthews at Cambridge read the manuscript. My thanks are also extended to W. M. Cady for his helpful suggestions for final revision of the manuscript.

REFERENCES CITED

Ahmad, F., 1968, Orogeny, geosyncline and continental drift: Tectonophysics, v. 5, p. 177–189.

Argand, E., 1922, La tectonique de l'Asie: Brussels, Internat. Geol. Cong. Comptes Rendus, p. 171–372.

Aubouin, Jean, 1965, Geosynclines: Amsterdam, Elsevier Pub. Co., 335 p.

Bucher, W. H., 1933, The deformation of the earth's crust: Princeton, Princeton Univ. Press.

Cady, W. M., 1950, Classification of geotectonic elements: Am. Geophys. Union Trans., v. 31, p. 780–785.

Coney, P. J., 1970, The geotectonic cycle and the new global tectonics: Geol. Soc. America Bull., v. 81, p. 739–748.

Dewey, J. F., and Bird, J. M., 1970, Mountain belts and the new global tectonics: Jour. Geophys. Research, v. 75, p. 2625–2647.

Dickinson, W. R., 1970, The new global tectonics: Geotimes, v. 15, p. 18–22.

Dietz, R. S., 1962, Ocean basin evolution by sea floor spreading, in Runcorn, S. K., ed., Continental drift: New York, Academic Press, p. 289–298.

—— 1963, Collapsing continental rises: an actualistic concept of geosynclines and mountain building: Jour. Geology, v. 7, p. 314–333.

Dietz, R. S., and Holden, J. C., 1966, Miogeoclines in space and time: Jour. Geology, v. 74, p. 566–583.

Drake, C. L., Ewing, M., and Sutton, J., 1959, Continental margins and geosynclines—the east coast of North America north of Cape Hatteras, in Ahrens, L. H., and others, eds., Physics and chemistry of the earth, Vol. 3: London, Pergamon Press, p. 110–198.

Emery, K. O., and others, 1970, Continental rise of eastern North America: Am. Assoc. Petroleum Geologists Bull., v. 54, p. 44–108.

Gilluly, James, 1949, Distribution of mountain building in geologic time: Geol. Soc. America Bull., v. 60, p. 561–590.

Glaessner, M. F., and Teichert, C., 1947, Geosynclines: a fundamental concept in geology: Am. Jour. Sci., v. 245, p. 465–483, 571–591.

Hess, H. H., 1962, History of ocean basins, in Engel, A. E., James, H. L., and Leonard, B. F., eds., Petrologic studies (Buddington volume): New York, Geol. Soc. America, p. 599–620.

Hsü, K. J., 1957, Isostasy and a theory for the origin of geosynclines: Am. Jour. Sci., v. 256, p. 305–327.

Isacks, Bryan, Oliver, Jack, and Sykes, L. R., 1968, Seismology and the new global tectonics: Jour. Geophys. Research, v. 23, p. 5855–5899.

Kay, G. M., 1944, Geosynclines in continental development: Science, v. 99, p. 461–462.

King, P. B., 1959, The evolution of North America: Princeton, Princeton Univ. Press.

Knopf, Adolph, 1960, Analysis of some recent geosynclinal theory: Am. Jour. Sci., v. 258, p. 126–136.

Krynine, P. D., 1951, A critique of geotectonic elements: Am. Geophys. Union Trans., v. 32, p. 743–748.

Mitchell, A. H., and Reading, H. G., 1969, Continental margin, geosynclines and ocean floor spreading: Jour. Geology, v. 77, p. 629–646.

Pettijohn, F. J., 1957, Sedimentary rocks (2d ed.): New York, Harper & Bros., 718 p.

Shepard, F. P., Dill, R. F., and Rad, U., 1969, Physiography and sedimentary processes of La Jolla submarine fan and fan valley, California: Am. Assoc. Petroleum Geologists Bull., v. 53, no. 2, p. 390–420.

Stille, Hans, 1940, Einführung in den Ban Amerikas: Berlin, Gebrüder Borntraeger.

Trümpy, Rudolph, 1960, Palaeotectonic evolution of the central and western Alps: Geol. Soc. America Bull., v. 71, p. 843–907.

Umbgrove, J.H.F., 1947, The pulse of the earth: The Hague, Martinus Nijhoff, 358 p.

Vening Meinesz, F. A., 1955, Plastic buckling of the earth's crust: the origin of geosynclines, in Poldervaart, A., ed., Crust of the earth: Geol. Soc. America Spec. Paper 62, p. 319–330.

Wang, C. S., 1968, Continental terrace: its initiation, growth and destruction: Acta Geol. Taiwanica, no. 12, p. 59–74.

Zen, E-an, White, W. S., Hadley, J. J., and Thompson, J. B., Jr., eds., 1968, Studies of Appalachian geology: New York, Interscience, 475 p.

Manuscript Received by the Society April 12, 1971

Revised Manuscript Received November 23, 1971

46

Reprinted from pages 1–13 of *Modern and Ancient Geosynclinal Sedimentation,*
R. H. Dott, Jr. and R. H. Shaver, eds., Society of Economic Paleontologists and
Mineralogists Spec. Pub. 19, 1974, 380 p.

THE GEOSYNCLINAL CONCEPT

R. H. DOTT, JR.

University of Wisconsin, Madison

ABSTRACT

Whether they formed under shallow or deep water, thick geosynclinal sediments were regarded until recently as essential precursors to mountains. Every orogenic belt presumably had evolved stage by stage from geosyncline to mountain, ultimately producing a peripheral accretion to some evergrowing continental craton. Conversely, by implication, thick sediment prisms along any present continental margin inevitably should lead to mountains. These long-standing deterministic generalizations were hardly justified, however, for modern orogenic belts are not consistently located at continental margins, nor do they all contain thick sediments. Moreover, it is impossible to designate any uniquely geosynclinal sediment type. Most geosynclinal sediments are results more than causes of orogenesis; an orthogeosyncline is simply a sediment-filled orogenic belt.

The early, strictly uniformitarian sea-floor spreading model for geosynclines was unacceptable because it regarded continental terrace sediment prisms (miogeoclines) formed on passive or nonorogenic continental margins as essential, evolutionary precursors to mountain building—a holdover from the venerable tectonic cycle. But most existing continental terraces are almost 200 million years old and still show practically no tectonic mobility. Genetically, these miogeoclines belong to a different genus than accumulations formed in active orogenic zones. Miogeoclines form on passive trailing edges of diverging continents, whereas orthogeosynclines form near active leading edges of converging lithosphere plates. Plate tectonics shows how these two genetically distinct sediment prisms may become coincidently crushed together in orogenic belts. Rather than being simple concentric accretions of successive orogenic belts, continents are mosaics of very complexly truncated, overprinted, and even rifted tectonic elements containing haphazard relics of former plate margins.

The geosynclinal concept has suffered from an analogue syndrome. Dogmatic generalizations were applied to all cases from an incomplete list of supposed modern analogues, and tectonic environments were confused with sedimentary ones. But intensive marine research over the past two decades has provided many more well-documented possible analogues for the testing of truly actualistic models. Armed with these, as well as with new tectonic insights and new vocabulary, geosynclinal studies can advance from a long descriptive phase to a more genetic one.

INTRODUCTION

Actualism and analogy.—In the two decades since publication of Marshall Kay's (1951) *North American Geosynclines,* students of both modern and ancient sediments have compiled an immense body of knowledge relevant to the geosynclinal concept. Moreover, the new theory of plate tectonics has seemed overnight to require a complete reassessment of the geosyncline as well as of orogenesis. The purpose of this symposium volume is to evaluate by comparison of modern and ancient sediments a number of depositional models applicable to the great variety of strata seen in orogenic belts and heretofore called "geosynclinal."

Studies of ancient sedimentary rocks have relied heavily upon analogies with modern features of known origin for genetic interpretations. Large-scale analogues, such as deltas and deep-sea trenches, have long been invoked in trying to understand geosynclines. Such comparisons represent familiar applications of the uniformitarian doctrine. Of course, what is involved is simply inductive reasoning from present, more or less ideal dynamic features to allow

inferences about some now-static and imperfect fossilized phenomenon. Being based upon simple morphologic comparisons, such inferences represent a *strict uniformitarianism* (of Dott and Batten, 1971, or *substantive uniformitarianism* of Gould, 1965). But if we are concerned with the full genesis of ancient sediments, then we must make more abstract inferences about processes as well as about morphology. These inferences involve more sophisticated analogies, but with only the single methodological assumption of the invariance of natural laws through geologic time. This restricted assumption was termed *actualisme* by Constant Prevost in 1825. seven years before *uniformitarianism* was coined. It allows the induction of ancient causes through knowledge gained from modern processes without also requiring a one-to-one identity of form, dimensions, rates, and the like between ancient and modern phenomena. The distinction urged here is not trivial because allegedly actualistic models are being invoked carelessly in a wide variety of contexts; a truly actualistic hypothesis or model is something other than merely a one-to-one, modern-ancient analogy (see Hooykaas, 1963). Long ago actualism was given a more precise meaning than was uniformitarianism. As the single most important philosophic doctrine for the study of earth history, that meaning should be strictly maintained.

Geology is greatly dependent upon comparative reasoning as is illustrated throughout this volume. But ". . . it will always be a hypothetical investigation. It cannot be in any way made a series of facts—its production must be *probable* analogies" (Davy, 1840, p. 185; italics mine). The history of geology is rife with abuses of analogical reasoning. As the phenomena being compared differ more and more in scale, details of similarity, or age, analogy becomes increasingly tenuous. Because analogical reasoning has no internal criterion of truth, "it is (either) the bane or the blessing of science" (Hooykaas, 1963, p. 157). Too often we either have chosen in ignorance the wrong modern analogue (the apple-and-orange fallacy) or have seized zealously upon only one of several possible analogues. Moreover, once a comparison is drawn, we have tended to dogmatize and fall victim to the "tyranny of the pigeonhole" (John Rodgers, 1970 oral presidential address, Geological Society of America). A classic example of such tyranny is the widespread use of Kay's 1951 Ordovician cross section through New England as *the* model for miogeosynclinal and eugeosynclinal relationships for all times and places, whereas it was intended simply as one illustrative case.

THE CLASSIC CONCEPT

Early American and European views.—Early literature on geosynclines reminds one of the Indian fable of "The Blind Men and the Elephant," for everyone was partly right, yet partly wrong about his example. Most writers, that is, those from 1859 to 1900, acknowledged a great thickness of sediments within orogenic belts, but agreement ceased there (fig. 1). Early writers on the Appalachians (e.g., James Hall and J. D. Dana) regarded all the geosynclinal sediments as shallow-marine deposits. Meanwhile, European writers such as Bertrand (1897), Haug (1900), and Suess (1909), considering the Alpine orogen, regarded the sediments most typical of geosynclines as relatively deep marine (bathyal) deposits (fig. 1). Haug, especially, thought bathymetry and biofacies more important characteristics than thickness—to him a geosyncline began as an elongated trough receiving pelagic sediments. Soon after 1900, most Europeans tended to assume that geosynclinal accumulation was initiated by the ophiolite sequence, consisting of oceanic ultramafic and mafic igneous rocks succeeded by deep-marine

siliceous and argillaceous sediments (Pantanelli, 1883; Steinmann, 1905). According to Hsü (1972), Suess even drew one-to-one analogies between such deep-marine sediments and mafic rocks on mid-Atlantic islands. Analogies also were drawn between deep-sea sediment samples from the *Challenger* expedition and Alpine strata by T. Fuchs, Emil Haug, G. Steinmann, and A. Heim (Bernoulli and Jenkyns, this volume). Ultramafic and mafic rocks also were recognized in the Appalachian belt, but Americans tended to interpret them as intrusions and extrusions within the orogen, that is, as emplaced during deformation. Were ophiolites extruded from the mantle up through sialic crust, or were they an oceanic basement beneath the sediments? To find deep-sea rocks now in mountains on continents seemed to be a threat to an American assumption of continental permanency, but, by repeatedly drilling through abyssal sediments into mafic basement, the recent Deep Sea Drilling Program has vindicated the European view.

Based upon the Appalachian example, Americans concluded that geosynclines and their apparent descendants, mountains, formed at the margins between continents and ocean basins. Hall (1859) challenged lateral compression due to earth shrinkage as the cause of deformation of strata, which was first elucidated by Elie de Beaumont in 1825. He believed, instead, that sediment loading was the sole cause of both the tenfold greater subsidence of the Appalachian region than of the continental interior and the subsequent deformation of the strata. From Hall (1859):

> "The line of greatest depression would be along the line of greatest accumulation (p. 70) . . . (thus) the course of the original transporting current (p. 73). By this process of subsidence . . . the diminished width of surface above, caused by this curving below, will produce wrinkles and folding of the strata (p. 70)."

J. D. Dana, however, postulated downbending of a *geosynclinal* complemented by an upwarped *geanticlinal* resulting from shrinkage of a supposedly cooling earth having a liquid or plastic interior (fig. 1). Unequal contraction had first given rise to continents and ocean basins; continuing contraction caused bending due "to lateral pressure from the contraction of that crust" (Dana, 1873, pt. V, p. 170). What could be more logical than for buckling to occur primarily at the juncture between two fundamentally different types of crust?

> ". . . The position of mountains on the borders of continents . . . is due to the fact that the oceanic

NORTH AMERICAN

EUROPEAN

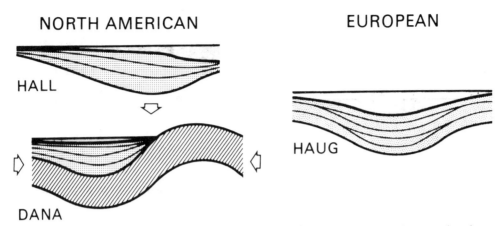

HALL

HAUG

DANA

F<small>IG</small>. 1.—Early conceptions of the geosyncline: American shallow versus European deeper marine views; James Hall's subsidence by sedimentation (upper left) versus J. D. Dana's subsidence due to crustal buckling with complementary upwarping of a geanticline (lower left). Emil Haug's view of a symmetrical, intercratonic trough is idealized at right.

areas were much the largest, and were the areas of greatest subsidence under continued general contraction of the globe . . . the oceanic crust had the advantage through its lower position of leverage, or more strictly speaking, of obliquely upward thrust against the borders of the continents." (Dana, 1873, p. 170–171.)

Meanwhile, again largely on the basis of the Alpine example, Europeans (e.g. Haug, 1900) inferred that geosynclines could form between two continents (fig. 1). Moreover, because of its apparent intercontinental position and the presence of continental (Hercynian) basement rocks in its core, it appeared that the Alpine geosyncline had been initiated by the foundering of continental crust (see Bernoulli and Jenkyns, this volume). Thus were planted the seeds of a long controversy about the ensialic-versus-ensimatic locus of geosynclines (Wells, 1949).

The tectonic cycle.—By the turn of the century the notion of distinct stages of sedimentation in all orogens had developed from the recognition in Europe of the vertical succession of chert and shale through flysch to molasse, the general significance of which was first advocated by Bertrand (1897) and Haug (1900). Daly (1912), Kossmat (1921), and Kraus (1927) broadened the concept to include sequential igneous and structural events. By mid-20th century, both in Europe and North America, the stages of the sedimentary-tectonic cycle had become *preorogenic* or *geosynclinal* (graywacke), *orogenic* (arkose), and *cratonic* (orthoquartzite) (e.g., Krynine, 1948). There are many variations, including the incorporation of metamorphism and ore deposition by many au-

thors (see Knopf, 1948; Aubouin, 1965; Coney, 1970; and Hsü, 1972).

Even before plate tectonics was formulated, incongruities revealed by detailed features of orogenic belts seemed glaring enough for one to question the validity of the geotectonic or geosynclinal cycle (e.g., Dott, 1964a; 1964b). With the advent of plate theory, Coney (1970) argued emphatically for its complete abandonment. The tectonic cycle implied that all orogens developed in just the same manner. There is, however, little consistency of stage of occurrence within the idealized cycle, either of sediment or of igneous types from belt to belt or even within a single belt. For example, flysch and molasse are diachronous, occurring at different times in different loci and grading laterally as well as vertically into one another; in some belts flysch may not be present at all. Such sediment differences are, after all, a function of local depositional environment, which is only indirectly related to stage of tectonic development (fig. 2). Another great shortcoming of the tectonic cycle is its implication of regularity of timing of tectonic events and the seductive implication of a simple evolutionary continuity from a tranquil continental terrace to an unstable orthogeosyncline through time.

In fact, there have been profound genetic discontinuities between overprinted tectonic patterns, and the developmental history of different orogenic belts is similar in only the most general way. To insist upon fitting each belt into one simplistic, man-constructed cycle obscures the element of uniqueness of such belts and thus impedes understanding more than it helps. The most serious defect of the tectonic cycle, how-

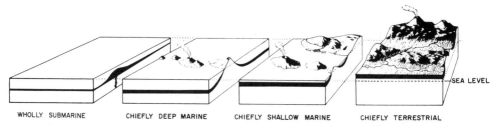

WHOLLY SUBMARINE   CHIEFLY DEEP MARINE   CHIEFLY SHALLOW MARINE   CHIEFLY TERRESTRIAL

Fig. 2.—Diagrammatic portrayal of great diversity of sedimentary environments represented by geosynclinal sediments and volcanic rocks found in orogenic belts. No implication of a temporal sequence (cycle) is intended. (After Dott and Batten, 1971, p. 251).

ever, was its strong implication that thick sediments caused orogenesis. Although there were a few doubters, such as Eduard Suess and H. H. Hess, a geosyncline was generally considered to be a prerequisite for orogenesis, and, conversely, the absence of thick sediments should preclude significant mountain building. "The greater the accumulation, the higher will be the mountain chain," said Hall (1859, p. 83).

Continental accretion.—A logical handmaiden of the tectonic cycle was the important concept of enlargement of continents through the development of successive, more or less concentric orogenic belts. In America this idea became firmly established through the writings especially of Dana (1873) and in Europe, of Suess (1909), Bertrand (1887), Haug (1900), and Stille (1936, 1941).

> "That each epoch of mountain-making ended in annexing the region upturned, thickened and solidified, to the stiffer part of the continental crust, and that consequently the geosynclinal that was afterward in progress occupied a parallel region more or less outside the former. . . ." (Dana, 1873, p. 171).

Accretion generally dictated that juvenile continental crust was generated by regional metamorphism, volcanism, and granitic plutonism accompanying orogenesis in former geosynclinal belts. Moreover, it became a common uniformitarian assumption that continental crust has increased in volume and area at a more or less linear rate through time, a view allegedly supported by geochronology (e.g., Wilson, 1949; Hurley and others, 1962; Engel, 1963). On the other hand, lead-isotope evolution (Patterson, 1964), as well as the increasing documentation of sharply discordant, isotopic-date provinces and of successive metamorphic overprintings or orogenic reworkings (Gastil, 1960; Dott, 1964b), suggested anything but constant-rate accumulation of continental crust and anything

but temporal and spatial regularity of orogenesis. Although limited concentric growth has occurred (cf. Okada, this volume), one was confronted with a very complex picture of continents as mosaics laced by orogenic belts of widely varying ages and patterns. Recognition today that plate spreading can extend from oceans into continents casts doubt upon a long-popular implication of irreversibility of accretion.

Continental accretion implied that most orogenic belts must be at least partially ensimatic and marginal to continents initially if new continental crust was to be generated therein. Early studies suggested that, in North America, at least, Phanerozoic orogens supported most of the implications of accretion. The theory became less popular abroad, however, because of the obviously sialic basement found in many belts, the clear truncations of Caledonian structures by Hercynian ones, the overprinting in turn of Alpine upon Hercynian trends, the present intercratonic position of the Urals and Himalayas, and the location of several modern arc-trench systems far from any continent in a wholly oceanic environment. It seemed to many workers that orogenic belts could be, after all, either ensialic or ensimatic and that they had not formed in any simple concentric fashion. Hess (1960) argued that volcanic arcs were oceanic and not precursors to Alpine-type mountains; the two represent different responses to the same tectonic regime in different geologic environments.

An additional objection to accretion exists if new granitic crust forms solely through granitization of sediments, an idea popular around 1940. The sediments themselves were formed largely by erosion of older continental crust and of volcanic rocks, which were commonly assumed to represent crustal melting. But without additions of juvenile material from outside the continental realm, obviously no increase of crustal volume would be possible. Recent proposals that andesitic eruptions in magmatic arcs probably originate by fractional melting well below the continent provides an appealing answer to this dilemma (e.g., Kuno, 1959; Dickinson and Hatherton, 1967; Dickinson, 1970). If granitic plutons as well as andesites do originate from such melting, a mechanism for accretion of new continental rocks by igneous differentiation is provided after all.

Sedimentation and paleogeography.—European workers have envisioned the geosyncline as a deep sea-floor trough or furrow receiving only fine pelagic sediments at first. Subsequently the sand of the flysch was introduced from terrigenous sources, which were interpreted as embryonic mountain ridges called *cordilleras* that were raised as elongate islands within the geosynclinal belt (Argand, 1916). Flysch was deposited between the cordilleras, but, with full uplift of most of the geosyncline, coarser, largely nonmarine molasse was deposited at the margins of the rising mountains (fig. 2).

Meanwhile in America, the prominence of sandstones and conglomerates in middle and late Paleozoic rocks of the Appalachian Mountains led to the view of geosynclinal sediments all being shallow-marine and nonmarine deposits. Eastward coarsening of these strata and the presence in adjacent western New England of crystalline rocks, which were assumed until the 1930's to be entirely pre-Paleozoic, gave rise to the concept of a persistent eroding borderland of ancient rocks lying next to the subsiding geosyncline. James Hall had already recognized vaguely an eastern land source for Ordovician and later clastic sediments, but Dana (1873, p. 171) was more specific in stating that:

". . . on the ocean side of the progressing geosynclinal referred to, there has been generally, as the first effect of the thrust against the continental border, a progressing geanticlinal, which usually disappeared in the later history of the region. . . ."

In 1882 Chamberlin showed such a borderland on an early Paleozoic paleogeographic map for North America (apparently the first of its kind). Williams (1897) coined the name "Appalachia" for the land, and Schuchert (1910) then generalized *borderland* (or geanticline) sources for all American geosynclines. Deltaic and fluvial deposits were recognized widely among the later Paleozoic strata of the Appalachians, and uniformitation analogies were drawn with the Mississippi River delta system (e.g., Barrell, 1913-1914; Storm, 1945), but the origin of Appalachian Ordovician and Devonian flyschlike deposits remained a mystery.

In 1937 Kay recognized *tectonic lands raised* *within* the Appalachian belt by episodic uplift of earlier geosynclinal rocks to provide sources of younger deposits laid in still-subsiding portions of the belt (fig. 3). These were much like the European cordilleras. Kay's interpretation was given strong impetus through the discovery by Harvard geologists of Paleozoic marine fossils in the alleged pre-Paleozoic crystalline complex of New England, that is, in the heart of old Appalachia (see Kay, this volume).

At the same time, Hans Stille noted the rather consistent arrangement of belts of different rock types within most ancient orogens. His subdivisions focussed upon two parallel zones, one having volcanic and plutonic rocks (including ophiolites)—the eugeosynclinal belt—and one lacking such rocks—the miogeosynclinal belt. Although Stille was European, his principal examples were chosen from the North American Cordillera (Stille, 1936; 1941). Meanwhile, Kay had compiled for teaching purposes the distributions of volcanic rocks and sedimentary facies for various rock systems in North America and found the same patterns to hold for long spans of history in every orogenic belt that he studied. The prominence of volcanic rocks in the eugeosynclinal zones and the physical continuity of several such belts with modern volcanic arcs then led to a general paleogeographic analogy between the two features (Hess, 1939; Kay, 1947; Eardley, 1947). Was this not the first actualistic geosynclinal model? Early in this century, Haug had compared geosynclines to the newly discovered deep troughs of Indonesia, but it was the Dutch studies of the 1930's that gave the greatest impetus to the volcanic arc-trench analogue (Umbgrove, 1933; Vening Meinesz, 1940; Van Bemmeln, 1949). Direct comparisons of Indonesia with the Alpine geosyncline were especially popular, for the trench seemed fully compatible with the long-standing European assumption of deep-marine geosynclinal sedimentation. An unfortunate tendency to equate trenches and eugeosynclines on a one-to-one basis developed and still persists (e.g., Hsü, 1972, p. 34-35). The analogy intended by Kay and others was more general and would encompass both the volcanic lands (the arc) *and* the trench as a possible modern counterpart for an ancient eugeosyncline (see Scholl and Marlow, also Dickinson, this volume). It must be stressed, however, that both "eugeosyncline" and "miogeosyncline" were descriptive terms referring to tangible subdivisions of ancient orogens rather than to any modern morphologic features, sedimentary facies, or hypothetical paleogeographic restoration. So viewed, geosynclines may be *elucidated*

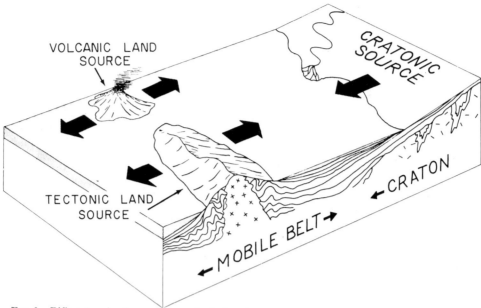

FIG. 3.—Different major types of source lands for clastic geosynclinal sediments; purely diagrammatic in terms of spatial dimensions and number of source types present at any given time. Note cannibalism of slightly older geosynclinal sediments so characteristic of tectonic lands. (After Dott and Batten, 1971, p. 248).

by reference to modern features, but they can not be *defined* by them (see also Aubouin, 1965, p. 3).

In Britain, E. B. Bailey (1930), with his distinction of graded from cross-stratified sandstone suites, was among the first to perceive that much coarse clastic sediment might be deposited in deep-marine environments by some kind of gravity processes. However, O. T. Jones (1938), in spite of his pioneering work with slump structures, maintained that no great depth necessarily prevailed during deposition of graywackes. Yet, in California Natland (1933) had noted that conglomerates as well as sandstones were interstratified with Cenozoic abyssal shales, and in 1950 Kuenen and Migliorini finally resolved the long-standing European-American bathymetric dilemma with their diagnosis of turbidity current transport. Through actualistic reasoning, ancient, coarse submarine fan deposits are being recognized widely in geosynclines (e.g., Sullwold, 1960; Dzulynski and others, 1959; Walker, 1966; and Kulm and Fowler, Mutti, Nelson and Nilsen, and Normark, all this volume). Now we know that a great deal of gravel as well as sand has reached deep-sea fans via submarine canyons, although the exact mechanism of transport of the coarsest materials is just beginning to be understood

(Stauffer, 1967; Aalto and Dott, 1970; Middleton, 1970; Hampton, 1972). Very coarse slide breccias (olistostromes and some wildflysch) also are recognized increasingly, but some are difficult to distinguish from tectonic melanges (see Hsü, Blake and Jones, and Wood, all this volume). Widespread application of paleocurrent analysis pioneered by F. J. Pettijohn and his students in the 1950's also has shed much light upon paleobathymetry and paleogeography in general over the past two decades. The importance of diverse source terranes for geosynclinal sediments—cratonic, volcanic, and tectonic lands (fig. 3)—has been confirmed beyond question through such investigations coupled with sandstone petrography (see Crook, this volume).

By at least 15 years ago, it had become clear that most geosynclines at different times and in different places have had practically every conceivable type of sediment deposited within them (table 1). Schwab (1971) noted that evaporites are the only sediments almost totally lacking. Carbonate rocks and mature sandstones are certainly more abundant in miogeosynclines, and graywacke and chert in eugeosynclines, but none are mutually exclusive. Nonetheless, by 1960 unfortunate dogmatic generalizations were entrenched. For example, "graywacke," "turbi-

TABLE 1.—PERCENTAGE PROPORTIONS OF ROCK TYPES IN SOME REPRESENTATIVE
NORTH AMERICAN ORTHOGEOSYNCLINES

| Rock type | Eugeosynclines | | | | Miogeosynclines |
| | Including volcanics | | Without volcanics | | |
	Dott[2]	Schwab[3]	Dott[2]	Schwab[3]	Schwab[3]
Shale	40	26	50	33	26
Volcanic	20	21	—	—	—
Graywacke[1]	10	30	13	38	2
Carbonate	10	5	13	6	44
Quartz arenite	9	4	11	5	12
Chert	9	7	11	9	1
Conglomerate	2	?	2	?	?
Arkose and subgraywacke	?	7	?	9	15

[1] Includes tuffaceous sandstones.
[2] Point counts of relative areas on geologic maps from eight regions (done for writer by L. J. Suttner, Univ. Wisconsin Research Assistant, 1965).
[3] Percentage of total thickness in two stratigraphic sequences from Appalachian region and two from Cordilleran region (Schwab, 1971).

dite," and "flysch" all became loosely equated with "eugeosyncline" even in the absence of associated volcanic rocks. In fact, many examples of clastic wedges within miogeosynclinal (or, strictly, exogeosynclinal) belts exist. Such wedges are comprised either of flyschlike graywacke deposited by turbidity currents (e.g., Ordovician of the Appalachians, upper Paleozoic of the Ouachita-Marathon region, and some classic flysch of the Northern Alps) or of molasselike deposits (e.g., Old Red Sandstone of Britain, Catskill and upper Paleozoic strata of the Appalachians, and the Cretaceous of the Rocky Mountains). Conversely, mature quartz arenites occur in volcanic eugeosynclinal belts. In the North American Cordillera, pure quartzites occur within Ordovician black slates, where they comprise as much as 25 percent of the sequence (Ketner, 1966; Churkin, this volume), and within Lower Jurassic carbonates (Stanley, and others, 1971). In the middle Precambrian orogen of the Great Lakes region, quartzites and quartz wackes are widespread (Dott, 1964b; 1972).

Environments of deposition, and thus also paleogeography, were as variable as the sediments themselves (fig. 2). Many workers have assumed unconsciously that most geosynclinal sediments must be marine; no rational reason exists, however, for excluding thick nonmarine ones. Great volumes of nonmarine sediments, most notably the famous molasse (Van Houten and Eisbacher, both this volume), are contained within every classic geosyncline. And it is the rule that simultaneously along a single belt nonmarine, shallow- and deep-marine environments coexisted. Moreover, sediments formed by the same processes may occur in diverse sedimentary or tectonic environments (Crowell, Harms and Pray, Normark, all this volume). For example, submarine fans built by turbidity currents can occur in cratonic basins, continental rises, arc-rear basins, and trenches (see Reading, 1972).

Stille and Kay's various subclasses of geosynclines were based upon tectonic setting and upon presence or absence of certain igneous rocks, rather than upon sedimentary rock type or inferred environment of deposition. "Geosynclinal facies are characterized by thickness rather than kind" (Kay, 1951, p. 1).

THE TECTONIC REVOLUTION

The causality myth.—As noted above, causality very early became a part of the geosynclinal concept, especially in North America. It was an article of faith that "thick sediments must invariably lead to mountains—in fact are a prerequisite—and mountain building leads to enlargement of continents." In reality, however, ancient orogenic belts and modern magmatic arcs, which clearly are genetically related species, by no means seem to be consistently marginal to continents. Considering the diversity, their role in crustal evolution seemed unclear to some workers a decade ago (e.g., Hess, 1960; Dott, 1964b). All that did seem definite was that the tectonically mobile belts of the earth, regardless of age or location, represent first and foremost a tectonic environment. The sediments of geosynclines—strictly, the *orthogeosynclines* of Stille and Kay—must be quite secondary! This crucial point was appreciated by Suess, Kay, Hess, and others, but not by many other workers. Modern belts such as the Marianas arc-trench system are impoverished of sediments and may always be so, yet surely they are experiencing a tectonic regime similar to, say Japan's or to that of the northeastern Mediterranean. This line of reasoning led to the suggestion that orthogeosynclines are best thought of as sediment-filled mobile belts (Dott, 1964a, 1964b; Dott and Batten, 1971); "sediments are but innocent bystanders" (Hsü, 1972. p. 33). Conversely, then, because structure is the key to orogens and controls sedimentation to a great extent, the Gulf Coast type of so-called "'geosyncline" must belong to a genus wholly different from the orthogeosyncline. The Gulf Coast type owes its origin to a tranquil progradation of a continental terrace and apparently has nothing causal to do with orogenesis. Ironically, this *paraliageosyncline* of Kay (1951) (synonymous with the *miogeocline-eugeocline* couplet of Dietz and Holden, 1966) fits closely James Hall's original, wrongly conceived model for sedimentary accumulation on the postulated site of future mountains.

The miogeocline paradox.—Counter to the above reasoning, Drake and others (1959) already had suggested a possible analogy between young shelf-rise (continental terrace) sediment prisms and ancient orthogeosynclines. Dietz (1963) then incorporated this view and that of Hess (1962) about sea-floor spreading into an inclusive model for geosynclines and mountain building (see also Dietz, this volume). Whereas to me a great degree of tectonic mobility had seemed to be a prerequisite for all true orthogeosynclines, Dietz argued that such belts began first as passive continental shelf-rise prisms of the Atlantic or Gulf Coast type. The key point is that his miogeocline was first conceived as an essential stage in the evolution of orthogeosynclines. But why was there so little tectonic mobility evidenced in modern miogeoclines? And where was there any compelling evidence that a continental rise (Dietz' eugeocline) on an Atlantic-type margin could be a youthful eugeosyncline? Submarine volcanism on such rises was a guess only. The entire model, although allegedly actualistic, was in fact strictly uniformitarian, for it involved a one-to-one analogy between a modern topographic entity, the continental terrace, and complexly deformed ancient rocks, which in fact share only limited lithologic similarities (Stanley, 1970; Moore and Curray, this volume). As outlined in the introduction, a truly actualistic model would imply a more complete analogy of tectonic and sedimentary processes in addition to primarily morphologic comparisons (e.g., Stanley, this volume). Moreover, the uniqueness of the Quaternary System makes modern shelf-rise sediments doubtful an-

alogues for ancient ones (Moore and Curray, this volume).

Salvation through plate tectonics?—The paradox between the sea-floor-spreading orogenic model of Dietz and the concepts of orthogeosynclines derived from studies of ancient rocks seems to have been resolved dramatically with the appearance of papers on the geologic implications of plate tectonic theory (e.g., LePichon, 1968; Isacks and others, 1968; Dewey and Bird, 1970a; Dickinson, 1970; North, 1971). These papers came close on the heels of Dietz' (1963) and Rodger's (1968) inference that strata of the Appalachian region ranging in age from earliest Cambrian through Early Ordovician represent a stable Atlantic-type continental-shelf sequence caught up by orogenesis, which began abruptly in medial Ordovician time (see Bird, this volume). The new global tectonic theory showed how a stable continental margin could become involved coincidentally at any time in orogenesis at an active convergent plate margin. Apparently, many ancient miogeosynclines were, in part, paraliageosynclines that had suffered exactly such a fate (fig. 4). The important Drake-Dietz-Rodgers analogy between youthful Atlantic-type continental terraces and some of the (early) deposits now seen within orogenic belts could be accepted so long as a genetic discontinuity separated such so-called "miogeoclinal" or nonorogenic rocks from subsequent, truly orthogeosynclinal or orogenic sequences. The crucial point revealed by plate theory is that the two do not represent a single, continuous evolutionary succession, but a coincidental association, an overprinting. The discontinuity typically is marked by an abrupt reversal of source direction and maturity of major clastic sediment contributions, and it may involve considerable lateral translation or telescoping of quite different terranes (fig. 4, VI). The significance of such temporal and spatial discontinuities can not be overemphasized. Even if many ancient orogens have one, a continental terrace record (or miogeocline in some usage) is not a requirement for every mobile belt—witness the Marianas arc built in a wholly oceanic realm. Miogeoclines are coincidental to orogenesis, but as Mitchell and Reading (1969) and Dickinson (1971) have stressed, one consequence of plate theory is that most passive continental margins are almost certain sooner or later to become tectonically crumpled by some active plate boundary.

The new global tectonics is so attractive because it seems, more than any older global theory, to generalize and simplify so many diverse phenomena. Moreover, some of those phenom-

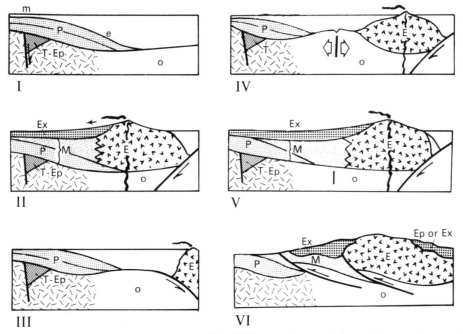

Fig. 4.—Possible genetic relations among some Kay-Stille geosynclinal species, especially showing ways that nonorogenic paraliageosynclines (P) [or miogeocline-eugeocline couplets (m and e)] can be coincidentally associated with orthogeosynclines. Not to scale: o, oceanic crust and mantle, including ophiolite suites; I, continental terrace on a passive margin following rifting [taphrogeosynclines (T) and certain epieugeosynclines (EP)]; II, subsequent development of an eugeosyncline-miogeosyncline-exogeosyncline complex (E-M-Ex) when the continental margin became orogenically active; III, a nonorogenic trailing-edge terrace destined to be involved in orogenesis by propagation toward it of an arc-trench system; note problem of distinguishing after the fact the polarity of this arc system from that of II and IV; IV, continental terrace facing a small-ocean basin having a local spreading site and an orogenic outer margin; V, marginal ocean basin after cessation of spreading therein followed by filling, which juxtaposed exogeosynclinal strata over continental terrace strata; VI, orogenic telescoping of rock suites and basements as might accompany the closing of an ocean basin (as in III, IV, V) or the termination of a continent-margin orogenic belt (as in II). Dual exogeosynclines shown characterize especially intercratonic orogens; in such belts, thrusting also may be bilateral. This figure is not intended to advocate perpetuation of the terms.

ena seemed almost totally unrelated before, for example, rifting in one region contemporaneously with orogenesis far away. Significantly, the new theory accommodates the several species recognized by Stille and Kay within the broad genus "orthogeosyncline." Twenty-five years ago species distinctions were necessarily descriptive and were conceived in order to focus upon what appeared to be tectonically significant differences. Now, at last, the new tectonics provides a rational genetic explanation for each species (fig. 4; see also Dewey and Bird, 1970b).

Loose ends.—The new tectonics seems to constitute a scientific revolution and to have produced a sweeping new *paradigm* in the sense of Kuhn (1962). Plate theory and its many implications—some entirely new—are shared by a large scientific community, which adopted them with unprecedented rapidity, presumably because of a so-called "crisis" of rapidly accumulating anomalies that were less satisfactorily explained by older theories. Moreover, the new tectonics has almost overnight provided a host of new problems for investigation as well as new ways of looking at old problems. Plate tectonics indeed seems to represent a quantum jump rather than a cumulative or inexorable evolutionary progression in geologic thought. It has provided a new tradition—a "reconstruction of the field from new fundamentals" (Kuhn, p. 85). For example, the mobilist view of the earth that involves breathtaking translations of tectonic elements, for which we were partially prepared by Wegener, Carey, and others, has rapidly become widely accepted as commonplace.

Significant as it seems to be, however, I doubt that plate tectonics has changed the goals and methods of geology to anything like the degree that the revolutions discussed by Kuhn changed astronomy, physics, and chemistry. Moreover, as Kuhn himself would predict (p. 79), the new tectonic paradigm has not completely resolved all old problems. Critics cannot condemn it (or any theory) for that, although they might scold its advocates for a disturbing degree of dogmatism acquired astonishingly quickly. We should remind ourselves that plate theory is not a catechism, but only a model, which must be held up for some time yet to the critical mirror of reality. The new theory is remarkably similar qualitatively to Taylor's (1910) and Holmes' (1930) early speculations during this century on continental drift as a mountain-building mechanism, and it bears some resemblance to the ideas of crustal downbuckle or tectogene that were popular in the 1930's. Finally, we must not forget that, like Charles Darwin's theory of natural selection, plate tectonics will remain an incomplete theory so long as a driving mechanism remains unverified.

There are some questions to ask about the new orthodoxy. Why do some orogenic belts seem to lack any clear oceanic relics? Perhaps they really were ensialic—"not organs of continental growth, but wounds healed by orogenesis," to paraphrase Aubouin (1965, p. 220). Were marginal and interarc ocean basins of such universal significance in the past as is currently suggested? If so, why are not clear suture zones representing their former positions more widely recognizable? Is it simply because of the varied structural levels that are exposed? How can we tell which chert-basalt assemblages are actually ocean-ridge deposits (Kanmera, this volume)? Why do modern ocean trenches not show more obvious commonalities with parts of ancient eugeosynclinal belts? Many of the latter, for example, lack tectonic mélanges and blueschists, which are assumed by many to be identifying characteristics of the trench tectonic environment. Can subduction zones flip so easily and frequently as some authors suggest? Lastly, must an orthogeosyncline now be viewed as an entire half-ocean basin riding on a conveyor belt, so to speak, away from its parental spreading site so that most of its rocks simply are scraped off in a trench?

Regardless of the ultimate fate of plate tectonics, which history tells us must be at least somewhat ephemeral, a number of concepts developed as handmaidens to it seem to stand on their own merits. Most important to the venerable geosynclinal concept is to distinguish the non-orogenic, miogeoclinal (or paraliageosynclinal or Atlantic-type continental marginal) sequences and some aulacogen sequences (Hoffman and others, this volume) that are now seen in orogenic belts from the rock sequences that developed within the genetic context of active orogenesis (or Pacific- and Mediterranean-type convergent plate margins). Also of great significance is the growing recognition of the enormous importance of submarine channels (Whitaker, also Picha; this volume) and fans to sedimentation within developing orogenic belts as well as on passive continental margins. But how, in ancient rocks, can we discriminate fans formed on stable margins from those formed either in more active marginal sea areas or in trenches? We see ever-increasing evidence of mass gravity failures both in the ancient and recent records, but their distinction from tectonic mélanges commonly presents serious problems. We have a number of bathymetric indicators (e.g., foraminiferal ratios, trace fossils, vesiculation in submarine flow rocks, and relative position of carbonate compensation), but still-more-accurate criteria would be welcome.

We need to distinguish among ancient turbidites formed on passive continental rises, arc-rear basins, arc fronts and trenches, and abyssal plains (see Crook, this volume for a partial key). Paleogeographic criteria for polarity of arc-trench systems to supplement the geochemical gradients for igneous rocks as proposed by Dickinson (1970) are needed as well, for in ancient rocks chemical alteration and metamorphism commonly have rendered that approach impossible. Difficulties of polarity interpretation are underscored, for example, by Stanley's (this volume) documentation of flysch derived from the stable craton of North Africa being sandier than flysch derived from the Hellenic-arc (opposite) side of the eastern Mediterranean. We need to distinguish among sea-floor, arc, and rift volcanics, all of which may be found together in the same belt. Ignimbrites, for example, are most common in arcs sited upon continental crust. Finer distinctions of volcanic products, including their trace-element chemistry, may be the most important means of unravelling the mysteries of the bewildering array of rock sequences found today jumbled together in orogenic belts (see Helwig, this volume).

CONCLUSION

"When I use a word, it means just what I choose it to mean—neither more nor less." (Humpty Dumpty in Lewis Carroll's *Through the Looking Glass*).

Classification and terminology are essential to all disciplines, especially if the phenomena under study are themselves both as complex and poorly understood genetically as was the geosyncline twenty years ago. Without some ordering of material, the human cannot cope. Whether nature really exists "out there" or only in our minds, classification is an indispensible part of thought. The earlier phase of study of orogenic belts necessarily was a descriptive or taxonomic one in which classifications of geosynclinal species served to focus upon important space-time differences. "The concern about the terms has resulted in more penetrating analyses of the history of the rocks; the very endeavor to classify has been rewarding" (Kay, 1967, p. 315). With the advent of a powerful, unifying explanation of orogenic belts, however, we move into a new, more genetic phase of study in which taxonomy should be far less important.

What is likely to be the fate of the geosyncline in the next 20 years? Some would argue that the entire concept is dead or dying and that the term itself should be put quietly to rest in the archives of science history. Others, however, have sought to rationalize every detail of older geosynclinal concepts with the new global tectonics (e.g., Wang, 1972). While most of the various descriptive species designated over the years can now be seen to have some genetic significance within the context of plate tectonics, there is little utility—and much potential danger—in perpetuating all of them. Long before the new global tectonics was formulated, one was confronted with such a multitude of contradictory and overlapping usages of terms that much of the utility of a scientific shorthand was lost. The most obvious anomaly is that noted by Aubouin (1965), whereby "geosyncline" is loosely applied to individual local depositional basins or furrows within an orogen as well as to the sedimentary and volcanic fill of the entire orogen.

If much of the old terminology is retained, then communication problems are likely to be aggravated with the advent of new hypotheses. Toulmin (1953) argued that major new breakthroughs in physical science involve the discovery chiefly of new modes of representation, thus new generalized models. The best model, which is "something more than a simple metaphor" (Toulmin, p. 39), should inspire new ways of drawing inferences and thus of formulating and testing hypotheses. It follows that a revolution in scientific theory need not involve discovery of any new evidence, but it must involve the discovery of a new model, or, in the approximately equivalent, but more colorful language of Kuhn, a new paradigm. It must provide a new conceptual and methodological framework. As a new general model is discovered, the bases for classification will change, which results in an important language shift (Toulmin, 1953, p. 13). In short, changing concepts require changing language! Terminology conceived in the context of, and useful to, an earlier general model or concept is almost certain to be constraining in the context of a new one (see also Coney, 1970). A language shift could, of course, take place in either of two ways: first, by a redefinition of all old terms, or, secondly, by the construction of an entirely new terminology. Different workers will have their own preferences, but the clutter of different shades of meaning attached during the past two decades to geosynclinal and tectonic nomenclature should give one considerable pause before he opts for the former course. So many of the premises associated with the older taxonomy are incompatible with the new idiom that, if perpetuated, such terminology would certainly act as a mental straitjacket. Conversely, exploitation of the new idiom's truly revolutionary implications could be greatly accelerated by a clean break with the old language.

ACKNOWLEDGMENTS

It is impossible to recall, much less to cite, all who have contributed materially over the years to the views expressed here. Most important, certainly, has been Marshall Kay's influence on my thinking. Previous historical reviews of the geosynclinal concept, such as those by Glaessner and Teichert (1947), Knopf (1948), Aubouin (1965), Coney (1970), and especially a recent one by Hsü (1972), were very helpful in composing this paper. The reader should consult them for a broader historical perspective. The manuscript was materially improved by criticism at various stages by I. W. D. Dalziel, James Helwig, W. R. Dickinson, J. F. Dewey, and H. G. Reading. Much of the substance of this paper was first presented orally as an invited paper in September 1971 at Edinburgh, Scotland, for a symposium on "Global Topics in Sedimentology."

REFERENCES

Aalto, K. R., and Dott, R. H., Jr., 1970, Late Mesozoic conglomeratic flysch in southwestern Oregon, and the problem of transport of coarse gravel in deep water: *in* Lajoie, J. (ed.), Flysch sedimentology in North America, Geol. Assoc. Canada Special Paper 7, p. 53–65.

Argand, E., 1916, Sur l'arc des Alpes occidentales: Eclogae Geol. Helvetiae, v. 14, p. 145–191.

Aubouin, Jean, 1965, Geosynclines: Amsterdam, Elsevier Pub. Co., 335 p.

Bailey, E. B., 1930, New light on sedimentation and tectonics: Geol. Mag., v. 67, p. 71–92.

BARRELL, JOSEPH, 1913–1914, The Upper Devonian delta of the Appalachian Geosyncline: Am. Jour. Sci., v. 36, p. 429–472; v. 37, p. 87–109.
BERTRAND, MARCEL, 1887, La chaîne des Alpes et la formation du continent européen: Soc. géol. France Bull., v. 15, p. 423–447.
———, 1897, Structures des Alpes francaises et recurrence de certains facies sedimentaires: 6th Cong. Geol. Internat. (1894), Compte Rendu, Lausanne, p. 161–177.
CHAMBERLIN, T. C., 1882, Geology of Wisconsin: Madison, Wisconsin, v. 4, 779 p.
CONEY, P. J., 1970, The geotectonic cycle and the new global tectonics: Geol. Soc. America Bull., v. 81, p. 739–748.
DALY, R. A., 1912, Geology of the North American Cordillera at the 49th parallel: Geol. Survey Canada Mem. 38, pt. 2, p. 547–857.
DANA, J. D., 1873, On some results of the earth's contraction from cooling, including a discussion of the origin of mountains, and the nature of the earth's interior: Am. Jour. Sci., ser. 3, v. 5, p. 423–443; v. 6, p. 6–14; 104–115; 161–172.
DAVY, JOHN (ed.), 1840, The collected works of Sir Humphrey Davy: Cornhill, London, Smith, Elder and Co., v. 8 (Johnson Reprint Edition of 1972; The sources of science No. 114), 365 p.
DEWEY, J. F., AND BIRD, J. M., 1970a, Mountain belts and the new global tectonics: Jour. Geophys. Research, v. 75, p. 2625–2647.
———, AND ———, 1970b, Plate tectonics and geosynclines: Tectonophysics, v. 10, p. 625–638.
DICKINSON, W. R., 1970, Relation of andesites, granites, and derivative sandstones to arc-trench tectonics: Rev. Geophysics and Space Physics, v. 8, p. 813–860.
———, 1971, Plate tectonic models of geosynclines: Earth and Planetary Sci. Letters, v. 10, p. 165–174.
———, AND HATHERTON, T., 1967, Andesitic volcanism and seismicity around the Pacific: Science, v. 157, p. 801–803.
DIETZ, R. S., 1963, Collapsing continental rises: an actualistic concept of geosynclines and mountain building: Jour. Geology, v. 71, p. 314–333.
———, AND HOLDEN, J. C., 1966, Miogeoclines (miogeosynclines) in space and time: ibid., v. 74, p. 566–583.
DRAKE, C. L., EWING, M., AND SUTTON, G. H., 1959, Continental margin and geosynclines: the east coast of North America, north of Cape Hatteras, in AHRENS, L. H., AND OTHERS (eds.), Physics and chemistry of the earth: London, Pergamon Press, v. 3, p. 110–198.
DOTT, R. H., JR., 1964a, Mobile belts and sedimentation (abs.): Geol. Soc. America Special Paper 76, p. 49–50.
———, 1964b, Mobile belts, sedimentation and orogenesis: New York Acad. Sci., Trans. ser. 2, v. 27, p. 135–143.
———, 1972, Implications of Precambrian quartz-rich sediments of the Lake Superior region for crustal evolution: 24th Internat. Geol. Cong. (Montreal), Proc., abs. v., p. 8–9.
———, AND BATTEN, R. L., 1971, Evolution of the earth: New York, McGraw-Hill Book Co., 649 p.
DZULYNSKI, S., KSIAZKIEWICZ, M., AND KUENEN, PH. H., 1959, Turbidites in flysch of the Polish Carpathian Mountains: Geol. Soc. America Bull., v. 70, p. 1089–1118.
EARDLEY, A. J., 1947, Paleozoic Cordilleran Geosyncline and related orogeny: Jour. Geology, v. 55, p. 309–342.
ENGEL, A. E. J., 1963, Geologic evolution of North America: Science, v. 140, p. 143–152.
GASTIL, R. G., 1960, Continents and mobile belts in the light of mineral dating: 21st Internat. Geol. Cong. Rept., pt. 9, p. 162–169.
GLAESSNER, M. F., AND TEICHERT, CURT, 1947, Geosynclines: a fundamental concept in geology: Am. Jour. Sci., v. 245, p. 465–482; 571–591.
GOULD, S. J., 1965, Is uniformitarianism necessary?: ibid., v. 263, p. 223–228.
HALL, JAMES, 1859, Description and figures of the organic remains of the lower Helderberg Group and the Oriskany Sandstone: New York Geol. Survey, Natural History of New York, pt. 6, Paleontology, v. 3, 532 p.
HAMPTON, M. A., 1972, The role of subaqueous debris flow in generating turbidity currents: Jour. Sed. Petrology, v. 42, p. 775–793.
HAUG, EMIL, 1900, Les géosynclinaux et les aires continentales: Soc. géol. France Bull., v. 28, p. 617–711.
HESS, H. H., 1939, Island arcs, gravity anomalies and serpentine intrusions, a contribution to the ophiolite problem: 17th Internat. Geol. Cong. Rept., v. 2, p. 263–282.
———, 1960, Caribbean research project, progress report: Geol. Soc. America Bull., v. 71, p. 235–240.
———, 1962, History of ocean basins, in ENGEL, A. E. J., AND OTHERS (eds.), Petrologic studies: a volume in honor of A. F. Buddington: Geol. Soc. America, p. 599–620.
HOLMES, ARTHUR, 1930, Radioactivity and earth movements: Geol. Soc. Glasgow Trans. (1928–29), v. 18, p. 559–606.
HOOYKAS, R., 1963, Natural law and divine miracle: the principle of uniformity in geology, biology and theology: Leiden, Brill, 237 p.
HSÜ, K. J., 1972, The concept of the geosyncline, yesterday and today: Leicester Lit. and Philos. Soc. Trans., v. 66, p. 26–48.
HURLEY, P. M., HUGHES, H., FAURE, G., FAIRBAIRN, H. W., AND PINSON, W. H., 1962, Radiogenic strontium-87 model of continent formation: Jour. Geophys. Research, v. 67, p. 5315–5334.
ISACKS, B., OLIVER, J., AND SYKES, L. R., 1968, Seismology and the new global tectonics: Jour. Geophys. Research, v. 73, p. 5855–5899.
JONES, O. T., 1938, On the evolution of a geosyncline: Geol. Soc. London Quart. Jour., v. 94, p. lx–cx.
KAY, MARSHALL, 1937, Stratigraphy of the Trenton Group: Geol. Soc. America Bull., v. 48, p. 233–302.
———, 1947, Geosynclinal nomenclature and the craton: Am. Assoc. Petroleum Geologists Bull., v. 31, p. 1289–1291.
———, 1951, North American geosynclines: Geol. Soc. America Mem. 48, 143 p.
———, 1967, On geosynclinal nomenclature: Geol. Mag., v. 104, p. 311–316.
KETNER, K. B., 1966, Comparison of Ordovician eugeosynclinal and miogeosynclinal quartzites of the Cordilleran Geosyncline: U.S. Geol. Survey Prof. Paper 550C, p. 54–60.

Knopf, Adolph, 1948, The geosynclinal theory: Geol. Soc. America Bull., v. 57, p. 649–670.

Kossmat, F., 1921, Die mediterranen Kettengebirge in ihrer Berichtigung zum Gleichgewichtszustande der Erdrinde: Sachsische Akad. Wiss., Abh., Math.-Phys. Kl., v. 38, p. 46–68.

Kraus, E., 1927, Der orogene Zyklus und seine Studien: Centralbl. für Minéralogie, Abt. B, p. 216–233.

Krynine, P. D., 1948, The megascopic study and field classification of sedimentary rocks: Jour. Geology, v. 56, p. 130–165.

Kuenen, Ph. H., and Migliorini, C. I., 1950, Turbidity currents as a cause of graded bedding: *ibid.*, v. 58, p. 91–127.

Kuhn, T. S., 1962, The structure of scientific revolutions: Chicago, Univ. Chicago Press, 2d ed. (1970), 210 p.

Kuno, H., 1959, Origin of Cenozozoic petrographic provinces of Japan and surrounding areas: Bull. volcanol., v. 20, p. 37–67.

LePichon, Xavier, 1968, Sea-floor spreading and continental drift: Jour. Geophys. Research, v. 73, p. 3661–3697.

Middleton, G. V., 1970, Experimental studies relating to problems of flysch sedimentation, *in* Lajoie, J. (ed.), Flysch sedimentology in North America: Geol. Assoc. Canada Special Paper 7, p. 253–272.

Mitchell, A. H., and Reading, H. G., 1969, Continental margins, geosynclines, and ocean floor spreading: Jour. Geology, v. 77, p. 629–646.

Natland, M. L., 1933, Depth and temperature distribution of some Recent and fossil Foraminifera in the southern California region: Scripps Inst. Oceanography Bull., Tech. Ser., v. 3, p. 225–230.

North, F. K., 1971, Alpine serpentinites, oceanic ridges, and continental drift: Geol. Mag., v. 108, p. 81–192.

Pantanelli, D., 1883, I diaspri della Toscana e loro fossili: Reale Accad. Nazl. Lincei, v. 7, p. 13–14 (not seen; *fide* Aubouin, 1965).

Patterson, Clair, 1964, Characteristics of lead isotope evolution on a continental scale in the earth, *in* Craig, H., and others (eds.), Isotopic and cosmic chemistry; Amsterdam, North Holland Publishing Co., p. 244–268.

Reading, H. G., 1972, Global tectonics and the genesis of flysch successions: 24th Internat. Geol. Cong. Proc., Sec. 6, p. 59–66.

Rodgers, John, 1968, The eastern edge of the North American continent during the Cambrian and Early Ordovician, *in* Zen, E., and others (eds.), Studies of Appalachian geology: northern and maritime: New York, Wiley-Inter-Science, p. 141–149.

Schuchert, Charles, 1910, Paleogeography of North America: Geol. Soc. America Bull., v. 20, p. 427–606.

Schwab, F. L., 1971, Geosynclinal compositions and the new global tectonics: Jour. Sed. Petrology, v. 41, p. 928–938.

Stanley, D. J., 1970, Flyschoid sedimentation on the outer Atlantic margin off northeast North America, *in* Lajoie, J. (ed.), Flysch sedimentology in North America: Geol. Assoc. Canada Special Paper 7, p. 179–210.

Stanley, K. O., Jordan, W. M., and Dott, R. H., Jr., 1971, New hypothesis of Early Jurassic paleogeography and sediment dispersal for western United States: Am. Assoc. Petroleum Geol. Bull., v. 55, p. 10–19.

Stauffer, P. H., 1967, Grain-flow deposits and their implications, Santa Ynez Mountains, California: Jour. Sed. Petrology, v. 37, p. 487–508.

Steinmann, G., 1905, Die geologische Bedeutung der Tiefseeabsätze und der ophiolithischen Massengesteine: Naturf. Gesell. Freiburg Ber., v. 16, p. 44–65.

Stille, Hans, 1936, Wege und Ergebnisse der geologisch-tectonischen Forschung: Gesell. Wiss. Förh., 25 Jahre Kaiser Wilhelm, Bd. 2, p. 84–85 (not seen; *fide* Kay, 1951).

———, 1941, Einfuhrung in den Bau Amerikas: Berlin, Borntraeger, 717 p. (not seen; *fide* Kay, 1951).

Storm, L. W., 1945, Resumé of facts and opinions on sedimentation in the Gulf Coast region of Texas and Louisiana: Am. Assoc. Petroleum Geol. Bull., v. 29, p. 1304–1335.

Suess, Eduard, 1909, Das Antlitz der Erde: Leipzig, Freytag, v. 3, pt. 2, 789 p.

Sullwold, H. H., Jr., 1960, Tarzana fan, deep submarine fan of late Miocene age, Los Angeles County, California: Am. Assoc. Petroleum Geol. Bull., v. 44, p. 433–457.

Taylor, F. B., 1910, Bearing of the Tertiary mountain belt on the origin of the earth's plan: Geol. Soc. America Bull., v. 21, p. 179–226.

Toulmin, Stephen, 1953, The philosophy of science: New York, Harper Torchbook, paper ed. (TB 513), 176 p.

Umbgrove, J. H. F., 1933, Verschillende typen van Tertiare geosynclinalen in den Indischen Archipel.: Leidsche Geol. Meded., v. 6, p. 33–43.

Van Bemmelen, R. W., 1949, The geology of Indonesia: The Hague, Nijhoff, 997 p.

Vening Meinesz, F. A., 1940, The earth's crust deformation in the East Indies: Prac. K. Nederlandsch Akad. Wetensch. Proc., ser. B, v. 43, p. 278–306.

Walker, R. G., 1966, Deep channels in turbidite-bearing formations: Am. Assoc. Petroleum Geol. Bull., v. 50, p. 1899–1917.

Wang, C. S., 1972, Geosynclines and the new global tectonics: Geol. Soc. America Bull., v. 83, p. 2105–2110.

Wells, F. G., 1949, Ensimatic and ensialic geosynclines (abs.): *ibid.*, v. 60, p. 1927.

Williams, H. S., 1897, On the southern Devonian formations: Am. Jour. Sci., v. 3, p. 393–403.

Wilson, J. T., 1949, The origin of continents and Precambrian history: Royal Soc. Canada Trans., v. 43, ser. 3, p. 157–182.

AUTHOR CITATION INDEX

SUBJECT INDEX

About the Editor

FREDERIC L. SCHWAB is professor of Geology at Washington and Lee University where he has taught since 1967. He received the A.B. from Dartmouth College in 1961, the M.S. from the University of Wisconsin in 1963, and the PhD. from Harvard University in 1968.

Professor Schwab is a member of the Geological Society of America, the Society of Economic Paleontologists and Mineralogists, and the International Association of Sedimentologists, and he serves as a member of the editorial board of *Geology*. From 1971 to 1972 he was a National Science Foundation Science Faculty Fellow at the University of Edinburgh in Scotland, and from 1977 to 1978 he was a N.A.T.O. senior scientist at the University of Grenoble, France.

Professor Schwab has served as the consulting editor for the McGraw-Hill *Dictionary of Science and Technology*, the *Encyclopedia of Energy*, and the *Encyclopedia of the Geological Sciences*. He has published several papers in the *Journal of Sedimentary Petrology*, the *Bulletin of the Geological Society of America*, *Sedimentology*, and *Geology*. His major research interests are geosynclinal development, the evolution of the earth's sedimentary rock record, and the origin of the late Precambrian stratified assemblages of the Appalachian Blue Ridge.